P9-DWN-939

CHEMISTRY OF CATALYTIC PROCESSES

McGraw-Hill Chemical Engineering Series

Building the Literature of a Profession

Fifteen prominent chemical engineers first met in New York more than fifty years ago to plan a continuing literature for their rapidly growing profession. From industry came such pioneer practitioners as Leo H. Baekeland, Arthur D. Little, Charles L. Reese, John V. N. Dorr, M. C. Whittaker, and R. S. McBride. From the universities came such eminent educators as William H. Walker, Alfred H. White, D. D. Jackson, J. H. James, Warren K. Lewis, and Harry A. Curtis. H. C. Parmelee, then editor of *Chemical and Metallurgical Engineering,* served as chairman and was joined subsequently by S. D. Kirkpatrick as consulting editor.

After several meetings, this committee submitted its report to the McGRAW-HILL Book Company in September 1925. In the report were detailed specifications for a correlated series of more than a dozen texts and reference books which have since become the McGRAW-HILL Series in Chemical Engineering and which became the cornerstone of the chemical engineering curriculum.

From this beginning there has evolved a series of texts surpassing by far the scope and longevity envisioned by the founding Editorial Board. The McGRAW-HILL Series in Chemical Engineering stands as a unique historical record of the development of chemical engineering education and practice. In the series one finds the milestones of the subject's evolution: industrial chemistry, stoichiometry, unit operations and processes, thermodynamics, kinetics, and transfer operations.

Chemical engineering is a dynamic profession, and its literature continues to evolve. McGRAW-HILL and its consulting editors remain committed to a publishing policy that will serve, and indeed lead, the needs of the chemical engineering profession during the years to come.

The Series

Bailey and Ollis: *Biochemical Engineering Fundamentals*
Bennett and Myers: *Momentum, Heat, and Mass Transfer*

CHEMISTRY OF CATALYTIC PROCESSES

Bruce C. Gates
James R. Katzer
G. C. A. Schuit

University of Delaware
Center for Catalytic Science and Technology
Department of Chemical Engineering

McGraw-Hill Book Company

New York St. Louis San Francisco Auckland Bogotá Düsseldorf
Johannesburg London Madrid Mexico Montreal New Delhi
Panama Paris São Paulo Singapore Sydney Tokyo Toronto

CHEMISTRY OF CATALYTIC PROCESSES

67890 HDHD 8987654

This book was set in Times Roman.
The editors were Rose Ciofalo, Douglas J. Marshall, and Bob Leap;
the production supervisor was Milton J. Heiberg.
The drawings were done by Lorraine Turner and Judy Katzer.

Library of Congress Cataloging in Publication Data

Gates, Bruce C
Chemistry of catalytic processes.

(McGraw-Hill series in chemical engineering)
Includes bibliographical references and index.
1. Catalysis. I. Katzer, James R., date joint author. II. Schuit, G. C. A., joint author. III. Title.
TP156.C35G37 660.2′9′95 77-16112
ISBN 0-07-022987-2

For
Eva, Judy, and Jutta

CONTENTS

PREFACE

Because of its economic importance, catalysis is one of the most intensely pursued subjects in applied chemistry and chemical engineering. It is complex, encompassing solid and surface structure, reaction mechanism, and analysis and design of chemical reactors. The complexity makes the subject difficult to teach and write about with depth and coherence, and most practitioners have learned their trade almost entirely from on-the-job training. We believe that there is need for a book about catalysis to convey what the science and practice of the subject are really like. We hope to have begun to meet this need by writing in detail about some of the most important industrial applications of catalysis, attempting to integrate the science and engineering in a way reflecting their integration in practice.

The book is not meant to be comprehensive but to provide a representative cross section of applied catalysis and some insight into catalytic practice. In particular, we have attempted to illustrate how the chemistry constrains the engineering design and how the design limitations, at the same time, restrict the choice of chemical variables such as catalyst composition. We hope that the book demonstrates the complexity of industrial catalysts, which have been developed through years of empirical testing to offer surfaces with combinations of functions just suited to the desired reactions.

There are five chapters, each concerned with an industrial process or class of processes, namely, catalytic cracking, transition-metal-complex catalysis, catalytic reforming, partial oxidation of hydrocarbons (as illustrated by ammoxidation), and hydrodesulfurization. The processess were chosen because they are industrially important and illustrate the major classes of catalysts: acids, transition metals, metal oxides, and metal sulfides. The sequence proceeds roughly from the best-understood to the least well understood chemistry. The coherence is intended to be provided by the chemistry rather than the engineering, and the engineering subjects are introduced as they arise in this context. We believe that there is value in the quantitative illustration of the engineering methods, and our intention is that this book, with its summaries of the available processing data, will complement the existing books on chemical reaction engineering.

Each chapter is arranged roughly along the following lines: the process is introduced with a brief statement of the catalytic chemistry and process engineering; the chemistry is then presented in detail; and the engineering follows, with quantitative examples included to illustrate design methods.

The final manuscript evolved from notes for a graduate course and an intensive one-week short course taught at the University of Delaware. Believing that others may find this useful as a textbook, we have included problems with each chapter. The background information required for understanding the book includes standard undergraduate chemistry and the basic concepts of catalysis and chemical reaction engineering. Graduate students of chemical engineering and of technical chemistry should be adequately prepared for it, although an instructor's guidance will be helpful in directing students to the appropriate fundamentals for review. Students of chemistry who lack any experience with chemical engineering would profit from working through an introduction to reaction engineering such as Denbigh and Turner's "Chemical Reactor Theory," Cambridge University Press, 1971. Russell and Denn's "Introduction to Chemical Engineering Analysis," Wiley, 1972, is also recommended.

Without the help and criticisms of our students and colleagues, this book could not have been written. The comments of the industrial chemists and engineers who attended our annual short course have been especially helpful in eliminating errors and correcting false impressions of commercial practice. Many colleagues have helped us, and we especially thank W. H. Manogue, who offered invaluable comments, and J. H. Olson, who prepared the final section of Chap. 1, which is concerned with the reaction engineering of catalytic cracking. We are very grateful to our department chairman, A. B. Metzner, for his encouragement and stimulation during the preparation of the manuscript. We also acknowledge the Fulbright-Kommission in Bonn for the fellowship that allowed B. C. Gates time to work the manuscript into final form.

Bruce C. Gates
James R. Katzer
G. C. A. Schuit

CHAPTER
ONE

CRACKING

INTRODUCTION

PROCESSES

Most industrial reactions are catalytic, and many process improvements result from the discovery of better chemical routes, usually involving new catalysts. One of the largest scale catalytic processes practiced is *cracking*, the conversion of large petroleum molecules into smaller hydrocarbons, primarily in the gasoline range. In the United States cracking capacity exceeds 5 million barrels per day, and because the process has such a large production volume, years of research and development giving incremental improvements in gasoline yields have been highly profitable.

Cracking processes were first carried out in the absence of catalysts, but in the last four decades a series of continuously improved cracking catalysts has been applied, all of them solid acids. The most important advance in cracking technology in the last three decades has been the development of zeolite catalysts. These catalyze cracking so much more rapidly than the earlier catalysts like silica-alumina that the processes have had to be essentially redesigned. Instead of a large fluidized bed, the reactor is now a small tube. Catalyst particles are conveyed through it by rapidly flowing oil vapors, which stay in contact with the catalyst for only about 5 s. Catalytic cracking is the process considered first in this book

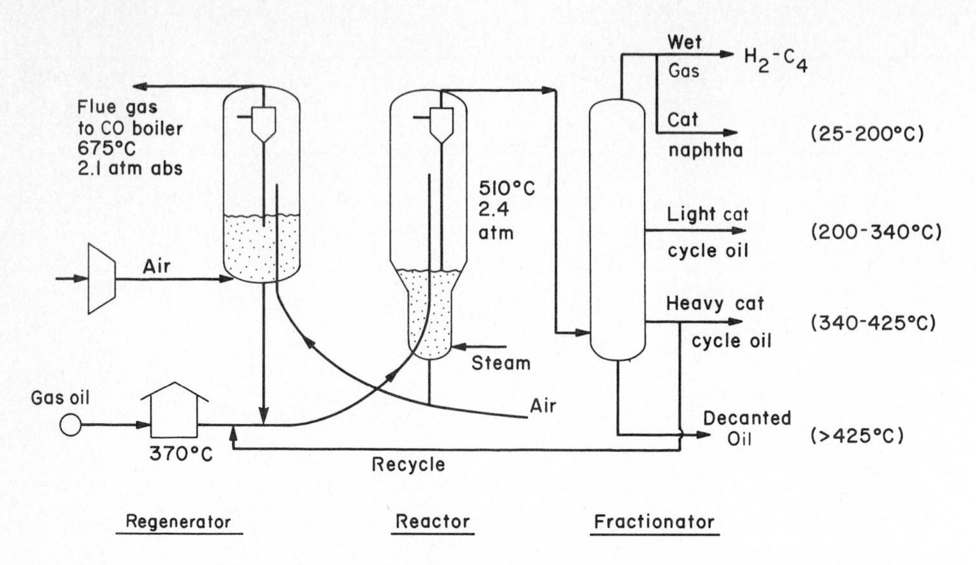

Figure 1-1 Flow diagram of catalytic-cracking process.

because cracking chemistry, unlike that of most catalytic processes, is well understood. It is the chemistry of strong acids, hydrocarbons, carbonium ions, and zeolites. The zeolite catalysts are familiar as molecular sieves, solids with crystal-

Table 1-1 Typical operating conditions for a catalytic cracking process

Riser-tube reactor	
Temperature, °C:	
Base	550
Top	510
Pressure, atm	3
Catalyst-to-oil ratio	6
Gas residence time, s	5–7
Regenerator	
Temperature in cyclone, °C	650–760
CO/CO_2 mole ratio	0.7–1.3 : 1
Pressure at bottom of fluidized bed, atm	3.5
Superficial gas velocity, cm/s	60
Solids residence time, s	30
Coke content of catalyst, wt %	
At entrance	0.8
At exit	< 0.1

line structures including uniform, molecular-scale pores. They have well-known surface structures, whereas most solid catalysts, being amorphous, have poorly understood surface structures.

The details of the chemistry of catalytic cracking follow, but before they are introduced, the process is outlined so that the chemistry can be understood in the context of industrial practice. The process (Fig. 1-1) consists of a riser-tube reactor, a fluidized-bed disengaging unit for separating catalyst particles from product vapors, and a fluidized-bed regenerator, in which high-molecular-weight carbonaceous products, called *coke*, are burned off the catalyst to restore its activity. A fractionator downstream of the reactor and disengaging unit separates the product into various boiling fractions, and the heavy oil which has not undergone sufficient cracking is recycled to the reactor.

Typical operating conditions for the reactor and regenerator are summarized in Table 1-1, and typical product yields are collected in Table 1-2. These data provide a preliminary comparison between silica-alumina and zeolite catalysts.

One version of a riser-tube catalytic cracking unit is illustrated in Fig. 1-2. Gas oil is introduced with dispersive steam at the base of the reactor and mixed with regenerated catalyst supplied from a standpipe at the base of the fluidized-bed regenerator. The reactor diameter increases with height in this unit to maintain a nearly uniform catalyst velocity as the hydrostatic head in the riser

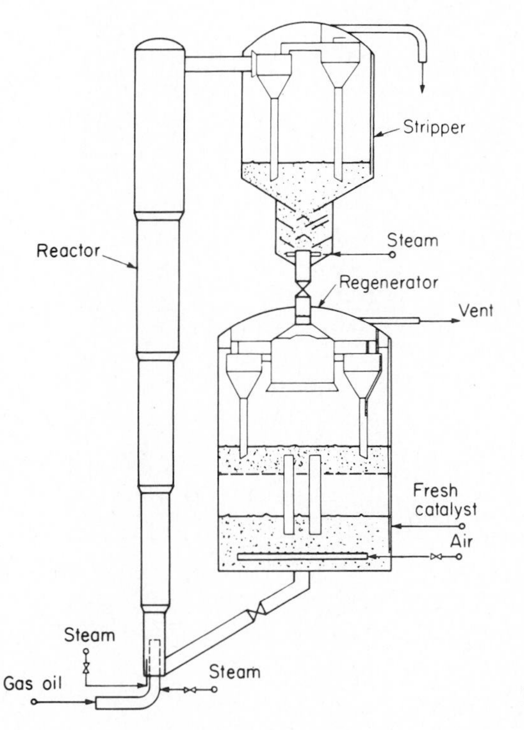

Figure 1-2 Riser catalytic-cracking unit.

Table 1-2 Performance of commercial cracking reactors with silica-alumina and zeolite catalysts [1]

Operating conditions	Durabead 5[a]		Durabead 1[b]	
Vapor inlet temperature, °C	476		476	
Catalyst inlet temperature, °C	548		549	
Vapor outlet temperature, °C	474		471	
Liquid hourly space velocity, vol/vol · h	1.0		0.9	
Catalyst-to-oil ratio, vol/vol	1.9		2.0	
Recycle ratio, vol recycle/vol fresh feed	0.84		0.82	
Steam content of feed, wt %	3.6		3.5	
Total reactor feed rate, bbl/day	12,900		13,400	
Catalyst circulation rate, kg/h	136,065		136,065	
Coke burnoff rate, kg/h	2,267		1,542	
Boiling range of recycle stream, °C	215–332		232–327	
Conversion, vol %	73.4		49.5	
Cracking efficiency, 100 × vol gasoline/vol converted	77.6		77.3	
	Yields			
	vol %	wt %	vol %	wt %
Synthetic tower bottoms	13.7	15.2	21.3	22.3
Distillate fuel oil	12.9	13.3	29.2	29.4
C_4-free gasoline	56.9	48.7	38.3	32.9
Butanes	13.4	8.5	8.5	5.4
Dry gas (C_3 and lighter)	...	8.9	...	6.6
Coke	...	5.4	...	3.4
Total	...	100.0	...	100.0
n-Butane	2.1	1.3	1.1	0.6
Isobutane	6.5	4.0	2.9	1.8
Butenes	4.8	3.2	4.5	3.0
Total C_4	13.4	8.5	8.5	5.4
iC_4/C_4 ratio	1.35	...	0.64	
Propane	3.8	2.1	2.4	1.3
Propylene	4.1	2.4	3.7	2.1
Total	7.9	4.5	6.1	3.4
Ethane	...	1.3	...	1.0
Ethylene	...	0.6	...	0.4
Methane	...	1.8	...	1.2
Hydrogen	...	0.1	...	0.1
Hydrogen sulfide	...	0.6	...	0.5
Total C_2 and lighter	...	4.4	...	3.2

[a] REHY zeolite in silica-alumina matrix.
[b] Silica-alumina.

decreases toward the exit. Downstream of the reactor, the catalyst is separated from most of the products in a two-stage cyclone. Steam stripping of hydrocarbons from the coked catalyst occurs in the baffled region of the stripper and in the dense bed below the two cyclone dip tubes. In the regenerator, coke is burned off the catalyst in a fluidized bed.

REACTIONS

Cracking reactions of hydrocarbons are catalyzed by acids and proceed through carbonium-ion intermediates. The details of the reaction chemistry are given later, but for purposes of engineering design it is sufficient to consider a greatly simplified reaction scheme which accounts for the essential character of product distributions in catalytic cracking reactors. In the following scheme, developed by Weekman and coworkers [3–5],†

$$\mathrm{O} \xrightarrow{k_1} \mathrm{G} \xrightarrow{k_2} \mathrm{X}, \qquad \mathrm{O} \xrightarrow{k_3} \mathrm{X} \tag{1}$$

gas oil O, gasoline G, and the undesired products X, including light, "overcracked" products, are treated as though they were true compounds rather than complex mixtures of compounds. The scheme indicates that overcracking of product (conversion of G to X) needs to be minimized. As will be shown in a later section of this chapter, a piston-flow reactor is superior to a well-mixed reactor for minimization of overcracking, and therefore the riser-tube reactors are designed for piston-flow operation. The reaction scheme shows that some undesired products X are formed not only from gasoline G, but also directly from gas oil O.

CATALYSTS

An industrial cracking catalyst is composed of 3 to 25 weight percent of roughly 1-μm-diameter crystallites of a zeolite embedded in a matrix of dense, amorphous silica-alumina. For satisfactory operation of fluidized beds, catalyst particles are chosen to be 20 to 60 μm in diameter. Zeolites are used in the silica-alumina matrix because alone they are expensive and catalytically too active to be used in units of practical dimensions without causing severe heat-transfer requirements. There are interactions between the zeolite and the surrounding matrix, and the mixture has an acidity distribution different from that of amorphous silica-alumina. These differences are responsible for the zeolite catalyst's substantially higher activity for gas-oil cracking and better selectivity for gasoline production than the amorphous silica-alumina catalysts.

† Numbers in brackets refer to References at the end of the chapter.

CRACKING REACTIONS

Introduction

Cracking reactions involve the rupture of C—C bonds, and since they are endothermic reactions, they are thermodynamically favored by high temperatures. Hydrocarbon cracking involves the following reactions:

1. Paraffins are cracked to give olefins and smaller paraffins:

$$\underset{}{C_nH_{2n+2}} \longrightarrow \underset{\text{Olefin}}{C_mH_{2m}} + \underset{\text{Paraffin}}{C_pH_{2p+2}} \qquad \text{where } n = m + p \tag{2}$$

2. Olefins are cracked to give smaller olefins:

$$C_nH_{2n} \longrightarrow \underset{\text{Olefin}}{C_mH_{2m}} + \underset{\text{Olefin}}{C_pH_{2p}} \qquad \text{where } n = m + p \tag{3}$$

3. Alkyl aromatics undergo dealkylation:

$$ArC_nH_{2n+1} \longrightarrow \underset{\text{Aromatic hydrocarbon}}{ArH} + \underset{\text{Olefin}}{C_nH_{2n}} \tag{4}$$

4. Instead of the foregoing dealkylation reaction, aromatic side-chain scission can occur:

$$ArC_nH_{2n+1} \longrightarrow \underset{\text{Aromatic with olefinic side chain}}{ArC_mH_{2m-1}} + \underset{\text{Paraffin}}{C_pH_{2p+2}} \qquad \text{where } n = m + p \tag{5}$$

Unsubstituted aromatics undergo relatively slow cracking under typical industrial reaction conditions because of the stability of the aromatic ring.

5. Naphthenes (cycloparaffins) are cracked to give olefins:

$$C_nH_{2n} \longrightarrow \underset{\text{Olefin}}{C_mH_{2m}} + \underset{\text{Olefin}}{C_pH_{2p}} \qquad \text{where } n = m + p \tag{6}$$

If the cycloparaffin contains a cyclohexane ring, however, the ring is not opened:

$$C_nH_{2n} \longrightarrow \underset{\text{Cyclohexane}}{C_6H_{12}} + \underset{\text{Olefin}}{C_mH_{2m}} + \underset{\text{Olefin}}{C_pH_{2p}} \qquad \text{where } n = m + p + 6 \tag{7}$$

Secondary reactions occurring after the initial cracking steps are important in determining the final product composition. These include the following:

6. Hydrogen transfer:

$$\text{Naphthene} + \text{olefin} \longrightarrow \text{aromatic} + \text{paraffin} \tag{8}$$

$$\text{Aromatic coke precursor} + \text{olefin} \longrightarrow \text{coke} + \text{paraffin} \tag{9}$$

7. Isomerization:

$$\text{Olefin} \longrightarrow \text{isoolefin} \tag{10}$$

8. Alkyl-group transfer:

$$C_6H_4(CH_3)_2 + C_6H_6 \longrightarrow C_6H_5(CH_3) + C_6H_5(CH_3) \tag{11}$$

9. Condensation reactions:

$$C_6H_5CH{=}CH_2 + R_1CH{=}CHR_2 \longrightarrow C_{10}H_6R_1R_2 + 2H \tag{12}$$

10. Disproportionation of low-molecular-weight olefins:

$$2H_2C{=}CHCH_2CH_3 \longrightarrow H_2C{=}CHCH_3 + H_2C{=}CHCH_2CH_2CH_3 \tag{13}$$

The principal cracking reactions are not limited by equilibrium under industrial reaction conditions; at equilibrium, the hydrocarbons would be almost completely degraded to graphite and hydrogen (Table 1-3). In contrast, such side reactions as isomerization, alkyl-group rearrangement, and dealkylation of aromatics can occur to only a moderate extent at equilibrium under cracking condi-

Table 1-3 Equilibria of hydrocarbon reactions under cracking conditions [6]

Reaction	Equilibrium constant at 420°C
$C_nH_m \longrightarrow \underset{\text{Graphite}}{nC} + \frac{m}{2} H_2$ (except for $n = 1$)	Very large
$C_nH_m \longrightarrow CH_4 + C_{n-1}H_{m-4}$ (except for $n = 1$)	Very large
Large paraffin $\longrightarrow$ paraffin + olefin (except olefin $\neq C_2H_4$)	Very large
Large olefin $\longrightarrow$ 2 olefins (except olefin $\neq C_2H_4$)	Very large
Paraffin $\longrightarrow$ aromatic + $4H_2$	Very large
Paraffin + H_2 $\longrightarrow$ 2 smaller paraffins	Very large
Hydroaromatic + olefin $\longrightarrow$ aromatic + paraffin	Very large
Cyclization of olefins to naphthenes (1-hexene $\longrightarrow$ cyclohexane)	Moderate (15.2)
Olefin isomerization (n-butene $\longrightarrow$ isobutene)	Moderate
Paraffin isomerization (n-butane $\longrightarrow$ isobutane)	Small (0.51)
Paraffin dehydrocyclization (n-hexane $\longrightarrow$ cyclohexane + H_2)	Small (0.07)

Source: From "Catalysis," vol. VI, P. H. Emmett (ed.).

tions. Paraffin-olefin alkylation, aromatic hydrogenation, and olefin polymerization (except for ethylene polymerization) cannot occur to any appreciable extent.

Cracking reactions are quite endothermic; isomerizations have very small heats of reaction; and hydrogen-transfer reactions are exothermic. In cracking processes the endothermic reactions always predominate, the magnitude of the heat effect depending on the feedstock, catalyst, and reaction conditions.

Thermal Cracking

Before the discovery of cracking catalysts, petroleum refining primarily involved noncatalytic cracking, and these processes are of current importance for converting naphtha into light olefins and for visbreaking of heavier feedstocks. When hydrocarbons in the absence of a catalyst are brought to high temperatures, they undergo thermal cracking via free-radical mechanisms. The initiation step in thermal cracking of a paraffin is the homolysis of a carbon-carbon bond:

$$R_1-\underset{H}{\overset{H}{C}}-\underset{H}{\overset{H}{C}}-R_2 \longrightarrow R_1-\underset{H}{\overset{H}{C}}\cdot + \cdot\underset{H}{\overset{H}{C}}-R_2 \tag{14}$$

The radicals formed can undergo scission to give ethylene and a primary radical which has two fewer carbon atoms. The empirical β rule states that C—C bond scission occurs at the C—C bond located β to the carbon atom having the unpaired electron:

$$RCH_2-CH_2-CH_2\cdot \longrightarrow RCH_2\cdot + H_2C=CH_2 \tag{15}$$

The newly formed primary free radical can continue to undergo β scission to give ethylene and a smaller radical until, ultimately, a methyl radical is formed.

The methyl radical abstracts a hydrogen radical from another hydrocarbon molecule, producing a random secondary radical and methane:

$$H_3C\cdot + RCH_2CH_2CH_2CH_2CH_2CH_2CH_3 \longrightarrow CH_4 + RCH_2CH_2CH_2CH_2\dot{C}HCH_2CH_3 \tag{16}$$

This radical can then undergo β scission, producing an α-olefin and a primary free radical:

$$RCH_2CH_2CH_2-CH_2\dot{C}HCH_2CH_3 \longrightarrow RCH_2CH_2\dot{C}H_2 + H_2C=CHCH_2CH_3 \tag{17}$$

The repetition of this reaction and reactions (15) and (16) leads to the formation of large amounts of ethylene, small amounts of methane, and small amounts of α-olefins.

Like the methyl radical, the $R\dot{C}H_2$ species is able to abstract a hydrogen radical from another paraffin to form a secondary free radical and a smaller paraffin, but it does so at a lower rate corresponding to its slightly greater stability. Only about 10 percent of all radical chains undergo such termination before forming $\dot{C}H_3$, as indicated by the small amounts of paraffins produced. At high hydrocarbon concentrations, such as those prevailing in high-pressure liquid-phase reactions, chain-transfer reactions become much more important, and the products of thermal cracking contain much higher concentrations of intermediate-molecular-weight paraffins and olefins, particularly C_3 to C_7 compounds:

$$R_1\dot{C}H_2 + RCH_2CH_2CH_2CH_2CH_2CH_2CH_3 \longrightarrow R_1CH_3 + RCH_2CH_2CH_2CH_2CH_2\dot{C}HCH_3 \qquad (18)$$

Free radicals do not undergo isomerization involving either migration of alkyl groups or shift of the radical center from one carbon atom to the neighboring carbon atom of the chain. Since a primary radical is less stable than a secondary or tertiary radical, however, a long-chain primary radical can coil back on itself and remove a hydrogen radical from a secondary or tertiary position:

$$\begin{array}{l} \qquad\quad \cdot CH_2CH_2 \\ RCHCH_3 \qquad \diagdown \\ \;| \qquad\qquad\quad CH_2 \\ CH_2CH_2{-}CH_2 \end{array} \longrightarrow \begin{array}{l} \qquad\qquad\quad CH_3 \\ \qquad\qquad\quad \diagdown \\ \quad\bullet \qquad\qquad CH_2 \\ R\dot{C}CH_3 \qquad \diagup \\ \;| \qquad\qquad\quad CH_2 \\ CH_2CH_2{-}CH_2 \end{array} \qquad (19)$$

This reaction and chain transfer [Eq. (18)] are important because they result in the production of less ethylene and more gasoline.

Chain termination by combination of free radicals [the reverse of Eq. (14)] happens infrequently because the free radicals are present in only low concentrations. Condensation and cyclization reactions also occur to a small extent in thermal cracking, as indicated by the presence of aromatic tars.

The relative rates of abstraction of a hydrogen atom from the various carbon atoms of a paraffin are predicted by the theory of Rice and coworkers [7–9] to be the following: for a primary carbon atom, 1; for a secondary carbon atom, 3.66; and for a tertiary carbon atom, 13.4. The relative rates of thermal cracking of straight- and branched-chain paraffins can be predicted from this theory if it is assumed that hydrogen-atom abstraction is slow. From experimental studies of thermal cracking of a series of paraffins, the relative rates of hydrogen-atom abstraction were found to be 1 : 3.66 : 11.4, in good agreement with theory [10].

Product distributions resulting from the thermal cracking of naphtha and of *n*-hexadecane are collected in Table 1-4. As expected, high yields of ethylene were observed with each feedstock. An even distribution of other products was observed in *n*-hexadecane cracking; at relatively high temperatures, the product containing C_5 and higher-molecular-weight compounds was highly aromatic. The Rice theory gives a good prediction of the product distribution for *n*-hexadecane cracking (Table 1-4).

Table 1-4 Thermal cracking of hydrocarbons

Commercial-scale naphtha cracking[a]		Laboratory cracking of *n*-hexadecane[b]		
			Mol/ 100 mol cracked	
Product	Mol %	Product	Obs.	Calc.
Hydrogen	16.2	C_1	53	61
Methane	25.7	C_2	130	139
Acetylene	1.4	C_3	60	50
Ethylene	30.7	C_4	23	27
Propadiene	2.5	C_5	9	15
Propylene	0.8	C_6	24	17
Propane	7.5	C_7	16	14
Butadiene	0.2	C_8	13	12
Butylene	2.4	C_9	10	11
Butane	1.0	C_{10}	11	10
C_{5+} liquid	11.5[c]	C_{11}	9	9
Total	100.0	C_{12}	7	8
		C_{13}	8	7
		C_{14}	5	7
H/C ratio of C_{5+} liquid = 1.0		iC_4/C_4 ratio = 0.07		

[a] Contact time in tube furnace = 0.05 to 0.1 s; temperature ≈ 900°C [11].
[b] Liquid hourly space velocity = 0.05. Reactor filled with quartz chips, temperature = 500°C, conversion 31.5 wt % [10].
c Estimated on the basis of C_5H_5.

In summary, the important characteristics of thermal cracking reactions are scission at a bond located β to the carbon atom having the unpaired electron, infrequent transfer of the radical from one hydrocarbon chain to another, and the inability of the unpaired electron to move from one carbon atom to another on the chain; i.e., the free radical cannot undergo isomerization reactions such as a methyl shift. The theory of thermal cracking satisfactorily predicts the high yields of ethylene, the low yields of methane, the low yields of evenly distributed α-olefins, the absence of isomerized products, and the high ratio of olefinic to paraffinic products [10].

Catalytic Cracking

Cracking reactions catalyzed by acidic surfaces proceed via surface carbonium-ion intermediates. This generalization is an inference from the known surface acid properties of cracking catalysts and the similarities in product distribution resulting from catalysis of many reactions by acidic surfaces and by acid solutions such as H_2SO_4. Carbonium-ion reactions in solution have been thoroughly studied, and the structures and reactivities of these species are well understood, as summarized in the following pages.

Properties of carbonium ions The heteropolar rupture of the C—H bond of a hydrocarbon molecule can lead to formation of either a carbonium ion or a carbanion:

$$\geq C-H \longrightarrow \geq C^+ + H^- - E_+ \tag{20}$$

$$\geq C-H \longrightarrow \geq C^- + H^+ - E_- \tag{21}$$

The energies E_+ and E_- associated with these reactions include the ionization energies and the electron affinities of the hydrogen and alkyl groups along with the dissociation energy of the C—H bond. The energy required for carbonium-ion formation increases with an increase in the number of H atoms attached to the carbon atom from which the hydride ion is abstracted. The stability of carbonium ions decreases in the order of increasing E_+:

Tertiary > secondary > primary > methyl

Values of E_+ based on measurements of appearance potentials for the gas-phase ions in a mass spectrometer are available (Table 1-5), but the corresponding data for solution- and surface-reactions are lacking. The tertiary carbonium ion is by far the stablest, and it is the easiest to form and the most prevalent whenever it can be formed. This pattern is valid for carbonium ions in solution and on surfaces, as inferred from product distributions of many reactions.

Table 1-5 Relative stabilities of gas-phase carbonium ions
[12, 13]

Type of ion	Relative value of E_+, kcal/mol
$(\geq C)_3C^+$ (tertiary)	0
$(\geq C)_2CH^+$ (secondary)	14
$\geq C-CH_2^+$ (primary)	21

Carbonium ions can be formed by several different routes, a common example being the interaction of an acid with an unsaturated hydrocarbon acting as a weak base:

$$H_2C{=}CHCH_3 + HX \rightleftharpoons H_3C{-}\overset{\displaystyle H}{\underset{+}{C}}CH_3 + X^- \tag{22}$$

In this reaction, a secondary rather than a primary carbonium ion is formed since it is the more stable.

Aromatic hydrocarbons can also act as proton acceptors, e.g.,

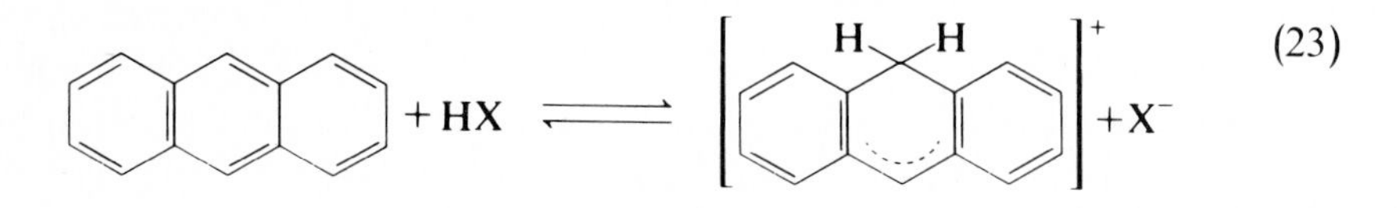

Carbonium-ion formation from a paraffin requires the abstraction of a hydride ion; e.g.,

$$RH + HX \rightleftharpoons R^+ + X^- + H_2 \tag{24}$$

or

$$RH + \underset{\text{Lewis acid}}{L} \rightleftharpoons LH^- + R^+ \tag{25}$$

A similar hydride-ion transfer can occur upon interaction of a carbonium ion with a saturated hydrocarbon to form a new carbonium ion:

$$R_1^+ + R_2H \rightleftharpoons R_1H + R_2^+ \tag{26}$$

Because of the relatively high stability of tertiary carbonium ions, a tertiary carbon atom readily donates a hydride ion to a primary or secondary carbonium ion; other transfers are slower.

Reactions of carbonium ions An important reaction of carbonium ions is their rearrangement by hydrogen-atom and carbon-atom shifts. The former leads to a double-bond isomerization of an olefin:

$$H_2C{=}CH{-}CH_2{-}CH_2{-}CH_3 \qquad CH_3{-}CH{=}CH{-}CH_2{-}CH_3$$

$$\searrow\!\!\nwarrow \; H^+ / -H^+ \qquad +H^+ / -H^+ \; \nearrow\!\!\swarrow \tag{27}$$

$$CH_3{-}\underset{+}{C}H{-}CH_2{-}CH_2{-}CH_3$$

The following skeletal rearrangement involves a methyl-group shift:

$$\underset{}{CH_3\overset{\overset{\displaystyle CH_3}{|}}{C}{=}CHCH_2CH_3} \underset{-H^+}{\overset{H^+}{\rightleftharpoons}} CH_3\underset{+}{\overset{\overset{\displaystyle CH_3}{|}}{C}}CH_2CH_2CH_3 \overset{\text{H shift}}{\rightleftharpoons}$$

$$CH_3\underset{\underset{\displaystyle H}{|}}{\overset{\overset{\displaystyle CH_3}{|}}{C}}\overset{}{\underset{+}{C}}HCH_2CH_3 \underset{\text{methyl-group shift}}{\rightleftharpoons} CH_3\underset{+}{C}H\underset{\underset{\displaystyle H}{|}}{\overset{\overset{\displaystyle CH_3}{|}}{C}}CH_2CH_3 \overset{\text{H shift}}{\rightleftharpoons}$$

$$CH_3CH_2\underset{+}{\overset{\overset{\displaystyle CH_3}{|}}{C}}CH_2CH_3 \underset{H^+}{\overset{-H^+}{\rightleftharpoons}} CH_3CH{=}\overset{\overset{\displaystyle CH_3}{|}}{C}CH_2CH_3 \tag{28}$$

Isomerization of saturated hydrocarbons takes place through these carbonium-ion intermediates, but it requires a hydride-ion abstraction as a first step, as illustrated below for *n*-pentane [14]:

$$CH_3CH_2CH_2CH_2CH_3 + HX \rightleftharpoons$$

$$CH_3CH_2\underset{+}{C}HCH_2CH_3 + X^- + H_2 \tag{29}$$

$$CH_3CH_2\underset{+}{C}HCH_2CH_3 \rightleftharpoons CH_3\overset{\overset{\displaystyle CH_3}{|}}{C}H\underset{+}{C}HCH_3 \tag{30}$$

$$CH_3\overset{\overset{\displaystyle CH_3}{|}}{C}H\underset{+}{C}HCH_3 \rightleftharpoons CH_3\underset{+}{\overset{\overset{\displaystyle CH_3}{|}}{C}}CH_2CH_3 \tag{31}$$

$$CH_3\underset{+}{\overset{\overset{\displaystyle CH_3}{|}}{C}}CH_2CH_3 + H_2 + X^- \rightleftharpoons CH_3\overset{\overset{\displaystyle CH_3}{|}}{C}HCH_2CH_3 + HX \tag{32}$$

Reaction (32) could be replaced by a hydride-ion transfer from a paraffin:

$$CH_3\underset{+}{\overset{\overset{\displaystyle CH_3}{|}}{C}}CH_2CH_3 + CH_3CH_2CH_2CH_2CH_3 \rightleftharpoons CH_3\underset{\underset{\displaystyle H}{|}}{\overset{\overset{\displaystyle CH_3}{|}}{C}}CH_2CH_3$$

$$+ CH_3CH_2\underset{+}{C}HCH_2CH_3 \tag{33}$$

In this way, the paraffin isomerization can proceed by a chain reaction, many *n*-pentane molecules being isomerized for each occurrence of reaction (29).

A high hydrogen partial pressure is expected to retard reaction (29) and enhance reaction (32), in agreement with experimental results for the isomerization of straight-chain paraffins in strong-acid solutions. Similarly, saturated hydrocarbons having relatively strong tendencies to donate hydride ions, such as methylcyclopentane or isopentane, inhibit the isomerization reaction.

Carbonium ions are also intermediates in reactions leading to the formation and breaking of C—C bonds. The polymerization of olefins catalyzed by acids is illustrated by

$$H_2C{=}CHCH_3 + HX \rightleftharpoons CH_3{-}\underset{+}{C}HCH_3 + X^- \tag{34}$$

$$CH_3{-}\underset{+}{C}HCH_3 + H_2C{=}CHCH_3 \rightleftharpoons \begin{array}{l} CH_3{-}CHCH_3 \\ \quad\;\; | \\ \quad CH_2\underset{+}{C}HCH_3 \end{array} \tag{35}$$

and so forth. The secondary carbonium ions shown above predominate over the less stable primary ions, with the consequence that highly branched polymers are formed in the reaction.

Cracking is essentially the reverse of polymerization, occurring at the bond located β to the carbon atom bearing the positive charge. Cracking of a straight-chain secondary carbonium ion results in the formation of a primary carbonium ion:

$$RCH_2\underset{+}{C}HCH_2{-}CH_2CH_2R' \rightleftharpoons RCH_2CH{=}CH_2 + \underset{+}{C}H_2CH_2R' \tag{36}$$

The primary ion can undergo a rapid hydrogen shift to give a more stable secondary carbonium ion:

$$\underset{+}{C}H_2CH_2R \rightleftharpoons CH_3\underset{+}{C}HR \tag{37}$$

The continued pattern of cracking of the straight chain at the β position leads to formation of propylene, in high yields; ethylene is not formed by this mechanism. High yields of ethylene are therefore indicative of free-radical (thermal) cracking, whereas high yields of propylene are indicative of catalytic cracking.

Carbon-carbon bond formation occurs not only in polymerization but in the industrially important processes of alkylation. Alkylation of an aromatic involves the electrophilic attack of a carbonium ion on the aromatic ring. For example, the propylation of benzene is catalyzed by H_3PO_4 held in the pores of kieselguhr:

$$H_2C{=}CHCH_3 + HX \rightleftharpoons CH_3\underset{+}{C}HCH_3 + X^- \tag{38}$$

$$CH_3{-}\underset{+}{C}H{-}CH_3 + C_6H_6 \rightleftharpoons \left[C_6H_6^{+} \text{(H)} {-}C(CH_3)_2{-}H \right]^{\ddagger} \xrightarrow{X^-} C_6H_5{-}CH(CH_3)_2 + HX \tag{39}$$

Alkylation of isoparaffins by olefins requires stronger acids, e.g., concentrated H_2SO_4 or HF; the reaction involves hydride-ion transfer and a chain reaction such as

$$H_2C{=}CHCH_3 + HX \rightleftharpoons CH_3{-}\overset{+}{C}HCH_3 + X^- \tag{40}$$

$$CH_3\overset{+}{C}HCH_3 + CH_3{-}\underset{H}{\overset{CH_3}{C}}{-}CH_3 \rightleftharpoons CH_3CH_2CH_3 + CH_3{-}\underset{+}{\overset{CH_3}{C}}{-}CH_3 \tag{41}$$

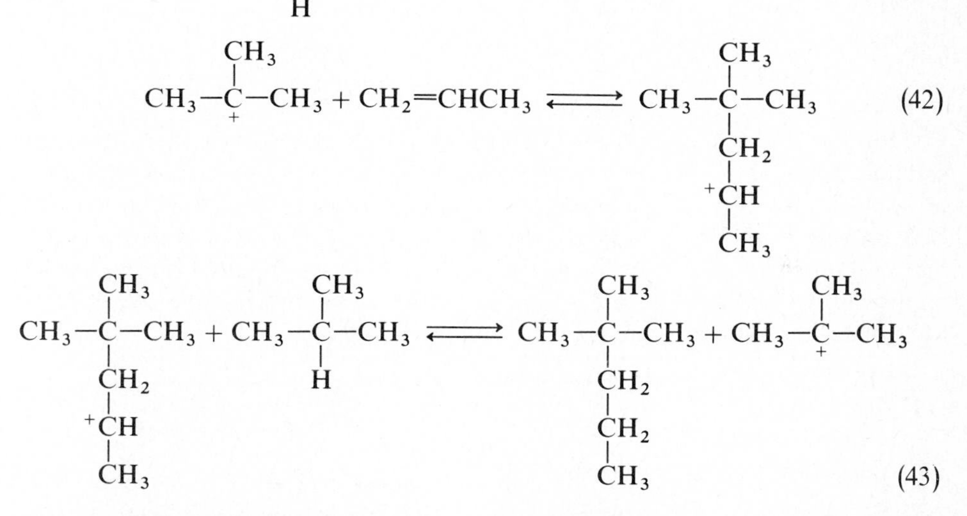

The *t*-butyl carbonium ion serves as a chain carrier. The process gives highly branched heptanes, and the yield is high, provided that polymerization is suppressed by use of high isobutane-to-olefin ratios. The primary product is 2,4-dimethylpentane, however, which indicates that substantial isomerization precedes reaction (43).

When hydrocarbons contact strong acids, a side reaction almost always occurs to give a high-molecular-weight, highly unsaturated material called *coke* or *tar*. Coke formed on surfaces of cracking catalysts remains there, causing severe loss of cracking activity. In practice, the coke is removed by oxidation in a fluidized-bed regenerator, as already mentioned. Coke can be formed from monoolefins by dehydrogenation and cyclization reactions proceeding through carbonium-ion intermediates. The hydrogens on the carbon atoms located α to the double bond of an olefin molecule are especially susceptible to hydride abstraction by a carbonium ion, since hydride-ion abstraction from this position results in the formation of a resonance-stabilized allylic carbonium ion:

$$>C{=}\overset{H}{C}{-}\underset{H}{\overset{H}{C}}{-}H \xrightarrow[-H^-]{SiO_2\text{-}Al_2O_3} >C{=}\overset{H}{C}{-}\overset{+}{C}\begin{matrix}H\\H\end{matrix} \rightleftharpoons \left(>C{\cdots}\overset{H}{C}{\cdots}\underset{H}{\overset{H}{C}}\right)^+ \tag{44}$$

Ultraviolet spectra assigned to an allylic carbonium ion have been observed for solutions of dienes in H_2SO_4 [16] and monoolefins adsorbed on SiO_2–Al_2O_3:

$$H_2C{=}CH{-}\underset{\displaystyle CH_3}{\underset{|}{C}}{=}CH_2 \underset{-H^+}{\overset{+H^+}{\rightleftharpoons}} \left(CH_2{\cdots}CH{\cdots}\underset{\displaystyle CH_3}{\underset{|}{C}}{-}CH_3\right)^+ \qquad (45)$$

Reactions of allylic carbonium ions can lead to the further unsaturation of hydrocarbons, as the carbonium ion back-donates a proton to the conjugate base to form a conjugated diene:

$$(R_2CH{\cdots}CH{\cdots}CH{-}CH_2R_3)^+ + X^- \rightleftharpoons R_2CH{=}CH{-}CH{=}CHR_3 + HX \qquad (46)$$

The overall reaction results in the saturation of one hydrocarbon molecule and increases the degree of unsaturation of another. Since the hydrogens on carbons located α to the double bond of the diene are now susceptible to further hydride-ion abstraction, reactions leading to higher degrees of unsaturation take place. Cyclization of trienes formed in this way occurs rapidly. The foregoing sequence of reactions therefore accounts for formation of aromatic compounds from olefins; the final reactions in the sequence are

$$R_1^+ + R_2{-}CH{=}CH{-}CH{=}CH{-}CH_2{-}CH_2CH_3 \rightleftharpoons R_1H + (R_2{-}CH{\cdots}CH{\cdots}CH{\cdots}CH{\cdots}CH{-}CH_2CH_3)^+ \qquad (47)$$

$$X^- + (R_2{-}CH{\cdots}CH{\cdots}CH{\cdots}CH{\cdots}CH{-}CH_2CH_3)^+ \rightleftharpoons R_2{-}CH{=}CH{-}CH{=}CH{-}CH{=}CHCH_3 + HX \qquad (48)$$

$$\begin{array}{ccc} & R_2{-}CH & \\ H_3C{-}CH & & {=}CH \\ \| & & | \\ CH & & CH \\ & CH & \end{array} \rightleftharpoons \begin{array}{ccc} & R_2 & \\ & | & \\ & CH & \\ H_3C{-}CH & & CH \\ | & & \| \\ CH & & CH \\ & {=}CH & \end{array} \qquad (49)$$

$$R_1^+ + \begin{array}{ccc} & R_2 & \\ & | & \\ & CH & \\ H_3C{-}CH & & CH \\ | & & \| \\ CH & & CH \\ & {=}CH & \end{array} \rightleftharpoons \begin{array}{ccc} & R_2 & \\ & | & \\ & C^+ & \\ H_3C{-}CH & & CH \\ | & & \| \\ CH & & CH \\ & {=}CH & \end{array} + R_1H \qquad (50)$$

$$X^- + \begin{array}{ccc} & R_2 & \\ & | & \\ & C^+ & \\ H_3C{-}CH & & CH \\ | & & \| \\ CH & & CH \\ & {=}CH & \end{array} \longrightarrow H_3C{-}C_6H_3{-}R_2 \text{ (benzene ring)} + HX \qquad (51)$$

Hydride-ion abstraction could alternatively take place at a position several carbons removed from the double bond, the result being the formation of an olefinic carbonium ion:

$$R_1^+ + R_2CH{=}CH{-}CH_2CH_2CH_2CH_2CH_3 \rightleftharpoons R_1H + R_2CH{=}CH{-}CH_2CH_2CH_2\overset{+}{C}HCH_3 \quad (52)$$

This carbonium ion could undergo intramolecular attack on the double bond:

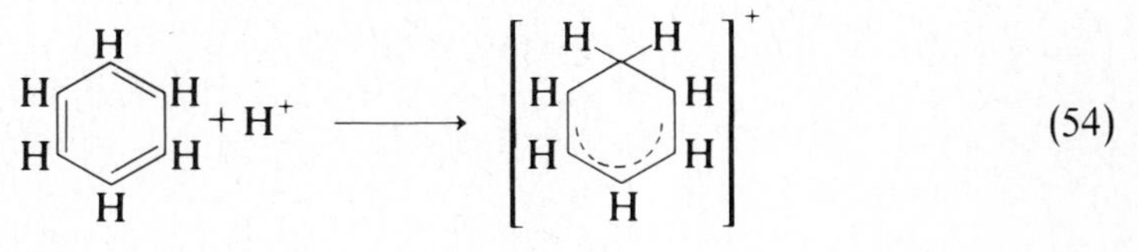

These reactions provide another route to cyclization and formation of aromatics.

Once aromatics are present, they can react to give higher-molecular-weight hydrocarbons and coke, which is formed by condensations such as reaction (12) and by combination of aromatics with other aromatics or coke. The formation of coke from benzene is illustrated by the following reaction sequence:

Step 1: initiation:

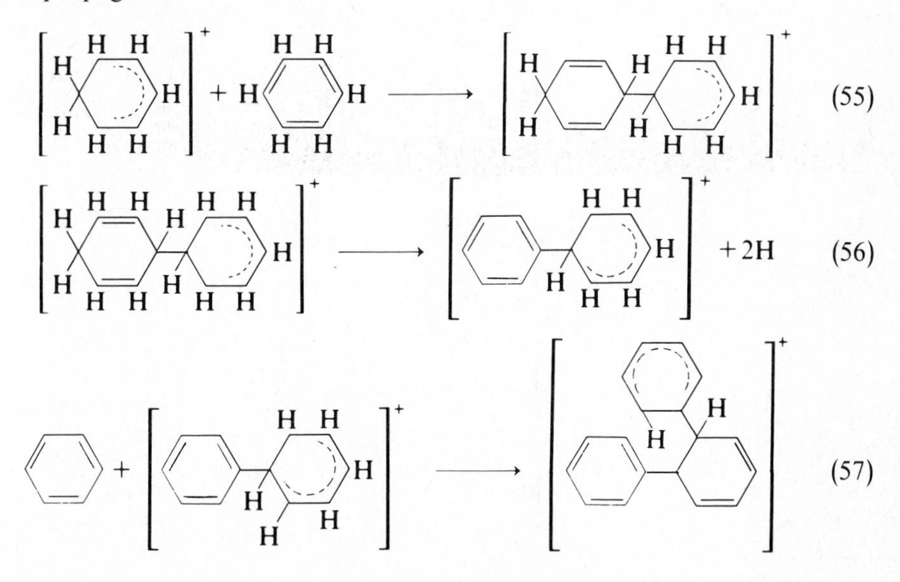

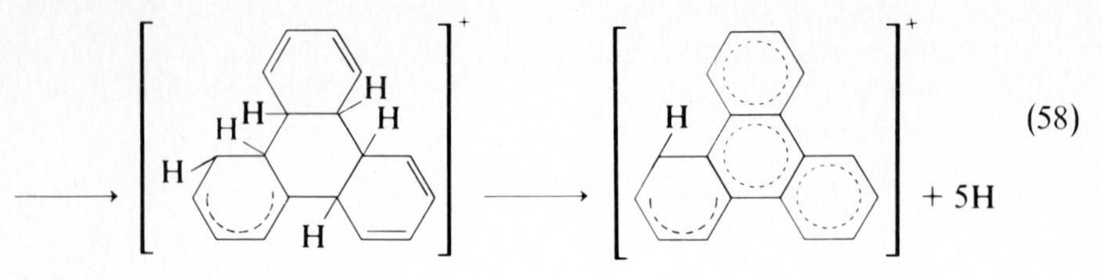

(58)

and so forth.

Step 3: termination:

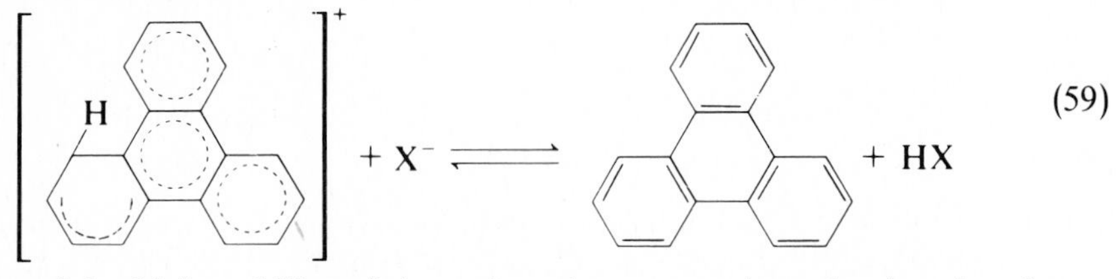

(59)

Because of the high stability of the polynuclear aromatic carbonium ion, it can continue to grow on the surface for a relatively long time before a termination reaction occurs through back donation of a proton.

Characterization of acidity To understand the mechanisms of cracking and related acid-catalyzed hydrocarbon reactions, it is necessary to understand the mechanism of carbonium-ion formation. The rate of protonation of a base (B) for the most part increases with increasing strength of the acid providing the proton. The strength of an acid may be defined as its ability to donate a proton to a neutral base:

$$H^+ + B \rightleftharpoons BH^+ \tag{60}$$

If

$$K_a = \frac{\hat{a}_{H^+} \hat{a}_B}{\hat{a}_{BH^+}}$$

where $\hat{a}_i$ is the thermodynamic activity of species i, then

$$K_a = \hat{a}_{H^+} \frac{f_B C_B}{f_{BH^+} C_{BH^+}} \tag{61}$$

where f_i is the activity coefficient of species i.

By definition,

$$pK_a = -\log K_a = -\log \left(\hat{a}_{H^+} \frac{f_B C_B}{f_{BH^+} C_{BH^+}} \right) \tag{62}$$

and

$$pK_a = -\log \left(\hat{a}_{H^+} \frac{f_B}{f_{BH^+}} \right) - \log \frac{C_B}{C_{BH^+}} \tag{63}$$

The Hammett acidity function is defined as

$$H_0 = -\log\left(\hat{a}_{H^+}\frac{f_B}{f_{BH^+}}\right) \tag{64}$$

Clearly, H_0 depends not only on the proton-donor strength of a solution but also on the base B. Hammett used a series of aniline bases as indicators to make f_B/f_{BH^+} nearly constant and obtain self-consistent data for a range of acid strengths, which required bases of very different strengths to give the required color changes.

It is evident from the definition of H_0 that as the limit of a dilute aqueous solution is approached, the value of H_0 approaches the pH. The Hammett acidity function is similar to pH in that it becomes more negative the greater the acidity of the solution. The value reaches -10 or less in the most concentrated solutions of H_2SO_4 and $HClO_4$.

The acidity of a solid surface can be determined by using the Hammett indicators [17]. The solid is dispersed in a nonpolar solvent, and a series of indicators requiring successively weaker acid strengths to become protonated is added until one becomes protonated, spectroscopically exhibiting a color change. When a color change is observed for a given indicator, it is concluded that some of the sites on the surface correspond to the pK_a value characterized by that indicator. The data obtained in this simple way are sufficient to determine only qualitatively the distribution of acid strengths of various surface groups, and correlations between rate constants of protonation reactions and acidity functions for a series of catalysts would not be expected for surfaces unless more quantitative measurements of the numbers and strengths of acid sites were made.

The strength and number of acid sites on a solid can be obtained more explicitly by determining quantitatively the adsorption of a base such as ammonia, quinoline, pyridine, or trimethylamine. This method is especially valuable because the experiments can be carried out under conditions similar to reaction conditions and infrared spectra of the surface can also be obtained. Monitoring of sample weight, system pressure, and infrared band intensity of the adsorbed base determines the number of acid sites combined with the base.

The infrared technique is powerful for studying both Brønsted and Lewis acidity of surfaces [18, 19], and cracking catalysts such as SiO_2–Al_2O_3 and the zeolites have been found to have both Brønsted and Lewis acid sites. For example, ammonia can adsorb on a surface physically as NH_3; it can be bonded to a Lewis acid site as coordinatively bound NH_3; or it can be adsorbed on a Brønsted acid site as NH_4^+. Each of these species is independently identifiable from its characteristic infrared absorption bands. Pyridine similarly adsorbs on Lewis acid sites as coordinatively bound pyridine and on Brønsted acid sites as the pyridinium ion. These two species can also be distinguished by their infrared spectra, allowing the numbers of Lewis and Brønsted acid sites on a surface to be determined quantitatively [20, 21].

The solid cracking catalysts are not unique in having both Brønsted and Lewis acid character. There is a comparable class of solutions formed from combinations of Brønsted and Lewis acids. Some examples are

Bronsted acids		Lewis acids
H_2O	+	BF_3
HF	+	BF_3
HF	+	SbF_5
FSO_3H	+	SbF_5
HCl	+	$AlCl_3$

The most strongly acidic of these solutions are referred to as *superacids* because they exhibit extraordinarily high proton-donor strengths; for example, $FSO_3H + SbF_5$ can even protonate methane [22]. Solids with analogous character can be formed from sulfonic acid ion-exchange resin and $AlCl_3$ and from sulfonated alumina and SbF_5. These initiate cracking of *n*-hexane at temperatures less than 100°C [23].

The high proton-donor strength of the superacids results in part from the action of the Lewis acid in forming a complex with the conjugate base of the Brønsted acid. For example, the equilibrium of reaction (65) is shifted to the right because the F^- ions are drained away by the SbF_5:

$$HF + B \rightleftharpoons (HB)^+ + F^- \tag{65}$$

$$F^- + SbF_5 \rightleftharpoons (SbF_6)^- \tag{66}$$

Reactions between the Brønsted and Lewis acids also take place; e.g., in $FSO_3H + SbF_5$, a very strongly acidic species, $H_2SO_3F^+$, can be formed along with the following complex, among others:

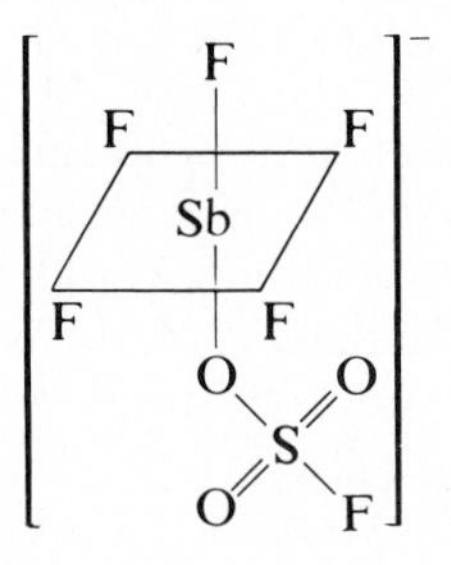

Mechanisms of carbonium-ion reactions The donation of a proton to an olefin or an aromatic molecule may often lead to the formation of a σ bond between C and H. The bond formation probably involves the π-bond electron pair of the hydrocarbon, the empty *s* orbital of the proton, and the full sp^3 orbital of carbon. The bond formation leaves a positive charge on the second carbon atom, which was initially involved in the double bond. The interaction may be represented as

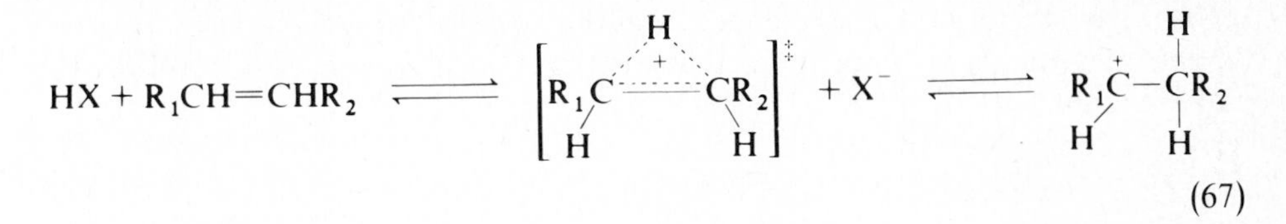

The carbonium ion produced in this reaction, often referred to as a *classical* carbonium ion,† is one having an sp^2-hybridized, electron-deficient central carbon atom, which, in the absence of constraining skeletal rigidity or steric interference, has a planar configuration with an empty p orbital perpendicular to the plane of the sp^2 bonds. The structure is illustrated in Fig. 1-3 for the secondary propyl carbonium ion.

The protonation of a saturated hydrocarbon requires an acid with very high proton-donor strength, such as $FSO_3H + SbF_5$. Paraffin protonation is believed to involve an intermediate containing a five-coordinated carbon atom [22].§ The formation of the *nonclassical* carbonium ion from methane is

$$CH_4 + H^+ \rightleftharpoons [CH_5]^+ \equiv \left[H_3C\cdots\begin{smallmatrix}H\\H\end{smallmatrix}\right]^+ \qquad (68)$$

(The right-hand structure is a simplified representation and is not to be mistaken for a 2-carbon species.) This ion involves three covalent two-electron bonds, the fourth bond being a two-electron three-center bond [24]. The dashed lines in the simplified representation indicate the interaction of a carbon valence orbital and a molecular orbital with two electrons of H_2. A similar configuration appears to exist in the BH_5 molecule [24].

The five-coordinate species can react further with cleavage of the

$$\left[H_3C\cdots\begin{smallmatrix}H\\H\end{smallmatrix}\right]^+$$

† In the system of nomenclature proposed by Olah [22], this species is referred to as a *carbenium ion*, corresponding to the trivalent state of the carbon atom bearing the positive charge.

§ This intermediate species is referred to by Olah as a *carbonium ion* [22].

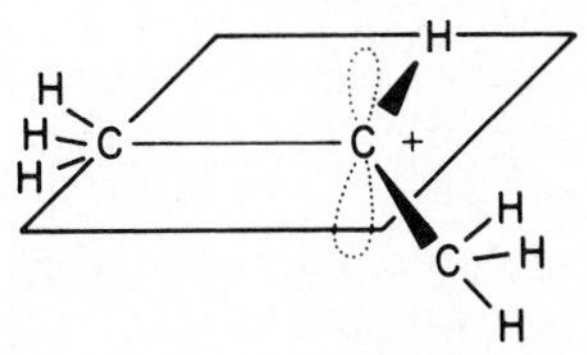

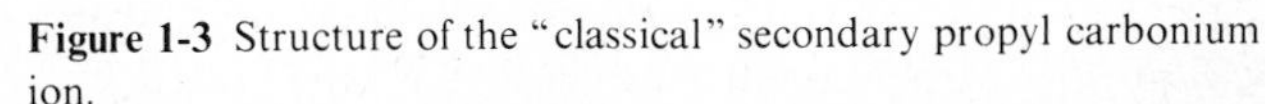

Figure 1-3 Structure of the "classical" secondary propyl carbonium ion.

bond to give the highly unstable methyl carbonium ion, H_3C^+, and H_2. A protonated paraffin containing more than one carbon can also experience C—H bond scission:

$$\mathrm{R_3{-}CH_2{-}CH_3 + H^+ \rightleftharpoons \left[R_3{-}\overset{H}{\underset{H\cdots H}{C}}{-}CH_3\right]^+ \rightleftharpoons R_3{-}\overset{H}{\underset{+}{C}}{-}CH_3 + H_2} \qquad (69)$$

Alternatively, a protonated paraffin can experience C—C bond scission:

$$\mathrm{R_3{-}CH_2{-}CH_3 + H^+ \rightleftharpoons \left[R_3{-}\overset{H}{\underset{H}{C}}\cdots\langle^{CH_3}_{H}\right]^+ \rightleftharpoons R_3{-}\overset{H}{\underset{H}{C^+}} + CH_4} \qquad (70)$$

The transition-state structures of the carbonium ion during alkyl or hydrogen shifts have been suggested to involve triangular species like those below:

$$\mathrm{RCH_2{-}CH_2{-}\overset{+}{\underset{H}{C}}{-}CH_2{-}CH_3 \rightleftharpoons \left[RCH_2{-}CH_2{-}\overset{\overset{CH_3}{\cdot+\cdot}}{CH{-}CH_2}\right]^{\ddagger} \rightleftharpoons RCH_2{-}CH_2{-}\overset{CH_3}{\overset{|}{C}}H{-}CH_2^+} \qquad (71)$$

and

$$\mathrm{RCH_2CH_2\overset{CH_3}{\overset{|}{C}}HCH_2^+ \rightleftharpoons \left[RCH_2CH_2\overset{CH_3}{\overset{|}{C}}{-}\underset{\underset{H}{\cdot+\cdot}}{CH_2}\right]^{\ddagger} \rightleftharpoons RCH_2CH_2\overset{CH_3}{\underset{+}{\overset{|}{C}}}CH_3} \qquad (72)$$

The mechanism has been refined by Brouwer [15] to account for observations involving isomerization of *n*-butane-1-^{13}C and of *n*-pentane catalyzed by superacid solutions of HF + SbF_5. *n*-Pentane undergoes rapid isomerization to isopentane, whereas *n*-butane undergoes only slow isomerization to isobutane under the same conditions; however, *n*-butane-1-^{13}C is rapidly isomerized to *n*-butane-2-^{13}C:

$$\mathrm{CH_3CH_2CH_2^{13}CH_3 \rightleftharpoons CH_3CH_2^{13}CH_2CH_3} \qquad (73)$$

The rate of carbon scrambling in the *n*-butane is approximately equal to the rate of isomerization of *n*-pentane to isopentane.

To account for this carbon-atom isomerization, Brouwer proposed a *protonated-cyclopropane intermediate*, which can be formed from the classical carbonium ion:

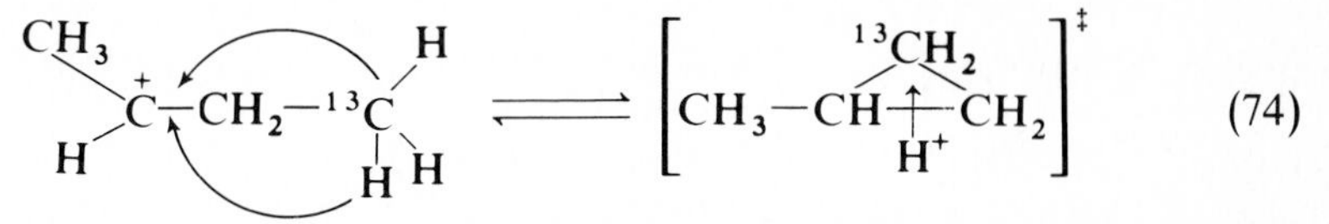

The location of the proton on the cyclopropane ring has not been determined, although quantum-mechanical calculations indicate that edge protonation of the ring gives the stablest form, shown as the species at the top of Fig. 1-4. The proton can be present in the three equivalent positions I, II, and III (Fig. 1-4). This intermediate can shift back to a classical carbonium ion if one of the C—C bonds is broken, provided that the proton bonds to the carbon atom having the free electron pair. There are three possible isomerization products, depending upon which ring C—C bond is broken (Fig. 1-4).

Breaking bond I gives back the original carbonium ion; breaking bond II would give skeletally isomerized butane, but this reaction does not readily occur because of the low stability of the primary carbonium ion which would have to be formed; breaking bond III gives the observed isomerization product. It is known that *n*-butane is isomerized to isobutane under somewhat more severe conditions

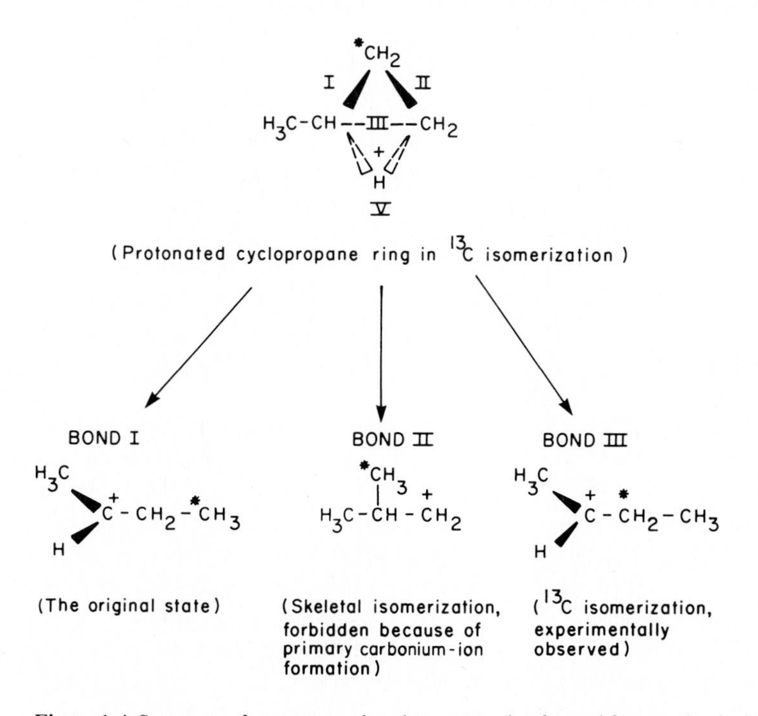

Figure 1-4 Structure of a protonated cyclopropane ring formed from a classical *s*-butyl carbonium ion and the carbonium ions formed by rupture of the indicated C–C bonds.

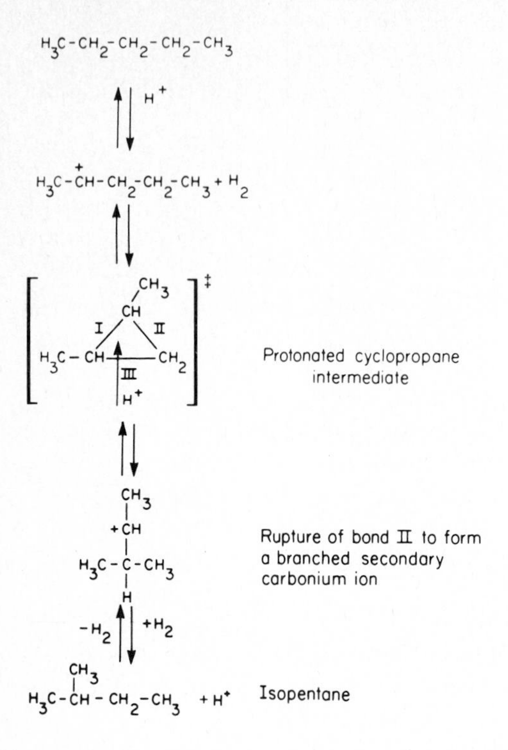

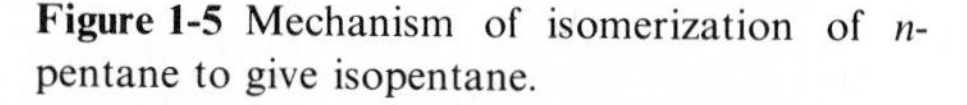
Figure 1-5 Mechanism of isomerization of *n*-pentane to give isopentane.

than applied by Brouwer; it appears that the mechanism of the isobutane formation may involve C—C bond scission, leading to formation of a methyl carbonium ion, which then recombines with a C_3 fragment [25, 26].

The mechanism illustrated in Fig. 1-4 readily explains the rapid isomerization of *n*-pentane when *n*-butane isomerization is slow; a $CHCH_3$ group swings and forms the protonated cyclopropane ring. Rupture of bond II of the protonated cyclopropane ring then leads to a secondary carbonium ion and skeletal isomerization, as shown in Fig. 1-5.

The final step in isomerization is a hydride abstraction converting the carbonium ion back into a paraffin. Since this carbonium ion has been formed by isomerization, it is usually a relatively stable tertiary carbonium ion. Hydride-ion exchange from the nonisomerized hydrocarbon (lacking tertiary carbon atoms) to the carbonium ion is energetically unfavorable and therefore slow. To increase the rate of isomerization it is useful to add an effective hydride donor (such as methylcyclopentane) or to operate under H_2 pressure. Too high a concentration of the hydride donor or too high a partial pressure of H_2, however, leads to conversion of the carbonium ion before it is isomerized. When reactions of paraffins are studied as a function of hydride-donor concentration, a pattern like the following emerges:

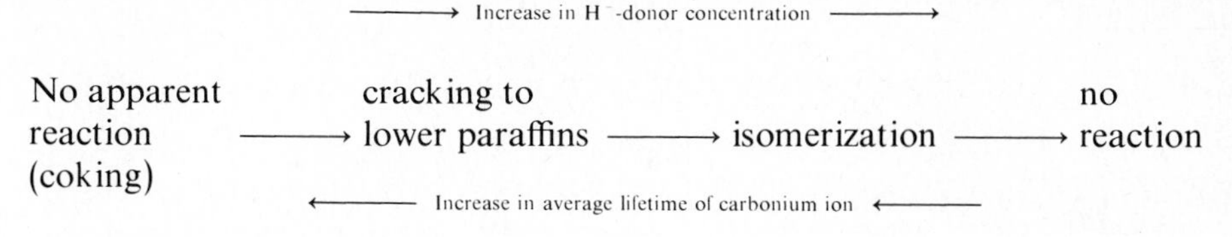

In this discussion of carbonium-ion reaction mechanisms, a variety of transition-state structures has been proposed (Fig. 1-6). Although these structures appear different, they are actually very similar because they all contain a delocalized electron-deficient bonding system of three centers and two electrons.

A single structural concept therefore suffices to explain qualitatively the complete set of reactions of carbonium-ion chemistry. Each of the structures has $2n$ electrons distributed over $2n + 1$ bonding atomic orbitals, where $n = 1, 2, 3, \ldots$. The interaction of a proton with a C—H bond (structure II) involves three orbitals (one of C, one of H, and one of the proton) and the two electrons from the C—H bond; hence $n = 1$. For the protonated cyclopropane ring, seven orbitals (two from each carbon atom and one from the proton) and six electrons (two from each C—C bond) are involved. In summary:

$$n = \begin{cases} 1 & \text{for proton + C—H bond (II)} \\ 2 & \text{for proton + C=C bond (I)} \\ 2 & \text{for carbonium ion + C=C bond} \\ 3 & \text{for proton + cyclopropane ring (IV)} \end{cases}$$

These intermediates are similar to the electron-deficient complexes B_2H_6 (BH_3 never occurs, only the dimer) and $Al_2(C_2H_5)_6$, which involve delocalized bond systems with four electrons and six orbitals. B_2H_6, for example, has a well-established structure with the two hydrogen atoms connecting the boron atoms above and below the plane containing the other hydrogens (Fig. 1-7).

Symmetry considerations show that such a system can be divided into two equivalent two-electron three-orbital systems. Cotton [27] has shown that if only "overlap" between 1 and 2 and 2 and 3 but not 1 and 3 (called the *open situation*) (Fig. 1-8) is taken into account, and if the orbitals are assumed to be similar, the

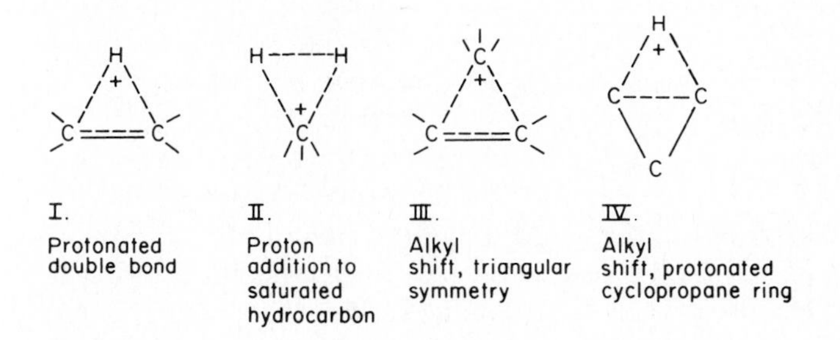

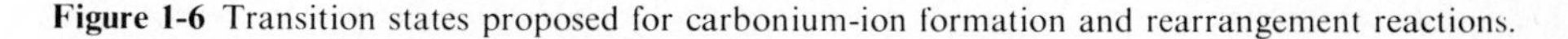

Figure 1-6 Transition states proposed for carbonium-ion formation and rearrangement reactions.

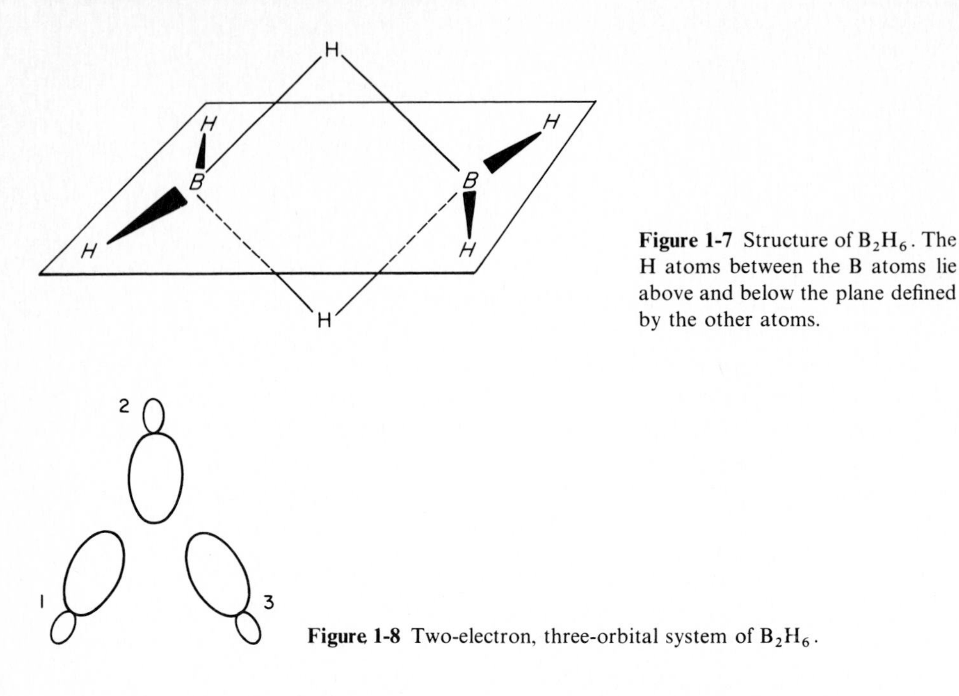

Figure 1-7 Structure of B_2H_6. The H atoms between the B atoms lie above and below the plane defined by the other atoms.

Figure 1-8 Two-electron, three-orbital system of B_2H_6.

energy-level scheme involves a bonding, a nonbonding, and an antibonding orbital, as shown in Fig. 1-9(*A*). This is the situation for the bonding bridge of B_2H_6 (Figs. 1-7 and 1-8). If, however, 1 and 2 and 2 and 3 overlap (the so-called *closed situation*), one more strongly bonding orbital and two antibonding orbitals occur, as shown in Fig. 1-9(*B*).

The transition states shown in Fig. 1-6 can now be discussed as a "normal" set of interacting atomic orbitals involving one or more localized bonds and a three-center two-electron delocalized system. The transition state for the alkyl-group shift (III), which involves four electrons and five orbitals, can be split into a localized bond with two electrons in a strongly bonding orbital and a high-energy empty antibonding orbital (two orbitals, two electrons) and into a delocalized

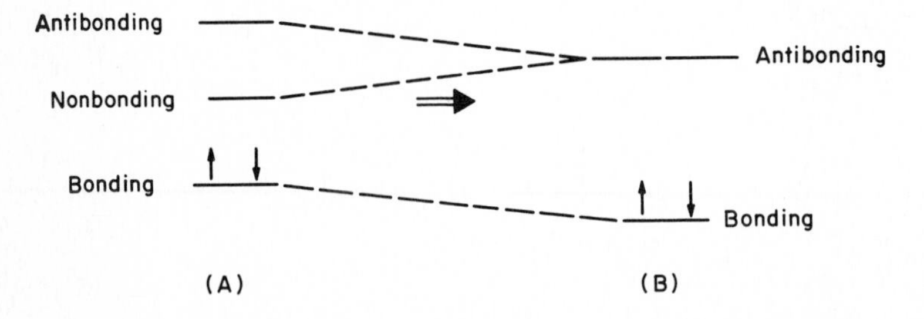

Figure 1-9 Energy levels in open three-center bonding (*A*) and closed three-center bonding (*B*).

three-orbital two-electron system. The three-center system is partly open (three-center bond) and partly closed (σ bond between two carbon atoms):

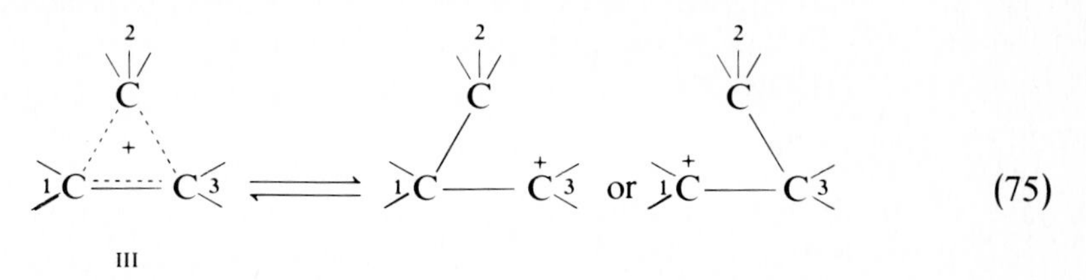

Similarly, the protonated cyclopropane ring (IV, Fig. 1-6) might be considered a combination of two localized C—C bonds (four orbitals and four electrons and a three-center two-electron delocalized system of two carbon atoms and a proton):

IV (76)

Both these intermediates are relevant to skeletal isomerization; they are similar to each other with respect to the electronic distribution in the delocalized state, and they represent possible transition states.

Carbonium ions on surfaces The nature of carbonium ions on surfaces is not as well understood as the nature of carbonium ions in solution, but the many product-distribution data for catalytic reactions catalyzed by solid acid surfaces leave little doubt that carbonium-ion mechanisms like those taking place in solution are also operative on surfaces. When a carbonium ion is formed, it becomes associated with an anion, becoming part of an ion pair. Reactions like the isomerizations just considered involve the movement of the positive charge of the carbonium ion along the carbon chain; this movement would result in the formation of a large potential energy if the negative charge were not to follow the positive charge. In solution, anion movement occurs readily, and only small charge separations exist. On a surface, however, movement of charge along a hydrocarbon chain cannot be accompanied easily by movement of the anion because it occupies a fixed position in the lattice. The mechanism whereby significant charge separation on surfaces is avoided is not clear. Perhaps reactions occur on surface sites having configurations allowing the positive charge to jump from one anion site to another separated by a certain distance, or perhaps the configuration of the hydrocarbon chain on the surface facilitates charge transfer from one carbon center to another with only a moderate energy barrier to be crossed. In a zeolite the charge may be distributed over the oxygen anions so that the molecular-scale pore acts as a solventlike envelope surrounding a carbonium ion.

Another explanation accounting for reaction without much charge separation is based upon the idea that carbonium ions may not actually be fully formed; instead, the molecules may become sufficiently polarized for the appropriate rearrangements to occur before a carbonium ion is formed. This idea is speculative for cracking and related reactions, but it is firmly based on many experimental results for reactions like eliminations catalyzed by solutions, enzymes, and surfaces. For example, elimination of water from alcohols and elimination of HCl from alkyl chlorides are reactions catalyzed by acids in solutions and on surfaces. Under certain conditions, these elimination reactions proceed through carbonium-ion intermediates, but under other conditions, nonionic intermediates are involved, as the catalyst removes a proton from the reactant molecule at one position and more or less simultaneously donates a proton and removes water at a neighboring position.

A whole spectrum of reaction mechanisms exists, characterized by the degree of charge development at a carbon atom. One limiting case is the carbonium-ion mechanism, and the other is a mechanism involving a hardly polarized reactant molecule, a concerted or synchronous mechanism involving simultaneous bond breaking and bond formation and the action of two catalyst functions. These reaction mechanisms can be studied with reactants having sufficient molecular complexity to permit the stereochemistry of the reaction to be inferred from the nature of the product. An example of a concerted mechanism proposed by Pines and Manassen [28], involving a nonionic intermediate bridged between opposite walls of a narrow alumina pore, is shown in Fig. 1-10. The predominant product, 1,9-octalin, originates by trans elimination and not from a carbonium ion. This example may be of particular interest because of the very narrow pore structure of zeolite cracking catalysts, which offer an excellent opportunity for such mechanisms. It may be significant that mordenite, a zeolite with especially narrow pores which provide solventlike surroundings to a molecule, shows superacidity and superactivity as a cracking catalyst.

It has been postulated that the active site on the surface of a zeolite cracking catalyst includes a combination of a Lewis acid site (an Al^{3+} ion with an empty

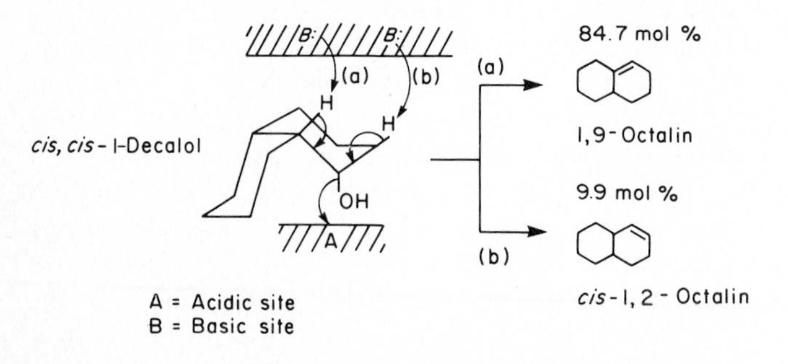

Figure 1-10 Proposed mechanism of dehydration of *cis,cis*-1-decalol catalyzed by alumina [28]. Reaction path (a), a trans-elimination, explains the predominant product. (Reprinted with permission from *Advances in Catalysis*. Copyright © by Academic Press.)

p orbital) and a Brønsted acid site (a proton-donating hydroxyl group), as will be discussed in detail later in this chapter. It is appropriate to suggest that cracking may proceed by multifunctional catalysis, involving these sites with surface anions and proceeding by concerted mechanisms. For the present, there is not sufficient information to allow us to determine whether it is an oversimplification to describe cracking reactions simply as carbonium-ion reactions.

Details of catalytic cracking chemistry *Cracking of pure hydrocarbons* Catalytic cracking reactions can now be considered in detail on the basis of the above discussion. The mechanistic interpretation is based on carbonium-ion reactions only, although the other possibilities mentioned above cannot be excluded.

Observed rates of cracking on silica-alumina are of the order of 0.01 to 0.5 molecule cracked per surface site per second; Mills et al. [29] reported rates of 0.2 molecule of cumene cracked per site per second and 0.03 to 0.2 molecule of gas oil cracked per site per second at 425°C. The rate of cracking is related to hydrocarbon structure and the concentration of adsorbed hydrocarbon. Figure 1-11 and Table 1-6 illustrate how cracking activity varies with carbon number and hydrocarbon structure for an alumina-zirconia-silica catalyst; the data were obtained at 500°C with a flow rate of 14 mol of hydrocarbon per liter of catalyst per hour [6].

When hydrocarbon reactants have tertiary carbon atoms, from which stable tertiary carbonium ions can be formed directly by hydride ion abstraction, they undergo rapid cracking; when the side chain has more than three carbon atoms, the presence of an aromatic ring also leads to relatively rapid cracking. Greensfelder et al. [10] estimated that tertiary carbons are 10 times as reactive as secondary carbons and 20 times as reactive as primary carbons. Correspondingly,

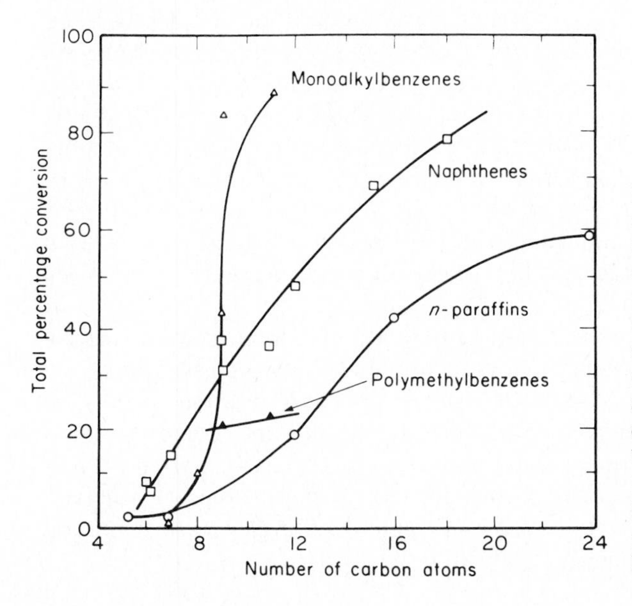

Figure 1-11 Comparison of various hydrocarbon classes in catalytic cracking. Reaction conditions: alumina-zirconia-silica catalyst; 500°C; 13.7 mol hydrocarbon feed/L catalyst · h; the conversions are integrated over 1 h of operation and therefore represent average activities for a deactivating catalyst [6]. (From "Catalysis," vol. VI, P. H. Emmett (ed.), copyright © 1958 by Litton Educational Publishing Company. Reprinted by permission of Van Nostrand Reinhold Company.)

Table 1-6 Cracking of various hydrocarbons[a] [6]

Compound	Number of carbon atoms	Conversion, %
n-Heptane	7	3
n-Dodecane	12	18
n-Hexadecane	16	42
2,7-Dimethyloctane	10	46
Decalin	10	44
Mesitylene	9	20
Isopropylbenzene	9	84
Cyclohexane	6	62
n-Hexadecene	16	90[b]

[a] Reaction conditions are specified in the legend to Fig. 1-11.
[b] Estimated from data obtained at 450°C.

Source: From "Catalysis," vol. VI, P. H. Emmett (ed.).

branched-chain paraffins, e.g., 2,7-dimethyloctane, crack more rapidly than *n*-paraffins, e.g., *n*-dodecane; and naphthenes, e.g., Decalin, crack more rapidly than straight-chain paraffins (Fig. 1-11). Decalin and 2,7-dimethyloctane crack at about equal rates, as there are tertiary carbons in both. Similar reactivity patterns have been observed with zeolite catalysts such as REHX.

Cracking of n-$C_{18}H_{38}$ catalyzed by REHX zeolite is 20 times faster than cracking of n-C_8H_{18} [30] (see also Fig. 1-11). The greater number of secondary carbons of the longer paraffin does not account for its higher reaction rate, and the result is inferred to mean higher surface concentration of the hydrocarbon with the greater chain length. Adsorption equilibrium data for hydrocarbons on zeolite catalysts (and on inert alumina) confirm the pattern of increased adsorption with increasing chain length.

The higher surface concentration of the long-chain hydrocarbon evidently corresponds to a higher rate of carbonium-ion formation and therefore a higher rate of cracking. This result is consistent with the idea that the formation of the carbonium ion is the slowest step in the surface-catalyzed reaction sequence. Similarly, for the solution-phase isomerization of paraffins catalyzed by superacids, hydride transfer from the paraffin to an alkyl carbonium ion has been demonstrated to be the rate-determining step [31].

Since protonation of a double bond is usually rapid compared with hydride-ion abstraction, it is to be expected that cracking of olefins would be more rapid than cracking of paraffins, provided that carbonium-ion formation is the slow step. This expectation has been confirmed experimentally; olefins crack at rates as much as two orders of magnitude greater than those observed for paraffin cracking [10, 30]. These observations and those showing the dependence of reactivity on hydrocarbon structure therefore support the supposition that carbonium-ion formation is the slow step in catalytic cracking.

Since formation of a surface carbonium ion from a paraffin is an apparently difficult step, it is appropriate to review the possible mechanisms:

1. Hydride-ion abstraction from the paraffin by a Lewis acid site on the catalyst surface.
2. Protonation of the paraffin by a Brønsted acid site on the surface via the nonclassical-carbonium-ion route to form hydrogen [reaction (69)], which is an observed product of catalytic cracking. This reaction might involve a proton-donor site located next to a Lewis acid site and therefore having an especially high acid strength, as will be discussed subsequently.
3. Olefin protonation; small amounts of olefin may be formed by thermal cracking at the elevated temperatures used in cracking, or olefins may be present as impurities in the feed to the reactor.

The role of olefins in initiating cracking of *n*-butane at 230°C catalyzed by the zeolite hydrogen mordenite has been shown by Weisz [32] (Fig. 1-12). The olefin content of the feed to a flow reactor containing the cracking catalyst was controlled by the temperature of the prereactor containing an active catalyst for olefin hydrogenation. The conversion of *n*-butane increased markedly with increasing olefin concentration, and therefore, for this example, the olefin protonation is inferred to have been the most important initiation mechanism.

Since catalytic cracking proceeds as a chain reaction, once carbonium ions are formed, the difficult step in paraffin activation probably becomes abstraction of a hydride ion by a carbonium ion. Data supporting this idea were reported by Aldridge et al. [20], who observed an induction period in the cracking of *n*-hexane catalyzed by LaX zeolite at 338°C in a recycle-flow reactor. The induction period is inferred to have been approximately the time required to form a steady-state concentration of surface carbonium ions from the paraffin. The reaction became much more rapid after the induction time had elapsed.

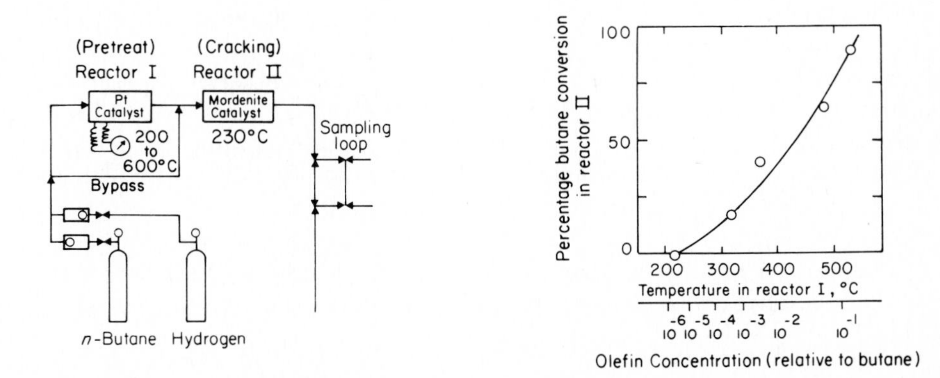

Figure 1-12 Apparatus and results establishing the role of olefins in initiating butane cracking [32]. (Reprinted with permission from *Chemtech*. Copyright by the American Chemical Society.)

As discussed previously, once a carbonium ion is formed, cracking occurs by scission of the carbon chain at a bond located β to the charged carbon atom to give an α-olefin and a smaller primary carbonium ion. The primary carbonium ion rapidly rearranges to the more stable secondary carbonium ion, and it then undergoes either β scission, to give more α-olefin, or further rearrangement. The catalytic cracking of long-chain paraffins gives high yields of C_3 to C_6 hydrocarbons, a maximum typically being observed for molecules containing 4 carbon atoms.

Instead of continuing to crack to form the smallest fragment possible, the carbonium ion can undergo chain transfer to form a paraffin and a new carbonium ion:

$$R_1CH_2CH_2CH_2CH_2CH_2R_2 + R_3\overset{+}{C}HCH_3 \rightleftharpoons R_3CH_2CH_3 + R_1CH_2CH_2\overset{+}{C}HCH_2CH_2R_2 \quad (77)$$

This chain transfer is frequent under commercial catalytic cracking conditions, and it has the advantage of giving increased yields of paraffins in the C_{5+} (gasoline) range.

The applicability of the foregoing carbonium-ion theory to catalytic cracking has been illustrated by prediction of the product distribution for the cracking of n-hexadecane. The product distribution calculation was based on the following rules [6, 10]:

1. Carbonium ions are formed from n-hexadecane by random loss of a hydride ion from any secondary position.
2. The secondary carbonium ion cracks at a position β to the charged carbon to form an α-olefin, C_nH_{2n}, and a primary carbonium ion, $C^+_{16-n}H_{33-2n}$. All bonds in the β position crack with equal probability unless a fragment smaller than C_3 would be formed, in which case cracking does not occur.
3. The carbonium ion isomerizes by a hydride-ion shift to give a secondary carbonium ion. A statistical distribution of secondary carbonium ions is assumed to be formed.† The carbonium ion then undergoes cracking again.
4. When a fragment contains 6 or fewer carbon atoms, it becomes a paraffin by hydride-ion abstraction.
5. The olefins formed by cracking react as they had been observed to react in experiments with pure olefin feeds. Specifically, half of the C_7 and higher olefins are protonated to form carbonium ions, which crack according to the rules given above. The remaining olefins are converted into paraffins (as explained later), and they do not crack.

The predicted product distribution is compared in Fig. 1-13 with that observed for n-hexadecane cracking catalyzed by an alumina-zirconia-silica catalyst

† In superacids, hydride-ion shifts along the chain occur several orders of magnitude faster than isomerization steps; therefore statistical redistribution of the carbonium-ion center would surely be expected in solutions. However, because of the possible need for charge separation on catalyst surfaces, redistribution may not be as rapid, and the assumption must be considered questionable.

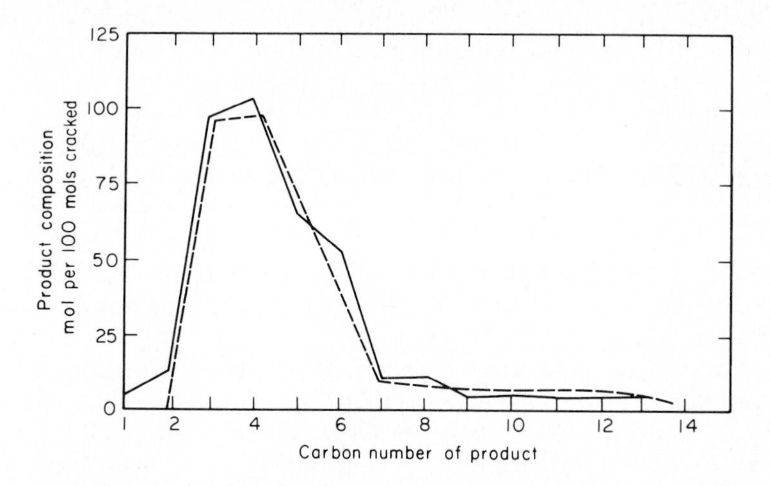

Figure 1-13 Catalytic cracking of *n*-hexadecane. The solid line represents experimentally observed products at 24% conversion in the presence of silica-zirconia-alumina at 500°C and LHSV = 10. The dashed line represents the calculated product distribution according to the rules given in the text [6, 10]. (From "Catalysis," vol. VI, P. H. Emmett (ed.). Copyright © 1958 by Litton Educational Publishing Company. Reprinted by permission of Van Nostrand Reinhold Company.)

at 500°C. The agreement is good for products containing between 3 and 14 carbon atoms. The major deviation is observed for products with fewer than 3 carbon atoms. Also formed were 12 mol of hydrogen per 100 mol of *n*-hexadecane cracked, possibly as a result of hydride-ion abstraction by the catalyst but probably from coke-forming reactions. Silica-alumina gave approximately the same product distribution [30, 33], whereas silica-magnesia gave a significantly different distribution, presumably because it was less active and less secondary cracking occurred [33]. The product distribution observed for reaction catalyzed by a rare-earth hydrogen-exchanged zeolite was also different, but the differences could be explained by considering secondary reactions and more rapid carbonium-ion removal by hydride-ion transfer [30].

Under commercial cracking conditions, the parent straight-chain paraffin never appears to undergo isomerization to an isoparaffin. In contrast, a parent straight-chain olefin, e.g., *n*-hexadecene [30, 34], is about half isomerized and half cracked under the same conditions. One might assume that the low rate of hydride-ion transfer to the tertiary carbonium ion formed from the *n*-paraffin would favor cracking over a return of the molecule to the gas phase as the isomeric paraffin. However, one should then expect the production of isomerized olefins of the same carbon number as the parent paraffin, since proton donation to olefins by carbonium ions is usually fast (compare olefin isomerization).

No such formation is reported to explain this anomaly, however, and therefore an explanation for the olefin isomerization different from that given before has been proposed by Nace [30]. The explanation is attractive, in particular as concerns the formation of cyclic hydrocarbons, but it has the disadvantage of introducing a complication as yet not needed in the discussion of carbonium-ion reactions. According to the new interpretation, both *n*-hexadecane and *n*-hexadecene undergo hydride-ion abstraction to form carbonium ions. Double-bond protonation to give the expected carbonium ion is also assumed to occur. The carbonium ion formed from *n*-hexadecane and that formed by protonation of the olefinic double bond behave equivalently, and the only reaction which readily occurs is cracking.

The carbonium ion formed from *n*-hexadecene by hydride-ion abstraction is an olefinic carbonium ion which can undergo self-alkylation to form a ring structure, as shown in Fig. 1-14 [reactions (52) and (53)]. The resultant cyclic carbonium ion can then undergo isomerization followed by β-scission reactions to form hexadecene isomers. It can also undergo other β-scission reactions to give smaller products; it can abstract a hydride ion to form a cyclic paraffin; or it can undergo proton loss to form a cycloolefin. Isoolefin protonation and hydride-ion abstraction, hydride-ion abstraction by the cyclic carbonium ion, and other hydrogen-transfer reactions result in isoparaffin formation. The cycloolefin can continue to undergo hydride-ion and proton-loss reactions via an allyl carbonium-ion mechanism until an aromatic product results.

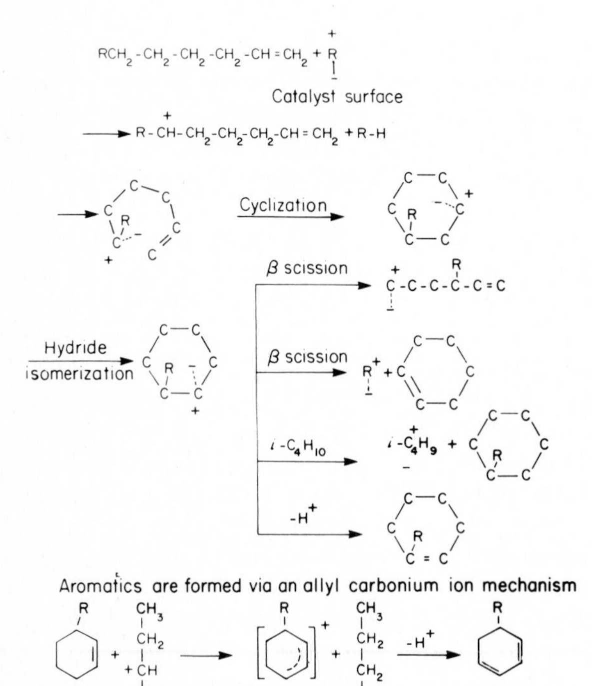

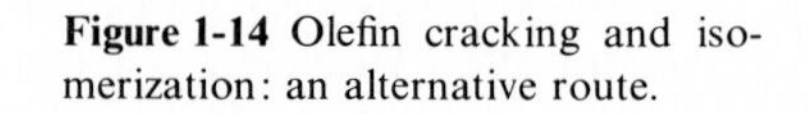

Figure 1-14 Olefin cracking and isomerization: an alternative route.

Table 1-7 Product distributions obtained from cracking of C_{12} hydrocarbons[a] [30]

	Hydrocarbon						
	n-Dodecane	*n*-Dodecene	Cyclododecane	Dodecylcyclohexane		Dodecylbenzene	
LHSV	650	1300	1300	1300		650	
Cracking conversion, %	9.0	14.8	8.71	19.9		16.0	
Isomerization conversion, %	...	25.3	3.05[b]	2.3[c]		0.0	
Cracking rate constant k, h^{-1}	500	1750	1020	1890		765	
C on catalyst, wt %	1.3	7.9	0.8	1.2		1.0	
Production distribution, mol %							
C_2	16.9	9.2	8.3	6.9		12.6	
C_4	31.0	17.7	20.7	16.4		26.4	
C_5	22.6	19.6	17.5	15.1		24.1	
C_6	17.6	18.2	16.5	15.9	88.7	18.6	48.4
C_7	7.7	16.2	15.1	14.7		9.1	
C_8	3.4	12.3	13.9	13.4		4.7	
C_9	1.0	5.4	6.7	10.5		2.2	
C_{10}	...	1.4	1.0	7.1		2.3	
iC_{12} or C_{11}	...	...	0.5	4.4		5.9	
C_{12}	...	...	...	2.8		9.8	
C_{13}	...	...	...	1.0		1.7	
C_{14}	...	...	...	0.4		2.5	
Benzene	...	...	...	2.7		21.8	
Toluene	...	...	...	...		6.0	
C_8 aromatic	...	...	...	...		3.8	

[a] 482°C, REHX catalyst, 2-min instantaneous samples.
[b] $iC_{12} + nC_{12}$.
[c] $iC_{18} + nC_{18}$.

Source: Reprinted with permission from *Industrial and Engineering Chemistry Product Research and Development*. Copyright by the American Chemical Society.

Nace [30] found that the isomer fraction for *n*-hexadecene cracking contained 6 to 14 percent cycloparaffins and 2 percent aromatics, consistent with proposed reactions for an olefinic carbonium ion. The olefin-to-paraffin ratio in this fraction was approximately 1 : 1.

Cracking of cycloparaffins has been shown to give the same product distribution as cracking of straight-chain olefins. For example, cracking of cyclododecane gave the product distribution shown in Table 1-7, which is about the same as that for *n*-dodecene [30]. The distribution is different from that observed for straight-chain paraffins, and it suggests that there are common intermediates in the cycloparaffin and olefin reactions.

The proposed mechanism involves formation of an olefinic carbonium ion, which can undergo the same cyclization and cracking reaction as *n*-dodecene:

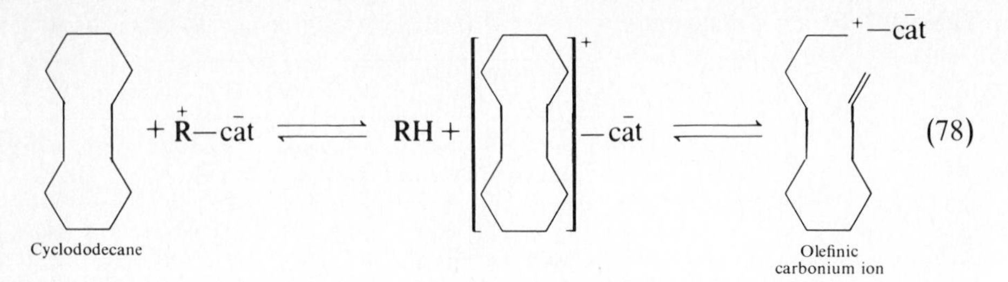

Cracking of gas oil PRIMARY REACTIONS The above considerations of the mechanism of acid-catalyzed cracking suggest that propylene, butenes, and higher-molecular-weight olefins, along with the paraffins formed by termination of carbonium ions which have resulted from the cracking step, should be the primary products of cracking. Figures 1-15 to 1-18 show the yield behavior of these products as a function of conversion for the fixed-bed cracking of a highly paraffinic gas oil [35]. The feed contained almost no olefins and less than 8 weight percent aromatics, mostly monosubstituted aromatics.

All the yield data shown in these figures fall on lines which can be extrapolated through the origin and have a positive slope there. Such behavior is a requirement for a primary product of cracking. The data are time-averaged for the deactivating catalyst. The data lying on the envelope which encloses the rest of the data are for very short on-stream times and represent the instantaneous yield (optimum yield) of the relatively fresh catalyst. The points lying beneath the curve are for longer times of operation and represent integrated yields corresponding to a condition of considerable deactivation of the catalyst.

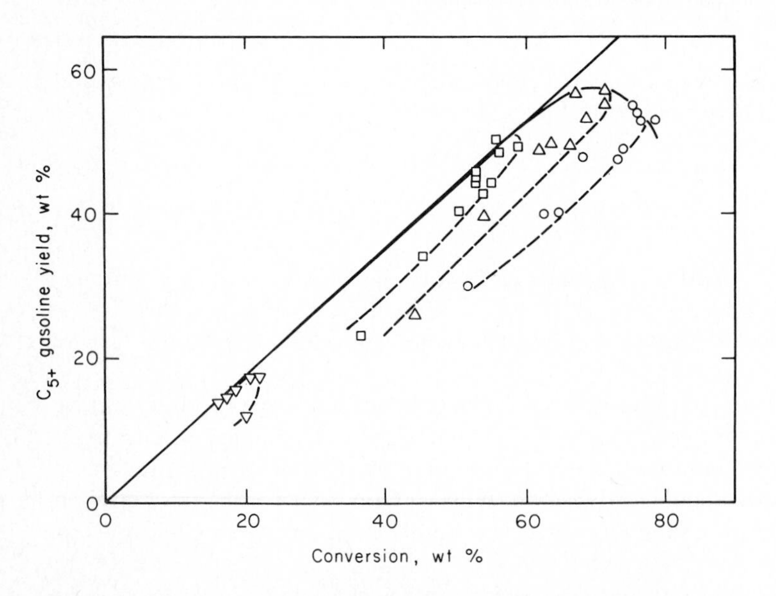

Figure 1-15 Gasoline selectivity with deactivating catalyst in a fixed-bed reactor. Conditions: temperature, 503°C; feedstock, highly paraffinic gas oil; catalyst/oil ratio: ○ 0.25, △ 0.05, □ 0.01, ▽ 0.0034 [35]. (Reprinted with permission from *Journal of Catalysis*. Copyright © by Academic Press.)

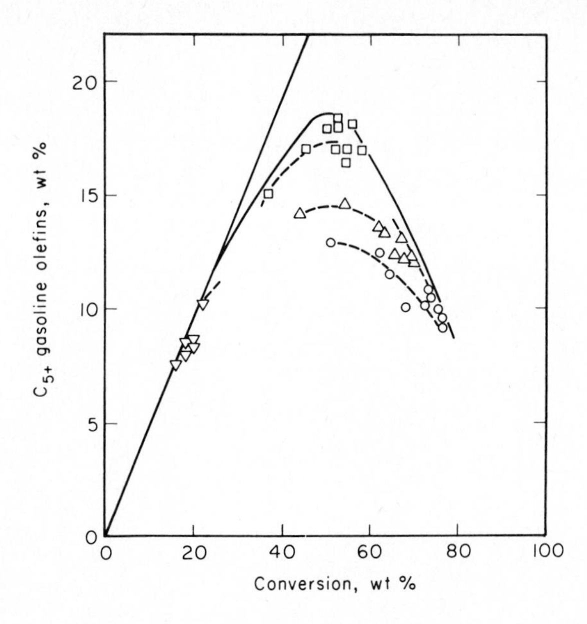

Figure 1-16 C_{5+} gasoline olefin selectivity with deactivating catalyst in a fixed-bed reactor. Conditions: temperature, 503°C; feedstock, highly paraffinic gas oil; catalyst/oil ratio: ○ 0.25, △ 0.05, □ 0.01, ▽ 0.0034 [35]. (Reprinted with permission from *Journal of Catalysis*. Copyright © by Academic Press.)

The data show that both gasoline-range (C_{5+}) paraffins and olefins were primary products, as were *n*-butane, butenes, and propylene. The gasoline-range paraffins (Fig. 1-15) began to undergo substantial further cracking (overcracking) at about 70 percent conversion, and therefore the optimum conversion envelope

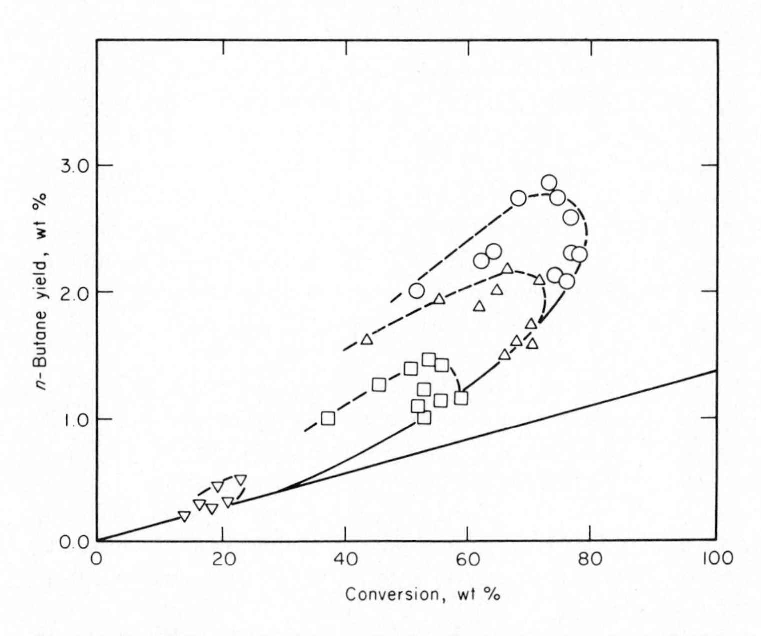

Figure 1-17 *n*-Butane selectivity with deactivating catalyst in a fixed-bed reactor. Conditions: temperature, 503°C; feedstock, highly paraffinic gas oil; catalyst/ratio ○ 0.25, △ 0.05, □ 0.01. ▽ 0.0034 [35]. (Reprinted with permission from *Journal of Catalysis*. Copyright © by Academic Press.)

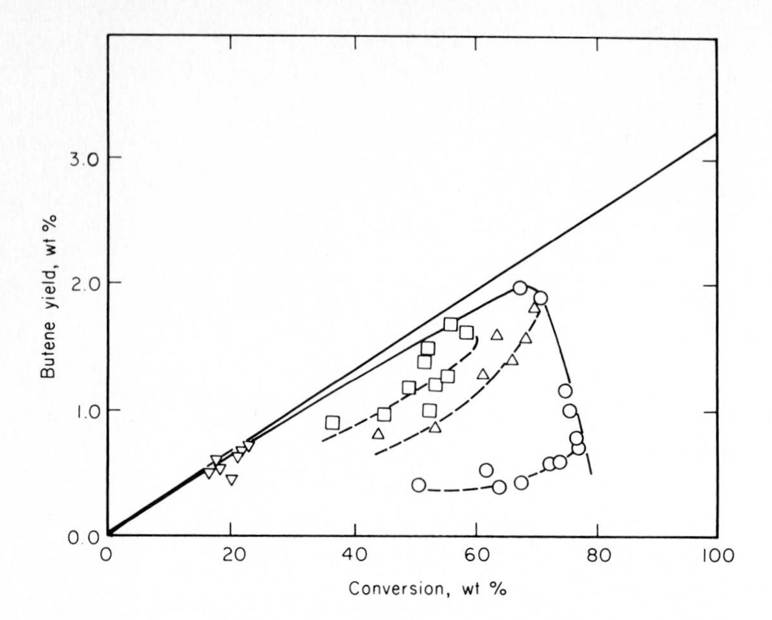

Figure 1-18 Butene selectivity with deactivating catalyst in a fixed-bed reactor. Conditions: temperature, 503°C; feedstock, highly paraffinic gas oil; catalyst/oil ratio: ○ 0.25, △ 0.05, □ 0.01, ▽ 0.0034 [35]. (Reprinted with permission from *Journal of Catalysis*. Copyright © by Academic Press.)

departs downward from the straight line extending from the origin; this same trend was observed for the more reactive gasoline-range olefins at about 40 percent conversion (Fig. 1-16). Similarly, butenes were unstable primary products which underwent further secondary reactions. Propylene, on the other hand, showed behavior similar to that of *n*-butane (Fig. 1-17), except that about $2\frac{1}{2}$ times as much was produced. Propylene is therefore concluded to be a stable olefin which was not substantially converted to other products at higher conversions; secondary cracking reactions, however, produced additional propylene, and correspondingly the optimum conversion envelope deviates above the extrapolated line from the origin, as does the envelope for *n*-butane.

SECONDARY REACTIONS The secondary reactions of importance in catalytic cracking also proceed through the carbonium-ion mechanisms mentioned previously. The α-olefins formed are protonated and rapidly isomerize [reactions (27) and (28)] in addition to undergoing cracking. A carbonium ion formed from a product olefin may abstract a hydride ion from another hydrocarbon molecule, resulting in its conversion to a paraffin. This reaction stabilizes these hydrocarbon products and largely prevents their further cracking; it therefore represents a desirable form of hydrogen transfer.

The ideal primary cracking reactions theoretically give high yields of olefins, one olefin molecule for each cracking event. Typically, about 3 mol of product is formed per mole of gas-oil-range hydrocarbon cracked; according to the simple theory, this process should result in a 2 : 1 olefin-to-paraffin ratio in the product. Cracking catalyzed by silica-alumina, however, gives much lower olefin yields, and cracking catalyzed by zeolites can result in products which are almost free of olefins. This lack of olefins is due to rapid hydrogen transfer (1) from cycloparaffins and cycloolefins, which are thereby converted to aromatics, e.g., the last two reactions of Fig. 1-14; (2) from olefins which undergo dehydrogenation and cyclization to give cycloolefins, e.g., reactions (45) to (49) or those of Fig. 1-14; and (3) from dehydrogenation of aromatic and olefinic species to yield coke. These last conversions may occur through many repetitions of reaction (55) accompanied by dimerization, condensation, and cyclization.

Hydrogen transfer usually occurs somewhat more slowly than cracking of gas oils at 500°C. At lower temperatures it is relatively more rapid, presumably because of a higher activation energy for cracking. Consequently, decreasing the cracking temperature results in a shift of the product distribution to higher molecular weights (1) because of increased hydride-ion abstraction by carbonium ions, which converts them into paraffins before they undergo further β scission, and (2) because of hydrogen transfer to α-olefins rather than cracking of them. Increased contact time, which allows additional time for hydrogen transfer, results in a less olefinic product.

Correspondingly, in the early production of aviation gasoline, which required products having high levels of saturation, cracking conditions were always relatively mild, and contact times were longer than those applied when a cracked stock containing higher olefin concentrations was permissible.

Hydrogen transfer does not involve a dehydrogenation step followed by a hydrogenation step. For example, dehydrogenatable naphthenes such as Decalin are effective in the saturation of olefins but not by a two-step process. Passing the naphthene alone over a cracking catalyst results in very little hydrogen formation, and passing a mixture of molecular hydrogen and olefin over the catalyst gives the same results as passing nitrogen and olefin over the catalyst. Passing the naphthene and the olefin over the catalyst simultaneously results in a marked increase in paraffin formation. Hydrogen transfer clearly occurs directly between naphthenes and olefins, which are thereby converted into paraffins.

Hydrogen transfer is important because it (1) reduces the amount of olefins in the product, (2) strongly influences the molecular-weight distribution of the product, and (3) contributes to coke formation and catalyst deactivation. Control of molecular-weight distribution occurs through (1) termination of carbonium ions by hydride-ion transfer from other molecules [reaction (77)] before the carbonium ions have cracked all the way to very short-chain fragments and (2) through saturation of olefins before they can undergo further cracking. Improved hydrogen-transfer rates can result in marked increases in conversion to gasoline-range hydrocarbons. Shifts in the source of the hydrogen transferred from coke precursors to naphthenic species can reduce rates of coke formation and improve gasoline yield and quality.

The data of John and Wojciechowski [35] (Figs. 1-19 to 1-21) show yields of

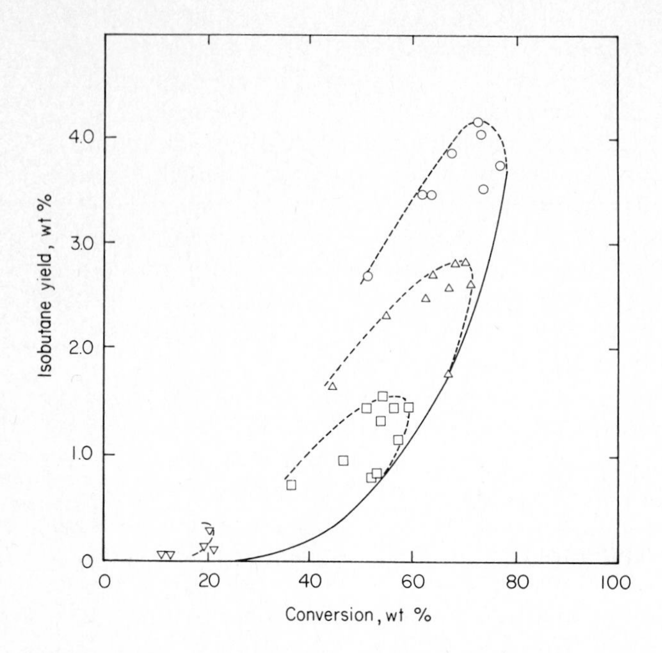

Figure 1-19 Isobutane selectivity with deactivating catalyst in a fixed-bed reactor. Conditions: temperature, 503°C; feedstock, highly paraffinic gas oil; catalyst/oil ratio: ○ 0.25, △ 0.05, □ 0.01, ▽ 0.0034 [35]. (Reprinted with permission from *Journal of Catalysis*. Copyright © by Academic Press.)

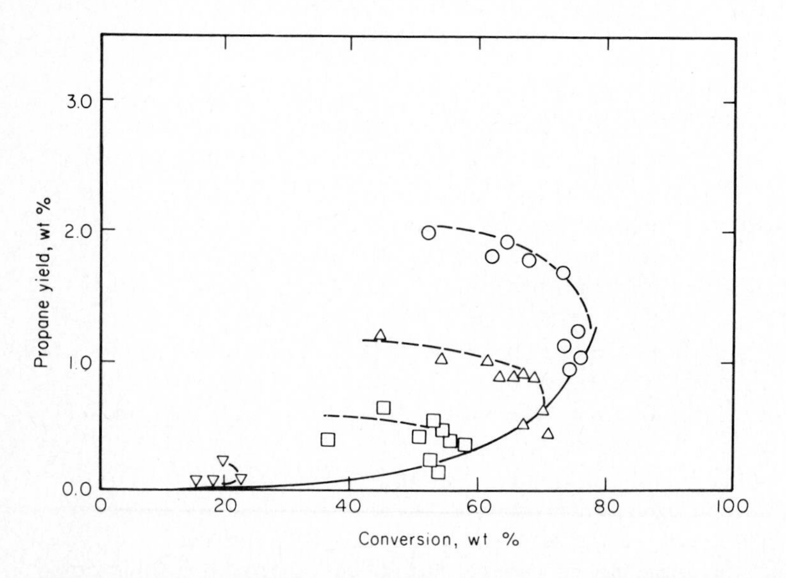

Figure 1-20 Propane selectivity with deactivating catalyst in a fixed-bed reactor. Conditions: temperature, 503°C; feedstock, highly paraffinic gas oil; catalyst/oil ratio: ○ 0.25, △ 0.05, □ 0.01, ▽ 0.0034 [35]. (Reprinted with permission from *Journal of Catalysis*. Copyright © by Academic Press.)

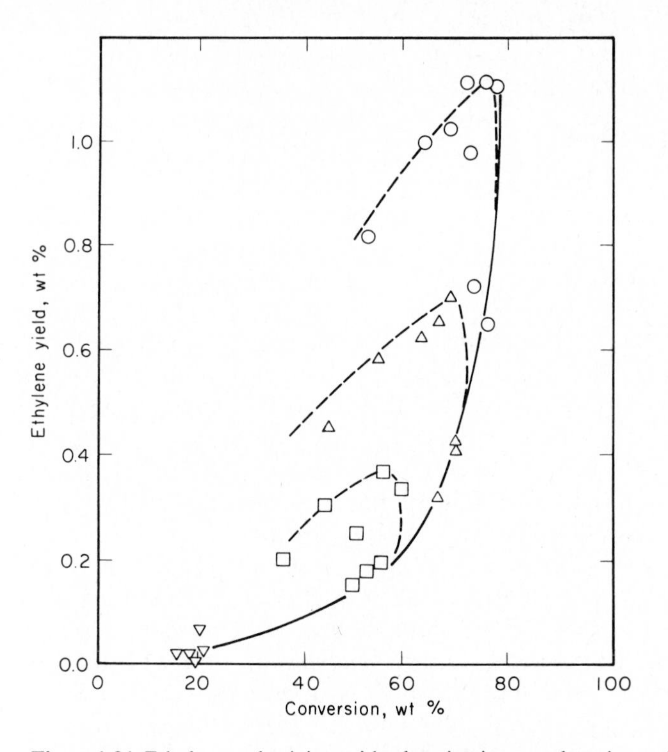

Figure 1-21 Ethylene selectivity with deactivating catalyst in a fixed-bed reactor. Conditions: temperature, 503°C; feedstock, highly paraffinic gas oil; catalyst/oil ratio: ○ 0.25, △ 0.05, □ 0.01, ▽ 0.0034 [35]. (Reprinted with permission from *Journal of Catalysis*. Copyright © by Academic Press.)

the secondary products of cracking. Isobutane was a stable secondary product which probably was formed from butenes by hydrogen transfer; similarly propane was formed from propylene or from the cracking of the C_4 fraction, since it appeared at about the same conversions as methane (conversions in excess of 50 percent) and since both are known products of butane and probably butene cracking [25]. Hydrogen transfer to, and cracking of, butenes explain their disappearance at high conversions (Fig. 1-18). Ethylene appeared at about 20 percent conversion (Fig. 1-21); ethane appeared only at conversions greater than 50 percent. The required high conversions before methane, ethane, or propane appeared suggest that these were tertiary products being formed either by saturation of olefins or cracking of chains having one or two more carbons than the tertiary product. The very light olefins could also have resulted from olefin disproportionation reactions in addition to secondary cracking; the literature provides little information about these reactions.

COKE FORMATION Coke formation is a mixed blessing: it causes rapid catalyst deactivation, but on the other hand coke combustion provides a source of heat for the endothermic cracking reactions, and coke is a source of hydrogen for stabilizing valuable lower-molecular-weight products. To maintain the overall hydrogen balance in the cracking process, the hydrogen which serves to saturate olefins must come from hydrocarbons which are converted into aromatics or coke.

The most probable reactions leading to coke formation are those given previously [(46) to (59)], but the coke-formation reactions must be considered the least well understood of those involved in catalytic cracking. Coke typically consists of a polyaromatic condensed-ring structure, which approaches the character of graphite, and it is formed in almost all catalytic hydrocarbon conversion processes. Highly unsaturated hydrocarbons having high molecular weights are strongly adsorbed on catalyst surfaces. Their presence on a surface in high concentration, their ease of protonation, and the stability of the resultant carbonium ion best explain the observation that aromatics have high coke-forming tendencies. For paraffins, both the rate of cracking and the rate of coke formation increase as the molecular weight of the reactant increases, and for paraffins of a particular carbon number, the coke-formation rate correlates well with paraffin reactivity [30]. This result suggests that the rate of coke formation may be tied to the rate of olefin formation and to the overall hydrogen balance for the system.

The rate of coke formation from a pure olefin feed was found to be high compared with that from a pure paraffin feed (Table 1-8), probably because of the high olefin concentration prevailing throughout the fixed-bed reactor when the former feed was used. Cyclododecane produced less coke than *n*-dodecane (Table 1-8), indicating that the olefin condensation reaction rather than the presence of the cyclic carbonium ion was of primary importance for coke formation; dodecylcyclohexane and dodecylbenzene produced about equivalent amounts of coke.

Table 1-8 Coke deposition on silica-alumina catalyst during cracking of pure hydrocarbons [36]

Reaction conditions: 445°C, 0.4 vol of hydrocarbon per volume of catalyst per hour; 13% Al_2O_3–87% SiO_2 catalyst

	Feed						
	Decalin	Tetralin	*n*-Butylbenzene	1-Methyl naphthalene	2-Methylbiphenyl	*n*-Hexadecane	1-Hexadecene
Time on stream, min	32	28	32	30	34	60	60
Conversion, wt %	98	91	78	86	67	71	100
Product distribution, wt % of feed:							
Gas	9	11	22	3	5	51	65
Liquid other than feed	64	74	51	67	54	13	23
Coke	25	6	5	16	8	7	12
Carbon on catalyst, wt %	7.80	1.72	1.20	4.50	2.77	2.39	4.64
Coke aromaticity[a]	1.03	0.92	1.35	1.23	1.55	0.51	0.47

[a] Defined in text.

Source: Reprinted with permission from *Industrial and Engineering Chemistry Process Design and Development*. Copyright by The American Chemical Society.

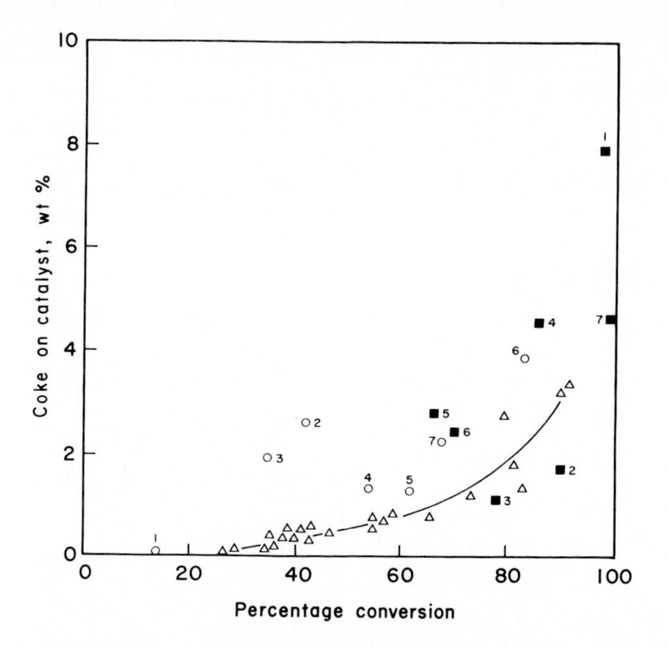

Figure 1-22 Extent of coke formation in the cracking of pure hydrocarbons. ■ SiO_2-Al_2O_3, 445°C, LHSV = 0.4 [36]: (1) Decalin, onstream time = 32 min, (2) Tetralin, onstream time = 28 min, (3) *n*-butylamine, onstream time = 32 min, (4) 1-methylnaphthalene, onstream time = 30 min, (5) 2-methylbiphenyl, onstream time = 34 min, (6) *n*-hexadecene, onstream time = 60 min, (7) *n*-hexadecene, onstream time = 60 min. △ SiO_2-Al_2O_3, 500°C, onstream time = 60 min, [36]; ○ SiO_2-Al_2O_3, 550°C, onstream time = 60 min [10]: (1) *n*-hexane, (2) *n*-octane, (3) *n*-dodecane, (4) Decalin (500°C), (5) *n*-hexadecene (400°C), (6) Cumene (isopropylbenzene), (7) *n*-hexadecane.

Table 1-8 summarizes results of coke-formation studies for several pure hydrocarbons, and Fig. 1-22 shows the extent of coke formation as a function of cracking conversion for a number of pure hydrocarbons. The data of Table 1-8 suggest that the coke-formation rate decreases in the order two-ring aromatic, one-ring aromatic, olefin, naphthene, paraffin. The data of Fig. 1-22 show the increase in coke formation with increasing conversion; they suggest that coke is a secondary product of *n*-hexadecane cracking and that its rate of formation is not as strong a function of molecular structure as indicated by the data of Table 1-8.

The aromaticity of the coke formed in cracking of various pure hydrocarbons, defined as the ratio of the absorbance of the infrared band characteristic of C—H ($3050\ cm^{-1}$) to that characteristic of CH_2 ($2930\ cm^{-1}$), was determined by Eberly et al. [36] (Table 1-8). The most highly aromatic coke resulted from aromatic feeds, naphthenes gave coke of intermediate aromaticity, and olefin and paraffin feeds gave coke of low aromaticity. These results suggest that cyclization was slow and that without an aromatic present, the coke produced was more of a high-molecular-weight, amorphous, nongraphitic deposit.

Figure 1-23 shows the coke concentration as a function of catalyst position in a fixed-bed reactor for *n*-hexadecane cracking [36]. The coke profile moved

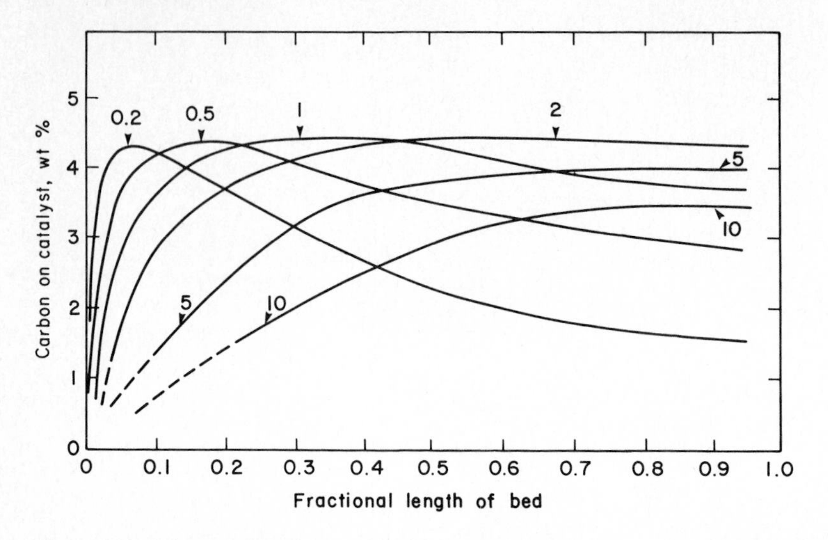

Figure 1-23 Coke concentration profiles for *n*-hexadecane cracking in a fixed-bed reactor at varying space velocities: 13% Al_2O_3-87% SiO_2 catalyst, 500°C, 60-min cycle time; the numbers on the curves represent liquid hourly space velocity [36].

toward the downstream end of the bed as the space velocity increased, which confirms that coke was the product of a secondary reaction and that its formation resulted from the reaction of olefins formed in the cracking step. Figure 1-24 shows that in the cracking of a paraffinic gas oil, coke was formed in secondary reactions involving the olefins formed from the primary cracking. In the absence

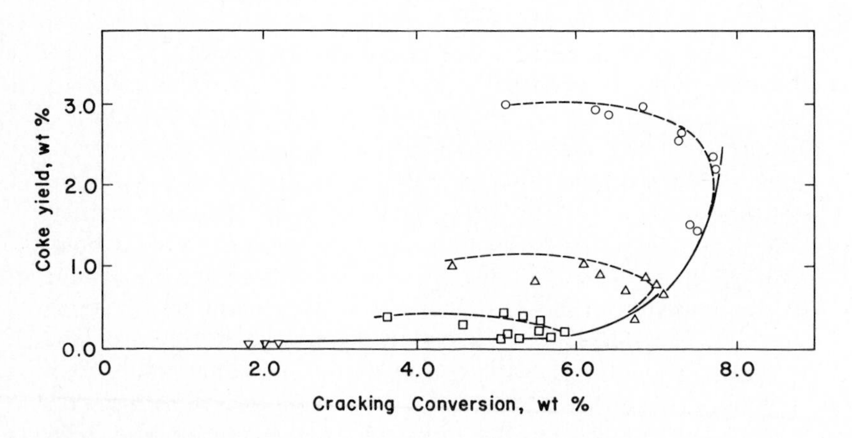

Figure 1-24 Coke formation during cracking of a gas oil. Conditions: temperature, 503°C; feedstock, highly paraffinic gas oil; catalyst/oil ratio: ○ 0.25, △ 0.05, □ 0.01, ▽ 0.0034 [35]. Reprinted with permission from *Journal of Catalysis*. Copyright by Academic Press.)

of aromatics, olefins are necessary for coke formation both as precursors to coke and as acceptors of hydrogen from coke precursors [37, 38]. If the feed had contained high-molecular-weight multiring aromatic hydrocarbons, coke formation would probably also have occurred as a primary reaction [reactions (54) to (59)]. Walsh and Rollman [36*a*] showed that in cracking a feed containing hexane and aromatics catalyzed by Y zeolite or mordenite, the contribution to coke formation by the aromatics exceeded that of the paraffin and that the alkylation of the aromatic was the initial step in coke formation, thus showing the great importance of interactions between the different compounds in coke formation.

The foregoing discussion of coke formation during catalytic cracking can be summarized as follows:

1. The rate of coke formation increases with increasing acid strength of the catalyst and with increasing base strength of the hydrocarbon reactant, i.e., with increasing ease of formation and stability of the carbonium ion.
2. The rate of coke formation increases with increasing acid-site density on the catalyst. Reactions (54) to (59) explain hydrogen transfer from growing coke species to olefins and require the proximity of two acid sites, one for the polynuclear aromatic and one for the carbonium ion. The rate of coke formation on silica-alumina increases as the alumina content increases and therefore presumably as the acid-site density increases [6]. Zeolite catalyst pretreatment which reduces acid-site strength and acid-site density also leads to decreased rates of coke formation [39, 40].
3. Pure-component studies may be misleading in that coke formation requires a source of hydrogen (coke precursor) and a sink for hydrogen (olefin). Mixed feeds containing small amounts of polynuclear aromatics and olefins may produce more coke than either would alone.

Summary of Catalytic Cracking Chemistry

Carbonium-ion chemistry explains the principal characteristics of catalytic cracking and the differences in product distribution between thermal and catalytic cracking. At times, however, it is necessary to introduce some details that are not readily apparent in the carbonium-ion chemistry deduced from low-temperature studies of reactions in superacid solutions. The expanded carbonium-ion theory explains the details of paraffin cracking behavior, including cyclization to naphthenes, dehydrogenation to aromatics, and hydrogen transfer between product species and from coke and coke precursors. From our understanding of carbonium-ion chemistry and the paraffinic gas-oil cracking results of John and Wojciechowski [35], we can now improve on the simplified reaction network given by Eq. (1). The detailed reaction network is shown in Fig. 1-25.

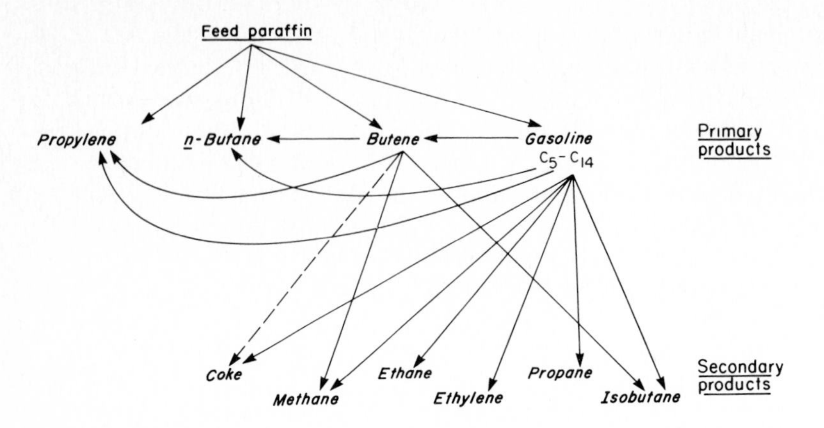

Figure 1-25 Detailed reaction scheme for catalytic cracking of paraffins.

CRACKING CATALYSTS

Introduction

The chapter so far has included many details of cracking chemistry, but the role of the catalyst has been glossed over. A complete picture of reaction mechanism requires inclusion of the catalyst surface, and the purpose of the following section is to summarize the structures of cracking catalysts and their influence on the cracking reactions. The discussion proceeds from the relatively well-known bulk structures of the solids to the surface structures and details of surface chemistry as they affect cracking reactions. The strategy of discussing catalyst structures in this way is applied repeatedly in the remaining chapters.

Amorphous Catalysts

Preparation and general properties Neither silica nor alumina nor a mechanical mixture of the two dry oxides is an active cracking catalyst, but a cogelled mixture of silica and alumina containing mainly silica is highly active. The incorporation of alumina into silica, even at very low concentrations, results in the formation of surface sites which catalyze cracking reactions. It is inferred from the preceding discussion that at least some of these sites must be Brønsted and/or Lewis acid sites.

One method of preparation of SiO_2–Al_2O_3 [40, 41] involves first the preparation of a porous silica hydrogel by acidifying a dilute aqueous solution of sodium silicate and allowing sufficient time for the silica in solution to undergo considerable polymerization and form a hydrogel. The hydrogel consists of a coherent aggregate of primary spherical particles about 30 to 50 Å in diameter. The primary particles consist of a three-dimensional network of interconnected SiO_4

tetrahedra, each silicon atom being linked to four oxygen atoms and each oxygen atom linking two silicons. The surfaces of the particles are terminated with hydroxyl groups in the form of silanol, Si—OH, groups, and the primary particles may be linked by Si—O—Si bridges. The pore system and large surface area ($\sim$ 500 m^2/g) of the dried hydrogel are associated with the open spaces between the aggregated primary particles and with the surface of the primary particles. This material is catalytically inactive for cracking.

After the silica hydrogel is formed, an aluminum salt is added to the solution and hydrolyzed so that the six-coordinate aluminum ions react with the surface of the primary silica particles. Reaction involves a condensation of the aluminum trihydrate and the surface OH groups of the silica with the elimination of water. An excess of reactive silica gel and a low pH (3.0) favor Al—O—Si rather than Al—O—Al bond formation. A schematic representation of the reaction is shown in Fig. 1-26. The result is that aluminum ions are incorporated into the surfaces of the silica particles.

The physical structure of silica-alumina is essentially the same as that of the silica resulting from drying and calcining the silica hydrogel, since the silica-alumina is derived from the same hydrogel. The 30 to 50-Å primary particles of silica-alumina are packed loosely together to form irregularly shaped aggregates ranging in size from 0.05 to 3 μm in diameter [42]. These aggregates are pressed together to form the pellets or particles available commercially. The small size and spherical shape of the primary particles are consistent with the fact that the material is amorphous to x-ray diffraction. In this respect, silica and silica-alumina differ from alumina, which contains larger primary crystallites and shows x-ray structure. The surface area of silica-alumina varies from 200 to about 600 m^2/g, which results directly from the small size of the primary particles.

Chemical properties The OH groups terminating the primary silica and alumina particles vary from weakly acidic to alcoholic and show no cracking activity. The incorporation of trivalent aluminum ions into the silica surface that produces an active cracking catalyst also produces surface OH groups having strong Brønsted

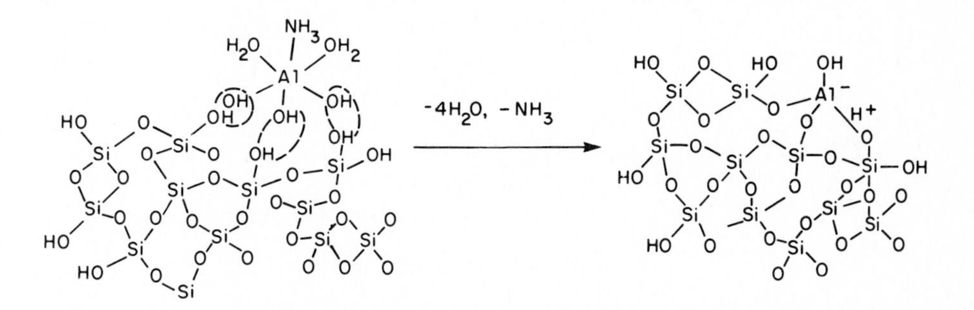

Figure 1-26 Condensation of aluminum trihydrate with the surface of an elementary silica hydrogel particle.

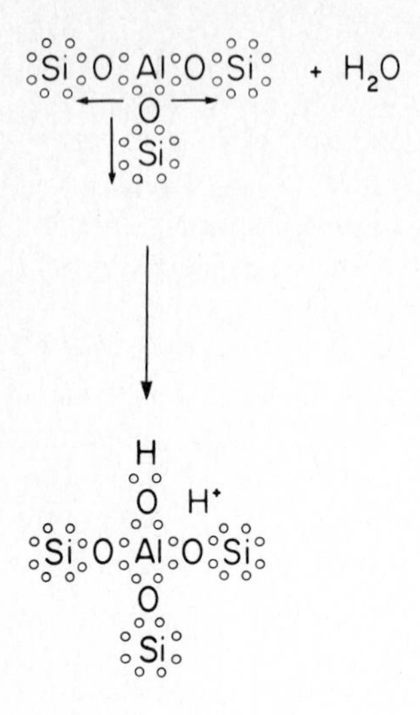

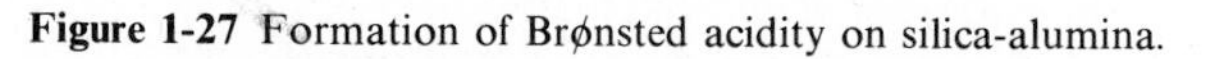

Figure 1-27 Formation of Brønsted acidity on silica-alumina.

acid character. The protonic acidity evidently arises from the dissociative adsorption of H_2O on the aluminum ion. It is associated with the proton which moves to the oxygen bonded to the silicon cation, as shown in Fig. 1-27. The strongly electrophilic nature of the aluminum ion can be traced to the asymmetry of its position in the surface, where it is surrounded by 4-valent silicon cations. This asymmetry could account for withdrawal of charge from the aluminum ion to make it more positive and develop a sufficiently strong field to allow it to acquire a hydroxyl group by splitting off hydrogen from water (Figs. 1-26 and 1-27). The resultant SiOOHAl group is strongly polarized by the asymmetry of the environment, which induces a strong acidity in the group.

When the silica-alumina surface is dehydrated by heating to high temperature, water is removed from the Brønsted site, exposing the aluminum ion with its

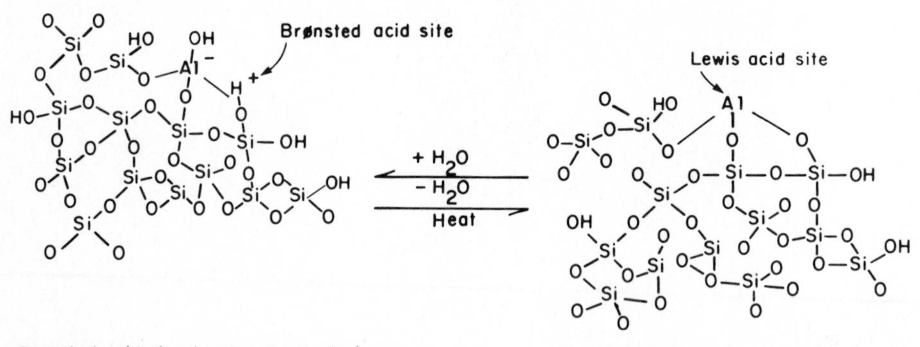

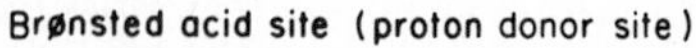

Figure 1-28 Conversion of a Brønsted acid site into a Lewis acid site by dehydration of alumina.

electron-pair-acceptor properties, thereby forming a Lewis acid site (Fig. 1-28). The dehydrated surfaces can show Brønsted acidity, Lewis acidity, or both, depending on pretreatment conditions.

Typical silica-alumina catalysts contain 10 to 12 percent alumina, but strong acidity is generated even at much lower alumina concentrations. These concentrations correspond to only partial surface coverage of the elementary silica particles by alumina. Higher alumina concentrations are usually not desirable, however, because alumina is more expensive than silica and because the resultant formation of many Al—O—Al bonds may decrease the acidity; Brønsted acidity has been shown to decrease for alumina concentrations exceeding about 20 percent [43].

A more specific picture of the surface symmetry and the catalytically active site for cracking has not been established, largely because the bulk structures of amorphous solids such as silica-alumina are not well known and correspondingly the surface structures are even less well known.

Crystalline (Zeolite) Catalysts

Introduction The zeolite cracking catalysts are unique among the many commercial oxide catalysts in that they are crystalline. Consequently, their bulk structures are susceptible to study with techniques like x-ray diffraction, and they are well understood.

There are 34 known natural zeolites, and about 100 zeolites which do not have natural counterparts have been synthesized [44, 45]. Of this large number of zeolites, only a few have found commercial application; they are mostly synthetic zeolites and synthetic analogs of natural zeolites.

The fundamental building block of all zeolites is a tetrahedron of four oxygen anions surrounding a smaller silicon or aluminum ion (Fig. 1-29). These tetrahedra are arranged so that each of the four oxygen anions is shared in turn with another silica or alumina tetrahedron. The crystal lattice extends in three dimensions, and the -2 oxidation state of each oxygen is accounted for.

Each silicon ion has its $+4$ charge balanced by the four tetrahedral oxygens, and the silica tetrahedra are therefore electrically neutral. Each alumina tetrahedron has a residual charge of -1 since the trivalent aluminum ion is bonded to four oxygen anions. Therefore each alumina tetrahedron requires a $+1$ charge

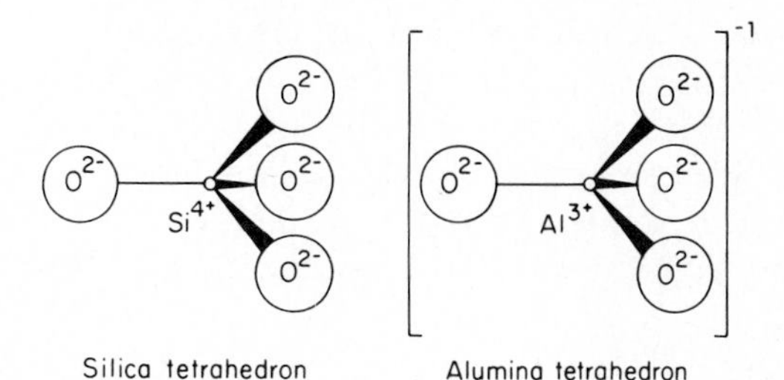

Figure 1-29 Primary building blocks of zeolites.

from a cation in the structure to maintain electrical neutrality. These cations are usually sodium in the zeolite as it is initially prepared, but they can readily be replaced by ion exchange. Ion exchange represents the most direct and useful method for the alteration of zeolite properties.

The silica and alumina tetrahedra are combined into more complicated secondary units, which form the building blocks of the framework zeolite crystal structures. The silica and alumina tetrahedra are geometrically arranged, with Al—O—Al bonds excluded [46]. The unit-cell formula is usually written $M_{j/n}(AlO_2)_j(SiO_2)_y \cdot zH_2O$, where M represents the exchangeable cation and j, y, z, and n are integers. The oxidation state of the cation is n.

The tetrahedra are arranged so that the zeolites have an open framework structure, which defines a pore structure with a high surface area. This surface area is different from that of amorphous solids such as silica-alumina in that it is a true part of the crystalline solid and not the area of termination of primary particles of the solid. Therefore, the chemistry of the zeolite surface is determined uniquely by the properties of the crystalline solid, which is a major advantage in elucidating the catalytic chemistry.

The pore structure varies greatly from one zeolite to another. In all zeolites, pore diameters are determined by the free aperture resulting from 4-, 6-, 8-, 10-, or 12-membered rings of oxygen atoms, and these have maximum values calculated to be 2.6, 3.6, 4.2, 6.3, and 7.4 Å, respectively [44, 45]. Because of puckering or elongation of the oxygen ring or the presence of cations near or within the pore apertures, the effective free aperture of the pore structure may be reduced somewhat, and the aperture may be elliptical.

Only zeolites with 8- and 12-membered oxygen rings have found major catalytic applications. The smaller apertures place unacceptable size limitations on the molecules to be adsorbed. The zeolites of industrial importance are zeolites A, X, and Y, erionite, and synthetic mordenite. Zeolites X and Y, erionite, and mordenite are the only ones of importance as catalysts in cracking and related reactions. The structures of the X and Y zeolites and mordenite are given in a following section.

Synthesis Zeolites are formed by hydrothermal synthesis, typically under mild conditions. The nature of the zeolite obtained is determined by the synthesis conditions, i.e., reactant concentrations, pH, time, temperature, and the nature and concentration of added promoters. Many of the phases formed are not equilibrium phases but only metastable ones, which in time are converted to other, more stable zeolitic phases or other mineral phases. Typical reactants are sodium aluminate, sodium silicate, silicic acid, and sodium hydroxide, the latter being used mainly to control pH. The sequence of events involves the formation of a hydrous aluminosilicate gel and then the crystallization of this gel under carefully controlled conditions to give the desired zeolitic phase. This sequence is illustrated schematically as follows:

$$NaOH(aq) + NaAl(OH)_4(aq) + Na_2SiO_3(aq) \xrightarrow{25°C} \underset{\text{Gel phase}}{Na_a(AlO_2)_b(SiO_2)_c \cdot NaOH \cdot H_2O} \xrightarrow{25\text{—}175°C} \underset{\text{Zeolite phase}}{Na_j[(AlO_2)_j(SiO_2)_y] \cdot zH_2O} + \text{solution} \quad (79)$$

The resultant zeolite must be removed from the mother liquor at the proper time and must be thoroughly washed to remove sodium silicate from the pore structure. Properly washed zeolite should have a Na/Al ratio of 1.00. Table 1-9 summarizes typical synthesis conditions for several important zeolites.

X and Y zeolite structures X zeolite, Y zeolite, and natural faujasite have topologically similar structures, which have been determined by x-ray diffraction [47–49]. They differ in their characteristic silica-to-alumina ratios and consequently differ in their crystal-lattice parameter, with a variation of about 2 percent over the range of permissible Si/Al ratios [50]. They also differ somewhat in such properties as cation composition, cation location, cation exchangeability, thermal stability, and adsorptive and catalytic character [51]. The typical unit-cell formula is $Na_j[(AlO_2)_j(SiO_2)_{192-j}] \cdot zH_2O$, where z is about 260. The value of j is between 48 and 76 (typically 57) for the Y zeolite and between 77 and 96 (typically 85) for the X zeolite [51, 52].

In the X and Y zeolites and faujasite, the silica and alumina tetrahedra are joined together to form a cuboctahedron, as shown in Fig. 1-30. This unit, referred to as a *sodalite unit* or *truncated octahedron*, contains 24 silica and alumina tetrahedra. The sodalite unit is the secondary building block of a number of

Table 1-9 Typical zeolite synthesis conditions [45]

Zeolite type	Reactant composition,[a] mol/mol Al_2O_3			Temp., °C	Time, h	SiO_2/Al_2O_3
	Na_2O	SiO_2	H_2O			
A	2	2	35	100	2–4	2
X	3.6	3	144	100	7	2.5 (2.0–3.0)
Y	8	20	320	100	7	5.0 (3.0–6.0)
Mordenite (large port)	6.3	27	61	100	168	9–12

[a] NaOH was used to adjust pH; the values were not specified.

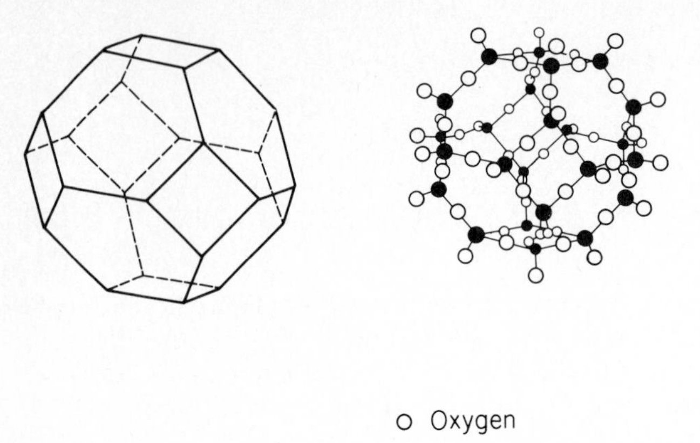

Figure 1-30 Sodalite cage structure. A formal representation of a truncated octahedron is shown on the left, and individual atoms are indicated on the right; the lines in the structure on the left represent oxygen anions, and the points of intersection represent silicon or aluminum ions.

zeolites, including sodalite, zeolite A, zeolite X, zeolite Y, and faujasite. Molecules can penetrate into this unit through the six-membered oxygen rings, which have a free diameter of 2.6 Å; the unit contains a spherical void volume with a 6.6-Å free diameter. Since the pore diameter is so small, only very small molecules, e.g., water, helium, or hydrogen, or ions can enter the sodalite cage.

When the truncated octahedra are stacked so that each four-membered ring is shared by two cages, sodalite is formed (Fig. 1-31). The maximum free-pore aperture of sodalite is only 2.6 Å, and sodalite can adsorb only small molecules at relatively high temperatures.

When the truncated octahedra are connected by bridge oxygen atoms between the four-membered rings, zeolite A is formed (Fig. 1-32). The free-pore

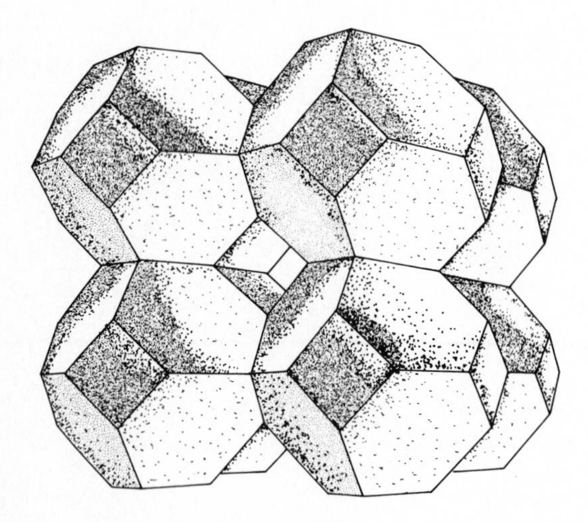

Figure 1-31 Arrangement of truncated octahedra in sodalite.

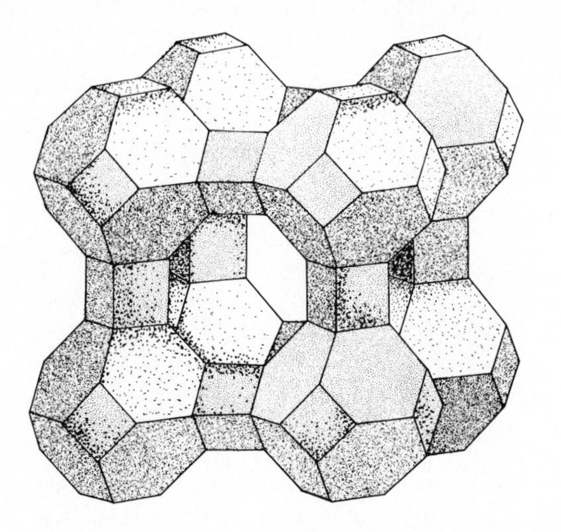

Figure 1-32 Arrangement of truncated octahedra in zeolite A.

aperture of zeolite A is determined by an eight-membered oxygen ring and therefore has a free-pore diameter of 4.2 Å. The resultant cavity enclosed by the eight sodalite units has a free dimension large enough to inscribe an 11.4-Å sphere. There are two three-dimensional pore structures, one involving the interconnected supercages separated by 4.2-Å minimum pore apertures and the other consisting of the sodalite units alternating with the supercages and having 2.6-Å free-pore apertures.

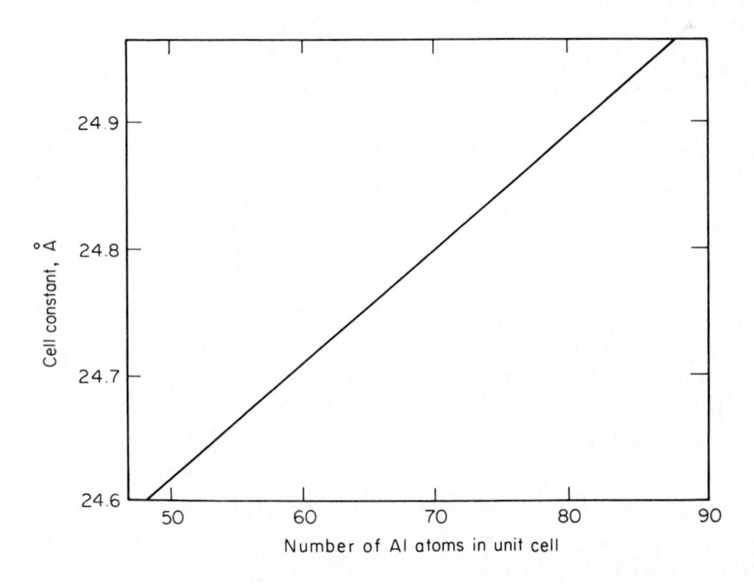

Figure 1-33 Experimental correlation of unit cell dimensions of faujasites. Unit cell dimension is the length of the unit cell in the *a* direction [45]. (Reprinted with permission of John Wiley & Sons, Inc. Copyright 1974 by John Wiley and Sons, Inc.)

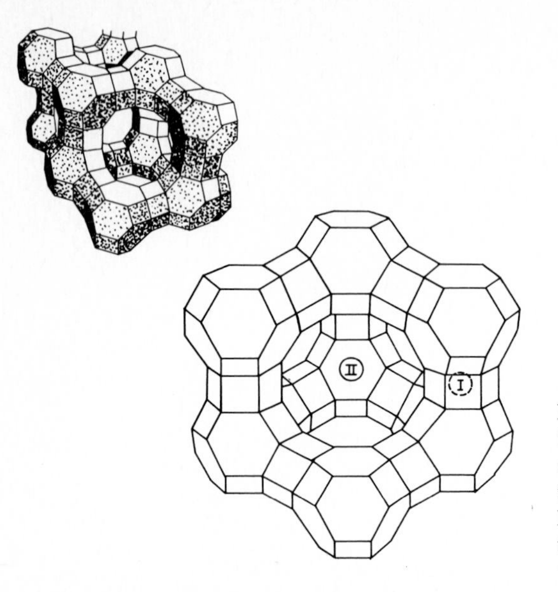

Figure 1-34 Perspective views of the faujasite structure. The silicon or aluminum ions are located at the corners and the oxygen ions near the edges. Type I and II sites are indicated; the supercage is in the center.

The unit cell of the faujasite-type zeolites is cubic with a unit-cell dimension of 25 Å, and it contains 192 silica and alumina tetrahedra. The unit-cell dimension varies with the Si/Al ratio, as shown in Fig. 1-33. Each sodalite unit in the structure is connected to four other sodalite units by six bridge oxygen ions connecting the hexagonal faces of two units, as shown in Fig. 1-34. The truncated octahedra are stacked like carbon atoms in a diamond. The oxygen bridging unit is referred to as a hexagonal prism, and it may be considered another secondary unit. This structure results in a supercage (sorption cavity) surrounded by 10 sodalite units which is sufficiently large for an inscribed sphere with a diameter of 12 Å. The opening into this large cavity is bounded by 6 sodalite units, resulting in a 12-membered oxygen ring with a 7.4-Å free diameter [46]. Each cavity is connected to four other cavities, which in turn are themselves connected to three additional cavities to form a highly porous framework structure.

This framework structure is the most open of any zeolite and is about 51 percent void volume, including the sodalite cages; the supercage volume represents 45 percent of the unit-cell volume. The main pore structure is three-dimensional and large enough to admit large molecules, e.g., naphthalene and fluorinated hydrocarbons. It is within this pore structure that the locus of catalytic activity resides for many reactions. A secondary pore structure involving the sodalite units exists, as in zeolite A, but its apertures are too small to admit most molecules of interest in catalysis. Because of the fineness of the major pore structure (pore aperture = 7.4 Å), transport restrictions associated with the counterdiffusion of reactant and product molecules are to be expected in catalysis by X and Y zeolites; this subject is considered subsequently.

Cation positions in X and Y zeolites The zeolite structure may be looked upon as a large matrix of oxygen anions, since the small silicon and aluminum ions are effectively shielded from interaction with species in the pore structure by the oxygen anions tetrahedrally surrounding them. This structure carries an effective charge per unit cell equivalent to the number of alumina tetrahedra per unit cell. The charge is partially delocalized over the entire structure, but the extent of delocalization is not known. The charge is neutralized by the sodium cations originally present in the structure. In the Y zeolite containing sodium ions (referred to as NaY) there are, depending on the Si/Al ratio, about 57 sodium cations per unit cell. These cations are easily exchangeable with other ions having charges of +1, +2, or +3.

Just as the framework structure is precisely defined, the positions of the cations within this structure are also precisely defined. The several cation sites which exist are filled to different degrees by the cations present, consistent with their preferred coordinations with oxygen [52–55]. These cations, even in the dehydrated form of the zeolite, are frequently exposed, essentially unshielded, on the surface and produce their own microchemistry within the zeolite pore, depending upon their charge, electronic structure, and surroundings.

In the X and Y zeolites, four distinct cation sites have been located, as shown in Figs. 1-35 and 1-36. The type I site is located at the center of the hexagonal prism formed by the six bridging oxygens; there are 16 per unit cell. The type I′ site is located in the sodalite unit just on the other side of the shared hexagonal face from the type I site (Fig. 1-35); there are 16 type I′ sites per unit cell. Type II sites are located on the unjoined hexagonal faces of the sodalite units and are

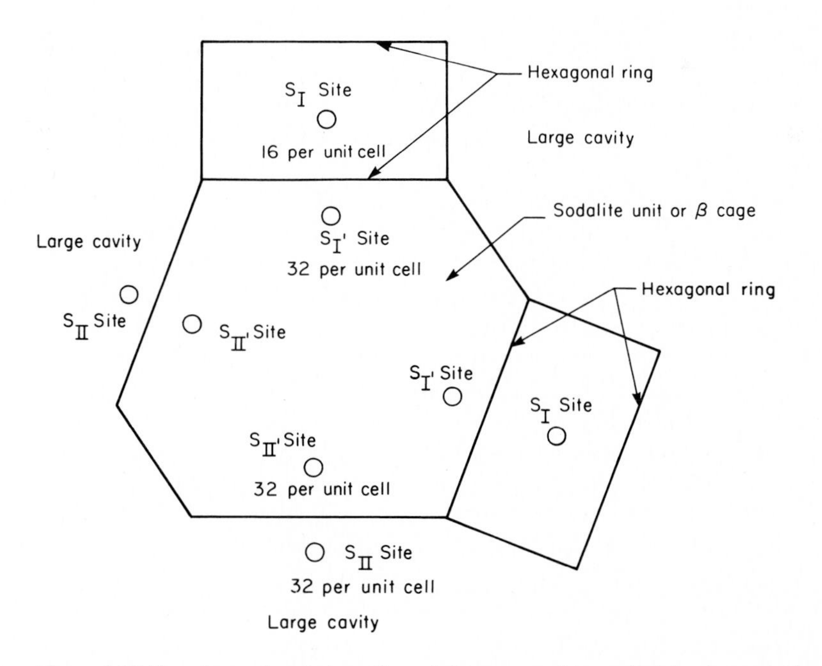

Figure 1-35 Two-dimensional view of a sodalite cage and two adjoining hexagonal prisms.

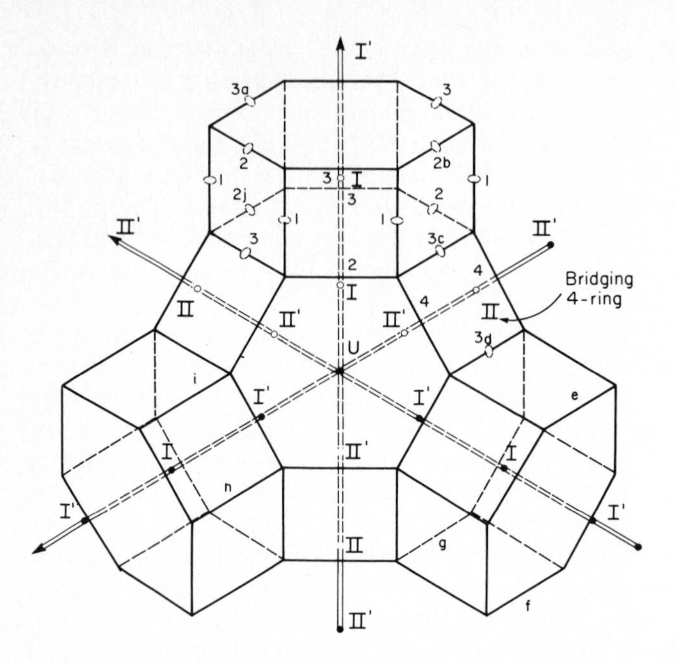

Figure 1-36 Locations of cation sites in X and Y zeolites [53]. (Reprinted with permission from *Advances in Chemistry Series*. Copyright by the American Chemical Society.)

slightly beyond the plane of the hexagonal face in the main sorption cavities. There are 32 such sites per unit cell. In the sodium form of the zeolite, all these sites contain sodium cations. The type II′ cation sites are located on the sodalite-cage side of the unshared hexagonal faces but are displaced considerably within the sodalite cavity.

Type III sites have been designated on the centers of the four-membered rings in the pore mouths. There are 48 such sites. Cation occupancy in these sites has never been observed, and it is likely that when the zeolite is in the hydrated form all the cations that are not located in the type I, I′, II and II′ sites are located in the main supercage and are solvated by water molecules.

Upon dehydration of the structure, the cations shift position, and there are also slight changes in the structural parameters of the zeolite. For the Y zeolite dehydrated at 350°C, Eulenberger et al. [56] found 7.5 sodium ions in the 16 possible type I sites per unit cell, corresponding to a so-called population parameter of 7.5/16 or 0.484. The population parameter for the type II sites was 0.947, and that for the type I′ sites was 0.612. Therefore, in the dehydrated form, there is a sodium cation protruding into the main sorption cavity from almost every unbridged hexagonal face of the sodalite cage.

Cation population parameters also depend on the cation type. For a dehydrated calcium faujasite, Pickert et al. [57] found that the calcium cations were located at type I and type II sites with population parameters of about 1.0

and 0.50, respectively. Upon dehydration, all cations of the catalytically active rare-earth-exchanged faujasite and X and Y zeolites move into the sodalite cages and hexagonal prisms and are located at type I, I′, and II′ sites [52, 58]. This configuration is stabilized by the presence of water molecules, OH groups, or oxygen atoms, which shield the cation charges from each other. Therefore, for the rare-earth faujasite-type zeolites no metal ions are left in the main pore structure to interact with guest molecules.

Typically, all but 16 Na^+ ions per unit cell are easily exchanged. These last 16 cations appear to be located in the type I sites, and their exchange requires their movement through the 2.6-Å six-membered oxygen rings, which requires stripping them of their hydration shells. This process needs considerable energy and occurs slowly even at 100°C. Essentially complete removal of Na^+ is best obtained by exchanging the zeolite several times with multivalent ions then calcining to 350°C and exchanging again. The calcination step replaces the 16 Na^+ ions in the type I sites with the multivalent cations that have been exchanged into the structure, and these remaining Na^+ ions are then easily exchanged out of the structure.

A hydrogen-form zeolite can be prepared by exchange with weak acid or by exchanging with NH_4^+ and decomposing the ammonium ion into $NH_3(g)$ and H^+, which maintains the charge balance in the structure. High degrees of H^+ exchange with acids typically result in structural collapse of the zeolite, but high degrees of H^+ exchange can be achieved without structural collapse if the NH_4^+-exchange technique is used. Such a treatment produces a zeolite free of metal ions; it is referred to as *decationization.*

Olson and Dempsey [58] determined the crystal structure of dehydrated hydrogen faujasite with sufficient accuracy to support the conclusion that there is one hydrogen cation (OH group) per six-membered ring projecting into the sodalite cavity and about one hydrogen per four-membered ring projecting into the main sorption cavity. They found no appreciable structural change upon decationization. Figure 1-37 illustrates this configuration. The literature does not reflect a consensus regarding the exact location of the H^+ ions in the structure.

The faujasite-type zeolite structure contains four crystallographically distinct types of oxygen (see Fig. 1-36). The hydrogen projecting into the supercage is usually placed on the O-1 oxygen in the four-membered ring; the hydrogen atom projecting into the sodalite cage is assigned to other oxygen atoms in the structure, usually the O-3 atom. It is probable that the hydrogens are distributed over all the oxygen types in the structure, oxygens 1 and 3 being the predominant ones with associated hydrogens.

Mordenite structure The mordenite crystal structure, which was determined by Meier in 1961 [59], consists of five-membered rings with each Si or Al tetrahedron associated with at least one such ring. The rings are interconnected to form chains, and the chains are cross-linked to identical chains to form the crystal structure. The relatively high thermal and acid stability of mordenite among zeolites is thought to result from mordenite's having the largest number of such five-membered rings; the mordenite chain consists of five-membered silica rings, which

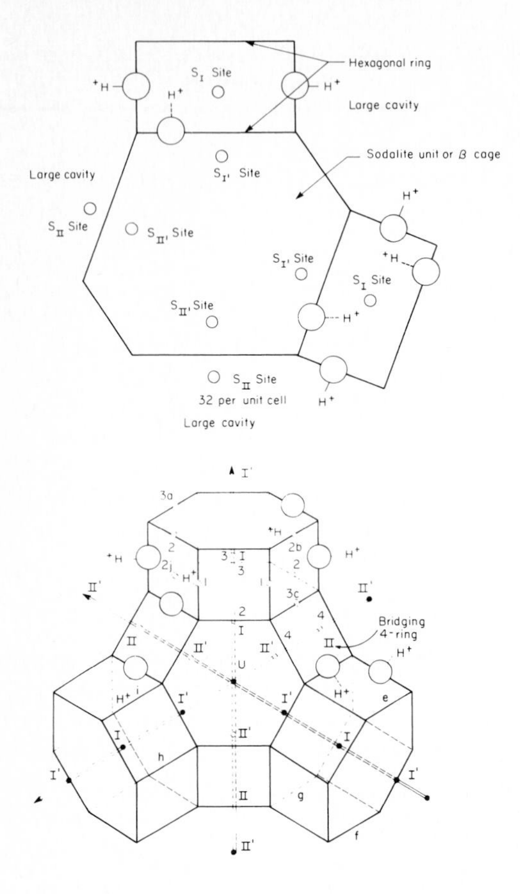

Figure 1-37 Proposed positions of H^+ ions (OH groups) in faujasite-type zeolites. Four different oxygen types are indicated. For clarity, only some of the H^+ ions are indicated. (Reprinted with permission from *Advances in Chemistry Series*. Copyright by the American Chemical Society.)

form the backbone of the chain, and individual alumina tetrahedra [46]. This structure is possible because the Si/Al ratio is high, 5.1. The ideal unit-cell formula is $Na_8 \cdot (AlO_2)_8 \cdot (SiO_2)_{40} \cdot 24H_2O$.

The mordenite pore structure consists of elliptical, noninterconnected channels parallel to the *c* axis, as illustrated in Figs. 1-38 and 1-39. The major channels are circumscribed by 12-membered oxygen rings, which have major and minor diameters of 7.0 and 6.7 Å, respectively [46, 59]. Side pockets open off the main channel along the *b* direction and have a free diameter of 3.9 Å. These side pockets do not interconnect the main channels because halfway between the two main channels the pore is constricted by two distorted eight-membered rings of 2.8 Å free diameter. In the sodium form of the zeolite, this pore is further blocked by a sodium cation located in the center of each distorted ring (Fig. 1-38). Consequently, each main channel is isolated from all other main channels for transport of even the smallest molecules. The remaining four sodium cations in the structure have not been located, but they are probably in the main channels [59]. Even in the hydrogen form, the main channels of mordenite are essentially isolated from each other for transport of all molecules except possibly He and H_2

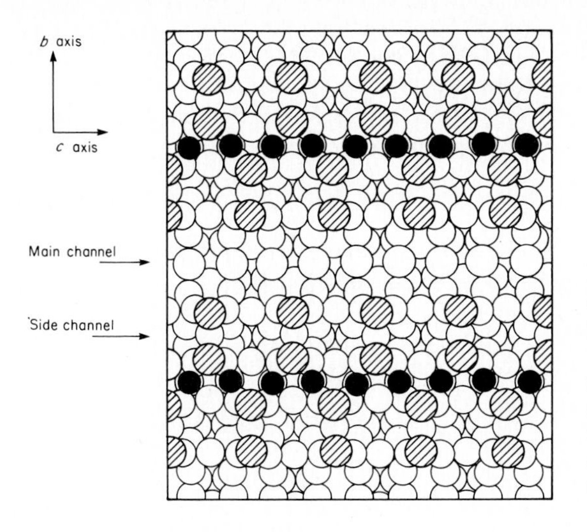

Figure 1-38 Cross section of a wide channel in mordenite showing the pockets lining each side of the channel: ○, O atoms; ⊘ O atoms in plane of paper; ●, Na ions. Some Na ions have not been located; Al and Si at the centers of each tetrahedron of O atoms are not shown [60]. (Reprinted by permission of the Royal Society of London.)

because of the small free diameter of the distorted eight-membered oxygen rings of the side pocket.

It follows that diffusion in the hydrogen mordenite pore structure is diffusion in parallel, noninterconnected channels having a minimum diameter of 6.7 Å. In sodium mordenite, diffusion is hindered by sodium cations in the main channels, but in hydrogen mordenite this cation hindrance is significantly reduced. The locus of catalytic activity is within this pore structure, but because of the transport limitations, particularly since counterdiffusion is required, it is not clear to what extent the internal surface area is accessible to reactants.

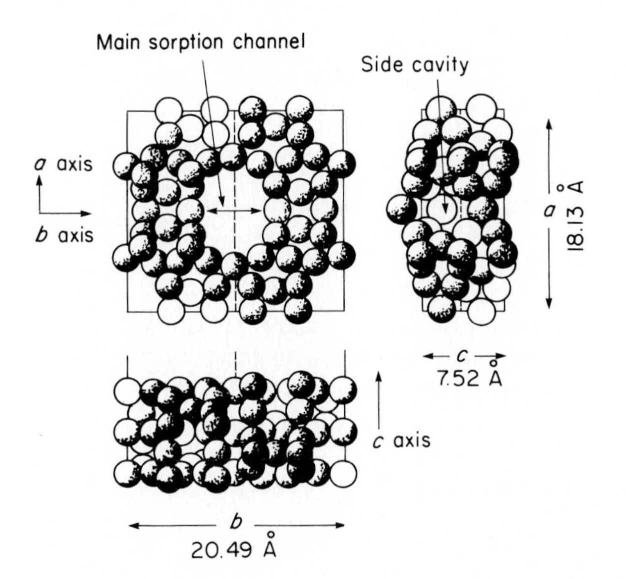

Figure 1-39 Crystal structure of mordenite. Only O atoms are shown. Shading indicates the vertical position of O atoms [59].

Because of the narrowness of the pores, the mordenite surface can exert stronger forces on sorbed reactants than are expected with other catalysts. This characteristic appears to be important in determining some of the unique catalytic properties of hydrogen mordenite, which are considered more fully in a following section.

Zeolite surface chemistry The surface of the zeolite structure can be represented in a simplified, *schematic* fashion, which for the sodium form is

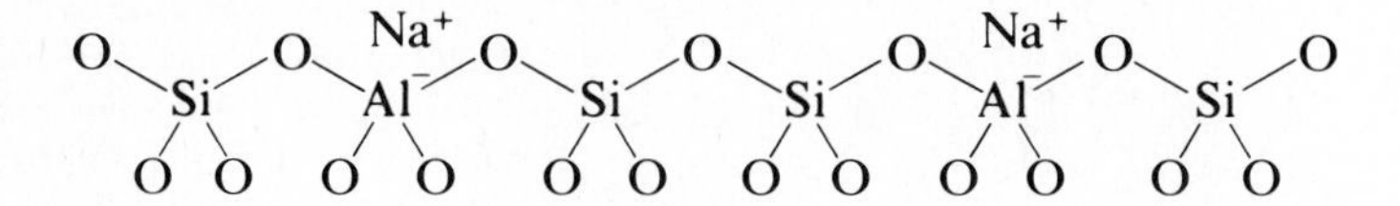

If Na^+ is replaced by Ca^{2+}, the structure is

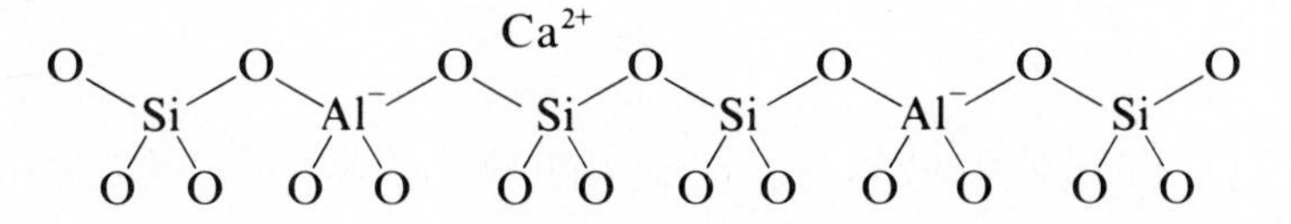

Here the single calcium ion balances the charges of two AlO_4 groups. In the true structure the negative charge is not localized on one tetrahedron but is at least partially distributed over a number of oxygen anions, which interact through either six- or twelvefold coordination with a cation.

The formation of the hydrogen Y zeolite by ion exchange with an ammonium salt to replace the Na^+ followed by heating to decompose the NH_4^+ into NH_3 and H^+ is shown schematically below:

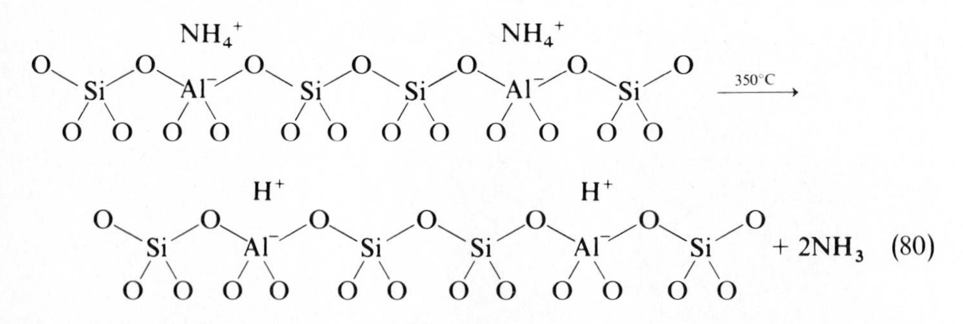

The protons probably bond immediately with one of the lattice oxygens to give OH groups as shown below; these sites represent potential Brønsted acid sites, and it is shown later that they have strong Brønsted acidity:

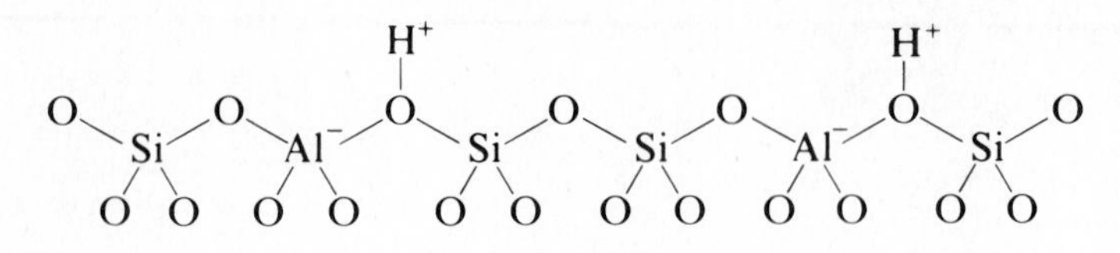

This structure is represented more precisely in Fig. 1-37. Upon further heating to temperatures exceeding 450°C, dehydroxylation of the above structure occurs, to give the following [61]

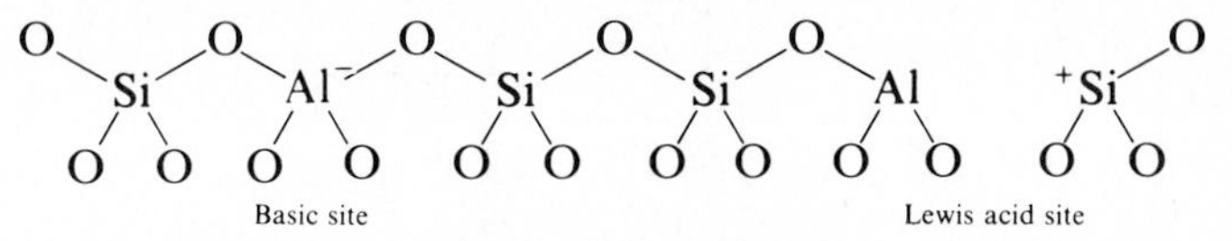

These structural considerations suggest that two Brønsted acid sites are lost for every Lewis acid site generated and that acid-base and positive-negative site pairs are generated. The dehydroxylation can be reversed if water is added back, starting at the highest temperature of dehydroxylation, and cooling is done in the presence of water vapor [61].

X-ray diffraction has been used to determine the exact framework structures of zeolites, thereby defining the interior surfaces. X-ray diffraction, thermogravimetric, and ion-exchange studies have provided quantitative information about the positions and interactions of cations in the structures and information about the behavior of these cations and the surfaces upon high-temperature activation. Our objective now is to consider relations between the structural information and the surface chemical and catalytic character. Structural details determined by infrared spectroscopy have been especially useful in defining the nature of the catalytically active sites in zeolites. A recent review summarizes most of the available literature [62]; below we present only the major elements of this work (Table 1-10).

Infrared measurements of sodium faujasites (X and Y) dehydrated at temperatures less than about 250°C show absorption bands at frequencies of 3750 and 3690 cm^{-1} [63, 64]. The 3690-cm^{-1} band has been observed for all monovalent zeolite forms (Table 1-10); it disappears when the zeolite is heated to temperatures exceeding 350°C, and it reappears as soon as water is added back to the zeolite at temperatures less than 300°C. No further water can be removed as temperature is increased beyond 300°C, and there is no observable weight loss upon heating beyond 300°C. The 3690-cm^{-1} band shifts with a change in the electric field strength of the cation (charge per cation radius). This band is therefore attributed to molecular water adsorbed on the monovalent cation, as shown below:

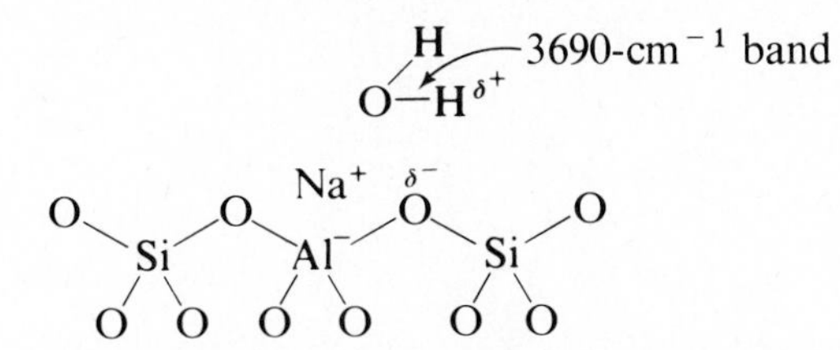

The 3750-cm^{-1} band, found in all cation forms of faujasite, is very weak and unaffected by any manipulation of the zeolite, including ion exchange and heating to very high temperatures. Exhaustive deuterium exchange causes the band to shift by the amount expected for converting an OH group into an OD group.

Table 1-10 Characteristic OH-band frequencies in Y zeolites

Cationic forms of Y	Characteristic frequency,[a] cm^{-1}	Effect of cation type	Effect of basic adsorbate	Effects of temperature and of other adsorbates	Band assignment
All that were investigated	3750	No effect	No effect (nonacidic)	Unaffected by heating to high temperature; deuterium-exchange shifts to ~ 2600–$2700\ cm^{-1}$; very weak	Silanol OH groups which terminate the exterior faces of the zeolite crystallites
Monovalent (Na^+)	3690	Frequency depends on cation	No interaction (nonacidic)	Disappears upon heating to $T > 300°C$; reappears with addition of H_2O at $T < 300°C$	OH of H_2O adsorbed on the cation, causing a field-dependent polarization
Divalent (Ca^{2+})	3600	Frequency depends on cation	No interaction (nonacidic)	Increases in intensity as zeolite is dehydrated to $\sim 400°C$; decreases in intensity at higher temperatures	OH associated with the cation, nonacidic, formed from hydrolysis of water
	3650	Frequency almost independent of cation	Interacts strongly with NH_3 and pyridine (Brønsted acid)	Increases in intensity as $3600\text{-}cm^{-1}$ band increases in intensity; decreases at $T > 500°C$ due to dehydroxylation	Brønsted-acidic, structural OH group attached to O-1 protruding into the supercage

	3540	Frequency almost independent of cation	Interacts only weakly at low temperatures; interacts strongly at high temperatures (proton mobility for $T > 250°C$)	Increases in intensity as 3600-cm^{-1} band increases in intensity; decreases at $T > 500°C$ due to dehydroxylation	Brønsted-acidic, structural OH group protruding into sodalite unit, not effective until proton mobility can occur; assigned to O-3 and also O-2 and O-4
Ammonium-exchanged (H^+)	3650		Interacts strongly with bases at all temperatures (Brønsted acid)	Appears upon decomposition of NH_4^+; decreases in intensity at $T > 500°C$; $OH \rightarrow OD$ shift occurs upon deuteration	Brønsted-acidic structural OH, as above; effectively defines structural OH band
	3540		Does not interact with bases at low temperatures; does at higher temperatures	Appears upon decomposition of NH_4^+; decreases in intensity at $T > 500°C$; $OH \rightarrow OD$ shift occurs upon deuteration	Structural OH, as above; effectively defines structural OH band
	3600, ~3700		No interaction (nonacidic)	Appear under conditions leading to formation of aluminum-lattice vacancies; reduced in intensity when conditions allow silicon migration into these vacancies	Nests of four structural Si—OH groups where aluminum vacancies are formed; non-Brønsted acid

[a] Other bands may exist; bands listed are those specifically characterizing the form for which they are given in the table.

Therefore this band is assigned to an OH-group stretch. The deuterium exchange of these OH groups [64] determined a number of OH groups approximately equal to the calculated number required to terminate the faces of the 1-μm crystallites, 0.2×10^{15} per gram. This band is therefore considered to be due to silanol, Si—OH, groups on the outer surfaces of the crystallites, i.e., OH groups which terminate the faces of the zeolite crystallites at positions where bonding in the interior would occur with adjacent tetrahedral Si or Al ions.

Infrared analysis of the dehydrated sodium form zeolite shows no strong OH bands, and so no protonic character or cracking activity would be expected. Correspondingly, NaY is not an active catalyst for reactions catalyzed by acids. Nonetheless, NaY is more active for catalyzing the cracking of *n*-hexane than amorphous silica-alumina is, but the reaction products are indicative of free-radical cracking rather than acid-catalyzed cracking [20, 65, 66]. It follows that the interactions of the paraffin with the zeolite pore structure are somehow strong enough to catalyze cracking, although by a free-radical rather than a carbonium-ion mechanism.

The CaY zeolite shows strong bands at 3750, 3650, 3540, and 3600 cm^{-1} (Table 1-10). All undergo the expected shift when extensive deuterium exchange is carried out, indicating that they all represent OH groups. The 3600-cm^{-1} band appears upon heating to relatively low temperatures, and then at higher temperatures it disappears. The band does not change in the presence of pyridine or ammonia and is therefore indicative of a nonacidic group. Its exact location depends on the divalent cation, and so it is attributed to OH groups which are localized on the divalent cations and which are formed by splitting water molecules as the zeolite is dehydrated.

As the 3600-cm^{-1} band appears upon heating, the bands at 3650 and 3540 cm^{-1} both increase in intensity. As the 3600-cm^{-1} band disappears upon further heating, the other two bands increase even more in intensity. At still higher temperatures, the 3650- and 3540-cm^{-1} bands are eventually eliminated. Olson [67] has shown by x-ray diffraction that under the conditions resulting in the decrease of the 3600-cm^{-1} band a Ca^{+}—O—Ca^{+} species appears in the zeolite structure.

These observations lead to the following explanation for the OH bands seen in the divalent cation-exchanged X and Y zeolites. The associated electrostatic field of the divalent cation is strong enough to induce dissociation of coordinatively bound H_2O molecules, as follows:

$$\underset{\text{Metal ion}}{M^{2+}} + H_2O \rightleftharpoons M^{2+}\cdots O\genfrac{}{}{0pt}{}{\diagup H}{\diagdown H} \qquad (81)$$

$$M^{2+}\cdots O\genfrac{}{}{0pt}{}{\diagup H}{\diagdown H} \rightleftharpoons M^{2+}\cdots \overset{\delta-}{O}\genfrac{}{}{0pt}{}{\diagup H^{\delta+}}{\diagdown H} \qquad (82)$$

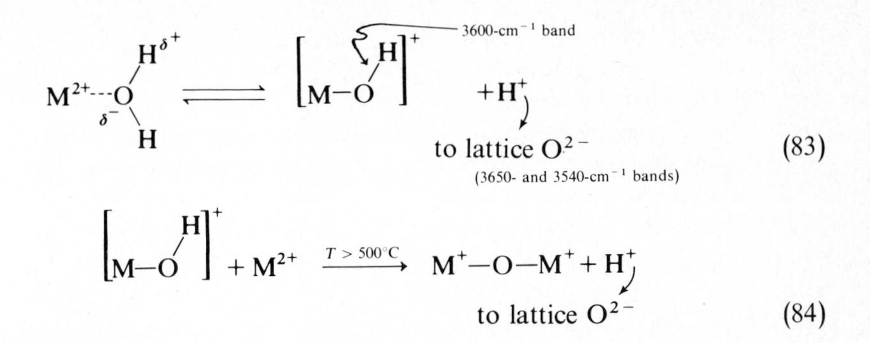

As shown, at higher temperatures, the H^+ splits off and associates with the lattice oxygen to give structural OH groups which are the same as those in the hydrogen-form zeolite.

To assign the 3650- and 3540-cm^{-1} bands, the completely exchanged HY zeolite is considered. These are the only strong bands observed for this zeolite, which indicates that they are associated with framework OH groups. Deuterium exchange gives the expected OH-to-OD shift. Therefore, in a divalent-cation-exchanged zeolite, these bands evidently arise from H^+ formed by the dissociation of water which had been localized on the cations during dehydration; the H^+ becomes located on lattice oxygens.

The 3650-cm^{-1} band is associated with a group showing Brønsted acid properties; it interacts strongly with pyridine and ammonia to give pyridinium ion and ammonium ion, respectively. It follows that the corresponding OH group protrudes into the supercage. The 3540-cm^{-1} band is associated with a group which does not interact with pyridine or ammonia at low coverages and low temperatures. Evidently this OH group protrudes into the sodalite cages or hexagonal prisms [62, 63] (Fig. 1-37). At higher temperatures, interaction between this group and polar bases is observed, which indicates proton mobility in the zeolite. When the HY zeolite is heated to 350 to 500°C, the 3650- and 3540-cm^{-1} bands first shift (also indicating proton mobility) and then at higher temperatures (500 to 700°C) disappear, doing so in accord with thermogravimetric indications of dehydroxylation. As expected, they return upon readdition of water [61].

Initial exchange of NaY with NH_4^+ results in formation of only the 3650-cm^{-1} band, consistent with exchange of the Na^+ ions in the type II sites by NH_4^+ [68]. Only at degrees of exchange greater than 55 to 60 percent does the 3540-cm^{-1} band appear, and it increases rapidly to the intensity found in HY with increasing degrees of exchange. Back-exchange of a highly exchanged HY with cesium ions, which are too large to enter the sodalite units or the hexagonal prisms, largely eliminates the 3650-cm^{-1} band without affecting the 3540-cm^{-1}-band intensity. Exchange results agree with adsorption results with respect to the location of these two OH groups. X-ray diffraction studies have located the two main OH groups on oxygens 3 and 1, leading to the assignment of the narrow 3650-cm^{-1} band to the O-1 hydroxyl protruding into the supercage and the assignment of the broad 3540-cm^{-1} band to the O-3 hydroxyl protruding into the hexagonal prism. The broader 3540-cm^{-1} band of HY has been resolved into five OH frequencies, the strongest of which is assigned to the O-3 hydroxyl. The other bands were assigned to O-2 and O-4 hydroxyls, along with OH groups in aluminum vacancy sites (Table 1-10) and cationic aluminum species [69].

Activation of ammonium-exchanged Y zeolite results in removal of up to 16 of the aluminum atoms from their tetrahedral sites with conversion of these aluminum atoms into cationic species which can be ion-exchanged out with sodium ions (0.1 *N* NaOH solution) [45]. This process, referred to as *dealumination*, results in a zeolite of somewhat increased stability; structural collapse occurs at about 600°C. Activation in the presence of NH_3 or water vapor produces a zeolite which is structurally stable at temperatures in excess of 1000°C; this material is referred to as *ultrastable zeolite* [45, 69, 70–74]. Cracking catalysts are of this general form, being produced as a result of the activation procedures, which may involve a steam treatment, or as a result of the continued steaming occurring in the regenerator.

Aluminum extraction from NaY can be accomplished at temperatures less than 100°C with a chelating agent such as ethylenediaminetetraacetic acid (H_4EDTA) [73]. Na_4EDTA does not remove aluminum, which suggests that the required first step is conversion into the HY form. The material remaining after the extraction has the same unit-cell constant as the parent NaY and is not necessarily ultrastable. Mild thermal treatment of NH_4^+Y followed by ion exchange produces similar results but not stabilization [75]. About 16 aluminum atoms per unit cell can be removed. These are particularly labile aluminums; additional aluminum removal is more difficult. It has been speculated that these aluminum atoms are from four-member rings containing two alumina tetrahedra which repel each other [76].

The initial step in stabilization involves hydrolysis of framework aluminum:

$$\begin{array}{c} \text{Si} \\ | \\ \text{O} \quad H^+ \\ | \\ -\text{Si}-\text{O}-\text{Al}^- -\text{O}-\text{Si}- \\ | \\ \text{O} \\ | \\ \text{Si} \end{array} + 3H_2O \longrightarrow \begin{array}{c} \text{Si} \\ | \\ \text{O} \\ | \\ \text{H} \\ -\text{Si}-\text{O}-\text{H} \quad \text{H}-\text{O}-\text{Si}- \\ \text{H} \\ | \\ \text{O} \\ | \\ \text{Si} \end{array} + Al(OH)_3 \qquad (85)$$

The OH nest formed at the defect site is responsible for the 3600-cm^{-1} infrared band (Table 1-10). As zeolite dehydration proceeds, hydroxoaluminum cations are formed by reaction with mobile protons, as follows [77, 78, 79]:

$$Al(OH)_3 + H^+ \longrightarrow Al(OH)_2^+ + H_2O \qquad (86)$$

These ions are probably located in the sodalite cages, 16 in all and thus 2 in each of the 8 sodalite cages per unit cell. Increased temperature results in reaction between hydroxoaluminum cations and framework protons to form $Al(OH)^{2+}$ ions; one $Al(OH)^{2+}$ ion can be located in each sodalite cage, and the other eight are probably located in type II sites with the OH groups in the supercages. The $Al(OH)^{2+}$ ions may contribute to the observed hydroxyl band [77]. Dehydroxylation at 550°C results in formation of eight Al—O—Al groups in the eight sodalite units per unit cell. The reaction sequence in dry air is summarized as follows [77]:

$$Na_8(NH_4)_{48}(Al_{56}Si_{136}O_{384}) \cdot 234H_2O \longrightarrow \tag{87}$$

$$Na_8(NH_4)_{32}[Al(OH)_3]_{16}[Al_{40}H_{64}Si_{136}O_{384}] + 16NH_3\uparrow + 186H_2O\uparrow \rightleftharpoons \tag{88}$$

$$Na_8(NH_4)_{16}[Al(OH)_2^+]_{16}[Al_{40}H_{64}Si_{136}O_{384}] + 16NH_3\uparrow + 16H_2O\uparrow \rightleftharpoons \tag{89}$$

$$Na_8[Al(OH)^{2+}]_{16}[Al_{40}H_{64}Si_{136}O_{384}] + 16NH_3\uparrow + 16H_2O\uparrow \longrightarrow \tag{90}$$

$$Na_8(Al{-}O{-}Al)_8[Al_{40}H_{64}Si_{136}O_{384}] + 8H_2O\uparrow$$

The unit-cell dimension is not changed, and the material is not stabilized.

The presence of water vapor or NH_3 during activation should inhibit reaction (89), permitting further hydrolysis of framework aluminum and allowing extraction of typically 24 Al atoms per unit cell. Dehydroxylation at 550°C leads to eight

$$\left(\overbrace{Al{-}O{-}Al{-}O{-}Al}^{O}\right)^{3+}$$

ions per unit cell, one in each sodalite cage [78]. Water vapor also increases the lability of framework oxygen, allowing silicon atoms to migrate and fill the vacancies left by removal of aluminum. Whether this process occurs by oxygen movement, with successive silicon atoms flipping in position (essentially a tetrahedral vacancy migration) or, alternatively, as movement of $Si(OH)_4$ species is not clear. X-ray, infrared, and ESR analyses confirm the occurrence of silicon migration [71], and the unit-cell dimension decreases to the value predicted for the new framework Si/Al ratio (Fig. 1-33). The 3600- and 3700-cm^{-1} bands both disappear, and the resultant zeolite shows ultrastability. Silicon migration and unit-cell shrinkage are essential to stabilization. The increased stability is probably best attributed to the increased Si/Al ratio and the unit-cell shrinkage. However, the cationic aluminum in the sodalite units may well contribute to increased stability, just as trivalent rare-earth ions do in the sodalite units; this stabilization may result from some sort of bridging action.

The ultrastable zeolites show far fewer hydroxyl groups in their infrared spectra than their precursors, and they show significant shifts in band positions. These results are consistent with the charge balance for eight $(Al_3O_3)^{3+}$ groups per unit cell balancing 24 of the remaining 32 alumina tetrahedra; only 8 structural hydroxyl groups remain per unit cell. But the exact results of producing ultrastable zeolites are strongly dependent on the treatment conditions; ultrastable Y is not a unique material.

The role of rare-earth cations in NH_4REY cracking catalysts may be to limit the extent of dealumination by filling some of the sodalite cages with RE—O—RE units, thereby increasing the concentration of acidic hydroxyl groups and improving the catalytic activity.

Relations between surface properties and catalytic activity

Silica-alumina Alumina shows no significant Brønsted acidity at any observed calcination temperature,† and infrared bands of NH_4^+ or pyridinium ion do not appear upon adsorption of ammonia or pyridine [80]. Although the OH groups on alumina do not show Brønsted acidity, alumina does show considerable Lewis acidity after calcination at relatively high temperatures. The Lewis acidity appears because of the dehydroxylation of the surface occurring at elevated temperatures to leave exposed electron-pair-accepting aluminum cations where OH groups have been removed. The concentration of the Lewis acid centers depends on treatment temperature [80, 81].

Silica gel shows neither Lewis nor Brønsted acidity [18, 80], and Lewis acidity is not generated by high-temperature calcination. Treatment of silica gel with HF produces Brønsted acid sites [82, 83], formed by charge induction through the oxygen-silicon bridge to the fluorine atom, resulting in a marked enhancement in the acidity of the neighboring OH group. This effect can be understood in terms of the asymmetry of the surface and the strong charge-withdrawing characteristics of the fluorine. The electronic charge is pulled toward the fluorine and away from the O—H bond, making the OH group a stronger proton donor. This situation is similar to that involving an aluminum ion on the surface of a silica particle.

Silica-alumina surfaces show both Lewis and Brønsted acidity, as has been mentioned, and the surface sites have been studied by infrared spectroscopy of adsorbed ammonia and pyridine. The ratio of the two types of sites depends upon the treatment temperature. The Brønsted acid sites are converted quantitatively to Lewis acid sites upon heating and reversibly back to Brønsted acid sites upon the addition of water; the surface contains predominantly Lewis sites after treatment at high temperatures [84]. The maximum Brønsted-acid-site concentration occurs at about 25 weight percent alumina [85]. This is the concentration at which uniformly distributed aluminum cations on the surface of the silica particles would begin to be interconnected as they are in an alumina particle.

These results are consistent with the observation that pure silica, which shows neither Brønsted nor Lewis acidity, is inactive for catalytic cracking, whereas silica-alumina is active. Alumina, which has only Lewis acidity, was shown by

† Some commercial forms of alumina show acidity associated with impurities.

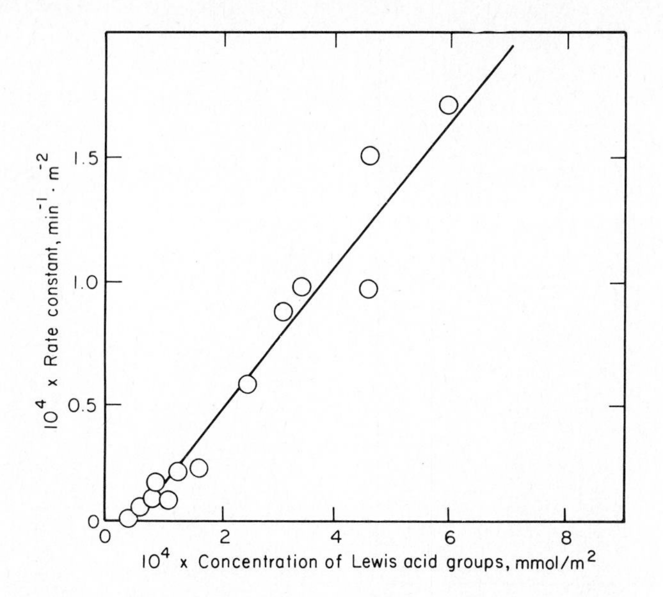

Figure 1-40 Isobutane cracking activity of SiO_2-Al_2O_3 catalysts of varying Lewis acid group concentration [43].

Hall et al. [83] to be very active in cracking but to deactivate almost immediately, which indicates that Brønsted acidity is necessary for the maintenance of activity.

Catalytic activity for carbonium-ion reactions has been correlated with catalyst surface acidity in several instances. Figure 1-40 shows the linear relationship observed between isobutane cracking activity and the Lewis-acid-site concentration on the surfaces of SiO_2–Al_2O_3 catalysts. The correlation suggests that Lewis acid sites are necessary for abstracting a hydride ion from isobutane to initiate the cracking reaction, but the possibility that the Lewis acid sites operate indirectly by enhancing the acidity of adjacent Brønsted sites through charge withdrawal from the O—H bond cannot be ruled out.

A correlation between the Brønsted-acid-site concentration of silica-aluminas (as determined by infrared measurements of pyridine adsorbed on Brønsted sites) and the activity for *o*-xylene isomerization is shown in Fig. 1-41. The results confirm the expectation that formation of a carbonium ion by protonation of the aromatic ring is the mechanism by which *o*-xylene isomerization occurs.

Zeolites Similar and somewhat more quantitative relationships between surface acidity and catalytic activity have been found for zeolites. Barthomeuf [86] presented a recent review. A qualitative summary of the results of several studies (Fig. 1-42) shows how the application of several techniques has provided information about the nature of the cracking sites in calcium-exchanged Y zeolite, the structure of which has already been given. With increasing replace-

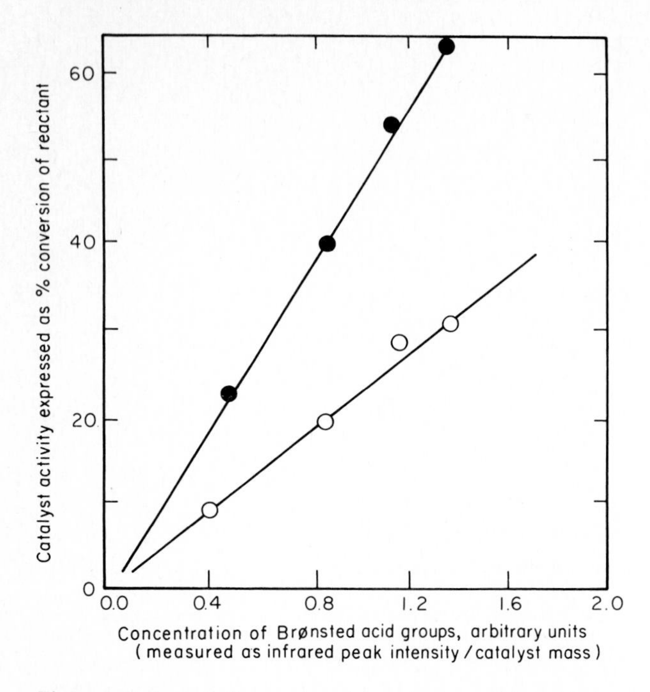

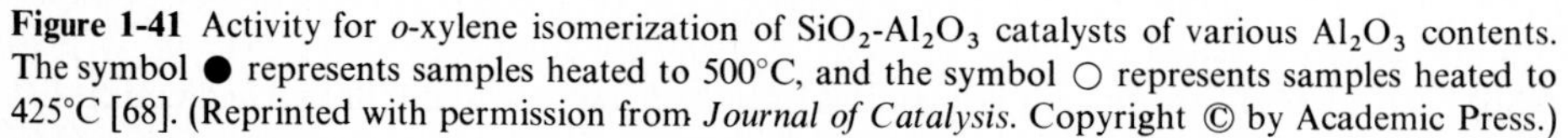

Figure 1-41 Activity for *o*-xylene isomerization of SiO_2-Al_2O_3 catalysts of various Al_2O_3 contents. The symbol ● represents samples heated to 500°C, and the symbol ○ represents samples heated to 425°C [68]. (Reprinted with permission from *Journal of Catalysis*. Copyright © by Academic Press.)

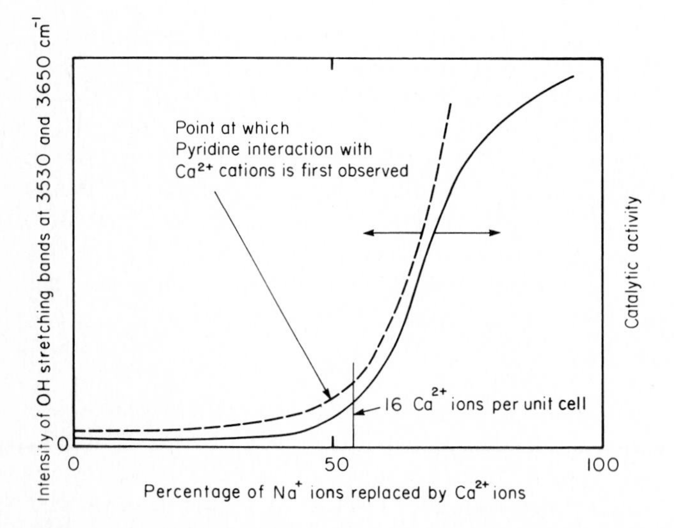

Figure 1-42 Qualitative representation of the correlation between cumene-cracking activity and concentration of accessible Brønsted acid sites in Y zeolite.

ment of Na^+ by Ca^{2+}, catalytic activity begins to appear at the point at which sufficient Ca^{2+} is present to fill all the type I sites (16 per unit cell). The result is explained as follows: Ca^{2+} ions preferentially occupy type I sites, and so until these sites are largely filled, no Ca^{2+} ions remain in the main pore structure (supercages) during the later stages of activation. When Ca^{2+} ions begin to appear in the type II sites, from which they protrude into the supercages, OH groups showing Brønsted acidity and cumene-cracking activity simultaneously appear. Since Ca^{2+} cations in type I sites are not in contact with water molecules, they cannot dissociate them to produce structural OH groups with acidic character; only partially coordinated Ca^{2+} cations in the type II sites generate a sufficient electric field to dissociate water. The similar behavior of the band indicating the OH stretch and the catalytic activity (Fig. 1-42) confirms that the catalytic activity is associated with the accessible surface OH groups.

This pattern of the dependence of catalytic behavior on exchange level has been demonstrated for benzene alkylation, toluene transalkylation, and cumene dealkylation [87–90] as well as for isomerization and cracking of *n*-hexane [66]. The behavior of the catalyst as a function of the degree of exchange is similar for both divalent and trivalent cations. In contrast, acid catalytic activity appears at very low degrees of NH_4^+ exchange [91].

The relative numbers of Brønsted and Lewis acid sites in a magnesium-hydrogen Y zeolite as determined from infrared spectra of adsorbed pyridine are shown in Fig. 1-43; the values depend strongly on the pretreatment temperature, as expected from the earlier discussion. As shown in Fig. 1-43*B*, the concentration of Brønsted acid sites plus twice the concentration of Lewis acid sites is constant for temperatures exceeding 400°C, which confirms that two Brønsted acid sites are converted to one Lewis acid site upon dehydroxylation of zeolites. In this respect, the zeolites differ from silica-alumina, for which a one-to-one relation has been observed (Fig. 1-28) [93].

A thorough study of the nature of the Brønsted and Lewis acidity and their relations to catalytic activity has been carried out by Ward [62, 63, 86, 92, 94–97]. For alkaline-earth- and rare-earth-cation forms of Y zeolite, gas-oil cracking and the *o*-xylene isomerization activities were found to be correlated with Brønsted acidity. The correlation for gas-oil cracking is shown in Fig. 1-44, and correlations for *o*-xylene isomerization are shown in Figs. 1-45 and 1-46.

Ward [97] recorded infrared spectra of HY during cumene cracking. At 250°C he observed a small effect of the reactant on the 3650-cm^{-1} band initially, with a slow decline in the band intensity with time of operation. There was no effect of the reactant on the 3540-cm^{-1} band, which confirms that the more accessible hydroxyl group at O-1 ($3650\ cm^{-1}$) plays the catalytic role at this temperature. The small initial effect of reactants on the 3650-cm^{-1} band intensity suggests that the surface groups were primarily Brønsted acid sites rather than adsorbed cations; this suggestion implies that carbonium-ion formation is rate-limiting and the carbonium ion a highly reactive, short-lived species. The band-intensity decline of the O-1 hydroxyl group was found to parallel the decline in catalytic activity, indicating the removal of individual acid sites. When the

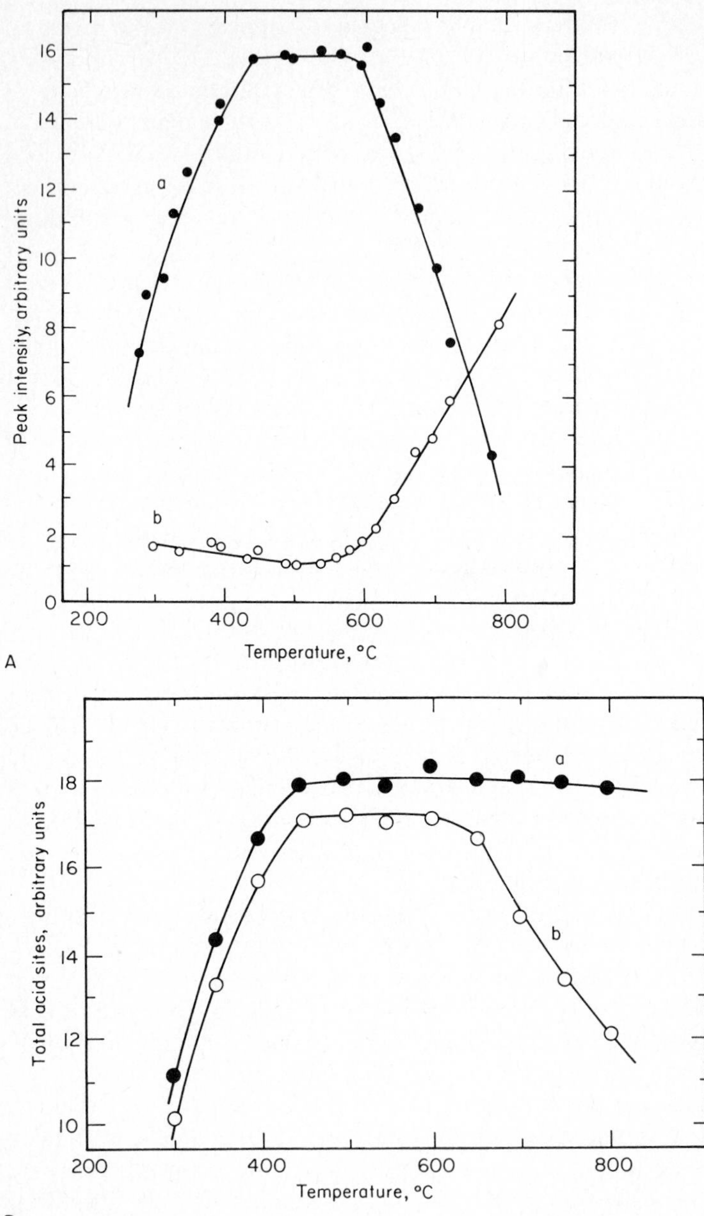

Figure 1-43 The relative Brønsted and Lewis acid site concentrations of Y zeolite as determined by infrared spectroscopy [92]. *A* Intensity of infrared absorption bands for pyridine chemisorbed on Brønsted and Lewis acid sites of magnesium-hydrogen zeolite Y as a function of the temperature of heat treatment. (a) Brønsted acid sites. (b) Lewis acid sites. *B* Total acid population as a function of the temperature of heat treatment. (a) Total Brønsted sites plus 2 times total Lewis sites, (b) total Brønsted sites plus total Lewis sites. (Reprinted with permission from *Journal of Catalysis*. Copyright © by Academic Press.)

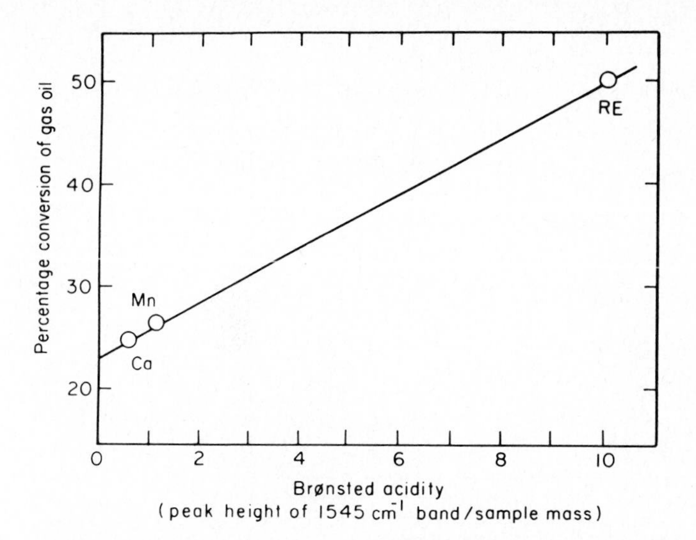

Figure 1-44 Correlation between catalyst activity for gas-oil cracking and Brønsted acidity of Ca, Mn, and REX zeolites. [94]. (Reprinted with permission from *Journal of Catalysis*. Copyright © by Academic Press.)

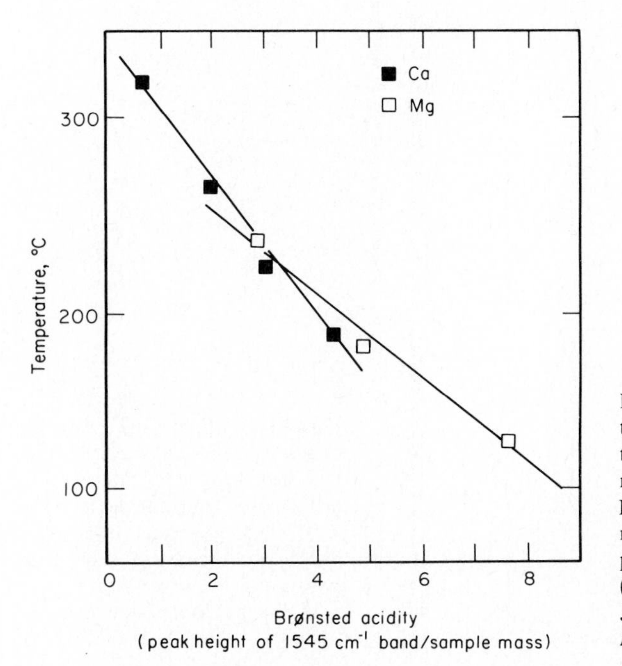

Figure 1-45 Correlation between catalyst activity (as measured by the temperature required for 20% isomerization of *o*-xylene) and the Brønsted acidity of calcium and magnesium faujasites; a lower temperature means a higher activity [95]. (Reprinted with permission from *Journal of Catalysis*. Copyright © by Academic Press.)

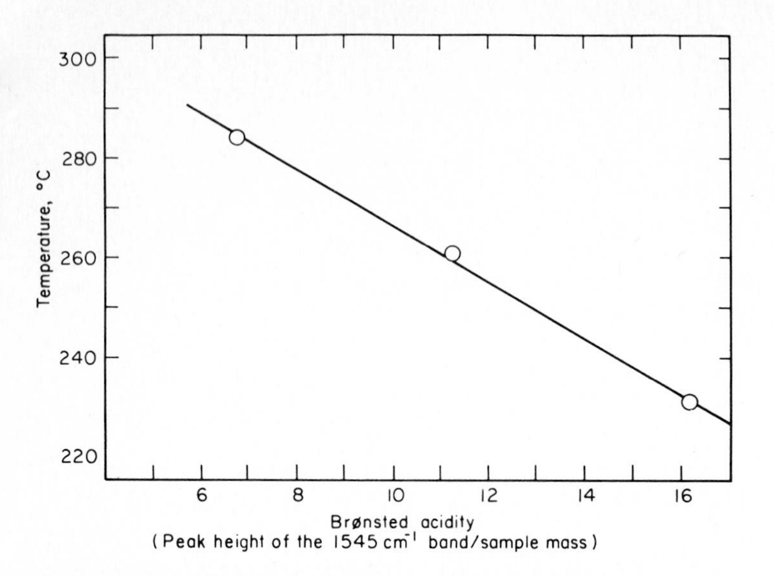

Figure 1-46 Correlation between catalyst activity (as measured by the temperature required for 25% conversion of *o*-xylene) and the Brønsted acidity of rare earth Y zeolite [96]. (Reprinted with permission from *Journal of Catalysis*. Copyright © by Academic Press.)

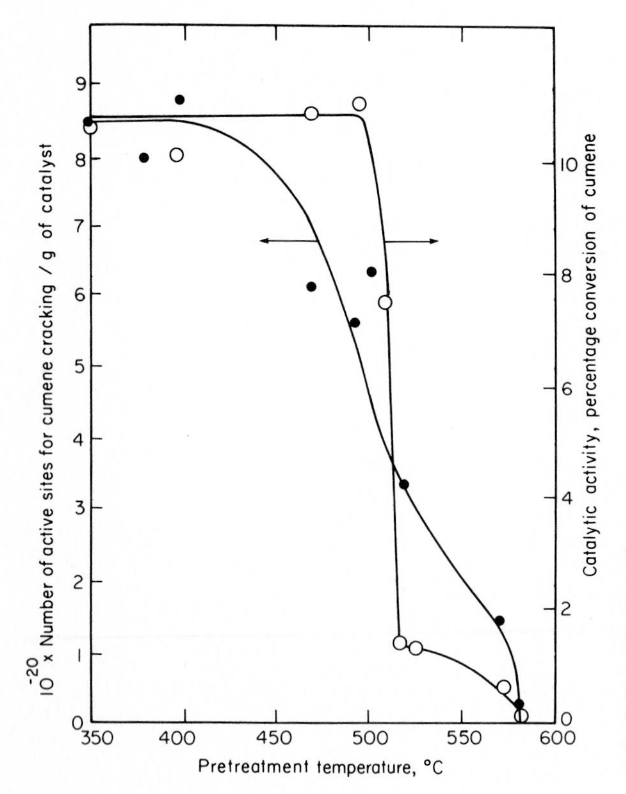

Figure 1-47 Effect of the treatment temperature of NaHY zeolite on the Brønsted acid site concentration and cumene cracking activity at 325°C. Adapted from [98]. (Reprinted with permission from *Advances in Chemistry Series*. Copyright by the American Chemical Society.)

reaction temperature was increased beyond 325°C, the 3540-cm^{-1} band was also affected, which shows that the acid sites associated with this band also became active when proton mobility began to be significant. In summary, Ward's data leave almost no room for doubt that the Brønsted acid sites are responsible for cracking and related reactions.

The effect of pretreatment temperature on the number of Brønsted acid sites within a NaHY (56 percent exchanged) zeolite and on the cumene-cracking activity of this zeolite at 325°C is illustrated in Fig. 1-47. The catalytic activity correlates with the number of Brønsted acid sites determined by quinoline poisoning, which confirms their role in the cracking. The number of active sites measured by poisoning (1.0×10^{21} sites per dry gram) is equal within 15 percent to the number of available OH groups in the zeolite, as calculated from the structure. The agreement and the direct relationship between the number of sites and the cracking activity indicate that each site is catalytically active. This conclusion is not accepted by all, however, and it is not expected to be valid for every reaction.

The effect of treatment temperature on the cracking of 2,3-dimethylbutane by the same catalysts at 400°C is shown in Fig. 1-48. A maximum cracking activity was observed at a temperature at which about 10 percent of the Brønsted acid sites had been converted to Lewis acid sites. These results clearly show the need for both Lewis and Brønsted sites in 2,3-dimethylbutane cracking, and they contrast strongly with the results for cumene cracking. Perhaps cracking of the rela-

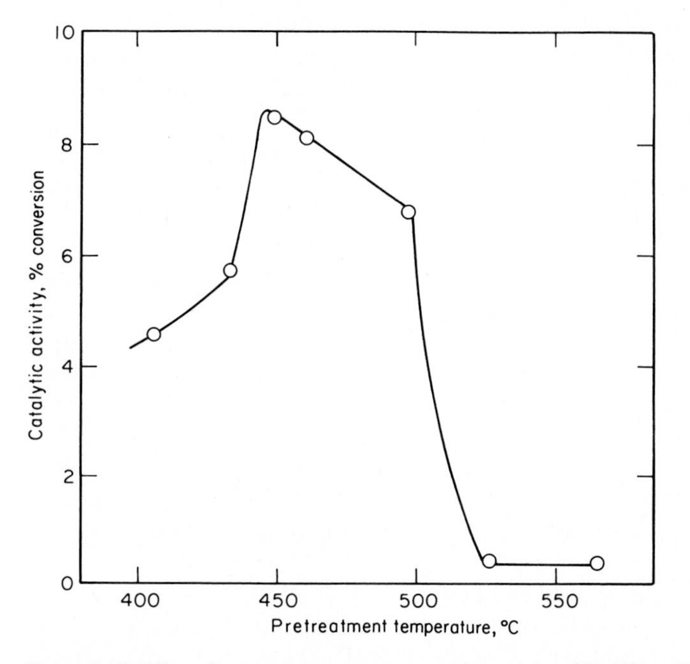

Figure 1-48 Effect of treatment temperature on the activity of NaHY zeolite for cracking 2,3-dimethylbutane at 400°C [98]. (Reprinted with permission from *Advances in Chemistry Series*. Copyright by the American Chemical Society.)

tively unreactive paraffin involves a concerted or dual-site mechanism, or perhaps the proton-donor strength of the Brønsted acid sites is markedly enhanced by the presence of Lewis acid sites. The decrease in catalytic activity at temperatures greater than the optimum is probably related to the loss of Brønsted acidity at the higher pretreatment temperatures. These results imply that although Lewis sites are needed for the catalytic reaction, they are not sufficient to form an adsorbed carbonium ion from the paraffin, which could then crack. Gas-oil cracking catalyzed by REHY and HY exhibits similar behavior, the maximum in activity occurring at a treatment temperature slightly greater than that for the maximum hydroxyl concentration.

One might be tempted to conclude from the preceding discussion that zeolites have uniquely identified catalytic sites for cracking and that simple correlations of catalytic activity with site concentrations are to be expected generally. Such a conclusion would be inappropriate, since many zeolites have significant nonuniformities in their acid sites due to different structural oxygens and OH environments. The HY zeolite is exceptional, having a rather narrow distribution of acid-site strengths with H_0 values from -4 to about -8 [90], which corresponds to a rather uniform set of acid groups.

The HY and partially exchanged forms derived from it exhibit a weak acidity (characteristic of solutions having $<3 \times 10^{-4}$ percent H_2SO_4) in addition to the stronger acidity (characteristic of solutions having >88 percent H_2SO_4). The weak acidity represents 30 percent of the total and is apparently associated with the easily extractable aluminum; it is eliminated upon extraction of about 16 aluminum atoms per unit cell [99–101]. Upon exchange of NH_4^+ into NaY only the weak acid sites appear at low (<30 percent) degrees of exchange. The rate of catalytic isomerization of *o*-xylene increases only slightly with increasing degree of exchange (NH_4^+ for Na^+) up to 30 percent, and it increases sharply and linearly with increasing degrees of exchange in excess of 30 percent; these results indicate that stronger acid sites are being introduced at degrees of exchange exceeding 30 percent [86]. Only this stronger acidity contributes to isooctane cracking [101]. Infrared data confirm the pattern defined by these acidity and catalytic-activity measurements. The integrated extinction coefficient of the 3650-cm^{-1} band increases by a factor of 1.6 between low degrees of NH_4^+ exchange (~ 30 percent) and high degrees of exchange [102]. The increasing extinction coefficient indicates increasing bond polarization, which indicates increasing Brønsted acidity of the OH group.

Rare-earth-exchanged zeolites have broader distributions of acid strengths than HY, the more strongly acidic sites appearing at higher degrees of exchange (Fig. 1-49). Some groups with acid strengths corresponding to $H_0 \leq -12.8$ occur; HY lacks groups which are so strongly acidic. The more strongly acidic sites can be preferentially removed by high-temperature calcination and by steaming, as shown in Fig. 1-50. The broad distribution of acid strengths may be indicative of cations in different sites interacting with OH groups involving different oxygens. The rare-earth ions could increase the acidity of the OH groups by withdrawing electrons from the O—H bonds [103].

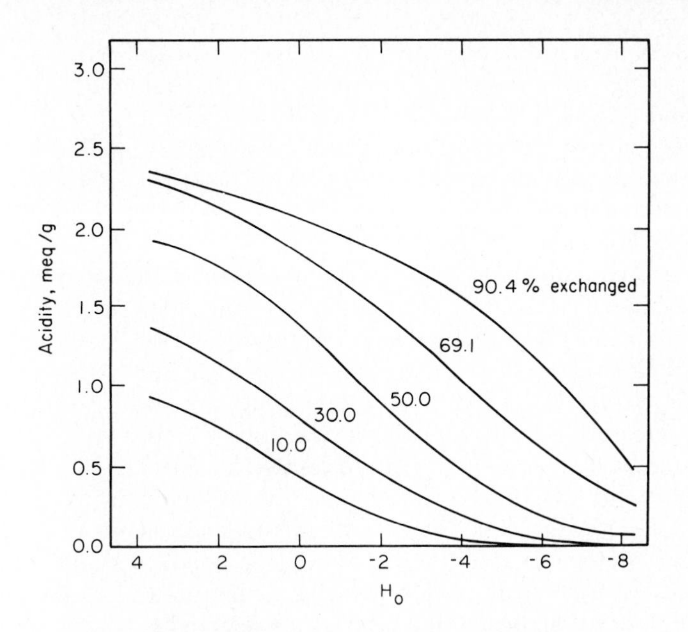

Figure 1-49 Distribution of acid-site strengths in NaLaX zeolite with increasing degrees of exchange (replacement of Na by La) [90].

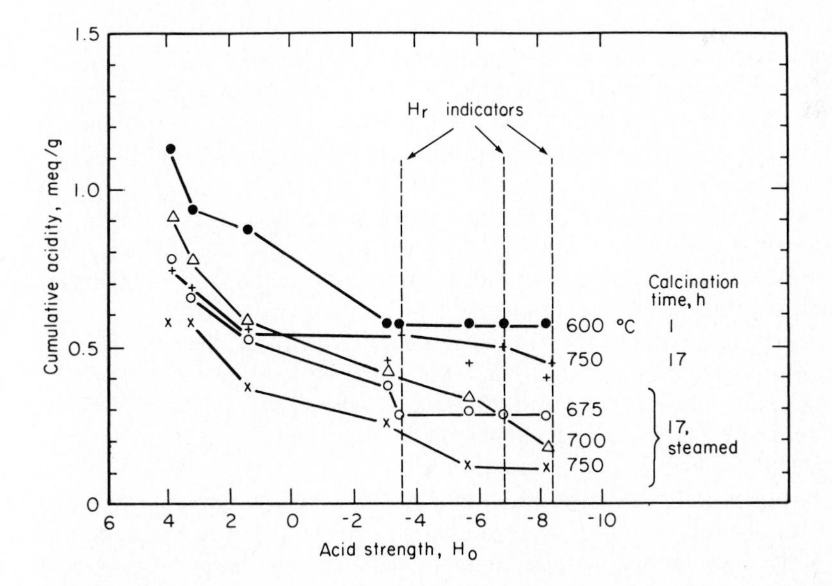

Figure 1-50 Acidity distributions in REY zeolite as a function of calcination and steaming conditions [39]. (Reprinted with permission from *Journal of Catalysis*. Copyright © by Academic Press.)

There is now a mass of evidence to confirm the existence of a distribution in acid-site strengths in zeolites. The evidence includes acidities determined by titration (cf. Figs. 1-49 and 1-50) as well as by adsorption of bases such as pyridine and ammonia [90, 99–105]; it also includes results of studies of isomerization [86] and cracking [101]. Unlike the zeolites, which often have broad distributions of acid-site strengths, silica-alumina has only very strong acid sites, with values of H_0 less than -8 and possibly less than -12 [90].

To this point we have considered only the acidic character of the interior zeolite surfaces, but it is important not to overlook the fact that the presence of unshielded cations and lattice charges in the fine pores creates strong electric fields. Dempsey [107] calculated field strengths of more than 4 eV/Å near Ca^{2+} ions in type II sites. Such strong fields can polarize bonds, as evidenced by the dissociation of water molecules adsorbed on calcium cations [reactions (82) to (84)]. The observation that NaY is much more active than silica-alumina in cracking *n*-hexane suggests involvement of these fields in catalytic reactions [20, 65, 66, 107]. Cracking patterns observed for NaY are consistent with free-radical reactions and (by inference) the induced homolytic rupture of C—H or C—C bonds to form the free radicals. The influence of these strong fields in the confining pore structures of zeolites cannot be overlooked in considering their catalytic activity.

PERFORMANCE OF SILICA-ALUMINA AND ZEOLITE CATALYSTS

Catalyst Activity

Zeolites are highly active cracking catalysts compared with silica-alumina, and therefore small riser-tube reactors have replaced large fluidized-bed reactors in practice. Nace [30] reported that the activity of REHX for *n*-hexadecane cracking was 17 times that of silica-alumina, and REX zeolite was found to be about 10 times as active for gas-oil cracking as amorphous silica-alumina [108]. Xylene isomerization, a reaction requiring Brønsted acidity [98], is 40 times as rapid in the presence of REHY and 10 times as rapid in the presence of HY as it is in the presence of amorphous silica-alumina [109]. Activities of a series of zeolite catalysts were reported to be more than 10,000 times the activity of amorphous silica-alumina for *n*-hexane cracking (Table 1-11).

The density of acidic sites in HY zeolites is compared with the density of acidic sites in amorphous silica-alumina in Table 1-12. Site densities of the zeolites are 10 to 100 times greater than those of silica-alumina. It follows that part of the higher activity of zeolites for *n*-hexadecane and gas-oil cracking and for *o*-xylene isomerization results from their higher acid-site densities.

The difference in densities of acidic sites is not sufficient to explain the much higher activities of zeolites for *n*-hexane cracking (Table 1-11). A full explanation of the high activity of zeolites compared with silica-alumina is lacking, but the following effects are probably the most important:

Table 1-11 Catalytic activity of various zeolites compared with silica-alumina in *n*-hexane cracking [110]

Catalyst	Major cations exchanged into zeolite	Analysis of zeolite, wt %					Activity[a] °C	α^b
		SiO_2	Al_2O_3	Na	Ca	RE		
Amorphous SiO_2-Al_2O_3 (standard catalyst)		...	...	...	...	...	540	1.0
Faujasite	Ca^{2+}	47.8	31.5	7.7	12.3	...	530	1.1
	NH_4^+	75.7	23.1	0.4	...	...	350	6,400
	La^{3+}	...	...	0.4	...	29.0	270	7,000
	RE	...	...	0.39	...	28.8	< 270	> 10,000
	RE, NH_4^+	40.0	33.0	0.22	...	26.5	< 270	> 10,000
	RE, NH_4^+	...	...	...	...	...	420	20
Zeolite A	Ca^{2+}	42.5	37.4	7.85	13.0	...	560	0.6
Zeolite ZK5	...	...	...	...	...	...	400	38
	H^+	76.8	23.1	0.47	...	...	340	450
Mordenite	Ca^{2+}	(~ 77)	...	1.01	...	...	520	1.8
	Ca^{2+}, H^+	82.0	14.0	0.4	...	...	360–400	40–200
	NH_4^+	...	...	...	...	...	< 270	> 10,000
	H^+	80.1	13.4	0.3	1.54	...	300	2,500
	NH_4^+	...	...	0.1	...	...	< 270	> 10,000
Gmelinite	NH_4^+	...	...	...	...	...	< 270(~ 200)	> 10,000
Chabazite	NH_4^+	...	...	...	...	...	< 270	> 10,000
Stilbite	NH_4^+	...	...	...	...	...	370	120
Offretite	NH_4^+	...	...	...	...	...	< 270	> 10,000

[a] Temperature required to achieve 5 to 20% conversion of *n*-hexane.
[b] Activity of zeolite extrapolated to 540°C divided by the activity of silica-alumina at 540°C.
Source: Reprinted with permission from *Journal of Catalysis*. Copyright by Academic Press.

1. Greater concentration of active sites in the zeolite, say by a factor of 50.
2. Greater effective concentration of hydrocarbon in the vicinity of a site resulting from the strong adsorption in the fine micropore structure of a zeolite. At 200°C, the concentration of *n*-heptane in NaX pores is about 0.4 g per cubic centimeter of pore volume [45], which is about 250 times greater than the gas density at this temperature. At the higher temperatures of cracking, the concentration in the zeolite pores may be roughly estimated to be 50 times that in the larger pores of silica-alumina. As explained previously (Fig. 1-11), a high surface concentration of reactant corresponds to a high rate of paraffin cracking. Hexane cracking has been shown to be first order in hexane concentration, and so a fiftyfold concentration increase could result in a fiftyfold rate increase. The effect of the zeolite pores in concentrating the reactant might not create such a difference between the zeolite and silica-alumina for the higher-molecular-weight *n*-hexadecane, since the silica-alumina itself may well be saturated with adsorbed reactant (Fig. 1-11).
3. The electric fields in the zeolite pores may enhance formation and reactions of carbonium ions through polarization of C—H bonds.

Table 1-12 Brønsted site densities of HY zeolite and silica-alumina

Method of acidity determinations	Cation form	10^{-21} × number of OH groups per gram	Ref.
	Y zeolite		
Structural:[a]			
Total H	H	2.9	
H protruding into supercage	H	~ 1.4	
Titration with *n*-butylamine	RE	0.70	39
	H	1.3	100
	H	1.6	90
	La	1.4	90
	H	1.2	106
Poisoning of cumene cracking activity	H	1.0	98
	H	1.0	111
	H	0.18[b]	111
Deuterium exchange with benzene	H	0.9–1.2	112
	Silica-alumina		
Absolute-rate theory	...	0.026	113
Titration	...	0.3	90
Deuterium exchange with benzene	...	0.012	112
Ethylene chemisorption	...	0.015	114

[a] SiO_2/Al_2O_3 ratio was 5.0.

[b] The poison was 2,6-dimethylpyridine; steric hindrance possibly influenced the result.

A combination of effects 1 and 2 could result in roughly a 2500-fold higher rate in a zeolite compared with silica-alumina, provided there was no hindrance of reaction by pore diffusion. This activity difference is about what is observed for *n*-hexane cracking, and so the relative importance of other effects, especially effect 3, remains in doubt.

Catalyst Selectivity

The most significant improvement offered by zeolites over silica-alumina is not increased activity but better selectivity. A comparison of the product distributions obtained in cracking of paraffins with REHX zeolite and silica-alumina is given in Fig. 1-51. Both distributions are characteristic of carbonium-ion cracking. The zeolite gives more products in the C_5 to C_{10} range and fewer in the C_3 to C_4 range.

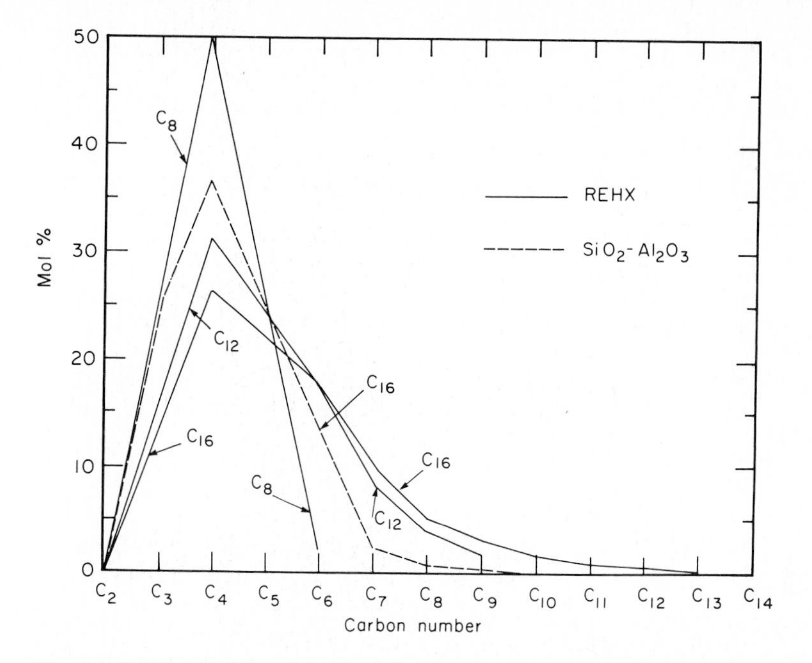

Figure 1-51 Distributions of products from cracking of paraffins catalyzed by REHX zeolite and SiO_2-Al_2O_3 [30]. (Reprinted with permission from *Industrial and Engineering Chemistry Product Research and Development*. Copyright by the American Chemical Society.)

The zeolite's better selectivity for gasoline-range products is attributed to its greater hydrogen-transfer activity relative to chain-scission activity. The cracking of carbon chains in the zeolite is stopped at a higher molecular weight by hydride-ion transfer to the carbonium ion and by hydrogen transfer to olefins. The high rates of these bimolecular transfer reactions in zeolites compared with silica-alumina can be explained by the high concentrations of reactants in the very small pores; a higher concentration would increase the rates of the presumably second-order transfer reactions more than the first-order cracking reactions.

Cracking catalyzed by zeolites results in the formation of much less olefinic product than cracking catalyzed by silica-alumina (Table 1-13), which implies that in the zeolite a larger fraction of the olefins initially formed is saturated with hydrogen before desorbing into the product stream. This difference corresponds directly to the higher rates of hydrogen transfer in the zeolite catalyst and to the stronger interaction of olefins with zeolite; this implies that desorption of a paraffin may be easier than desorption of an olefin, which is retained until hydrogenation occurs. The sources of the hydrogen are the coke and coke precursors which form in the zeolite during cracking, the naphthenes which undergo dehydrogenation to aromatics, and possibly even the hydroxyl groups of the zeolite [115].

Product-distribution data for cracking catalyzed by silica-alumina and REHX and REHY zeolites at a fixed conversion are summarized in Table

Table 1-13 Comparison of gasoline compositions from gas-oil cracking catalyzed by silica-alumina and zeolite [1]

	Feed					
	California virgin gas oil		California coker gas oil		Gachsaran gas oil	
Gasoline	Durabead 5[a]	Durabead 1[b]	Durabead 5[a]	Durabead 1[b]	Durabead 5[a]	Durabead 1[a]
Paraffins, %	21.0	8.7	21.8	12.0	31.9	21.2
Cycloparaffins, %	19.3	10.4	13.4	9.5	14.3	15.7
Olefins, %	14.6	43.7	19.0	42.8	16.3	30.2
Aromatics, %	45.0	37.3	45.9	35.8	37.4	33.1

[a] Durabead 5 = early-generation zeolite (REHX).
[b] Durabead 1 = silica-alumina.

1-14. These results confirm that the zeolites yield more gasoline than silica-alumina and less coke and light products ($< C_4$). The low zeolite coke yield is explained by relatively high rates of hydrogen transfer between cracked product molecules rather than by hydrogen transfer from coke precursors. This conclusion is illustrated by the product distributions shown in Fig. 1-52. The reaction scheme is inferred to be

$$\text{Olefins} + \text{naphthenes} \longrightarrow \text{paraffins} + \text{aromatics} \qquad (91)$$

The conversion of olefins to paraffins and of naphthenes to aromatics occurs almost stoichiometrically, based on H transferred.

Table 1-14 Product distributions in cracking of cycle stocks with silica-alumina and zeolite catalysts [1]

	Augusta light catalytic fuel oil[a]			Beaumont heavy catalytic fuel oil[b]		
	Silica-alumina	Durabead 5[c]	Durabead 7[d]	Silica-alumina	Durabead 5[c]	Durabead 7[d]
Conversion, vol %	35.6	35.6	35.6	42.5	42.5	42.5
C_{5+} gasoline, vol %	22.1	25.9	29.2	24.5	26.3	30.6
Total C_4, vol %	8.7	7.9	6.2	9.4	9.4	8.2
Dry gas, wt %	5.2	4.1	3.5	6.5	5.2	4.7
Coke, wt %	4.3	2.2	1.4	8.7	7.8	4.9

[a] Properties: 27.3°API; aniline no. 60°C; ASTM boiling range 269–352°C.
[b] Properties: 19.5°API; aniline no. 70°C; ASTM boiling range 210–404°C.
[c] Contains REHX in silica-alumina.
[d] Contains REHY in silica-alumina.

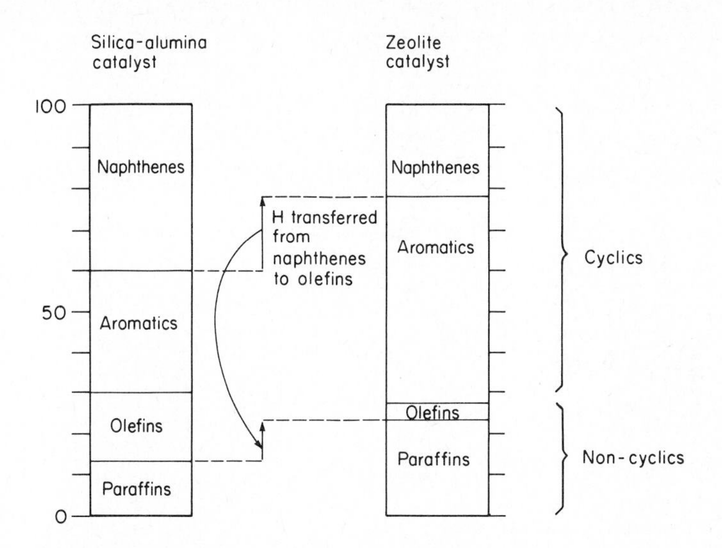

Figure 1-52 Comparison of product distributions in gas-oil cracking catalyzed by silica-alumina and Y zeolite [32]. (Reprinted with permission from *Chemtech*. Copyright by the American Chemical Society.)

The cracking selectivity of zeolites is sensitive to the distribution of acid-site strengths. Figure 1-53 shows how cracking selectivity to gasoline and coke is related to the concentration of very strong acid sites as measured by indicator titration. The changes in acid-site distributions referred to in Fig. 1-53 were achieved by high-temperature calcination and steaming, and these processes might be expected to have introduced significant structural changes. But back-addition of Na (as NaOH) to the REY, which presumably removed the strongest acid sites first (Fig. 1-49) without inducing any structural changes, resulted in a reduction in coke and gas yield and an increase in gasoline production. It is therefore clear that very strong acid sites are not desirable; they cause increased coke formation and production of light gases. Any influence of structural changes caused by the calcination and steaming of the catalysts appears to be of secondary importance.

Topchieva et al. [116] reported that dealumination of zeolite cracking catalysts results in increased activity but lower yields of gasoline and higher yields of gas and coke. These results are consistent with the results mentioned above, since dealumination leads to the removal of weaker acid sites and the formation of stronger acid sites [104]. Topchieva et al. also reported that steaming of the zeolite gave higher yields of gasoline and lower yields of coke and gas, which is in accord with the results of Figs. 1-50 and 1-53.

It appears then that the selectivity advantage of zeolites over silica-alumina is related to the increased concentration of hydrocarbons in the pores and the presence in the zeolite of weaker acid sites than those of silica-alumina. This idea finds some support in results of studies of other mixed-oxide catalysts, but definitive data are lacking.

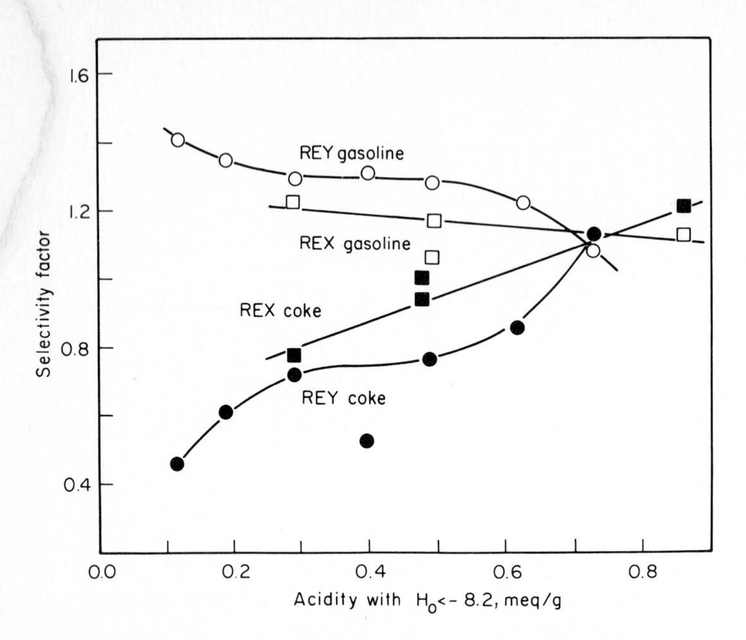

Figure 1-53 Dependence of cracking selectivity on very strong acid site concentration. The selectivity factor is defined as the yield obtained with the zeolite catalyst divided by the yield obtained with silica-alumina. The distributions of strengths of acid sites in the zeolites were changed by treatment at various conditions [39]. (Reprinted with permission from *Journal of Catalysis*. Copyright © by Academic Press.)

We speculate that weaker acid sites catalyze relatively slow cracking compared with hydrogen transfer thereby giving high selectivity to gasoline; the very strong acidity of silica-alumina may localize the reaction on the species tightly bound to the surface site, causing cracking to be relatively rapid compared with hydrogen transfer. This tight binding to the site may also promote coke formation by high-molecular-weight aromatics. Hall et al. [83] observed that alumina with Lewis acid sites was highly active for cracking but was deactivated very rapidly by coke formation on the surface; they speculated that the rapid coke formation was a consequence of the strong binding of reactants with the Lewis acid sites. The more moderate acidity of the zeolites could mean that binding and localization of reaction are not so strong and that hydrogen transfer can be rapid relative to cracking. It is also possible that zeolites are selective for hydrogen transfer because of a dependence of hydrogen-transfer rates on the extent of polarization of the C—H bond, which could be influenced by the high electric field strength in the pores.

The activity of a zeolite changes significantly as coke deposits are formed. The rapid activity decline of REHX as a function of on-stream time for hexadecane cracking is shown in Fig. 1-54. These data, combined with analyses of aged catalysts, show that most of the coke was formed very rapidly initially. The rates of coke formation decreased as coke was deposited; it is inferred that the initial rapid

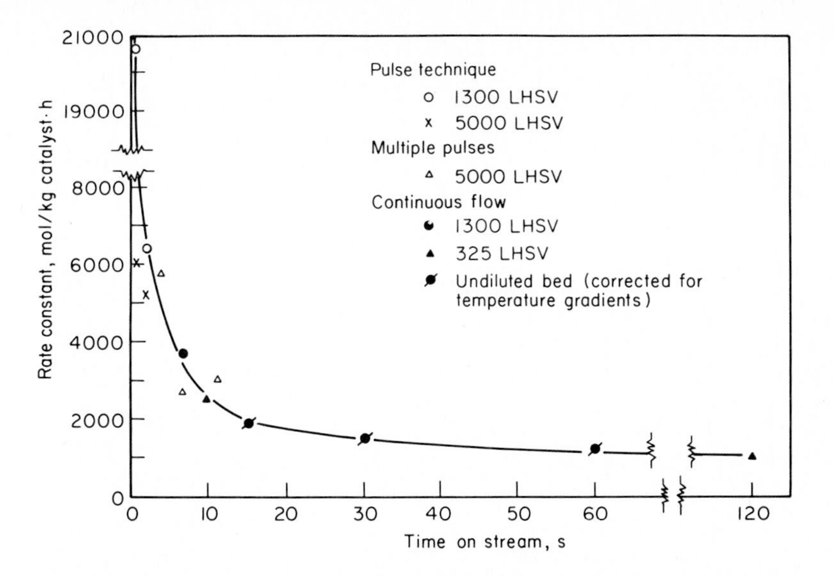

Figure 1-54 Deactivation of REHX catalyst by coke formation during *n*-hexadecane cracking at 482°C [30]. (Reprinted with permission from *Industrial and Engineering Chemistry Product Research and Development*. Copyright by the American Chemical Society.)

coke formation is associated with strong adsorption of hydrocarbons on fresh, active sites or that the rate of coke formation is related to the rate of the cracking reaction (which is also very high initially) as determined by the hydrogen balance on the system. The data of Thakur and Weller [117] indicate that the degree of catalyst deactivation is better related to integral conversion (a hydrogen balance) than to time on stream. It is possible that the slower coke buildup after a short time on stream also indicates that coke already formed is being further dehydrogenated by hydrogen transfer to unsaturated product molecules.

Diffusional Limitations and Shape-selective Catalysis

Weisz and coworkers [118] established what has been assumed throughout this chapter, that when reactants can enter into the microscopic intracrystalline pores of a zeolite, the locus of catalytic activity lies within this structure. Table 1-15 shows that Linde 5A zeolite (so-called because it has a 5-Å effective pore diameter) readily catalyzed the cracking of linear paraffins, which can enter the pore structure, but did not catalyze the cracking of isoparaffins, which are too large to enter; no isoparaffins were produced in the reaction (Table 1-15). In contrast, silica-alumina catalyzed cracking of both paraffins, and 3-methylpentane (with its tertiary carbon) was converted to a larger extent. Correspondingly, in the presence of the zeolite catalyst, *n*-butanol (but not isobutanol) was readily dehydrated at temperatures between 230 and 260°C. At considerably higher temperatures, a

Table 1-15 Demonstration of catalytic activity within the intracrystalline pore structures of zeolites [118]

Reaction conditions: 500°C, contact time = 7 s.

	Catalyst			
	None[a]	Linde 4A zeolite	Linde 5A zeolite	Silica-alumina
Reactant	Conversion, %			
n-Hexane	1.1	1.4	9.2	12.2
3-Methylpentane	< 1.0	< 1.0	< 1.0	28.0

[a] Conversion determined with reactor packed with quartz chips.

small fraction of the isobutanol was converted, which suggests that there were a few catalytic sites on the exterior surfaces of the zeolite crystallites.

Because of the narrowness of the pores and the high activity of zeolites under cracking conditions, one might expect that only a fraction of the interior surface would actually be used. The data needed to evaluate the difficulty of getting reactants to interior surface sites and products back out are unavailable and hard to obtain. Because of the very fine pore structure of zeolites, counterdiffusion rates are not predictable as would be rates of, say, Knudson diffusion in larger pores such as those of silica-alumina. Counterdiffusion rates are markedly affected by the nature of the cation within the zeolite structure, the extent of ion exchange, the nature of the pretreatment, the size and polarity of the counterdiffusing species, and the presence of impurities [119]. The data of Fig. 1-55 show a range of four orders of magnitude in the effective diffusion coefficient for counterdiffusion in NaY. Because of the extreme sensitivity of the diffusion rate to the size of the

Table 1-16 Unidirectional diffusion of hydrocarbons into Y zeolite (adapted from [121], [121*a*])

Compound diffusing into zeolite	Temp., °C	$D_{eff} \times 10^{17}$, m^2/s,[a]		
		NaY	HY(I)[b]	HY(II)[c]
1,2,3-Trimethyl benzene	0	13	> 100	
1,3,5-Triisopropyl benzene	30	0.047	3.2	9.3
1,3,5-Triisopropyl cyclohexane	30	4.9	3.2	8.7

[a] as fractional approach to equilibrium approaches zero.
[b] HY(I) = 75% ion-exchanged to hydrogen, remainder sodium.
[c] HY(II) = 97.7% ion-exchanged to hydrogen, remainder sodium.

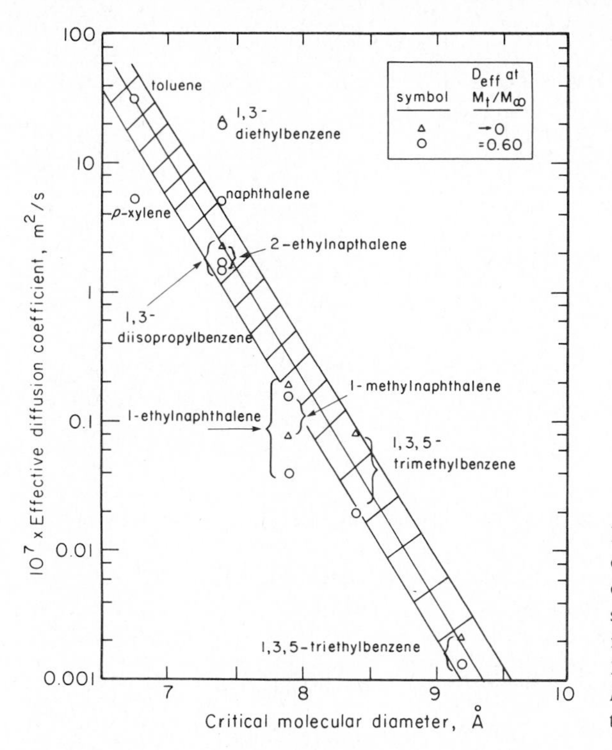

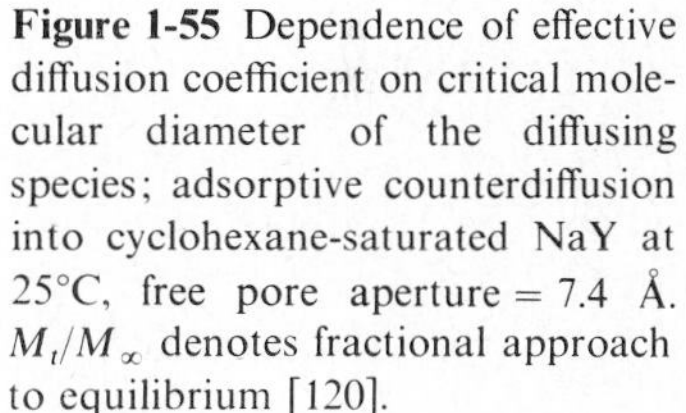
Figure 1-55 Dependence of effective diffusion coefficient on critical molecular diameter of the diffusing species; adsorptive counterdiffusion into cyclohexane-saturated NaY at 25°C, free pore aperture = 7.4 Å. M_t/M_∞ denotes fractional approach to equilibrium [120].

diffusing molecule relative to the pore size, small amounts of deposited coke can have marked effects on diffusion rates and catalytic properties, as discussed later. The data of Table 1-16 further show some of the effects of structure of the diffusing molecule and cationic composition of the zeolite. Elimination of acid-base interactions between zeolite and hydrocarbon by removal of the π electrons of the hydrocarbon or by exchanging out the Na^+ ions of the zeolite can result in a hundredfold increase in the diffusion rate.

Weisz and coworkers [110, 122] observed that the apparent activation energy for *n*-hexane cracking catalyzed by several of the catalysts listed in Table 1-11 was between 27 and 30 kcal/mol; nearly the same value was observed for silica-alumina, which suggests the absence of diffusional limitations at the relatively low temperatures of the study. Cracking of *n*-hexane catalyzed by a zeolite with smaller pores was characterized by an apparent activation energy of about 15 kcal/mol, which is suggestive of intracrystalline diffusion limitations [110, 122]. These results do not imply that diffusional limitations are absent in the cracking of C_6 hydrocarbons catalyzed by the large-pore zeolites at the higher temperatures normally needed for cracking.

Because of the complexity of the diffusion phenomena, it would be desirable to evaluate diffusion rates by making direct counterdiffusion measurements of reactants and products (or similar molecules) in the active form of the catalyst under reaction conditions. For most nonzeolitic catalysts, such an evaluation can be made by measuring reaction rates with catalyst particles of various sizes, but this standard technique has not often been applied to zeolites because zeolite crystallites of varying sizes are not readily available.

Large molecules can be excluded from the zeolite pore structure entirely, the most severe form of diffusional limitation. Such exclusion, evidenced by the small-pore zeolite A, provides a means of selectively converting only those molecules which can enter the pore structure (Table 1-15). This shape-selective catalysis is the basis of the so-called Selectoforming process for cracking straight-chain paraffins from mixtures including aromatics, naphthenes, or branched-chain paraffins, all of which are excluded from the catalyst pore structure.

Nace [30] observed that for the cracking of progressively larger ring structures catalyzed by silica-alumina, the rate constant increased continuously with increasing reactant size. When a REHX catalyst was used, however, the rate constant

Table 1-17 Cracking rate constants at 482°C [108] (2-min on-stream instantaneous values in units of h^{-1})

	SiO_2–Al_2O_3		REHX		Ratio
Reactant hydrocarbon	Rate constant	C on catalyst, %[a]	Rate constant	C on catalyst, %[a]	$k_{REHX}/k_{SiO_2-Al_2O_3}$
n-$C_{16}H_{34}$	60	0.1	1000	1.4	17
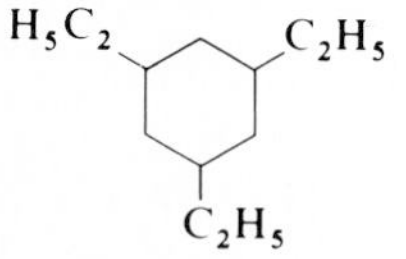 H_5C_2, C_2H_5, C_2H_5	140	0.4	2370	2.0	17
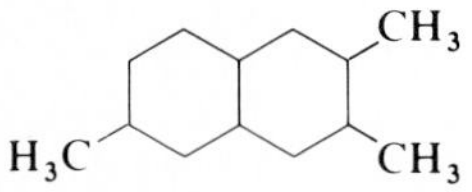 CH_3, CH_3, H_3C	190	0.2	2420	0.7	13
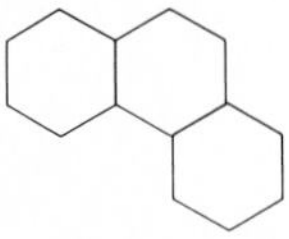	205	0.2	953	1.0	4.7
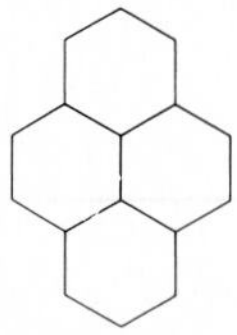	210	0.4	513	1.6	2.4

[a] The amount of coke on the catalyst after the 2-min test.

Source: Reprinted with permission from *Industrial and Engineering Chemistry Product Research and Development*. Copyright by the American Chemical Society.

increased to a maximum with increasing reactant size then decreased drastically for three- and four-ring reactants (Table 1-17). This behavior is indicative of diffusional limitations in the intracrystalline pore structure of the zeolite, which become more important as the molecular size increases. It is inferred that in cracking of gas oil catalyzed by zeolite, some fraction of the reactants experiences severe diffusional resistances, even to the limit of being excluded from the pore structure. These hindered species could even cause diffusional problems for species which alone would diffuse rapidly.

Thomas and Barmby [125] suggested that many of the gas-oil range molecules in a commercial feedstock are too large to enter the zeolite pore structure and are cracked instead on the exterior surfaces of the zeolite crystallites or on the surface of the silica-alumina matrix in which the crystallites are embedded. The gasoline-range molecules produced would diffuse into the zeolite pore structure, where they would undergo secondary reactions, including rapid hydrogen transfer. This suggestion is consistent with the available information, but quantitative diffusion-reaction data are needed to test it.

Gorring [126] reported data showing a strong dependence of the rate of diffusion of straight-chain paraffins in erionite on the paraffin chain length. The rate at first decreased markedly with increasing chain length, passed through a minimum, reversed, passing through a maximum, repeated the pattern, and then decreased. The cracking of a long straight-chain paraffin catalyzed by erionite produced a bimodal product distribution; the products present in highest concentrations were those which had been found to diffuse most rapidly in the erionite. Products with some carbon numbers were almost absent, and they were the ones which exhibited minima in the diffusion rate. These data evidently show a unique selectivity effect caused by diffusional properties of the zeolite catalyst, but an explanation of the complex behavior remains to be found.

Diffusional limitations caused by coke formation have been observed for zeolite catalysts. They are indicated by a decrease in the activation energy and a concomitant decrease in product selectivity as the coke level increases. The behavior is that expected for pore-mouth poisoning of the zeolite crystallite by coke. Therefore, reactor design and operation must be directed toward minimizing coke deposits, and regenerator design and operation must be directed toward maintaining low coke levels on the regenerated catalyst. These subjects are considered further in the next section.

REACTION ENGINEERING OF CATALYTIC CRACKING†

The purpose of this concluding section is to demonstrate how the catalytic chemistry is related to the performance of the cracking reactor and constrains the process design. A simplified representation of the process is given in Fig. 1-56. The

† This section was written by Jon H. Olson.

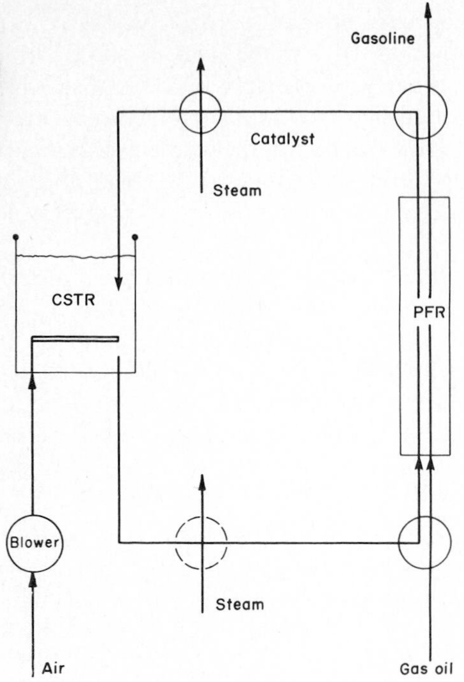

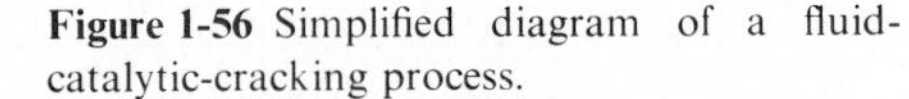
Figure 1-56 Simplified diagram of a fluid-catalytic-cracking process.

catalyst circulates between the riser-tube reactor and the regenerator, and the gas-oil stream and air separately fluidize the catalyst particles in the reactor and in the regenerator. An analysis of the operation of *fluid catalytic cracking* (FCC) requires the development of models for the riser-tube reactor and the regenerator and accounting for their cooperative operation, in particular a balancing of the heat produced in the regenerator with that consumed in the cracking reactor.

THE RISER-TUBE REACTOR

A rough summary of the fluid mechanics prevailing in the riser tube is given in Fig. 1-57. Gas oil and dispersing steam carry the freshly regenerated catalyst upward in two-phase (gas-solid) flow. The gas and catalyst particles do not move upward at the same velocity; the particles lag behind the gas. Therefore the ratio of solids to gas in the flowing mixture is greater than the ratio of solids to gas in the feed, and the residence time of the solids is longer than that of the gas.

The previously discussed, cracking and coke-forming reactions take place in this reactor, and it has been shown that coke formation involves a complex set of reactions in the pores of the zeolite and on the surface of the silica-alumina matrix. From a reaction-engineering standpoint, the loss of catalytic-cracking activity

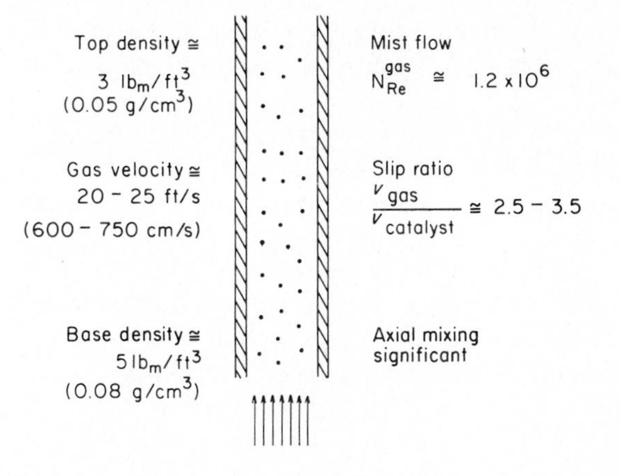

Figure 1-57 Riser-tube fluid-flow characteristics.

caused by the coke formation can be represented by a simple empirical equation. Voorhies [127] found that for many deactivating catalysts the fraction of the initial activity remaining can be expressed as

$$\Phi = e^{-\alpha t} \tag{92}$$

where Φ = fraction of initial activity remaining
α = empirical coking parameter
t = time solids are exposed to the gas oil

In riser-tube cracking the equation is written to account for the unburned coke remaining on the regenerated catalyst fed to the reactor:

$$\Phi = \Phi_0 e^{-\alpha t} \tag{93}$$

It is assumed generally that the fraction of activity remaining is linearly dependent on the coke content of the catalyst, at least when the weight fraction of coke on the catalyst is low; the initial activity is therefore given as [5]

$$\Phi_0 = 1 - \beta\omega_c \tag{94}$$

where β is an empirical deactivation parameter and ω_c is the weight fraction of coke on the catalyst.

If it is assumed that most of the coke is produced in side reactions during cracking, the coke level and the cracking activity remaining should be correlated with the amount of gas oil cracked rather than with time, and evidence to support this approach has been mentioned previously. Froment [127*a*] has argued that catalytic activity should be related to coke concentration on the catalyst (an argument similar to that above) and has provided detailed kinetics of coking, which would allow the coke concentration on the catalyst to be predicted for any point in the reactor. On the other hand, plant experience suggests an apparent independence of cracking activity and coking rate, and therefore the conventional

Voorhies equation is used for plant simulation; correspondingly, the residence time of the catalyst in the riser (typically 5 to 7 s) is an important parameter in the operation of an FCC unit.

Weekman and Nace [3, 4, 108, 128] developed a simple reaction-engineering model for catalytic cracking. Recently Jacob et al. [128*a*] have published a more detailed model of catalytic cracking, but only the simple Weekman-Nace model will be discussed here. As described in the simple reaction scheme (1), the vapors in the riser are considered to consist of three components, gas oil, gasoline, and other gases, primarily light gases. Although pure-compound studies have shown that cracking is a first-order reaction in the hydrocarbon concentration (at least for low-molecular-weight paraffins) the pseudocomponent gas oil cracks according to second-order kinetics.†

Although the fluid-flow pattern of the gas and catalyst in the riser is complex, it is useful to approximate the gas flow as idealized piston flow. The following simple mass-balance equation can then be written, which incorporates the assumption of second-order reaction of gas oil:

$$\frac{d\omega_O}{dZ} = -\frac{k_O}{\text{LHSV}} \Phi \omega_O^2 \tag{95}$$

where ω_O = mass fraction of gas oil in vapor stream
$Z = z/l$ = fractional distance downstream of riser entrance
$\Phi = \Phi_0 e^{-\alpha t}$ = fraction of cracking activity remaining
LHSV = liquid hourly space velocity = volume of liquid fed to reactor per gross volume of reactor per hour

$$k_O = k'_O \frac{(\rho_O^0)^2}{\rho_L} \frac{\rho_F}{\rho_C} \quad \text{pseudo-first-order rate constant,§ h}^{-1}$$

k'_O = second-order rate constant unit volume of catalyst per hour¶
ρ_O^0 = gas-oil density at inlet of reactor
ρ_L = liquid density used to define LHSV
ρ_F = density of fluidized catalyst in reactor
ρ_C = density of catalyst

† A similar result has been found for petroleum hydrodesulfurization, and it is explained in Chap. 5.

§ The Weekman definition of k_O includes the density of the catalyst in the reactor in the rate parameter. To convert from the moving-bed reactor used by Weekman and Nace to a fluidized-bed reactor, k_O should be adjusted by the ratio of the catalyst densities in the reactors, as follows:

$$k_O = k_O^{WN} \frac{\rho_F}{\rho_F^{WN}}$$

where k_O^{WN} = reported value (Weekman-Nace) of cracking parameter
ρ_F^{WN} = density of catalyst in moving-bed reactor
$\approx 50\ \text{lb/ft}^3$ or $0.8\ \text{g/cm}^3$

¶ Values of this parameter are not reported in the literature.

If the reactor is operated isothermally, the fraction of gas oil remaining in the product can be found by direct integration of Eq. (95):

$$\omega_O = \frac{1}{1 + (k_O/\text{LHSV})U} \tag{96}$$

where $U = \int_0^1 \Phi \, dZ$ = distance parameter modified to account for catalyst deactivation

$= \Phi_0 \dfrac{1 - e^{-\alpha t_c Z}}{\alpha t_c}$ when slip velocity (ratio of gas to catalyst velocity) in riser is constant

and where t_c is the contact time of the reactant with the catalyst in the riser. The foregoing equations account for gas-oil conversion, but the expressions for production of gasoline are more complex because gasoline can be formed from gas oil and lost by overcracking:

$$\frac{d\omega_G}{dZ} = \frac{k_1}{\text{LHSV}} \Phi \omega_O^2 - \frac{k_2}{\text{LHSV}} \Phi \omega_G \tag{97}$$

The first right-hand term of this equation represents the production of gasoline from gas oil; Nace et al. [5] found that k_1 is a fraction of k_O which depends primarily upon the aromatic-to-naphthene ratio of the gas oil. The second term represents the overcracking of gasoline to light gases. This equation can be integrated easily for isothermal operation and yields the following for ω_G

$$\omega_G = \theta_1 \theta_2 e^{-\theta_2/\omega_O} \left[\frac{e^{\theta_2}}{\theta_2} - \frac{\omega_O}{\theta_2} e^{\theta_2/\omega_O} - E(\theta_2) + E\left(\frac{\theta_2}{\omega_O}\right) \right] \tag{98}$$

where $\theta_1 = k_1/k_O$

$\theta_2 = k_2/k_O$

$E(x) = \int_{-\infty}^{x} \frac{e^u}{u} \, du$ = exponential integral†

A plot of the calculated production of gasoline as a function of the conversion of gas oil is given in Fig. 1-58. (These results are restricted to a particular value of k_1/k_O.) This figure shows that the overcracking ratio k_2/k_O must be very low if good gasoline yields are to be obtained. Further, the maximum gasoline yield is sensitive to the conversion, and therefore plant operators must be careful to limit the extent of reaction in the riser. As shown in Fig. 1-59, the maximum yield depends only upon the ratio k_2/k_O. Again, this plot shows that the overcracking ratio must be less than 0.025 if a 90 percent yield to gasoline is to be obtained.

Weekman et al. measured the apparent activation energies of the pseudoreactions. From the parameters $(E_{\text{act}})_O = (E_{\text{act}})_1 = 10{,}000$; $(E_{\text{act}})_2 = 18{,}000$;

† Values are collected in standard tables of functions.

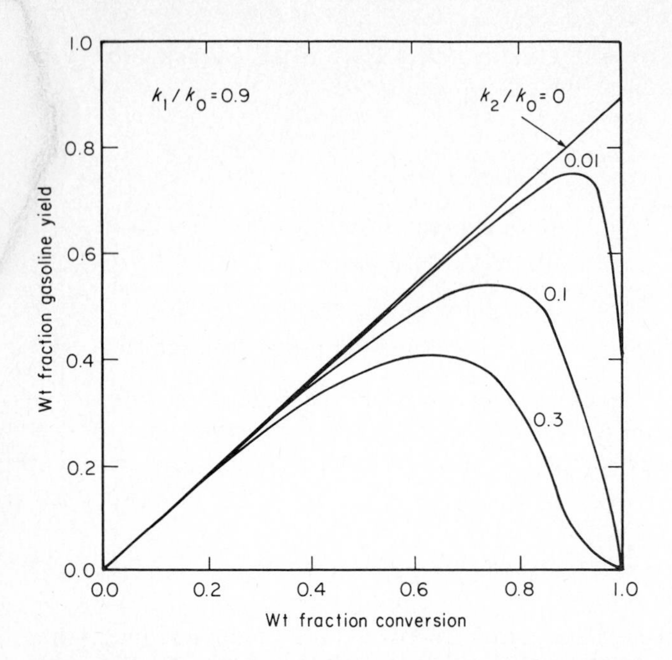

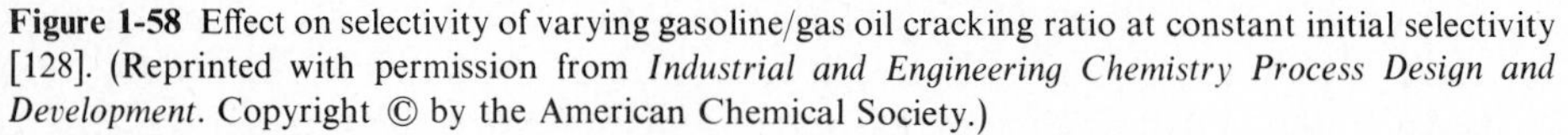

Figure 1-58 Effect on selectivity of varying gasoline/gas oil cracking ratio at constant initial selectivity [128]. (Reprinted with permission from *Industrial and Engineering Chemistry Process Design and Development*. Copyright © by the American Chemical Society.)

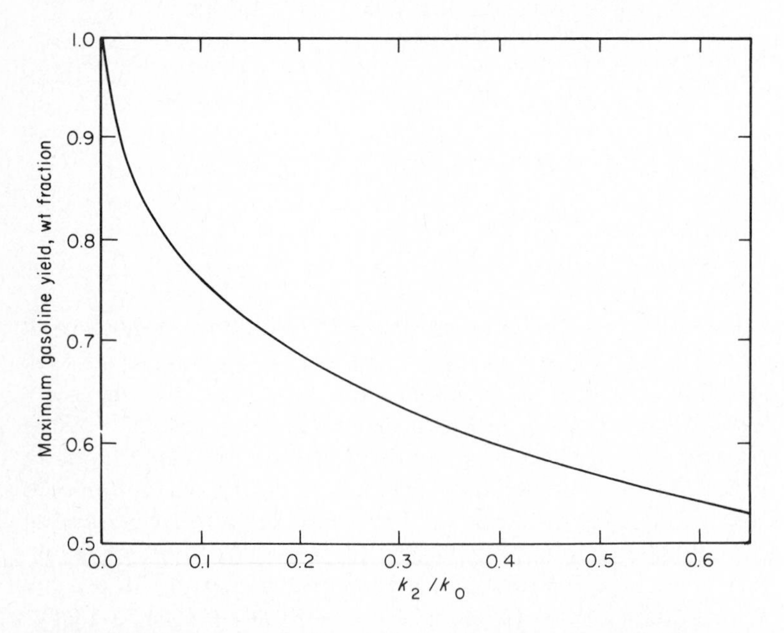

Figure 1-59 Maximum gasoline yield as a function of gasoline/gas oil cracking ratio [128]. (Reprinted with permission from *Industrial and Engineering Chemistry Process Design and Development*. Copyright © by the American Society.)

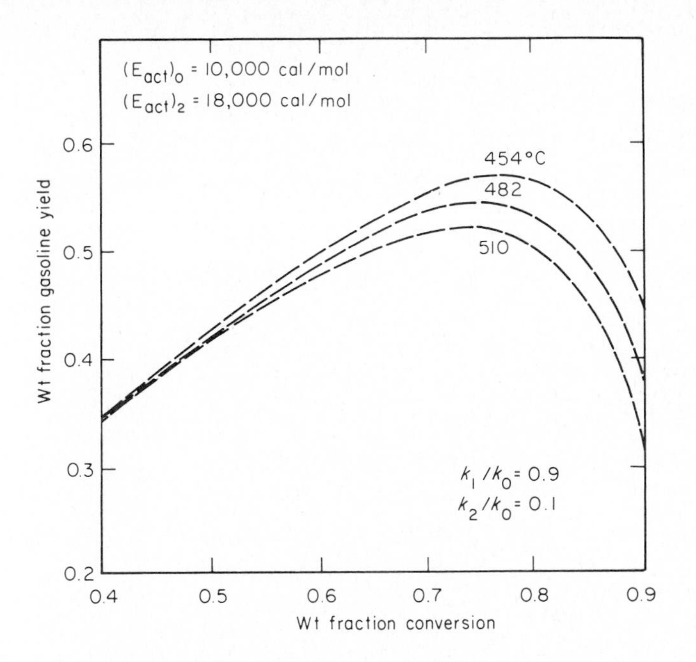

Figure 1-60 Effect of temperature on selectivity to gasoline formation in catalytic cracking [128]. (Reprinted with permission from *Industrial and Engineering Chemistry Process Design and Development*. Copyright © by the American Chemical Society.)

$(E_{act})_{deact} = -1700$ cal/mol, the isothermal yield-conversion plot shown in Fig. 1-60 was constructed. It is clear that the maximum yield increases with decreasing reaction temperature and that the reactor size for the maximum conversion also increases with decreasing temperature. It is also true, but not apparent from the kinetics, that the octane number of the gasoline decreases with reaction temperature.

Nace et al. [5] developed a set of correlations for k_0, k_1, and α for cracking of a series of feedstocks catalyzed by Durabead 5, an early-generation REHX zeolite in silica-alumina. These parameters are independent of the paraffin and olefin content. The aromatic-to-naphthene ratio is a key correlating parameter. Figure 1-61 shows that as the aromatic-to-naphthene ratio decreases, the cracking rate constant increases and the yield to gasoline increases. Since zeolites act to transfer hydrogen from naphthenes to form more paraffins [reaction scheme (91)], the cracking rates correlate with the supply of reactive hydrogen. Similarly, the rate constant for coke formation has been found to decrease with increased availability of reactive hydrogen. Although Nace et al. were not able to find a simple correlation for the overcracking parameter k_2, they found that its value was bounded in the narrow range of 1.5 to 2.5 h^{-1} at 482°C for the range of feedstocks examined. Therefore, there is a fairly complete and simple description of the gross kinetics for gas-oil cracking catalyzed by the zeolite Durabead 5.

A slightly more detailed analysis accounts specifically for the hydrogen-transfer activity of zeolite catalysts. Gasoline is represented as two fractions, an

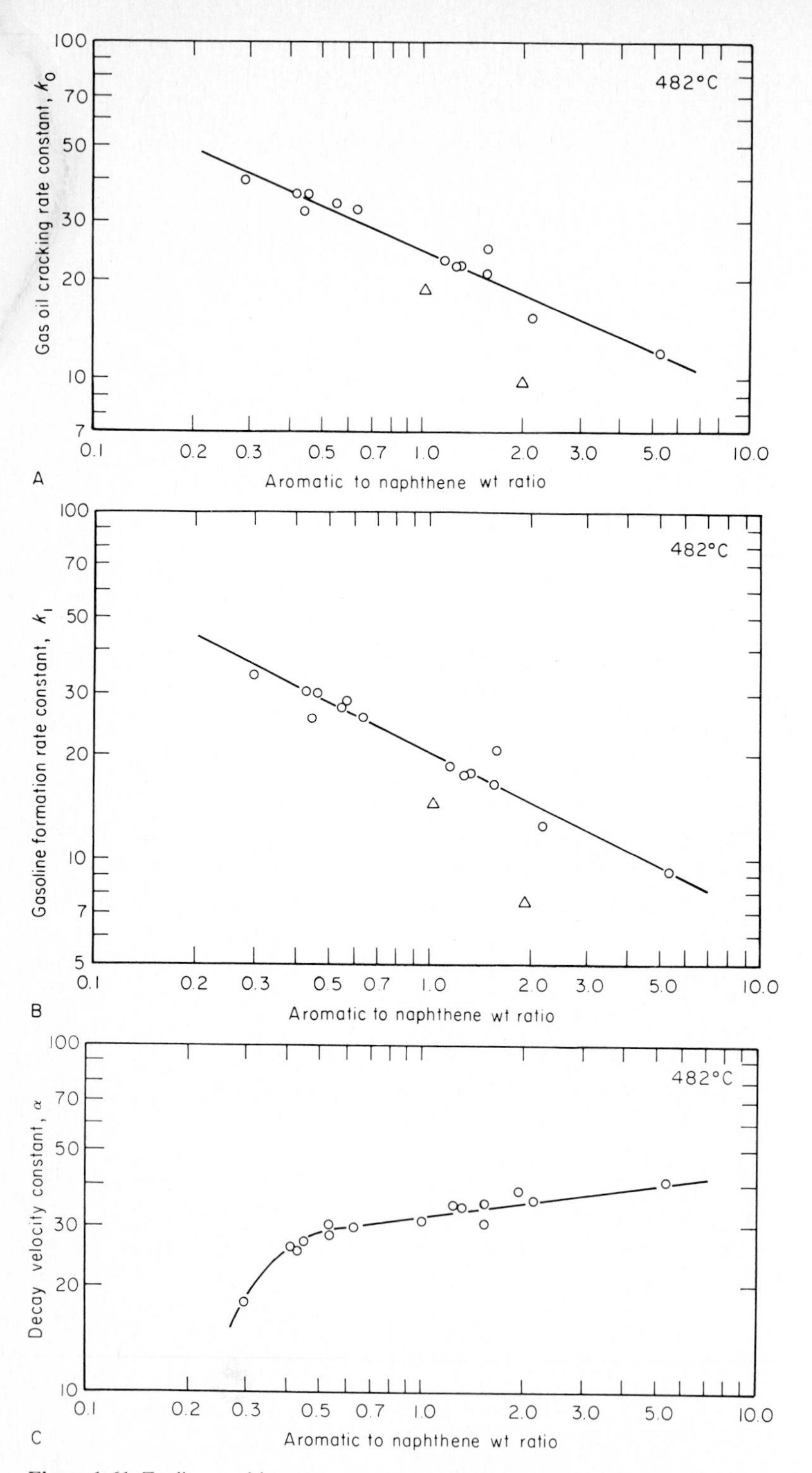

Figure 1-61 Zeolite cracking rate parameters for Durabead 5 [5]. (Reprinted with permission from *Industrial and Engineering Chemistry Process Design and Development*. Copyright © by the American Chemical Society.)

unstable (olefinic) fraction and a stable fraction, which can be formed from the unstable fraction. The mass-balance equations for isothermal reaction in a piston-flow reactor have been suggested to take the form

$$\frac{d\omega_U}{dZ} = \frac{k_1'}{\text{LHSV}} \Phi\omega_O^2 - \frac{k_2 + k_{II}}{\text{LHSV}} \Phi\omega_G \tag{99}$$

$$\frac{d\omega_S}{dZ} = \frac{k_1''}{\text{LHSV}} \Phi\omega_O^2 + \frac{k_{II}}{\text{LHSV}} \Phi\omega_G \tag{100}$$

where k_1' = rate constant for direct cracking to gasoline
k_1'' = rate constant for direct cracking to stabilized gasoline
$k_1 = k_1' + k_1''$ = total cracking rate constant for gasoline formation
k_{II} = first-order rate constant for conversion of unstabilized gasoline (olefins) to stabilized gasoline
ω_S = weight fraction of stabilized gasoline
ω_U = weight fraction of unstable gasoline

Using the terms of these equations, we can summarize how cracking catalysts can be improved, e.g., (1) by increasing the ultimate yield ratio k_1/k_0, (2) decreasing the overcracking-rate parameter k_2, and (3) increasing the stabilization-rate parameter k_{II}. These gross parameters presumably can be related to the structures of the zeolites and the silica-alumina matrix, but the relations are not available in the literature.

The reaction engineering of the riser reactor is actually more complex than it has been represented so far, although the results are not bad approximations of industrial cracking. Two complications need to be considered; (1) the reactor is not isothermal but is more nearly adiabatic, and (2) gas-phase axial dispersion (mixing in the direction of flow, a deviation from piston flow) affects the conversion and the yields significantly at high conversion levels. Both these complications can be accounted for straightforwardly.

A well-designed injector provides very rapid thermal equilibration between the gas oil and the regenerated catalyst it is mixed with. If it is assumed that the equilibration is achieved before a significant conversion has occurred, the temperature of the mixture at the inlet of the riser can be shown to be

$$T_0 = \frac{c_p^0 T_{O,0} + \bar{R}^C c_p^C T_{C,0} + \bar{R}^W c_p^W T_{W,0}}{c_p^0 + \bar{R}^C c_p^C + \bar{R}^W c_p^W} \tag{101}$$

where c_p^0 = heat capacity of gas oil per unit mass
c_p^C = heat capacity of catalyst
c_p^W = heat capacity of dispersing steam
$\bar{R}^C$ = ratio of catalyst-to-oil circulation rates (an important control parameter)
$\bar{R}^W$ = steam-to-oil flowrate ratio = 0 if dispersing steam not used
$T_{O,0}$ = inlet (absolute) temperature of gas oil
T_0 = resulting mixing-cup inlet temperature of gas oil
$T_{C,0}$ = (absolute) temperature of the catalyst returning from the regenerator
$T_{W,0}$ = the inlet (absolute) temperature of the dispersing steam

The riser can be considered an adiabatic reactor, and the energy balance on the reactor leads to the following expression for the temperature of the gas-solid mixture:

$$T - T_0 = \frac{-(h_g - h_O)(1 - \omega_O) + (h_g - h_G)\omega_G}{c_p^0 + \overline{R}^C c_p^C + \overline{R}^W c_p^W} \tag{102}$$

where h_O = heat of formation of gas oil per unit mass evaluated at inlet conditions
h_G = heat of formation of gasoline per unit mass
h_g = heat of formation of light gases per unit mass

The first term in the numerator represents the energy required to convert all the gas oil which is reacted into light gases. The second term represents the energy which would be recovered if the light gases were converted back into gasoline. This simple equation does not account for the small heat of reaction associated with the formation of coke. The values of the heat-of-formation terms depend significantly upon the composition of the gas oil, gasoline, and light gases; by comparison, the error resulting from neglect of the variations in the heat capacities with temperature and composition is small. Although T_0, the temperature of the mixture at the inlet, can be eliminated from the formulation, it is useful to retain it since operators often use T_0 as a control point.

The effect of axial dispersion on conversion in an isothermal reactor without deactivation can be accounted for by the following mass-balance equation, which has one more term than Eq. (95) to account for axial dispersion of gas oil:

$$\frac{d\omega_O}{dZ} = -\frac{k_O}{\text{LHSV}}\omega_O^2 + A\frac{d^2\omega_O}{dZ^2} \tag{103}$$

A is a reactor dispersion parameter which falls in the range of 0.02 to 0.10 [129]. Since A is small, a simple iterative solution of Eq. (103) is possible, giving

$$\omega_O = \frac{1}{1 + (k_O/\text{LHSV})(1 - 2A\zeta)\omega_O^0} \tag{104}$$

where $\zeta = 1 - \omega_O \approx \dfrac{k_O/\text{LHSV}}{1 + k_O/\text{LHSV}}$

= approximate extent of reaction;

$\omega_O^0 = (1 + Ak_O/\text{LHSV})^{-1}$ = inlet concentration modified to account for axial dispersion

The approximate solution confirms that axial mixing decreases the extent of reaction and that the effect is larger for higher conversions.

In a similar way, the effect of axial dispersion on the rate of production of gasoline can be evaluated. The result is a complicated expression of exponential integrals showing that axial dispersion decreases the maximum gasoline yield. To a good approximation, the maximum gasoline yield still depends primarily upon a ratio of the rate constants. The maximum yield in the absence of axial dispersion is shown in Fig. 1-59; the same figure applies approximately with the following substitution when axial dispersion is important:

$$\theta_2' = \frac{k_2}{k_0}\left(1 + \frac{2Ak_0}{\text{LHSV}} - \frac{Ak_2}{\text{LHSV}}\right) \tag{105}$$

Since k_2 is always smaller than $2k_0$, the effect of axial dispersion is to decrease the maximum gasoline yield.

The solutions just presented account for axial dispersion in isothermal rather than adiabatic operation of the riser. Adiabatic operation can be expressed in terms of the appropriate differential mass-balance equations for reactants and products, and another equation is needed to represent the energy balance.

THE REGENERATOR

The regenerator is a fluidized-bed reactor in which the deposited coke is burned off the catalyst (regenerating its activity). The energy needed for the endothermic cracking reaction is supplied by the exothermic combustion of the coke to give CO and CO_2. The operation of the regenerator is limited by the temperature limit of the materials of construction and by the capacity of the air compressor; since large compressors are expensive, the units are designed to operate near maximum capacity.

The kinetics of combustion of coke from a cracking catalyst has been determined by Weisz and Goodwin [130]:

$$\frac{dC_{O_2}}{dt} = -kC_{O_2}C_c \tag{106}$$

where C_c is the concentration of coke (wt %) on the catalyst. The rate of burnoff of a particular coke is a function of the feedstock from which it is formed. For example, low concentrations of organometallic compounds in gas oil decompose to leave metal deposits on catalysts, and these deposits may accelerate coke combustion.† It follows that burning-rate parameters must be determined for each feedstock. Burning kinetics, however, is apparently the same for coke deposited on dense silica-alumina and on commercial zeolite cracking catalysts.

The fluidized-bed reactor appears to contain two phases, a bubble phase consisting of bubbles ranging from about 0.1 to 1 m in diameter and a so-called emulsion phase, which is really a mixture of gas and fluidized particles. The bubbles rise rapidly through the emulsion phase, and there is a rapid transfer of gas between the two phases [131] (Fig. 1-62).

† Metal may be incorporated in low concentration in cracking catalysts specifically to accelerate the combustion of CO in the regenerator.

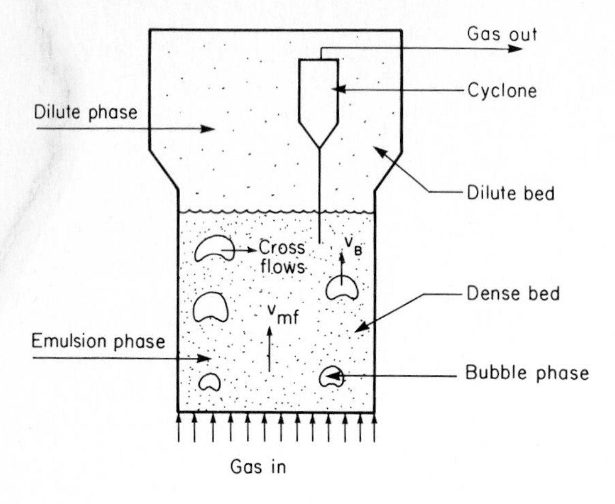

Figure 1-62 Flow of air through the regenerator; gas in the emulsion phase flows at a rate close to that of minimum fluidization; bubbles rise through the bed at a much higher rate; contact is good because catalyst particles continually rain down through the bubbles and exchange of gas between the bubble and emulsion phases is rapid.

Gas flows through the emulsion phase at a velocity nearly equal to the minimum velocity required for fluidization of the particles; the minimum fluidization velocity is low ($\sim$ 5 cm/s), and a fluidized bed could not be operated profitably with a throughput corresponding to this velocity. Fortunately, however, the bubbles flow through the bed at a much higher velocity, which depends upon the bubble diameter.

A review of the operating parameters of fluidized-bed reactors [131] shows that both the gas and the solid particles in the emulsion phase are well mixed by the passage of the bubbles through the phase. Consequently, there is no positional variation in composition in the emulsion phase. In contrast, there is little mixing in the bubble phase, and it can be represented as though it were in piston flow. The rate of interchange of gas between the emulsion and bubble phases is large, so that there is no real difference between the exit compositions of the two phases. To put these observations into a quantitative form, the compositions of the bubble phase and the emulsion phase are found from component-mass balances on carbon and oxygen. A carbon balance on the emulsion phase provides a third equation. These three equations then can be solved to determine the oxygen consumption in a fluidized-bed regenerator. The external parameters needed are the burning kinetics, the void fraction bubbles in the reactor [131], and the minimum velocity for fluidization [131]. The operation of the regenerator is stable and straightforward.

The energy balance in the regenerator can be developed easily if some of the details of the disengaging section and the cyclones downstream of the regenerator are neglected. The *disengaging section* of the regenerator is the section above the dense bed where particles carried out of the dense bed by the bubbles slow down and eventually return to the dense bed. The cyclones are needed to control the rate at which fines escape in the flue gas, later to be collected in dust-precipitation equipment. The particles in the disengaging section still contain carbon, and if the gas leaving the dense bed is too rich in oxygen, further burning of carbon or CO to CO_2 is possible. Consequently, the temperature difference between the dense bed and the disengager is used as measure of the oxygen concentration in the disengager. This temperature is used to control the air rate to the regenerator.

Assuming that the dense bed is well mixed, a steady-state energy-balance equation on the dense bed can be written and solved to determine (for a given inlet coke level, catalyst temperature, and flow rate) the exit temperature and coke level as functions of the flow rate of air from the blowers (which is usually close to the maximum capacity) and the inlet air temperature. The other variables for the regenerator are of less importance.

The older FCC units coupled the fluidized reactor and the regenerator through the overall energy balance. The coke concentration on the catalyst then rose to a value at which most of the oxygen in the air was consumed, and the catalyst was maintained at temperatures below the limits of the regenerator. Lee and Weekman [132] described a method for decoupling the reactor and the regenerator operation with zeolite catalysts, which operate most effectively at low coke levels. The decoupled-control method was asserted to give improved gasoline yields for FCC units. The decoupling of the reactor and the regenerator was

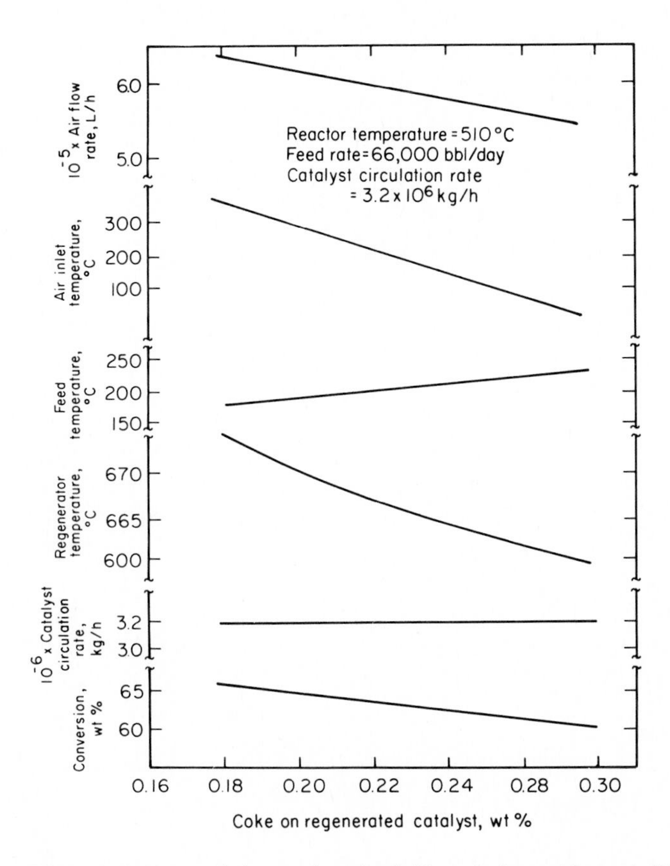

Figure 1-63 Uncoupled operation of a fluid-catalytic-cracking unit [132].

achieved by putting separate heat exchangers on the inlet airstream and the gas-oil stream, allowing the regenerator temperature to be controlled independently of the reactor temperature. The calculated performance is illustrated in Fig. 1-63.

In operation the catalyst circulation rate and gas-oil rate are maintained constant; the catalyst circulation rate is limited by the size of the transfer lines and the pressure drop. The reaction temperature also is fixed at the optimum level for the particular feedstock. As the regenerator temperature is increased to reduce the coke level, the gas-oil feed temperature is decreased to keep the reactor operating at the desired conditions. The independent control of the gas-oil temperature allows the catalyst temperature from the regenerator to vary over a 42°C range. The lower coke concentration on the catalyst requires additional oxygen, and so as the coke concentration is reduced the air flow must be increased.

NOTATION

A	type A zeolite
A	reactor dispersion parameter
Ar	aromatic
a	integer
$\hat{a}$	(thermodynamic) activity
aq	aqueous
B	base
b	integer
c	concentration, M/l^3 or mol/l^3
C_{n+}	hydrocarbon with n or more carbon atoms per molecule
c_p	specific heat, E/mol T
c	integer
CSTR	continuous stirred-tank reactor
D_{eff}	effective diffusion coefficient, l^2/t
E_{act}	activation energy, E/mol
E_+	energy of reaction for carbonium-ion formation, E/mol
E_-	energy of reaction for carbanion formation, E/mol
f	activity coefficient
G	gasoline
H_0	Hammett acidity function
h	heat of formation, E/mol
j	integer
K_a	acid dissociation (equilibrium) constant
k, k', k''	reaction rate constant, variable dimensions
L	Lewis acid
LHSV	liquid hourly space velocity, volumes of liquid feed per volume of catalyst per hour
l	length of riser-tube reactor
M	exchangeable cation in zeolite, mol
N_{Re}^{gas}	gas Reynolds number
n	integer, oxidation state of cation
O	gas oil
PFR	piston-flow reactor
R	gas constant, circulation rate per circulation rate of oil

R	alkyl group
$\bar{R}^c$	wt ratio of catalyst-to-oil circulation rate
$\bar{R}^w$	wt ratio of steam-to-oil flow
RE	rare earth
r	rate
T	temperature, °C or K
t	time
U	$\int_0^1 \Phi \, dZ$
v	velocity, l/t
v_B	bubble velocity, l/t
v_{mf}	minimum fluidization velocity, l/t
X	products of cracking other than gasoline
y	integer
Z	dimensionless length, z/l
z	integer; distance, l

Dimensional

E = energy
F = force
l = length
m = mass
t = time
T = temperature

Greek

α	empirical coking parameter, t^{-1}
β	empirical deactivation parameter
ζ	approximate extent of reaction
θ_1	k_1/k_0 in Eq. (98)
θ_2	k_2/k_0 in Eq. (98)
θ_2'	k_2/k_0 in Eq. (98) modified to account for axial dispersion
ρ	density, m/l^3
Φ	fraction of initial catalytic activity remaining
ω	weight fraction
ω_O	weight fraction of gas oil in vapor stream
ω_s	weight fraction stabilized gasoline

Subscripts and Superscripts

B	base, bubble
C	catalyst
c	coke, contact
F	fluidized catalyst
G	gasoline
g	gas
H	hydrogen transfer
j	integer
L	liquid
m	integer
mf	minimum fluidization
n	integer
O	gas oil

0	inlet or initial
p	integer
S	stabilized gasoline
U	unstable gasoline
w	water, steam
y	integer
z	integer

REFERENCES

1. Eastwood, S. C., C. J. Plank, and P. B. Weisz, *Proc. 8th World Pet. Cong. Moscow, 1971.*
2. Cartmell, R. R., U.S. Patent 3,785,782 (1974).
3. Weekman, V. W., Jr., *Ind. Eng. Chem. Process Des. Dev.*, **7,** 90 (1968).
4. Nace, D. M., S. E. Voltz, and V. W. Weekman, Jr., *Ind. Eng. Chem. Process Des. Dev.*, **10,** 530 (1971).
5. Voltz, S. E., D. M. Nace, and V. W. Weekman, Jr., *Ind. Eng. Chem. Process Des. Dev.*, **10,** 538 (1971).
6. Voge, H. H., in P. H. Emmett (ed.), "Catalysis," vol. VI, p. 407, Reinhold, New York, 1958.
7. Rice, F. O., *J. Am. Chem. Soc.*, **55,** 3035 (1933).
8. Rice, F. O., and E. Teller, *J. Chem. Phys.*, **6,** 489 (1938); **7,** 199 (1939).
9. Kossiakoff, A., and F. O. Rice, *J. Am. Chem. Soc.*, **65,** 590 (1943).
10. Greensfelder, B. S., H. H. Voge, and G. M. Good, *Ind. Eng. Chem.*, **41,** 2573 (1949); B. M. Fabuss, J. O. Smith, and C. N. Satterfield, *Adv. Pet. Chem. Refining*, **9,** 157 (1964).
11. Prescott, J. H., *Chem. Eng.*, July 7, 1975, p. 52.
12. Franklin, J. L., *J. Chem. Phys.*, **21,** 2029 (1953).
13. Franklin, J. L., and F. H. Field, *J. Chem. Phys.*, **21,** 550 (1953).
14. Oelderik, J. M., unpublished results cited in Ref. 15.
15. Brouwer, D. M., and H. Hogeveen, *Prog. Phys. Org. Chem.*, **9,** 179, (1972).
16. Tamaru, K., S. Teranishi, S. Yoshida, and N. Tamura, *Proc. 3d Int. Cong. Catal.*, p. 282, North-Holland, Amsterdam, 1965.
17. Tanabe, K., "Solid Acids and Bases," Academic, New York, 1970.
18. Mapes, J. E., and R. P. Eischens, *J. Phys. Chem.*, **58,** 809 (1954).
19. Pliskin, W. A., and R. P. Eischens, *J. Phys. Chem.*, **59,** 1156 (1955).
20. Aldridge, L. P., J. R. McLaughlin, and C. G. Pope, *J. Catal.*, **30,** 409 (1973).
21. Forni, L., *Catal. Rev.*, **8,** 65 (1973).
22. Olah, G. A., *Angew. Chem. Int. Ed. Engl.*, **12,** 173 (1973).
23. Magnotta, V. L., B. C. Gates, and G. C. A. Schuit, *J. Chem. Soc. Chem. Commun.*, **1976,** 342; V. L. Magnotta and B. C. Gates, *J. Catal.*, **46,** 266 (1977).
24. Olah, G. A., *J. Am. Chem. Soc.*, **94,** 808 (1972).
25. Fejes, P., and P. H. Emmett, *J. Catal.*, **5,** 193 (1966).
26. Greensfelder, B. S., and J. Samaniego, *Scuola Azione*, pt. I, no. 15, p. 61; pt. II, no. 16, p. 3; pt. III, no. 17, p. 3 (1960–1961).
27. Cotton, F. A., "Chemical Applications of Group Theory," 2d ed., p. 160, Wiley-Interscience, New York, 1971.
28. Pines, H., and J. Manassen, *Adv. Catal.*, **16,** 49 (1966).
29. Mills, G. A., E. R. Boedeker, and A. G. Oblad, *J. Am. Chem. Soc.*, **72,** 1554 (1950).
30. Nace, D. M., *Ind. Eng. Chem. Prod. Res. Dev.*, **8,** 24, 31 (1969).
31. Brouwer, D. M., and J. M. Oelderik, *Rec. Trav. Chim.*, **87,** 721 (1968).
32. Weisz, P. B., *Chemtech.*, August 1973, p. 498.
33. Gladrow, E. M., R. W. Drebs, and C. N. Kimberlin, Jr., *Ind. Eng. Chem.*, **45,** 142 (1953).
34. Egloff, G., J. C. Morrell, C. L. Thomas, and H. S. Block, *J. Am. Chem. Soc.*, **61,** 3571 (1939).
35. John, T. M., and B. W. Wojciechowski, *J. Catal.*, **37,** 240 (1975).

36. Eberly, P. E., Jr., C. N. Kimberlin, W. H. Miller, and H. V. Drushel, *Ind. Eng. Chem. Process Des. Dev.*, **5,** 193 (1966).
36*a*. Walsh, D. E., and L. D. Rollmann, *J. Catal.*, **49,** 369 (1977).
37. Blue, R. W., and C. J. Engle, *Ind. Eng. Chem.*, **43,** 494 (1951).
38. Thomas, C. L., *J. Am. Chem. Soc.*, **66,** 1586 (1944).
39. Moscou, L., and R. Moné, *J. Catal.*, **30,** 417 (1973).
40. Okkerse, C., in B. C. Linsen (ed.), "Physical and Chemical Aspects of Adsorbents and Catalysts," p. 213, Academic, New York, 1970.
41. Ryland, L. B., M. W. Tamele, and J. N. Wilson, in P. H. Emmett (ed.), "Catalysis," vol. VI, p. 1, Reinhold, New York, 1958.
42. Adams, C. R., and H. H. Voge, *J. Phys. Chem.*, **61,** 722 (1957).
43. Sato, M., T. Aonuma, and T. Shiba, *Proc. 3d Int. Cong. Catal.*, p. 396, North-Holland, Amsterdam, 1965.
44. Barrer, R. M., *Chem. Ind.* (*Lond.*), **1968,** 1203.
45. Breck, D. W., "Zeolite Molecular Sieves," Wiley, New York, 1974.
46. Meier, W. M., in "Molecular Sieves," pp. 10–27, Society of Chemical Industry, London, 1968.
47. Bergerhoff, G., H. Koyama, W. Koyama, and W. Nowacki, *Experientia*, **12,** 418 (1956).
48. Bergerhoff, G., W. H. Baur, W. Nowacki, *Neues Jahrb. Mineral. Monatsh.*, **1958,** 193.
49. Broussard, L., and D. P. Shoemaker, *J. Am. Chem. Soc.*, **82,** 1041 (1960).
50. Dempsey, E., *J. Phys. Chem.*, **73,** 387 (1969).
51. Breck, D. W., and E. M. Flanigen, in "Molecular Sieves," pp. 47–61, Society of Chemical Industry, London, 1968.
52. Smith, J. F., J. M. Bennett, and E. M. Flanigen, *Nature*, **215,** 241 (1967).
53. Smith, J. V., *Adv. Chem. Ser.*, **101,** 171 (1971).
54. Barry, T. I., and L. A. Lay, *J. Phys. Chem. Solids*, **29,** 1395 (1968).
55. Barry, T. I., and L. A. Lay, *Nature*, **208,** 312 (1965).
56. Eulenberger, G. R., G. P. Shoemaker, and J. G. Keil, *J. Phys. Chem.*, **71,** 1812 (1967).
57. Pickert, P. E., J. A. Rabo, E. Dempsey, and V. Schomaker, *Proc. 3d Int. Cong. Catal.*, p. 714, North-Holland, Amsterdam, 1965.
58. Olson, D. H., and E. Dempsey, *J. Catal.*, **13,** 221 (1969).
59. Meier, W. M., *Z. Kristallogr.*, **115,** 439 (1961).
60. Barrer, R. M., and D. L. Peterson, *Proc. R. Soc.*, **A280,** 466 (1964).
61. Bolton, A. P., and M. A. Lanewala, *J. Catal.*, **18,** 154 (1970).
62. Ward, J. W., in J. A. Rabo (ed.), "Zeolite Chemistry and Catalysis," p. 118, American Chemical Society, Washington, 1976.
63. Ward, J. W., *Adv. Chem. Ser.*, **101,** 380 (1971).
64. Uytterhoeven, J., L. G. Christner, and W. K. Hall, *J. Phys. Chem.*, **69,** 2117 (1965).
65. Weisz, P. B., and V. J. Frilette, *J. Phys. Chem.*, **64,** 382 (1960).
66. Tung, S. E., and E. McIninch, *J. Catal.*, **10,** 166 (1968).
67. Olson, D. H., *J. Phys. Chem.*, **72,** 1400 (1968).
68. Ward, J. W. and R. C. Hansford, *J. Catal.*, **13,** 364 (1969).
69. Jacobs, P. A., and J. B. Uytterhoeven, *J. Chem. Soc. Faraday Trans. 1*, **69,** 359, 373 (1963).
70. Maher, P. K., and C. V. McDaniel, U.S. Patent 3,293,192 (1966).
71. McDaniel, C. V., and P. K. Maher, in J. A. Rabo (ed.), "Zeolite Chemistry and Catalysis," p. 285, American Chemical Society, Washington, 1976.
72. McDaniel, C. V., and P. K. Maher, in "Molecular Sieves," p. 186, Society of Chemical Industry, London, 1968.
73. Kerr, G. T., *J. Phys. Chem.*, **72,** 2594 (1968); **73,** 2780 (1969).
74. Kerr, G. T., *Adv. Chem. Ser.*, **121,** 219 (1973).
75. Jacobs, P., and J. B. Uytterhoeven, *J. Catal.*, **22,** 193 (1971).
76. Dempsey, E., *J. Catal.*, **33,** 497 (1974).
77. Breck, D. W., and G. W. Skeels, *Proc. 6th Int. Cong. Catal.*, p. 645, The Chemical Society, London, 1977.
78. Breck, D. W., and G. W. Skeels, in J. R. Katzer (ed.), "Molecular Sieves II," p. 271, American Chemical Society, Washington, 1977.

79. Kerr, G. T., *J. Catal.*, **15,** 200 (1969).
80. Parry, E. P., *J. Catal.*, **2,** 371 (1963).
81. Pines, H., and W. O. Haag, *J. Am. Chem. Soc.*, **82,** 2471 (1960).
82. Chapman, I. D., and M. L. Hair, *J. Catal.*, **2,** 145 (1963).
83. Hall, W. K., F. E. Lutinski, and H. R. Gerberich, *J. Catal.*, **3,** 512 (1964).
84. Basila, M. R., and T. R. Kanther, *J. Phys. Chem.*, **71,** 467 (1967).
85. Shiba, T., M. Sato, H. Hattori, and K. Yoshida, *Shokubai* (*Tokyo*), **6**(2), 80 (1964).
86. Barthomeuf, D., in J. R. Katzer (ed.), "Molecular Sieves II," p. 453, American Chemical Society, Washington, D.C., 1977.
87. Minachev, Kh. M., *Kinet. Katal.*, **11,** 342 (1970).
88. Minachev, Kh. M., and Ya. I. Isokov, *Bull. Acad. Sci. USSR Div. Chem. Sci.*, **1968,** 903.
89. Tsutsumi, K., and H. Takahashi, *J. Catal.* **24,** 1 (1972).
90. Otouma, H., Y. Arai, H. Ukihashi, *Bull. Chem. Soc. Jap.*, **42,** 2449 (1969).
91. Tung, S. E., and E. McIninch, *J. Catal.*, **10,** 175 (1968).
92. Ward, J. W., *J. Catal.*, **11,** 251 (1968)
93. Schwarz, J. A., *J. Vac. Sci. Technol.*, **12**, 321 (1975).
94. Ward, J. W., *J. Catal.*, **14,** 365 (1969).
95. Ward, J. W., *J. Catal.*, **17,** 355 (1970).
96. Ward, J. W., *J. Catal.*, **13,** 321 (1969).
97. Ward, J. W., *J. Catal.*, **11,** 259 (1968).
98. Turkevich, J., and Ono, Y., *Adv. Catal.*, **20,** 135 (1969).
99. Beaumont, R., and D. Barthomeuf, *J. Catal.*, **26,** 218 (1972).
100. Beaumont, R., and D. Barthomeuf, *J. Catal.*, **27,** 45 (1972).
101. Barthomeuf, D., and R. Beaumont, *J. Catal.*, **30,** 288 (1973).
102. Bielanski, A., and J. Datka, *J. Catal.*, **37,** 383 (1975).
103. Richardson, J. T., *J. Catal.*, **9,** 182 (1967).
104. Beaumont, R., P. Pichat, D. Barthomeuf, and Y. Trambouze, *Proc. 5th Int. Cong. Catal.*, p. 343, North-Holland, Amsterdam, 1973.
105. Benson, J. E., K. Ushiba, and M. Boudart, *J. Catal.*, **9,** 91 (1967).
106. Ikemoto, M., K. Tsutsumi, and H. Takahashi, *Bull. Chem. Soc. Jap.*, **45,** 1330 (1972).
107. Dempsey, E., in "Molecular Sieves," p. 293, Society of Chemical Industry, London, 1968.
108. Nace, D. M., *Ind. Eng. Chem. Prod. Res. Dev.*, **9,** 203 (1970).
109. Hansford, R. C., and J. W. Ward, *J. Catal.*, **13,** 316 (1969).
110. Miale, J. C., N. Y. Chen, and P. B. Weisz, *J. Catal.*, **6,** 278 (1966).
111. Jacobs, P. A., and C. F. Heylen, *J. Catal.*, **34,** 267 (1974).
112. Haag, W. O., paper presented at *Am. Chem. Soc. Meet., Dallas, 1973.*
113. Prater, C. D., and R. M. Lago, *Adv. Catal.*, **8,** 293 (1956).
114. Amenomiya, Y., and R. J. Cvetanovic, *J. Catal.*, **18,** 329 (1970).
115. Bolton, A. P., and R. L. Bujalski, *J. Catal.*, **23,** 331 (1971).
116. Topchieva, K. V., L. M. Vishnevskaya, and H. S. Thoang, *Dokl. Akad. Nauk. SSSR*, **213,** 1368 (1973).
117. Thakur, D. K., and S. W. Weller, *Adv. Chem. Ser.*, **121,** 596 (1973).
118. Weisz, P. B., V. J. Frilette, R. W. Maatman, and E. B. Mower, *J. Catal.*, **1,** 307 (1962).
119. Satterfield, C. N., and J. R. Katzer, *Adv. Chem. Ser.*, **102,** 193 (1971).
120. Moore, R. M., and J. R. Katzer, *AIChE J.*, **18,** 816 (1972).
121. Cheng, C. S., Sc.D. thesis, Massachusetts Institute of Technology, Cambridge, Mass., 1970.
121*a*. Satterfield, C. N., and C. S. Cheng, *AIChE Symp. Ser.* **67**(117), 43 (1971).
122. Chen, N. Y., *Proc. 5th Int. Cong. Catal.*, p. 1343, North-Holland, Amsterdam, 1973.
123. Chen, N. Y., J. Maziuk, A. B. Schwartz, and P. B. Weisz, *Oil Gas J.*, **66**(47), 147 (1968).
124. Burd, S. D., and J. Maziuk, *Hydrocarbon Process.*, **51**(5), 97 (1972).
125. Thomas, C. L., and D. S. Barmby, *J. Catal.*, **12,** 341 (1968).
126. Gorring, R., *J. Catal.*, **31,** 13 (1973).
127. Voorhies, A., *Ind. Eng. Chem.*, **37,** 318 (1947).

127*a*. Froment, G. F., *Proc. 6th Int. Cong. Catal.*, p. 10, The Chemical Society, London, 1977.
128. Weekman, V. W., Jr., and D. M. Nace, *Am. Inst. Chem. Eng. J.*, **16,** 397 (1970).
128*a*. Jacob, S. M., B. Gross, S. E. Voltz, and V. W. Weekman, Jr., *AIChE J.*, **22,** 701 (1976).
129. Pavlica, R. T., and J. H. Olson, *Ind. Eng. Chem.*, **62**(12), 58 (1970).
130. Weisz, P. B., and R. D. Goodwin, *J. Catal.*, **2,** 397 (1963).
131. Kunii, D., and O. Levenspiel, "Fluidization Engineering," Wiley, New York, 1969.
132. Lee, W., and V. W. Weekman, Jr., U.S. Patent 3,769,203 (1971).
133. Butter, S. A., C. D. Chang, A. T. Jurewicz, W. W. Kalding, W. H. Lang, A. J. Silvestri, and R. L. Smith, Belgian Patent 14,339 (1973).

PROBLEMS

1-1 Ethylene is produced by pyrolysis of ethane at 650°C at short contact times. It is believed that the reaction proceeds by the following sequence of steps (the Rice-Herzfeld mechanism):

$$C_2H_6 \xrightarrow{k_1} 2CH_3\cdot$$

$$CH_3\cdot + C_2H_6 \xrightarrow{k_2} CH_4 + C_2H_5\cdot$$

$$C_2H_5\cdot \xrightarrow{k_3} C_2H_4 + H\cdot$$

$$H\cdot + C_2H_6 \xrightarrow{k_4} H_2 + C_2H_5\cdot$$

$$2H\cdot \xrightarrow{k_5} H_2$$

(*a*) Derive the rate expression for ethylene production.

(*b*) Compare the activation energy of the overall reaction with the activation energies of the individual steps.

(*c*) How would you determine the chain length and relate it to the overall stoichiometry?

1-2 In this chapter three routes to carbonium ions from hydrocarbon molecules have been suggested. What are these routes? Which route(s) to carbonium-ion formation is (are) probably most important in catalytic cracking of a highly paraffinic gas oil? Discuss your reasons for this choice.

1-3 Using the available mechanistic information about catalytic cracking and skeletal isomerization, predict the order of each of these reactions in hydrocarbon.

1-4 Use the stability of the carbonium ions formed from *n*-dodecane, 2,7-dimethyloctane and Decalin to predict the relative rates of cracking of these three hydrocarbons at 450°C if carbonium-ion formation is rate-limiting. How do these relative rates compare with relative rates estimated from Table 1-6?

1-5 Even though cracking reactions are endothermic, the equilibrium conversions in cracking processes may be high. What does this imply about the entropy changes in cracking reactions?

1-6 (*a*) Assume that an equilibrium distribution of carbonium ions is formed from 2-methylheptane and estimate the relative concentrations of primary, secondary, and tertiary carbonium ions at 450°C. Assume that no skeletal isomerization occurs.

(*b*) If each carbonium ion in this distribution undergoes β scission at the same rate, what is the predicted distribution of primary cracked products? Assume that all carbonium ions from a cracking event are terminated by hydride transfer and that the olefins formed do not undergo reaction.

1-7 The data in Fig. 1-12 show the effect of olefin concentration on butane cracking.

(*a*) Calculate and plot the first-order rate constant for butane cracking as a function of olefin concentration. (Assume that the reaction is first order in butane.)

(*b*) Assuming that the reaction is a chain reaction involving transfer of the carbonium-ion center to a butane molecule after each butane molecule has been cracked, derive the expression for the rate in terms of rate constants for each step and the concentrations of the reactants.

(*c*) Show how you would predict, if possible, the chain length for butane cracking and estimate the chain length. Is it independent of olefin concentration? What experiments can be used to determine the chain length in such a cracking reaction?

1-8 *n*-Butane appears to be a primary product of gas-oil cracking (Fig. 1-17) whereas isobutane appears to be a secondary product (Fig. 1-19). Explain this behavior in terms of carbonium-ion chemistry.

1-9 The formation of ultrastable HY (Si/Al ratio = 2.5) results in the removal of 24 aluminum cations per unit cell. A large fraction of the cation vacancies which are created are filled with silicon, giving a reduction in the unit-cell dimension and a stabilized zeolite. Assume that a zeolite crystallite is 1.5 μm in diameter and spherical and that the required silicon is derived from the surface region. What fraction of the crystallite must be destroyed to provide the required silicon assuming that all silicon in the exterior region migrates into the central crystalline region?

1-10 Consider a sample of hydrogen mordenite powder to be composed of particles such that a pore is 10 μm long. There are approximately 4.0×10^{16} pores per gram of sample. It has been observed that 0.087 g of benzene or 0.068 g of cumene can be adsorbed per gram of mordenite.

(*a*) Considering the molecular geometry, show whether these values can be realistically considered to represent saturation of the pore structure. Discuss the stacking of the aromatic molecules in the hydrogen mordenite pores.

(*b*) What tentative conclusion can you draw concerning the counterdiffusion of these two species in hydrogen mordenite pores?

(*c*) How much krypton would you expect to be adsorbed in the hydrogen mordenite pore structure at saturation?

1-11 Catalytic cracking of heavy gas oil depends on *space rate*, which is defined as the volume of oil feed per hour per unit volume of catalyst. The following data show the behavior.

Cracking with silica-alumina catalyst at 452°C and 1.68 atm

	Space rate, vol/vol · h			
	0.1	0.5	1.0	2.0
Product, wt %:				
$H_2 + C_1$ to C_4 paraffin + olefin	22.9	15.7	11.9	8.9
Gasoline (C_5 + higher-molecular-weight hydrocarbon)	43.2	33.7	30.4	25.0
Coke	11.0	3.9	2.5	1.8
Total product	77.1	55.3	44.8	35.7
Unconverted gas oil, wt %	22.9	44.7	55.2	64.3

(*a*) What kinetics expression best fits the data for the coke-formation reaction?
(*b*) What kinetics expression best fits the data for cracking?

1-12 Plot the data in Fig. 1-54 to test the correlation for deactivation (rate = At^n) and estimate the initial rate and the deactivation parameter n. Next plot the data in terms of Eq. (92) and evaluate the parameters. Finally plot the data according to the proposal of Thakur and Weller [117]. Evaluate the correlations.

1-13 (*a*) Assume that the rate constant for cracking each of the compounds in Table 1-17 catalyzed by REHX is 17 times greater than that for silica-alumina. Estimate the effectiveness factor for reaction of the trimethyl Decalin, of the saturated three-ring compound, and of the saturated four-ring compound catalyzed by REHX.

(*b*) From the data of Table 1-17 estimate the effective diffusion coefficient of each of the above compounds in the REHX zeolite. Plot D_{eff} vs. critical molecular diameter of each reactant molecule and compare with the diffusion behavior shown in Fig. 1-55.

1-14 Diffusion-reaction phenomena in crystallites of hydrogen mordenite catalyst have been studied by Swabb and Gates [*Ind. Eng. Chem. Fundam.*, **11,** 540 (1972)]. They measured rates of the dehydration reaction of methanol to give dimethyl ether using catalyst crystallites of various sizes. Interpret the data below in terms of the Thiele model, assuming first-order reaction. What is the value of the effective diffusion coefficient of methanol in the mordenite pores?

Catalyst pore length, μm	Rate of methanol dehydration at 205°C and 1 atm, moles of methanol converted per gram of catalyst per second
5.9 ± 3.2	$(7.33 \pm 0.94) \times 10^{-4}$
11.3 ± 4.3	$(6.17 \pm 0.48) \times 10^{-4}$
16.6 ± 5.7	4.94×10^{-4}

1-15 The Mobil Oil Corporation recently announced [133] a process for converting methanol into highly aromatic gasoline-range hydrocarbons using a zeolite catalyst. The first reaction in the sequence is the dehydration of methanol to give dimethyl ether, which then undergoes further reaction to give substituted benzenes. The temperature is less than 350°C, so that chain formation is thermodynamically allowed.

(*a*) Suggest a sequence of reactions to explain this conversion. Consider the possibility that ethylene is an intermediate in the conversion of dimethyl ether into aromatics. Show how aromatics can be formed from ethylene. What other chemical routes can you suggest to aromatic hydrocarbons?

(*b*) Table 1-18 gives the product distribution for the conversion of methanol into aromatic hydrocarbons under the same reaction conditions. The constraint index for different zeolite catalysts, also shown, is determined from the conversion of each of the components in an equimolar mixture of n-hexane and 3-methylpentane flowing over the catalyst at a temperature giving 10 to 60 percent overall conversion. The constraint index is defined as

$$\frac{\log\,(\text{fraction of } n\text{-hexane unconverted})}{\log\,(\text{fraction of 3-methylpentane unconverted})}$$

(1) Explain the relationship between the constraint index and the product selectivity. Why do the C_6 to C_{10} and C_{11+} yields vary as they do? Interpret in terms of zeolite structure and molecular size. (2) What zeolite properties would be required in a catalyst to make p-xylene for a chemicals feedstock?

(*c*) Propose a reactor design for a commercial-scale plant and discuss its integration into a process. What are the critical elements in reactor design, and what do you expect to be particularly troublesome problems in any reactor scheme? Do you expect the catalyst life to be long?

Table 1-18 Product distribution in methanol conversion catalyzed by different zeolites

Hydrocarbon concentrations in weight percent

	Zeolite type							
Product	ZSM-4	Hydrogen mordenite	β	ZSM-12	ZSM-5	ZSM-11	TEA mordenite	ZSM-2
C_6	0.2	1.7	0.6	12.3	2.1	2.7	1.8	3.5
C_7	1.5	4.5	3.5	5.0	16.4	22.2	11.4	14.6
C_8	3.8	4.9	2.7	16.3	38.8	40.6	53.5	42.5
C_9	6.9	4.8	3.2	14.0	28.3	29.4	22.3	31.6
C_{10}	9.5	9.6	6.5	26.9	13.0	5.1	10.5	3.0
C_6–C_{10}	21.9	25.5	16.5	74.5	98.6	100.0	~ 100.1	~ 100.0
C_{11+}	78.1	74.5	83.5	25.5	1.4	~ 0.0	~ 0.0	~ 0.0
Constraint index	0.52	0.53	0.56	1.8	6.0	2.2	1.7	38

1-16 A process has been suggested for separating the three isomers of xylene (dimethylbenzene) by selective diffusion through zeolite pores considered to have diameters of 6 to 8 Å.

(*a*) Estimate the minimum diameters of *o*-, *m*-, and *p*-xylene using van der Waals radii of the atoms.

(*b*) Estimate roughly their diffusion coefficients at 25 and at 425°C using the Knudsen diffusion equation. Compare with Fig. 1-55. Discuss.

1-17 Weekman's data for X zeolites can in principle be extended to riser-tube cracking, but more realistically one expects that the parameters will need adjusting to represent the more active catalysts of today. Assume the following inlet conditions at the base of the riser:

$$T = 524°\text{C} \qquad \text{catalyst/oil ratio} = 6 \text{ by weight}$$

$$\text{Catalyst density} = \begin{cases} 0.08 \text{ g/cm}^3 & \text{at base} \\ 0.05 \text{ g/cm}^3 & \text{at top} \end{cases}$$

$$\text{Pressure} = \begin{cases} 2.4 \text{ atm} & \text{at base} \\ 1.6 \text{ atm} & \text{at top} \end{cases}$$

Straight riser   inlet velocity = 670 cm/s

Coke on catalyst = 0.2 wt %

(*a*) Develop the profiles of G, O, light gases and coke through the reactor, first neglecting axial dispersion and then considering it.

(*b*) Repeat the development for a system with hydrogen transfer to form stabilized gasoline. One-sixth of the initial gasoline formed is stable gasoline, and the rate of conversion of gasoline into stabilized product is a parameter of the analysis.

1-18 When *n*-hexane cracking was catalyzed by zeolites, high activities were observed (Table 1-11). For each zeolite the apparent activation energy for cracking was about 30 kcal/mol except for offretite, for which it was about 15 kcal/mol. For the other zeolites diffusion rates were sufficient to allow a rate of cracking of the order of 10^{-6} mol/s per cubic centimeter of zeolite. If the lower activation energy for offretite was due to severe diffusional restrictions, estimate the upper limit on the diffusivity in offretite. Assume that the zeolite crystallites are spherical with a radius of 2 μm.

1-19 Consider the operation of a fluidized-bed regenerator (Fig. 1-62) for which the following parameters are known:

Catalyst recirculation rate = 60 tons/min
Catalyst inlet temperature = 500°C nominal
Bed temperature range = 468 to 565°C
Bed density = 0.56 g/cm^3
Coke on catalyst = 1 wt %
Coke composition = $CH_{0.2}$
Ambient temperature = 0°C

In the coke burning in the dense bed, one half of the carbon goes to CO and the other half to CO_2; hydrogen goes to H_2O rapidly, corresponding to fast CO in the dense bed; carbon on the catalyst goes to CO_2 in the dilute bed. Assume that all the coke remaining on the catalyst is burned in the cyclone and that coke burning is first order in carbon and in oxygen. Assume the blower rate is variable. Assume a stoichiometric ratio of oxygen to carbon (air to coke).

$$\frac{v}{v_{mf}} = 25 \qquad \text{residence time of solids} = 30 \text{ s} \qquad v = 2.5 \text{ ft/s at STP}$$

Air inlet temperature (variable): assume first a value of 150°C.

(*a*) Derive an energy balance for the regenerator.

(*b*) Assuming a fixed air inlet temperature, find the coke remaining, dense- and dilute-bed temperatures, and the oxygen partial pressure in the dilute bed in terms of blower throughput and other parameters. Show how these values change for different air inlet temperatures.

(*c*) Show how the above scheme permits independent adjustment of the catalyst temperature and the amount of coke on the catalyst.

1-20 Revise the regenerator analysis of Prob. 1-19 for burning kinetics which is zero order in carbon.

$$r(\text{mol } O_2 \text{ consumed/cm}^3 \text{ catalyst} \cdot s) = k_0 P_{O_2} e^{-E_{act}/RT}$$

where

$$E_{act} = 37 \text{ kcal/mol} \cdot \text{K} \qquad k_0 = 2.96 \pm 0.4 \text{ mol/cm}^3 \text{ catalyst} \cdot \text{atm}^{-1} \cdot \text{s}$$

and where P_{O_2} is in atmospheres. The density of cracking catalyst is 1.12 g/cm^3. Note that the carbon balance on the dense phase is found from a mass balance on oxygen.

1-21 Derive mass balances for coke burning from cracking catalyst for Probs. 1-19 and 1-20.

CHAPTER

TWO

CATALYSIS BY TRANSITION-METAL COMPLEXES: THE WACKER, VINYL ACETATE, OXO, METHANOL CARBONYLATION, AND ZIEGLER-NATTA PROCESSES

INTRODUCTION

CATALYSTS AND PROCESSES

In the last two decades there has been a renaissance in inorganic chemistry inspired by understanding of structure and bonding in the broad class of coordination complexes of transition metals. Simultaneously there has been a surge of interest in the chemistry of catalysis by certain transition-metal complexes and by surfaces of many of the same metals. Some of the classes of homogeneously catalyzed reactions are summarized in Table 2-1; the catalysts are complexes of iron, cobalt, rhodium, iridium, palladium, platinum, and other metals. Some transition metals, such as iron, also play important roles in catalysis by enzymes, such as the cytochromes.

In this chapter our attention is directed to the most important industrial reactions catalyzed by transition-metal complexes, which are listed in Table 2-2. The subject is unified by the process engineering. In general, the reactions proceed under mild conditions, with a typical temperature of 150°C and a typical pressure of 20 atm. The reactants include small molecules, such as CO, H_2, O_2, and olefins, which are usually gaseous at ambient conditions. The catalysts are either

Table 2-1 Some reactions of olefins catalyzed by transition-metal complexes [1]

Reaction	Typical catalysts ($L = PPh_3$)							
	Ru(II)	Co(II)	Fe(0)	Co(I)	Rh(I)	Ir(I)	Pd(II)	Pt(II)
Hydrogenation	$RuCl_6^{4-}$	$Co(CN)_5^{3-}$	$Fe(CO)_5$	$CoH(CO)_4$	$RhClL_3$	$IrI(CO)L_2$	...	$Pt(SnCl_3)_5^{3-}$ [a]
Hydroformylation	$RuCl_2L_4$ [a]	...	...	$CoH(CO)_4$	$RhCl(CO)L_2$	$IrCl(CO)L_2$	...	...
Double-bond migration	...	...	$FeH(CO)_4^-$	$CoH(CO)_4$	$RhCl_3(olefin)^{2-}$	...	$PdCl_4^{2-}$	$Pt(SnCl_3)_5^{3-}$ [a]
Dimerization	...	...	...	...	$RhCl_2(C_2H_4)_2^-$	...		...
Oxidation	...	...	...	...	...	...	$PdCl_4^{2-}$	...

[a] Active form of catalyst uncertain.

Source: Reprinted with permission from *Advances in Chemistry Series*. Copyright by the American Chemical Society.

Table 2-2 Industrial processes involving transition-metal-complex catalysis

Process	Reaction	Typical catalyst	Approximate temperature, °C	Approximate pressure, atm
Wacker	$C_2H_4 + \frac{1}{2}O_2 \longrightarrow CH_3CHO$	$PdCl_2/CuCl_2(aq)$	110	5
Vinyl acetate	$C_2H_4 + \frac{1}{2}O_2 + CH_3CO_2H \longrightarrow CH_3CO_2CH{=}CH_2 + H_2O$	$PdCl_2/CuCl_2(aq)$	130	30
Oxo	$RCH{=}CH_2 + CO + H_2 \longrightarrow RCH_2CH_2CHO + RCH(CHO){-}CH_3$	$HCo(CO)_4$ in organic solution	150	250
		or $RhCl(CO)(PPh_3)_2$ in organic solution	100	15
Methanol carbonylation	$CH_3OH + CO \longrightarrow CH_3CO_2H$	$RhCl(CO)(PPh_3)_2$ with CH_3I promoter in organic or aqueous solution	175	15
Ziegler-Natta polymerization	$C_2H_4 \longrightarrow \frac{1}{n}(C_2H_4)_n$	α-$TiCl_3(s) + Al(C_2H_5)_2Cl$ suspended in organic liquid	70	5
	$C_3H_6 \longrightarrow \frac{1}{n}(C_3H_6)_n$		100	10

dissolved or suspended in a liquid, so that all the reactor designs involve introduction of reactant gas into a liquid phase, through which the reactants must be transported to reach the catalyst. Since the reactions are exothermic, liberating roughly 30 to 50 kcal per mole of reactant, the well-mixed liquid medium allows rapid heat removal from the reactor. The reacting solutions are usually corrosive, and attempts to control the corrosion have led to process designs involving application of a catalyst confined within a single reactor, thus minimizing the use of expensive corrosion-resistant alloys. Since most of the processes operate with soluble catalysts, it is difficult and expensive to separate reaction products from catalysts, and a typical process flow diagram indicates a single reactor and an array of purification devices for catalyst recovery and recycle.

The individual processes are discussed in detail in later sections of the chapter, which concludes with consideration of process-design methods, exemplified by quantitative prediction of gas-liquid reactor performance. Alternative catalyst designs are summarized, illustrating methods for extending the chemistry of homogeneous catalysis to a reactor containing a catalyst phase separate from the reactant phase.

The subject of catalysis by transition-metal complexes is unified not only by these common features of the process engineering but also by the general patterns of the reaction chemistry, in particular by the structures of the transition-metal complexes involving bonded reactants as ligands and by the general occurrence of the cis-insertion mechanism. Therefore, we give an introductory summary of bonding, structure, and reaction mechanism for the class of coordination compounds of transition metals before discussing details of specific processes, which include the reaction kinetics, mechanisms, and process designs.

CHEMICAL BONDING IN TRANSITION-METAL COMPLEXES

The chemistry of transition-metal complexes has been reviewed thoroughly by many authors, including Cotton and Wilkinson [2], and only a brief summary is given here as an introduction to the complexes of importance to catalysis.

Organometallic compounds used as catalysts have transition-metal atoms or ions bonded to atoms or groups of atoms called *ligands*. The ligands in a complex surround the metal and form a polyhedron with the metal in the center. The most frequently observed surroundings are illustrated in Fig. 2-1; they are the regular six surrounding (octahedral), two types of five surrounding (tetragonal pyramidal and trigonal bipyramidal), and two types of four surrounding (tetrahedral and square planar).

The metal uses its *d* orbitals, which are partially filled, and the next higher *s* and *p* orbitals for the formation of the metal-ligand bonds in the complex. Bonds can be formed by a combination of a metal orbital and a ligand orbital, each having one electron, or by a combination of an empty metal orbital (Lewis acid) and a filled ligand orbital (Lewis base), or vice versa. Since in transition-metal

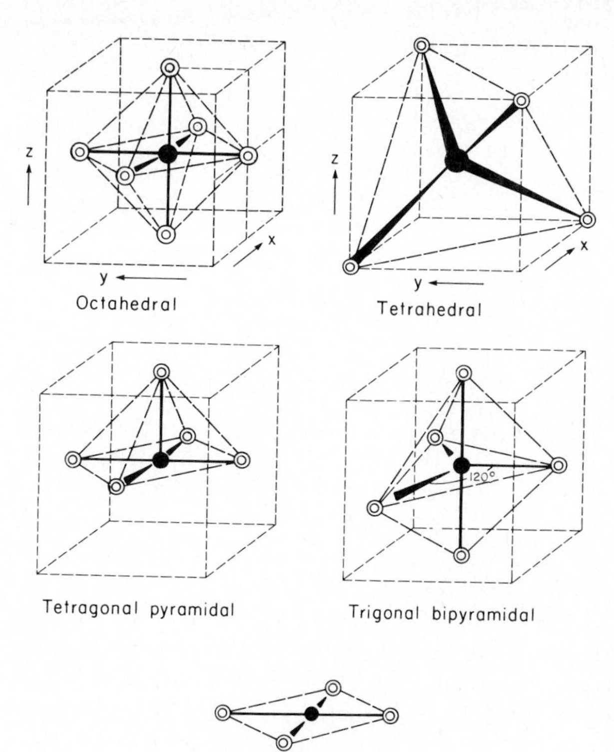

Figure 2-1 Ligand arrangements in organometallic compounds. The ligand polyhedra are presumed to contain only a single type of ligand. Similar surroundings occur with more than one kind of ligand in the complex, but then symmetry is generally lower although derived from ideal surroundings as given here.

atoms or ions there are both empty and filled orbitals, a ligand which also has both filled and empty orbitals can be doubly bonded.

To understand the chemical bonding in organometallic complexes and in particular the catalytic action of these complexes, which usually results from interaction of ligands in the coordination sphere of the metal, a short survey of ligands and their bonds with the metal is helpful.

CLASSIFICATION OF LIGANDS

1. *Ligands, including* NH_3 *and* H_2O, *which have only a filled orbital (lone pair) for interaction with the metal.* They are combined only via interaction with empty *d*, *s*, or *p* orbitals of the metal; i.e., they are Lewis bases, and the metal is a Lewis acid. The bond formed is rotationally symmetric around the metal-ligand axis and is therefore designated as a σ bond.
2. *Ligands, such as* H *and alkyls, with only one orbital and with one electron in this orbital.* These also interact with a metal through σ bonding, but they need a half-filled metal orbital to form an electron-pair bond. Therefore, bond formation is accompanied by transfer of a metal electron from a nonbonding to a bonding orbital, i.e., by oxidation of the metal.

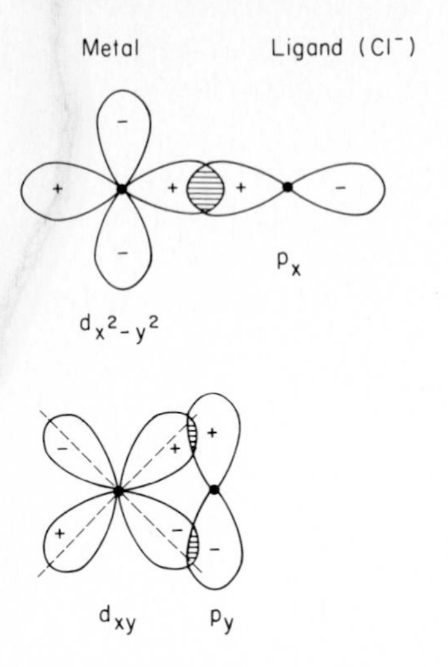

Figure 2-2 Double bond involving two filled ligand orbitals and two empty metal orbitals (donor π-bonding). The ligand may be Cl^-, Br^-, I^-, or OH^-.

3. *Ligands, including* Cl^-, Br^-, I^-, *and* OH^-, *with two or more filled orbitals which can interact with two empty metal orbitals, as illustrated in Fig.* 2-2. One of the ligand orbitals (p_x) forms a σ bond, but the second (p_y), which is necessarily oriented perpendicular to the metal-ligand axis, can only form a bond having no rotational symmetry; it is therefore called a π bond. For both σ and π bonds, the electrons are furnished by the ligand. To distinguish these ligands from those of the following group and to account for the electron donation by the π-bond-forming orbital, these ligands are called π-donor ligands.
4. *A large group of ligands, such as* CO, *olefins, and phosphines, having filled as well as empty orbitals.* The diversity of these ligands and their importance to organometallic catalysis justify a somewhat more detailed discussion. Figure 2-3

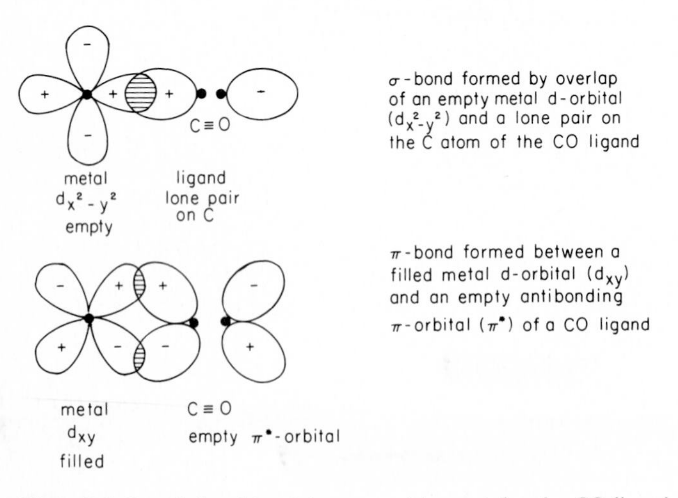

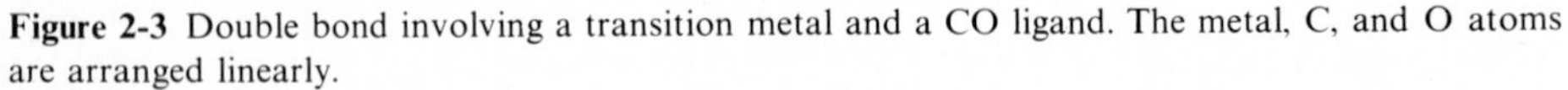

Figure 2-3 Double bond involving a transition metal and a CO ligand. The metal, C, and O atoms are arranged linearly.

shows the interaction of a CO ligand and a metal. The C atom of CO has a lone pair, which is the filled orbital, and an antibonding π^* orbital, which is the empty orbital. The combination of the lone pair and the empty *d* orbital furnishes a filled bonding molecular orbital (σ bond), and the combination of an empty antibonding π^* orbital and a filled *d* orbital furnishes another filled molecular orbital for a π bond. A related example with a C_2H_4 ligand is shown in Fig. 2-4. Here the filled orbital of C_2H_4 is the bonding π orbital, and the empty orbital is the antibonding π^* orbital. For this kind of double bonding to occur, the C—C axis of the olefin must be perpendicular to the metal-ligand axis.

In each of these examples, the bond formation between an empty *d* orbital and a filled ligand orbital involves a partial transfer of electrons from the ligand to the metal. Simultaneously, formation of a bond between the filled metal orbital and the empty ligand orbital involves partial transfer of electrons from metal to ligand. This second interaction is referred to as *backbonding* or *acceptor π bonding*. The net transfer of electric charge is small. The electron transfer to the ligand, however, serves to weaken the ligand double bond. Consequently, atomic distances such as C—O or C—C become greater upon bonding of the ligand, as shown by x-ray structural determinations. Weakening of the bonds is also indicated by decreased force constants of the stretching vibrations shown by band shifts in the infrared and Raman spectra.

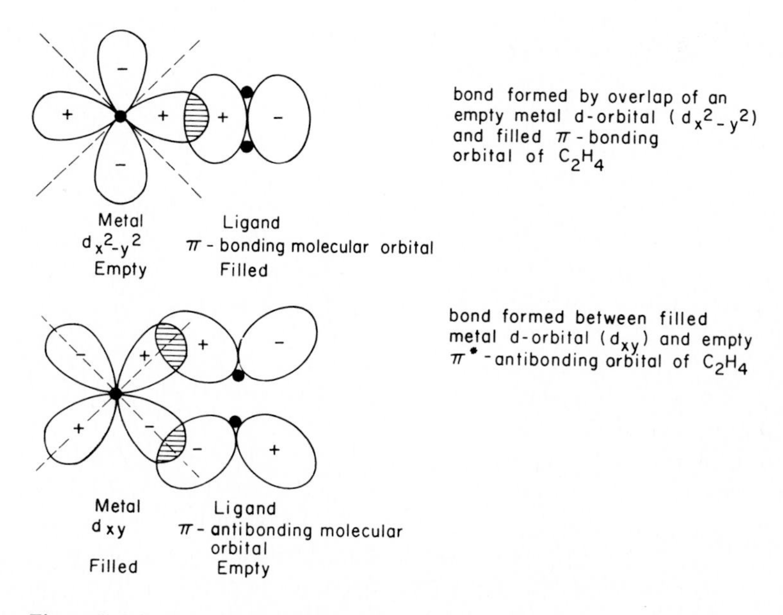

Figure 2-4 Double bond involving a transition metal and a C_2H_4 ligand. The C—C axis is perpendicular to the metal-ligand bond. The point group symmetry is O_h.

The reason for the weakening of the double bond is the introduction of electrons into the antibonding orbital of the π bond combined with the removal of electrons from the bonding orbital. Electrons are therefore transferred via the metal from a bonding to an antibonding orbital. The antibonding orbital points outward from the carbon atoms in the direction of other ligands; acceptor double bonding is a necessary prelude to the insertion reaction, discussed later.

Ligands such as butadiene and aromatics have molecular orbitals which are resonating systems of p orbitals. Their symmetry is related to the symmetry of π bonds, and they can serve as backbonding ligands. A special case is presented by unsaturated radicals such as allyl and cyclopentadienyl. These can bind either as other radicals do, via σ bonds, or as resonating polyenes, via acceptor π bonding. Some of them appear to show rapid changes from one to the other position, and they are then referred to as *fluxional ligands*.

Phosphines are among the most common ligands in catalyst complexes. They are considered to be backbonding ligands. A phosphine's filled σ-bond-forming orbital is easily identified as the sp^3 lone pair on P:

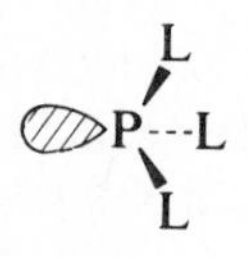

The empty d orbitals on the P atom are usually considered to cause the π-backbonding interaction.

LIGAND SURROUNDINGS OF A METAL

In the foregoing discussion it has been assumed that bond formation between metal and ligand orbitals occurred when the overlap (crosshatched areas in Figs. 2-2 to 2-4) was nonzero. The total number of electrons involved in each interaction was assumed to be not more than 2. These assumptions require some clarification. Overlap is defined by the integral over space of $\phi_M \cdot \phi_L$, where ϕ_M is the metal orbital and ϕ_L the ligand orbital. Nonzero overlap occurs when the two orbitals are of similar symmetry with their mutual positions in space taken into account. For instance, the two d orbitals in Figs. 2-2 to 2-4 have the same symmetry, and they interact with two ligand orbitals which differ in symmetry. The orbital referred to as $d_{x^2-y^2}$ was found to lead to overlap with the lone pairs of CO or the bonding π orbital of the olefin, provided that the ligands were arranged appropriately with respect to the coordinate axes of the system. The d_{xy} orbital could then overlap with the π-antibonding ligand orbitals. However, the overlap of d_{xy} with the former ligand orbitals and of $d_{x^2-y^2}$ with the latter is then zero, and it follows that only some of the d orbitals can be used for σ bonding and others for π bonding. The coordinate system used in discriminating between the two was defined with the axes pointing toward the ligands and the origin at the central metal atom. For a correct choice of the σ- and π-forming orbitals, a knowledge of the ligand positions with respect to the metal atom is required; in other words the symmetry of the ligand surroundings, i.e., the point-group symmetry, must be known.

The mathematical apparatus to handle the relations between orbital and point group symmetry is called *group theory*. For the ligand surroundings given in Fig. 2-1, group theory gives the following:

Surroundings and point groups	Orbitals suited to σ bonding	Hybridization scheme
Octahedral	$d_{x^2-y^2}$, d_{z^2}	d^2sp^3
O_h	s, p_x, p_y, p_z	
Tetrahedral	d_{xy}, d_{yz}, d_{xz}	d^3s
T_d	s, p_x, p_y, p_z	sp^3
Trigonal bipyramidal	d_{z^2}	dsp^3
D_{3h}	s, p_x, p_y, p_z	
Tetragonal pyramidal	$d_{x^2-y^2}$	dsp^3
D_{4d}	s, p_x, p_y, p_z	
Square planar	$d_{x^2-y^2}$	dsp^2
D_{4h}	s, p_x, p_y	

These results take on importance with regard to the second assumption mentioned previously, since this requires that metal orbitals for σ bonding be empty. The maximum electron occupation for octahedral symmetry is therefore d^6, provided that the ligand orbitals contain two electrons; the d^8 configuration fits either a square-planar, tetragonal-pyramidal, or trigonal-bipyramidal surrounding, and a d^{10} situation can only be tetrahedral. Consequently, the following sequence of surroundings can be easily understood, since in each complex the total number of bonding electrons involved is 18, corresponding to the stable noble-gas configuration of argon:

d^6 d^7 d^8 d^9 d^{10} ←—— Number of d electrons per metal atom

$Cr(CO)_6$ (Octahedral) $Fe(CO)_5$ (Trigonal bipyramidal) $Ni(CO)_4$ (Tetrahedral)

Similarly, the bonding of one-electron ligands such as H can also be understood. Bonding involving an atom or radical (see page 115, case 2) adds only one electron to the total number of electrons, and the 18-electron rule requires that the number of electrons on the metal be odd. The configuration of $HCo(CO)_4$ should then correspond to a five surrounding (tetragonal pyramidal or trigonal bipyramidal), but it should be distorted because there are two types of ligands involved. The structure is indeed the distorted trigonal bipyramidal. This line of reasoning leads to the suggestion that the manganese hydridocarbonyl configuration should be distorted octahedral:

d^7 d^9

$HMn(CO)_5$ $HCo(CO)_4$

The assumption of only two electrons per bond is not a hard-and-fast rule since electrons can be allocated to antibonding orbitals provided that antibonding is not pronounced, i.e., provided that bonding is weak. A surrounding of six can still be attained with more electrons than 18, as demonstrated by coordination complexes like $Ni^{2+}(NH_3)_6$. It is important to realize, however, that for a square-planar d^8 complex such as $Pt^{2+}(Cl^-)_4$ the statement that there are two empty coordination sites should be interpreted with caution; filling these sites with ligands entails considerable loss in energy.

In the following chapter we discuss ligand exchange reactions, which can occur by filling empty coordination sites by a new ligand and removal of one already bonded. For example, the following reactions take place in the Wacker process:

$$Pd^{2+}(Cl^-)_3(C_2H_4) + H_2O \longrightarrow \underset{\text{(Unstable)}}{Pd^{2+}(Cl^-)_3(H_2O)(C_2H_4)} \tag{1}$$

$$Pd^{2+}(Cl^-)_3(H_2O)(C_2H_4) \longrightarrow Pd^{2+}(Cl^-)_2(H_2O)(C_2H_4) + Cl^- \tag{2}$$

For reasons that are not clear, these are fast processes for Pd^{2+} complexes but slow for the analogous Pt^{2+} complexes. As a consequence, the Pd^{2+} complex is a good catalyst for the Wacker reaction, and the Pt^{2+} complex is not.

FRONTIER THEORY

Symmetry determines whether a bond can or cannot be formed, but the strength of the bond is determined by such considerations as the relative positions of the energy levels of the interacting orbitals (Fig. 2-5). If $E_A = E_B$, the bonding orbital is at a position $E_B - \beta$, where β is the exchange integral $\int \phi_A H_0 \phi_B \, d\tau$. If the energy levels E_A and E_B are separated by a magnitude $E_A - E_B$, the lowering of the bonding orbital energy with respect to that of the lowest "separate atom" orbital (here B) is approximately equal to $\beta/(E_A - E_B)$. Hence bond strength rapidly decreases with the energy difference between the separated-atom levels.

In double-bond formation there are four interacting orbitals; for acceptor π bonding two of these are empty (excited states, see Fig. 2-6). There are two simultaneous interactions, each of which comprises a filled orbital (*highest occupied*

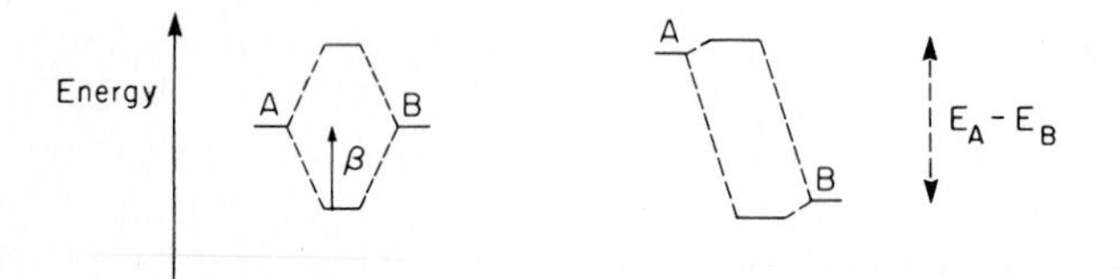

Figure 2-5 Interaction of two orbitals ϕ_A and ϕ_B. If $E_A = E_B$, the position of the bonding orbital is at a distance β, equal to the integral over space of $\phi_A H_0 \phi_B$, where H_0 is the Hamiltonian. If the energy level positions differ by $E_A - E_B$, the bonding orbital position as referred to B is approximately equal to $\beta/(E_A - E_B)$.

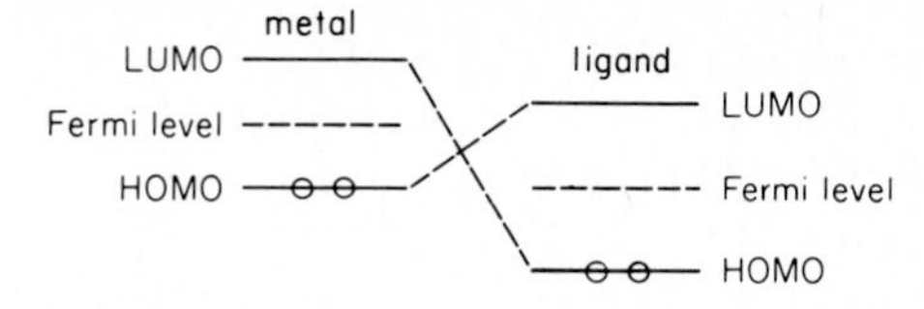

Figure 2-6 Double bond formation for acceptor π-bonding involving four orbitals, two of which are empty (LUMO) and two others of which are filled (HOMO). The degree of interaction depends on the relative positions of the Fermi levels, located between the HOMOs and LUMOs.

molecular orbital or HOMO) and an empty orbital (*lowest unfilled molecular orbital* or LUMO). The energy distance between the two depends on the positions of the energy levels midway between HOMO and LUMO: these are Fermi levels. The importance of these positions was first realized by Fukui [3], who named the HOMOs and LUMOs *frontier orbitals*. The theory was subsequently refined by Salem and Wright [5] and Pearson [6], following earlier work by Bader [4] to introduce the necessary considerations of symmetry.

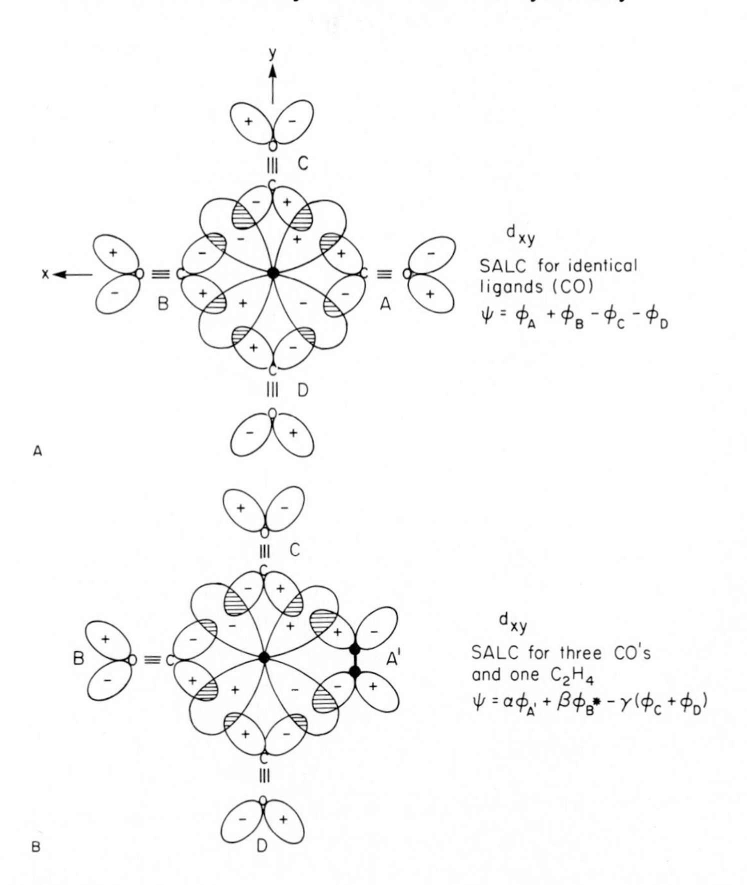

Figure 2-7 Illustration of symmetry adapted linear combinations (SALC) of orbitals.

In Fig. 2-6 if the relative positions of the energy levels are changed so that the metal HOMO approaches the ligand LUMO (the two orbitals involved in π bonding), the π bonding becomes stronger but at the same time the metal LUMO moves away from the ligand HOMO and the σ bonding becomes weaker. Such a change can be effected either by moving the metal energy level upward or the ligand energy level downward. Movements in the other direction cause the σ bonds to become strengthened and the π bonds weakened. Changes in the metal energy level can be made by substituting one metal for another or by varying the electron density of a given metal by changing its oxidation state.

The energy-level positions of a particular metal-ligand combination can also be changed by variation of the auxiliary ligands. Each metal orbital is necessarily associated with more than one ligand. According to ligand field theory, metal-ligand interactions can be deduced by the construction of symmetry-adapted linear combinations (SALC) of the ligand orbitals. Figure 2-7*a* gives a SALC for d_{xy} and four antibonding π^* orbitals of four CO molecules; Fig. 2-7*b* gives another SALC for three COs and one C_2H_4. The significance of a SALC is that it indicates a delocalization of the electrons over the various ligand orbitals. If the ligands are different, the average electron population over the ligand orbitals is different and π backbonding of C_2H_4 can therefore be influenced by the choice of the auxiliary ligands such as CO or phosphines. Qualitative ligand field theory does not give much information about the degree of this delocalization, and a quantitative method is necessary to shed more light on the problem.

Recent computations by Johnson and coworkers [7] provide quantitative confirmation of the action of auxiliary ligands. Zeise's salt, $K^+(Pt^{2+}Cl_3^-C_2H_4)$, contains the anion

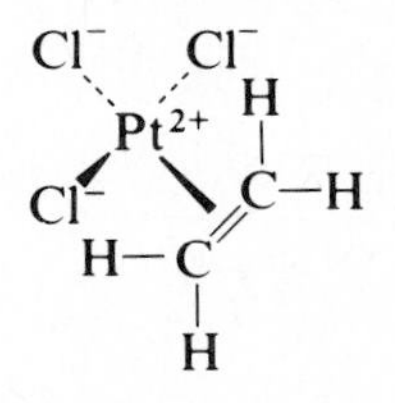

Inspection of the orbital arrangement with respect to π interaction along the line Cl^-–Pt^{2+}–C_2H_4 shows that the ligand SALC must be a linear combination of a Cl^- *p* orbital (filled) and an empty C_2H_4 orbital (π^* antibonding) as shown in Fig. 2-8. If π-bond formation between the Cl^- and the filled Pt^{2+} level were considerable, the bonding molecular orbital would be pushed into the middle of the Cl–Pt interval, i.e., away from the Cl^- ion. There would therefore be electron transfer from Cl to Pt, but because the Pt orbital is already filled, electron transfer to the empty C_2H_4 level is expected to be enhanced.

The method used by Johnson and coworkers [7] was the self-consistent-field Xα scattered-wave method. The result, given in Fig. 2-9, shows the contraction of the orbital density near the Pt^{2+} ion. In simpler but less precise words, the Cl^- pushes electrons toward Pt^{2+} and raises its energy level; as a consequence π bonding with C_2H_4 becomes stronger.

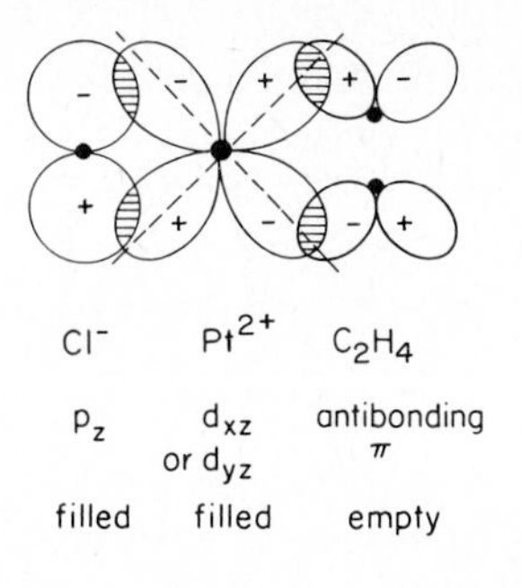

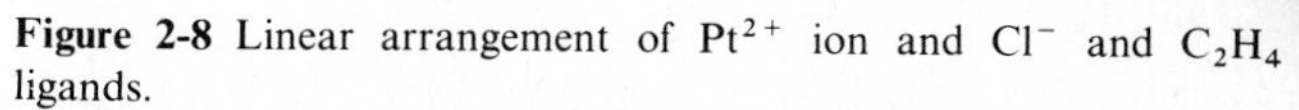
Figure 2-8 Linear arrangement of Pt^{2+} ion and Cl^- and C_2H_4 ligands.

The general impression acquired from catalytic results is that of a necessity for pushing the energy level as far upward as possible by drastically decreasing the oxidation state of the metal. This is readily understood: to cause olefins or CO to become reactive, electrons should be transferred to their antibonding π levels. Evidently there is an undesired limit, the precipitation of zero-valent metal. To prevent metal formation, ligands are needed to occupy coordination sites which are potentially able to interact in the formation of bonds with the metal but which are not themselves engaged in reactions in the coordination sphere of the metal. Most appropriate would be ligands having strong σ-donor but weak π-acceptor tendencies. This reasoning perhaps explains why phosphines are ligands in so

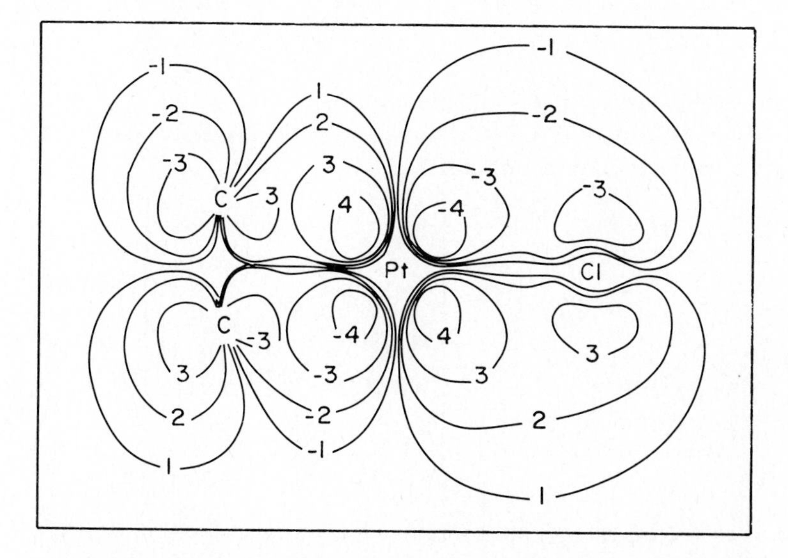

Figure 2-9 Contour map of Zeise's anion. The $2b_2$ π-bonding orbital wave function of $[Pt(C_2H_4)Cl_3]^-$ is plotted in a plane containing the central Pt atom, the *trans*-Cl ligand, and the C_2H_4 ligand. The H atoms lie outside of this plane and their contribution to the chemical bonding is not shown. The contour values are given in units of Rydbergs and the sign gives the sign of the wave function. The interior nodes of the wave function near each atom are omitted for clarity [7]. (Reprinted with permission from the *Journal of the American Chemical Society*. Copyright by the American Chemical Society.)

many of the best organometallic catalysts, e.g., the olefin hydrogenation catalyst $RhH(CO)(PPh_3)_3$, discovered by Wilkinson [8], who was awarded the Nobel prize in chemistry for his work with organometallic compounds.

STRUCTURES OF CATALYTIC COMPLEXES

Defining the structure of an organometallic complex acting as a catalyst is to some extent arbitrary because the ligands are normally changed during the cycle of elementary catalytic reaction steps. Even so, it is desirable to choose a key structure, starting from which the mechanism of the catalytic reaction can be explained. Sometimes such a structure must remain an inspired or convenient guess.

The key structure for the Wacker reaction, involving interaction of H_2O and C_2H_4 in Pd complexes containing Cl^-, is assumed to be related to that of Zeise's salt, which was discussed earlier:

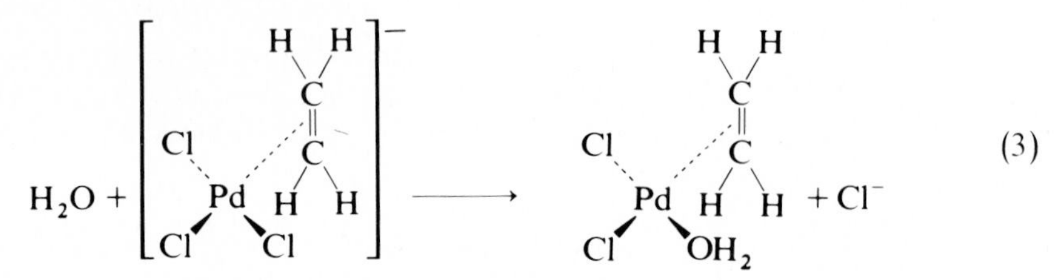

This reaction is the exchange of one Cl^- ligand with H_2O.

Besides Zeise's salt, the dimer $Pt_2Cl_4(C_2H_4)$ is also known to exist; assuming similar Cl^-–H_2O exchange for Pd, we obtain

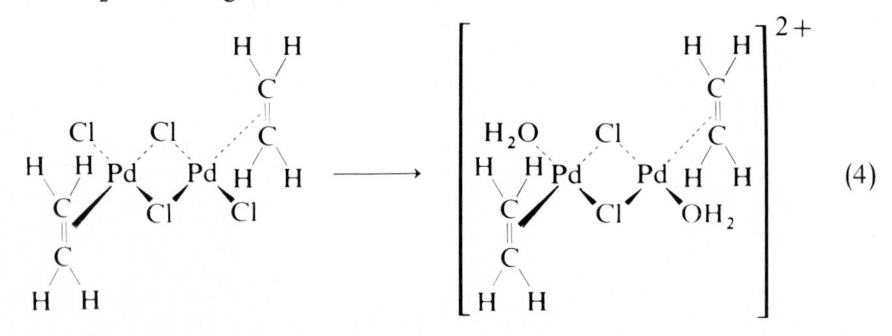

The Cl^- ions between the Pd cations are bridging chlorines like those in Al_2Cl_6; there is no Pd—Pd bond.

It is worthy of note that Ni^{2+} also can be bonded to Cl^- and H_2O, having a preference for octahedral surroundings and undergoing rapid exchange of Cl^- and H_2O. No C_2H_4 complexes of Ni^{2+}, however, are known to exist in the presence of H_2O and Cl^-, and Ni^{2+} complexes do not catalyze the Wacker reaction. Zeise's salt of Pt^{2+} is stable, but exchange of Cl^- and H_2O is exceedingly slow. In view of the position of Pd^{2+} in the periodic table, intermediate between

Ni and Pt, one might suggest from these results that a transient octahedral surrounding facilitates the Cl^{-}–H_2O exchange and allows the Wacker reaction to occur.

The key structure in the oxo reaction, involving CO, H_2, and olefins, is less readily identified, since two carbonyls and one hydridocarbonyl are known to exist. These are:

1. $HCo(CO)_4$, which has a distorted trigonal bipyramidal structure,

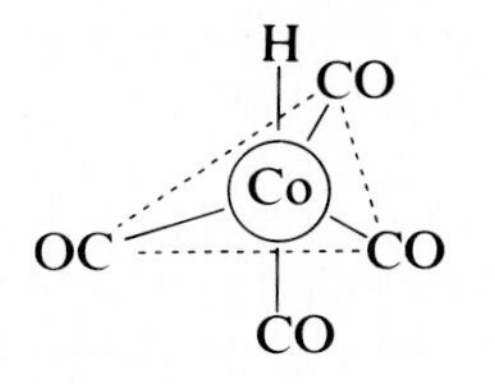

2. $Co_2(CO)_8$, a compound occurring in two forms, both having Co—Co bonds and one having two "ketonic" bridging CO ligands:

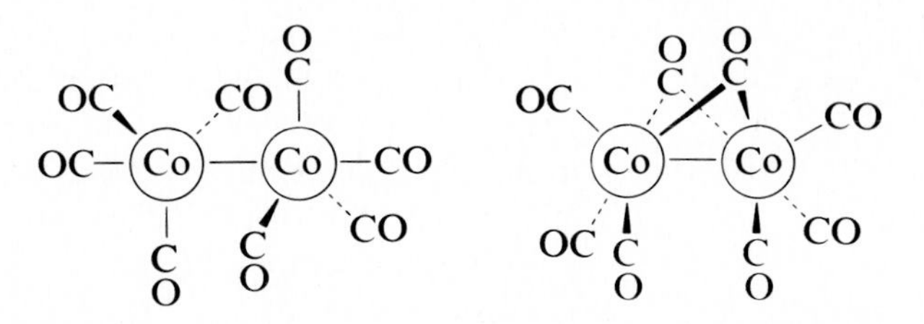

3. $Co_4(CO)_{12}$, a compound having a cluster of four Co atoms in a tetrahedral arrangement, with both bridging and terminal CO groups pointing outward:

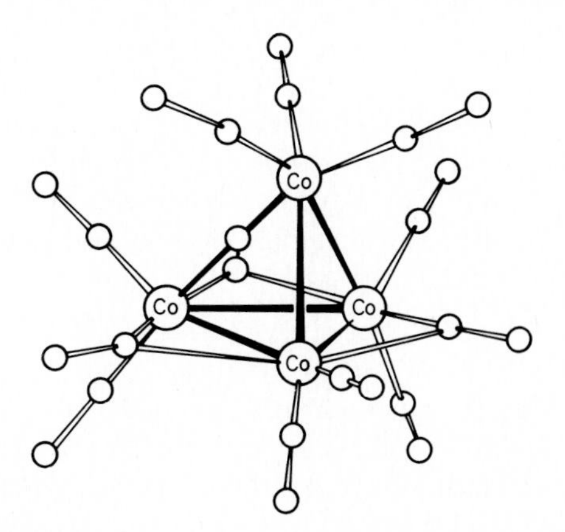

There is a rapid reaction giving the cobalt hydridocarbonyl from the dimeric compound:

$$H_2 + Co_2(CO)_8 \longrightarrow 2HCo(CO)_4 \tag{5}$$

and a slower reaction resulting in the breakup of the clusters:

$$Co_4(CO)_{12} + 4CO \longrightarrow 2Co_2(CO)_8 \tag{6}$$

The cluster carbonyl compound is similar to a metallic cluster of four Co atoms with CO adsorbed on the outer surface. In the presence of high CO partial pressures [Eq. (6)], the metal-metal bonds become separated, and electrons in bonding orbitals become nonbonding while electrons are simultaneously transferred to ligand antibonding orbitals. Evidently CO serves not only as a reactant in hydroformylation but prevents the involvement of metal electrons in undesired side reactions. The nonbonding electrons are available to bind hydrogen, and for the hydroformylation reaction to occur, the olefin must be introduced into a position similar to that of CO. The role of CO is therefore complex, involving the inhibition of metal-metal bond formation in favor of Co—H bond formation but at the same time allowing the olefin to become a ligand.

When Rh instead of Co is used as a hydroformylation catalyst, the second role assigned to CO, that is, preventing metal-cluster formation, is taken over by the phosphine ligand. Its acceptor bonding properties appear to be more exactly fitted to the purpose than those of CO; it prevents metal clustering but is more easily replaced by an olefin. It can even be part of a five-coordinate complex (as CO can), from which it is easily dissociated. Since there are many phosphine ligands with a range of bonding properties, a chemical variable is available, allowing the possibility of fine tuning the environment around the catalyst.

REACTIONS AND CATALYTIC PROPERTIES OF ORGANOMETALLIC COMPLEXES

Ligand Exchange

We consider a general reaction of a molecule A with another molecule B catalyzed by an organometallic complex ML_n, where L represents a ligand. The catalysis begins as A and B become coordinated to the metal next to each other, which requires two empty coordination sites on the metal. These can be formed by dissociation of two ligands L

$$ML_n \longrightarrow ML_{n-2}\square_2 + 2L$$

where $\square$ represents an empty coordination site. Subsequently the molecule $ML_{n-2}\square_2$ can react with A and B to form $ML_{n-2}AB$. The dissociation of two ligands L is usually thermally initiated, and the M—L bond strength should not be too great or else it would prevent formation of empty sites. Further, the M—A and M—B bonds, which may be similar in strength to the M—L bonds, should not be too weak, since then A and B would not be bonded in sufficient concentrations. This reasoning indicates the necessity of an optimum bond strength. Similar considerations apply to surface catalysis, as discussed in Chap. 3 and illustrated by the Balandin volcano plots.

Oxidative Addition

The bonding of neutral molecules such as CO or C_2H_4 does not appreciably change the electron density on a metal, i.e., does not change its oxidation state. If, however, molecules such as H_2 are to become bonded with dissociation of the H—H bond, the metal must donate two electrons when accepting the H atoms as ligands. Consequently, the metal is converted to a higher oxidation state. This type of reaction is known as *oxidative addition*, and the reverse process is known as *reductive elimination:*

$$\underset{\mathrm{L}\;\square}{\overset{\square\;\mathrm{L}}{\mathrm{L{-}M^{n+}{-}L}}} + \mathrm{AB} \underset{\text{reductive elimination}}{\overset{\text{oxidative addition}}{\rightleftharpoons}} \underset{\mathrm{L}\;\mathrm{L}}{\overset{\mathrm{B}\;\mathrm{L}}{\mathrm{L{-}M^{(n+2)+}{-}A}}} \tag{7}$$

Since two electrons must be given up by the metal ion, oxidative addition requires two vacant coordination sites and a metal such as Rh having a tendency to occur in oxidation states separated by 2 units. Many molecules such as H_2, HI, and CH_3I undergo oxidative addition reactions with metal complexes, and some of these are important in the catalytic reactions listed in Table 2-2.

The Insertion Reaction

The most important reaction which can occur in the coordination sphere of a transition-metal ion is the insertion reaction, which is the basic reaction of all the catalytic processes discussed in this chapter. It was first postulated by Cossee to account for Ziegler-Natta polymerization, as discussed later, and it was subsequently recognized to be involved in many other reactions.

An insertion reaction is defined as one in which an atom or group of atoms is inserted between two atoms initially bound together. Specifically, in this chapter, the insertion reaction is understood to be a reaction taking place in the coordination sphere of a transition-metal atom or ion, the result of which is the insertion of one ligand between the metal and another ligand.

The insertion reaction can be described in simplified terms on the basis of two different transition states. A three-center transition state explains the commonly observed CO insertion:

$$\overset{\mathrm{R_3C}}{\underset{|}{\overset{|}{-\mathrm{M}}}}{-}\mathrm{CO} \rightleftharpoons \left[\overset{\mathrm{R_3C}}{-\mathrm{M}\cdots\cdots\mathrm{CO}}\right]^{\ddagger} \rightleftharpoons \overset{\square}{\underset{|}{-\mathrm{M}}}{-}\mathrm{C}\overset{\mathrm{CR_3}}{\underset{\mathrm{O}}{\diagdown}} \tag{8}$$

A four-center transition state explains reactions such as the following, which occurs in the Ziegler-Natta polymerization:

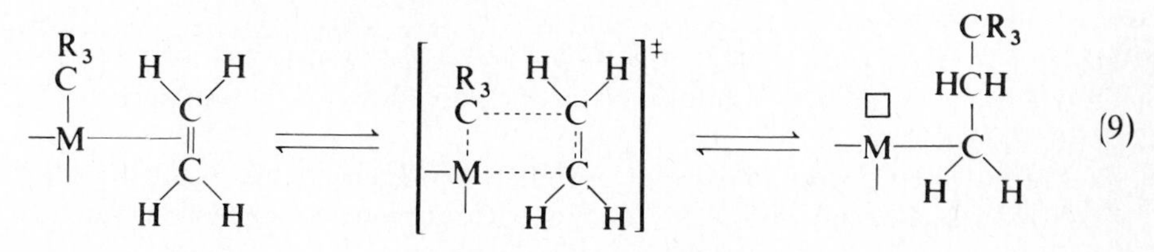

These reactions are referred to as cis insertions since the reactants are bonded adjacent to each other in the coordination sphere. The four-center reaction clearly shows that the attack from radical and metal on the two C atoms forming the double bond occurs in the cis position. The movement of the reactants takes place in the plane formed by σ- and π-bonding orbitals and is therefore restricted in space as well as in conformation. Although reasonably well described, the cis-insertion reaction is only partially understood. Cossee had suggested that during reaction step (9) the alkyl group moves to one of the carbon atoms of the ligand to be inserted by passing from one metal d orbital (e_g) to another (t_{2g}) under constant nonzero overlap [9].

PROCESSES

THE WACKER PROCESS: ETHYLENE OXIDATION TO ACETALDEHYDE

Reactions and Catalysts

Among the first to recognize the practical catalytic value of transition-metal complexes were Smidt and Hafner and their coworkers at the Consortium für elektrochemische Industrie, who in 1959 described the commercial Wacker process for single-step oxidation of ethylene to produce acetaldehyde [10]. The Wacker reactions include oxidation of ethylene to acetaldehyde in aqueous solution, accompanied by the reduction of dissolved $PdCl_2$ to give palladium metal:

$$H_2C{=}CH_2 + H_2O + PdCl_2 \longrightarrow CH_3CHO + Pd + 2HCl \qquad (10)$$

If the reactant solution contains the oxidizing agent $CuCl_2$, the palladium is reoxidized:

$$Pd + 2CuCl_2 \longrightarrow PdCl_2 + 2CuCl \qquad (11)$$

The CuCl is reoxidized by dissolved oxygen:

$$2CuCl + \tfrac{1}{2}O_2 + 2HCl \longrightarrow 2CuCl_2 + H_2O \qquad (12)$$

When the three reactions take place in a single reactor, the palladium and copper salts need be present only in small amounts and the overall stoichiometry corresponds to the oxidation of ethylene:

$$C_2H_4 + \tfrac{1}{2}O_2 \longrightarrow CH_3CHO \qquad (13)$$

We refer to $PdCl_2$ and $CuCl_2$ as catalysts and to the sequence of reactions as catalytic in a broad sense, although there is no single catalytic reaction in the traditional sense.

The only metal with activity for the Wacker reaction is Pd, in the presence of which reactions occur with many olefins, yielding ketones as well as aldehydes.

Table 2-3 is a summary of recent results taken from the review of Stern [11]. Our attention is focused on ethylene, which is the olefin used in the commercial process and the one for which the reactions have been most thoroughly studied.

In the industrial Wacker process, ethylene and oxygen flow into a reactor holding a dilute aqueous phase containing H_3O^+, $PdCl_4^{2-}$, Cu^{2+}, and Cl^-, among other ions. The reactants are absorbed into the liquid phase, where reactions (10) to (12) take place.

Reaction Kinetics

Reaction kinetics results have been reported by several investigators who measured volumetrically the uptake of ethylene into aqueous $PdCl_2$ solutions [11, 11a, 12].

Table 2-3 Reactions of monoolefins with Pd(II) in aqueous solution

	Reaction conditions				
Olefin	Catalyst	Temp., °C	Time, min	Product	Yield, %
Ethylene	$PdCl_2/CuCl_2$	20	5	Acetaldehyde	85
Propylene	$PdCl_2/CuCl_2$	20	5	Acetone	90
	K_2PdCl_4	70		Propionaldehyde	15.3
	$PdCl_2/CuCl_2$	70		Acetone	92–94
				Propionaldehyde	0.5–1.5
1-Butene	$PdCl_2/CuCl_2$	20	10	Methyl ethyl ketone	80
	$PdCl_2/CuCl_2$			Methyl ethyl ketone	85–88
				Butyraldehyde	2–4
	K_2PdCl_4	70		Butyraldehyde	8.9
1-Pentene	$PdCl_2/CuCl_2$	20	20	2-Pentanone	81
	K_2PdCl_4	70		1-Pentanal	20
1-Hexene	$PdCl_2/CuCl_2$	30	30	2-Hexanone	75
	K_2PdCl_4	70		1-Hexanal	3.8
3,3-Dimethyl-1-butene		20	15	3,3-Dimethyl-2-butanone	66
1-Heptene	$PdCl_2/CuCl_2$	50	30	2-Heptanone	65
	K_2PdCl_4	70		1-Heptanal	5
1-Octene	$PdCl_2/CuCl_2$	50	30	2-Octanone	42
1-Nonene	$PdCl_2/CuCl_2$	70	45	2-Nonanone	35
1-Decene	$PdCl_2/CuCl_2$	70	60	2-Decanone	34
1-Dodecene	$PdCl_2/CuCl_2$	60–70		2-Dodecanone	78–85
Cyclopentene	$PdCl_2/CuCl_2$	30	30	Cyclopentanone	61
Cyclohexene	$PdCl_2/CuCl_2$	30	30	Cyclohexanone	65
Styrene	$PdCl_2/CuCl_2$	50	180	Acetophenone	57
	K_2PdCl_4	70		Phenylacetaldehyde	75
Allylbenzene	$PdCl_2/CuCl_2$	40	30	Benzyl methyl ketone	76
n-Propenyl-benzene	$PdCl_2$	100	60	Benzyl methyl ketone	61
2-Phenyl-2-butene	$PdCl_2$	100	60	2-Phenyl-3-butanone	14.5
Stilbene	$PdCl_2$			Benzyl phenyl ketone	30

Source: Reprinted from [11] by courtesy of Marcel Dekker, Inc.

Recent kinetics studies of olefin oxidation by aqueous palladium chloride solutions reported by Moiseev, Vargaftik, and their coworkers have been based on potentiometric measurement of reacting solutions [13 and references cited therein]. The earlier kinetics data appear to have been influenced by mass transport at the higher rates and are therefore not entirely reliable. The recommended equation is [13]

$$r = \frac{k_1[PdCl_4^{2-}][C_2H_4]}{[Cl^-]^2[H_3O^+]} + \frac{k_2[PdCl_4^{2-}]^2[C_2H_4]}{[Cl^-]^3[H_3O^+]} \tag{14}$$

where brackets are used for concentrations instead of the C notation to avoid three-level subscripts.

At $PdCl_4^{2-}$ concentrations less than about 0.04 M, the second term on the right-hand side of the equation is usually negligible. The equation is valid over a wide range of concentrations and ionic strengths of the reactant solution, as demonstrated by the results of Fig. 2-10 for reaction at 15 to 40°C [13]. The linearity of the plots demonstrates consistency with Eq. (14), which can be rearranged to

$$\frac{r[Cl^-]^2[H_3O^+]}{[PdCl_4^{2-}][C_2H_4]} = k_1 + \frac{k_2[PdCl_4^{2-}]}{[Cl^-]} \tag{15}$$

The data show that k_1 and k_2 are dependent on ionic strength and temperature, as illustrated in Figs. 2-11 and 2-12.

Since the regeneration reaction (11) is fast compared with the acetaldehyde formation (10), the addition of small amounts of $CuCl_2$ to the solution does not significantly change the rate of acetaldehyde formation. In practice, palladium-metal formation is prevented by the presence of excess $CuCl_2$, and therefore the reaction kinetics of the Wacker process can be predicted approximately by Eq. (14) when the effects of $CuCl_2$ on the concentration of H_3O^+ and Cl^- are accounted for; complications may arise from interactions of Cu(II) with the Pd(II) complex [11].

Product Distribution

Yields of acetaldehyde from ethylene in the process are about 95 percent. Side products include acetic acid (~ 2 percent), carbon dioxide (~ 1 percent), and chlorinated products (~ 1 percent), including methyl chloride, ethyl chloride, and chloroacetaldehyde [14]. With other olefins as feeds, yields vary considerably (Table 2-3).

Reaction Mechanism

Zeise's salt, the anion of which is represented in Fig. 2-8, can undergo hydrolysis to give acetaldehyde. There is abundant evidence [11, 11a, 15, 16] that $PdCl_4^{2-}$ in the

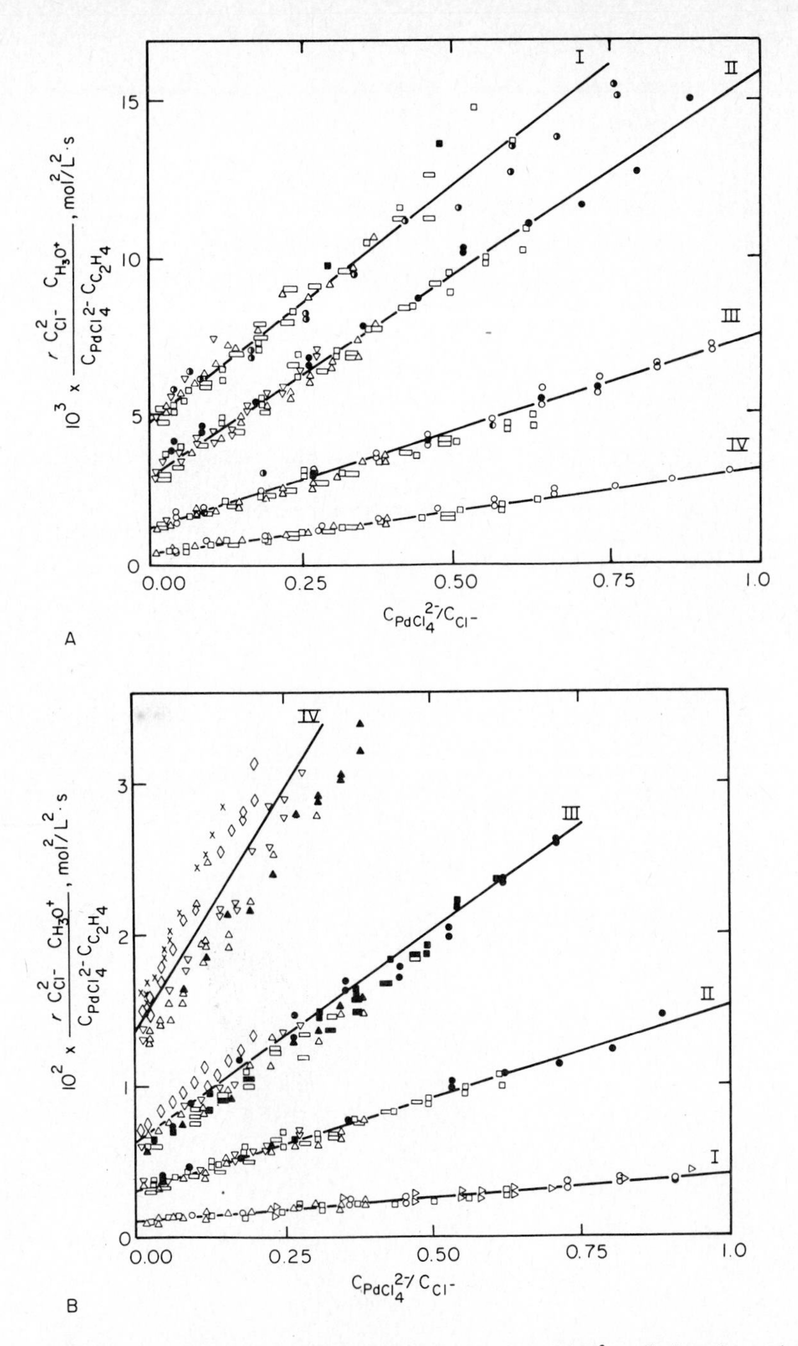

Figure 2-10 Kinetics of ethylene oxidation by aqueous $PdCl_4^{2-}$ solutions for various ionic strengths (Figure 2-10A) and temperatures (Figure 2-10B). Concentration (M) of Cl^- ions: ○, 0.2; □, 0.3; ▭, 0.4; △, 0.5; ▽, 0.7; ◇, 1.0; X, 1.4. All unfilled circles correspond to $C_{H_3O^+} = 0.2M$; all solid circles correspond to $C_{H_3O^+} = 0.5M$; ◐, $C_{H_3O^+} = 0.1M$, $C_{Cl^-} = 0.2M$; ◑, $C_{H_3O^+} = 0.8M$, $C_{Cl^-} = 0.2M$; ◓, $C_{H_3O^+} = 1.0M$, $C_{Cl^-} = 0.2M$; and ◒, $C_{H_3O^+} = 1.5M$, $C_{Cl^-} = 0.2M$ [13]. (Reprinted with permission from the *Journal of the American Chemical Society.* Copyright by the American Chemical Society.)

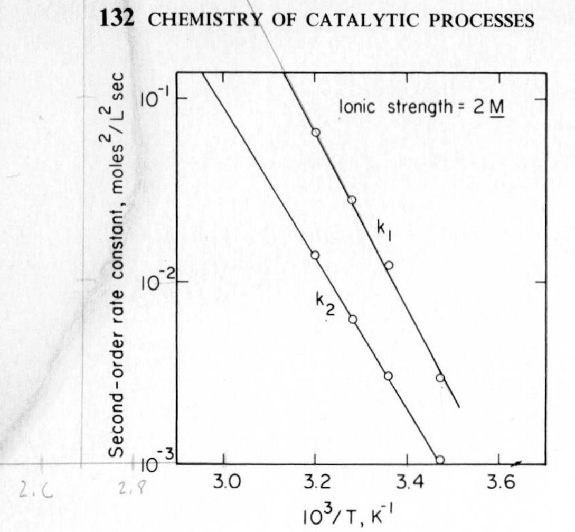

Figure 2-11 Temperature dependence of rate constants for ethylene oxidation by aqueous $PdCl_4^{2-}$; the data are from Fig. 2-10.

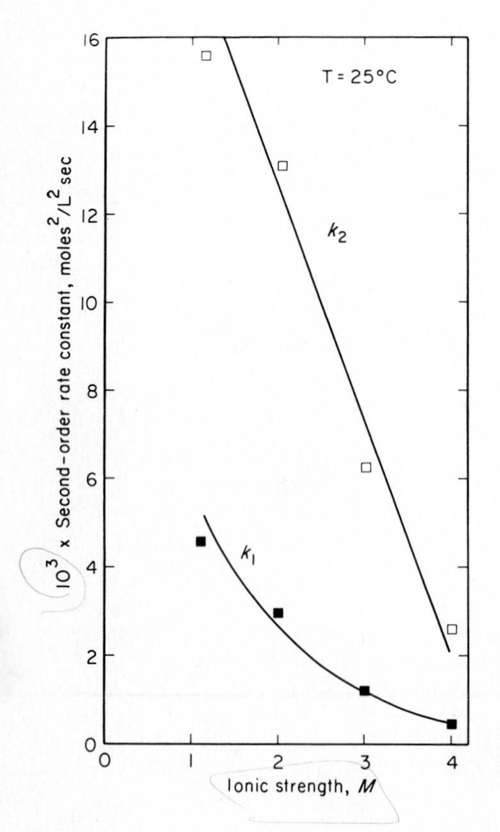

Figure 2-12 Dependence of rate constants for ethylene oxidation by $PdCl_4^{2-}$ on ionic strength; the data are from Fig. 2-10.

presence of ethylene forms a π-bonded Pd(II)–olefin complex very similar in structure to Zeise's salt:

$$\begin{array}{c}\mathrm{Cl^-}\\ \mathrm{Cl^-{-}Pd^{2+}{-}Cl^-}\\ \mathrm{Cl^-}\end{array} + \begin{array}{c}\mathrm{CH_2}\\ \|\\ \mathrm{CH_2}\end{array} \overset{K_{16}}{\rightleftharpoons} \begin{array}{c}\mathrm{Cl^-}\\ \mathrm{Cl^-{-}Pd^{2+}{-}}\\ \mathrm{Cl^-}\end{array}\!\!\begin{array}{c}\mathrm{CH_2}\\ \|\\ \mathrm{CH_2}\end{array} + \mathrm{Cl^-} \tag{16}$$

This complex reacts in the following sequence of elementary steps to give acetaldehyde:

$$\begin{array}{c}\mathrm{Cl^-}\\ \mathrm{Cl^-{-}Pd^{2+}{-}}\\ \mathrm{Cl^-}\end{array}\!\!\begin{array}{c}\mathrm{CH_2}\\ \|\\ \mathrm{CH_2}\end{array} + \mathrm{H_2O} \overset{K_{17}}{\rightleftharpoons} \begin{array}{c}\mathrm{OH_2}\\ \mathrm{{}^-Cl{-}Pd^{2+}{-}}\\ \mathrm{Cl^-}\end{array}\!\!\begin{array}{c}\mathrm{CH_2}\\ \|\\ \mathrm{CH_2}\end{array} + \mathrm{Cl^-} \tag{17}$$

$$\begin{array}{c}\mathrm{OH_2}\\ \mathrm{{}^-Cl{-}Pd^{2+}{-}}\\ \mathrm{Cl^-}\end{array}\!\!\begin{array}{c}\mathrm{CH_2}\\ \|\\ \mathrm{CH_2}\end{array} + \mathrm{H_2O} \overset{K_{18}}{\rightleftharpoons} \begin{array}{c}\mathrm{OH^-}\\ \mathrm{{}^-Cl{-}Pd^{2+}{-}}\\ \mathrm{Cl^-}\end{array}\!\!\begin{array}{c}\mathrm{CH_2}\\ \|\\ \mathrm{CH_2}\end{array} + \mathrm{H_3O^+} \tag{18}$$

$$\begin{array}{c}\mathrm{OH^-}\\ \mathrm{{}^-Cl{-}Pd^{2+}{-}}\\ \mathrm{Cl^-}\end{array}\!\!\begin{array}{c}\mathrm{CH_2}\\ \|\\ \mathrm{CH_2}\end{array} \longrightarrow \left[\begin{array}{c}\mathrm{OH^-}\\ \mathrm{{}^-Cl{-}Pd^{2+}\cdots}\\ \mathrm{Cl^-}\end{array}\!\!\begin{array}{c}\mathrm{CH_2}\\ \vdots\\ \mathrm{CH_2}\end{array}\right]^{\ddagger} \xrightarrow[\mathrm{H_2O}]{k_{19}} \left[\begin{array}{c}\mathrm{OH_2}\\ \mathrm{{}^-Cl{-}Pd^{2+}{-}CH_2CH_2OH}\\ \mathrm{Cl^-}\end{array}\right] \tag{19}$$

$$\left[\begin{array}{c}\mathrm{OH_2}\\ \mathrm{{}^-Cl{-}Pd^{2+}\cdots CH_2CH_2OH}\\ \mathrm{Cl^-}\end{array}\right] \xrightarrow{k_{20}} \mathrm{Pd^0 + 2Cl^- + H_3O^+ + CH_3C(=O){-}H} \tag{20}$$

Reaction steps (16) to (18) are well-established ligand exchanges, which have been demonstrated to proceed rapidly [16]. The rate-determining step is (19), which is the cis insertion. The details of this step are not entirely understood, but it is well established that the Pd^{2+} ion serves as a template and the two adjacent ligands, OH^- and $H_2C{=}CH_2$ (bonded cis to each other in the complex), react, forming a σ bond between the template ion and a carbon atom as well as a bond between the OH group and the other carbon atom; simultaneously the C—C double bond becomes a single bond, which no longer interacts with the *d* electrons of the template ion. The net result is the conversion of a π olefin complex to a σ complex with the insertion of the CH_2CH_2 unit between the template ion and the OH group.

If step (19) is rate-determining, then (if water is present in large excess)

$$r = k_{19}[cis\text{-}\mathrm{C_2H_4PdCl_2(OH)^-}] \tag{21}$$

The assumption that this step is rate-determining implies that the preceding steps in the sequence are virtually in equilibrium, so that the concentration of the complex in Eq. (19) can be related directly to the concentrations of reactants through the appropriate equilibrium constants. Consequently, the rate equation is the following (where the water concentration is incorporated appropriately in the equilibrium constants):

$$r = \frac{k_{19}K_{18}K_{17}K_{16}[PdCl_4^{2-}][C_2H_4]}{[Cl^-]^2[H_3O^+]} \tag{22}$$

This equation has the form of the observed kinetics, Eq. (14), when the second term on the right-hand side of Eq. (14) is negligible; this is usually the case when the concentration of $PdCl_4^{2-}$ is less than about 0.04 M.

At higher $PdCl_4^{2-}$ concentrations, there are more kinetically significant reaction steps than those shown in steps (16) to (19). The remaining steps postulated by Moiseev [13, 16] are the following, which involve a bridged species:

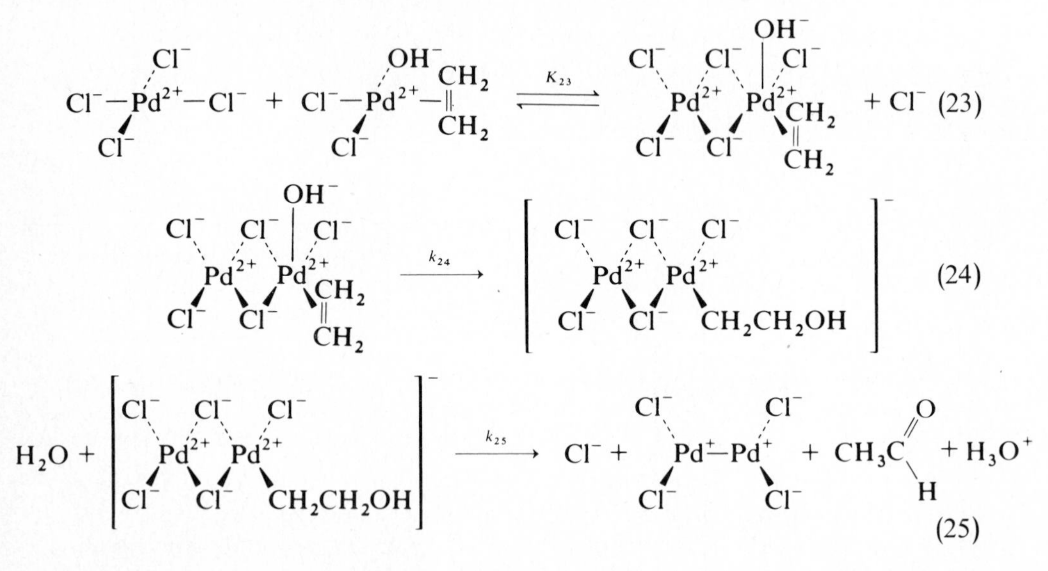

Again in this sequence, if the cis-insertion step (24) is assumed to be rate-determining, the kinetics can be stated directly; for the sequence of steps (16) to (19) followed by (23) to (24), the rate equation is

$$r = \frac{k_{24}K_{16}K_{17}K_{18}[PdCl_4^{2-}]^2[C_2H_4]}{[Cl^-]^3[H_3O^+]} \tag{26}$$

When both sequences of steps take place simultaneously, the rate is given by the sum of Eqs. (22) and (26). This result is in agreement with the observed kinetics, Eq. (14).

Process Design

Smidt et al. described three designs for the Wacker process [10, 17]. In the process shown schematically in Fig. 2-13, the acetaldehyde formation and reoxidation of metallic palladium take place in one reactor, and Cu^+ is reoxidized in the other. Ethylene and an aqueous solution containing palladium and copper salts flow concurrently as two phases in a tubular reactor, which contains inert packing material to facilitate fluid mixing. The ethylene is absorbed into the solution, where it reacts almost quantitatively to give acetaldehyde and small amounts of the previously mentioned side products. The reaction temperature is about 100 to 110°C, and pressure is about 10 atm. Reactor effluent is distilled, and the column heat load is supplied by the heat of reaction, −52 kcal/mol of acetaldehyde [14]. The overhead (organic) product is purified by further distillation to remove side products from the acetaldehyde. The bottoms (aqueous) product is continuously recycled to the second reactor. Air is mixed into this stream, which flows concurrently into the second reactor; oxygen is transferred to the liquid phase, where reaction (12) takes place, providing reoxidized copper. The conversion of oxygen is so high that the vented products can be used as inert gas.

A major engineering concern is the corrosive nature of the aqueous solution of chlorides of palladium and copper. The catalyst recirculation system was originally designed with a pump and lines constructed of costly titanium alloys [17]. The investment in these materials was probably a large fraction of the processing cost.

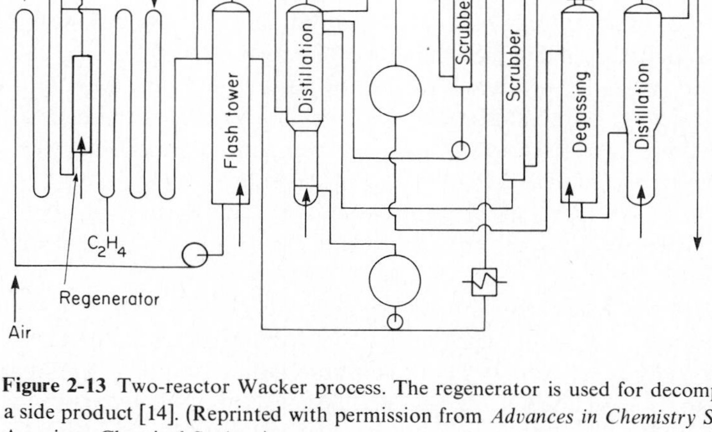

Figure 2-13 Two-reactor Wacker process. The regenerator is used for decomposition of Cu oxalate, a side product [14]. (Reprinted with permission from *Advances in Chemistry Series*. Copyright by the American Chemical Society.)

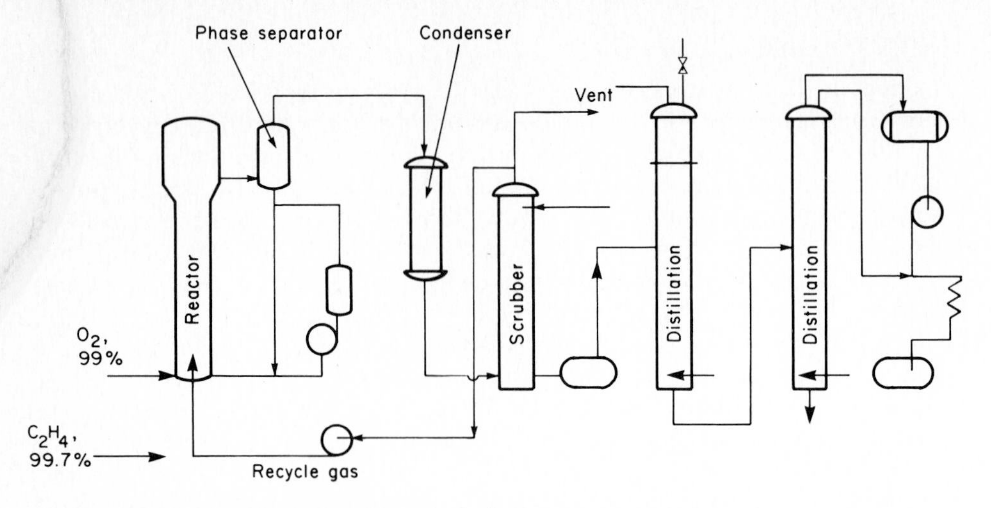

Figure 2-14 Single-reactor Wacker process [14]. (Reprinted with permission from *Advances in Chemistry Series.* Copyright by the American Chemical Society.)

An alternative process design suggested by Smidt [10, 14] and shown schematically in Fig. 2-14 differs from the preceding design in having one reactor instead of two and in having gas recycle but no liquid recycle. The air feed stream is now replaced by 99 percent oxygen. All three reactions in the sequence take place in the single reactor, to which the palladium and copper are confined. The evaporated water carried overhead from the reactor contains much of the energy liberated in the reaction; makeup water is added to the reactor. The product gases are water-scrubbed, and the resulting acetaldehyde solution is purified by distillation. Inerts are removed from the recycle gas in a continuous bleed stream, which flows to a separate reactor for further ethylene conversion. The reactor is operated at 120 to 130°C and 3 atm [14].

According to Smidt et al. [18], the concentration of palladium in the reactant solution is only 0.02 to 0.2 M; copper is usually present in large excess, as much as 100 mol per mole of palladium. The pH range may be about 0 to 2.

Both the one- and two-stage processes apparently have been used successfully. In the single-stage process, the advantages of one fewer reactor and elimination of noncorrosive materials of construction (except in the reactor) are offset by the need for a gas-recycle loop and a feed of pure oxygen rather than air; there is danger of forming explosive gas mixtures, especially with the pure oxygen feed. The feed ratio of oxygen to ethylene must be outside the explosion limits, which are shown in Fig. 2-15.

Smidt et al. [10] briefly described a fixed-bed process in which an ethylene-oxygen mixture saturated with water vapor is passed over a fixed bed of particles holding catalytically active palladium-copper solution within the pores. Evnin et al. [20] described a solid V_2O_5 catalyst containing Pd^{2+} which is highly active for ethylene oxidation to acetaldehyde. The catalyst will be discussed in Chap. 4.

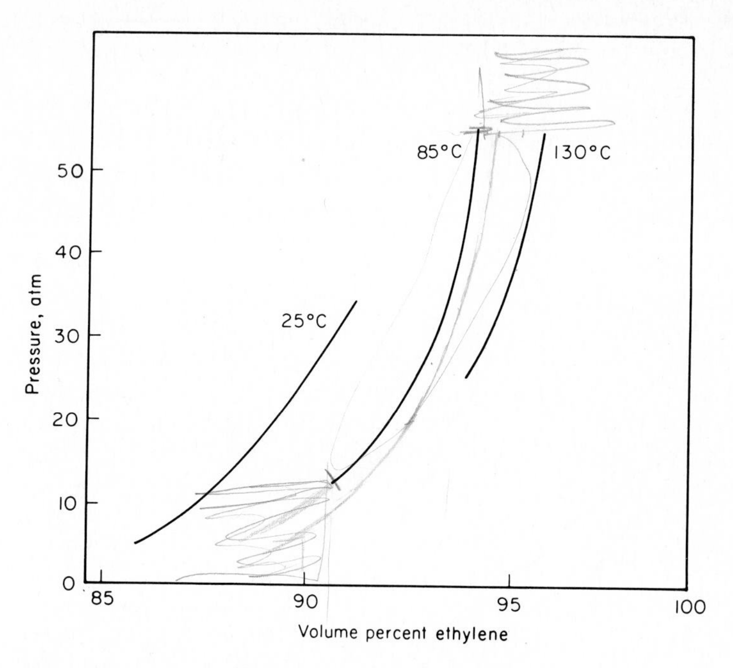

Figure 2-15 Explosive limits of ethylene-oxygen mixtures [19].

VINYL ACETATE SYNTHESIS

Reactions and Catalysts

Closely related to the Wacker process is a newer commercial process for the synthesis of vinyl acetate, which is used on a large scale in the production of poly(vinyl acetate) and a product of its reaction with water or methanol, poly(vinyl alcohol). The vinyl acetate formation from ethylene and acetic acid is based on the following reaction cycle, first observed by Moiseev et al. [21]:

$$H_2C{=}CH_2 + CH_3{-}C({=}O)OH + PdCl_2 \longrightarrow CH_3{-}C({=}O)OCH{=}CH_2 + Pd + 2HCl \qquad (27)$$

$$Pd + 2CuCl_2 \longrightarrow PdCl_2 + 2CuCl \qquad (11)$$

$$2CuCl + \tfrac{1}{2}O_2 + 2HCl \longrightarrow 2CuCl_2 + H_2O \qquad (12)$$

The overall reaction produces vinyl acetate and water:

$$H_2C{=}CH_2 + \tfrac{1}{2}O_2 + CH_3{-}C({=}O)OH \longrightarrow CH_3{-}C({=}O){-}O{-}CH{=}CH_2 + H_2O \qquad (28)$$

Since water is a product, the Wacker reaction (10) also occurs, giving acetaldehyde as the major side product; hydrolysis of vinyl acetate ultimately leads to further formation of acetaldehyde [19]. Ethylidene diacetate, $CH_3CH(O_2CCH_3)_2$, and the Wacker by-products are also formed, among other by-products.

Reaction Kinetics

The kinetics of the process is poorly defined because there are many products and their distribution varies strongly with reaction conditions [11, 15]. The reaction is approximately first order in ethylene concentration and second order in acetic acid concentration [11, 22], although the latter result is doubtful [11]. Rate depends in a complicated way on the concentrations of acetate and Cl^- ions [23]. The rate increases about 1.5-fold for a 10°C temperature increase in the industrial process [19].

Reaction Mechanism

Details of the reaction mechanism in vinyl acetate synthesis are not well established, but the weight of evidence suggests that this reaction is similar to the Wacker reaction. Both are believed to proceed via π-olefin-complex formation, incorporation of a nucleophile into the complex, rearrangement from a π to a σ complex (cis insertion), and decomposition of the complex into products [11]. The following steps adequately explain the reaction:

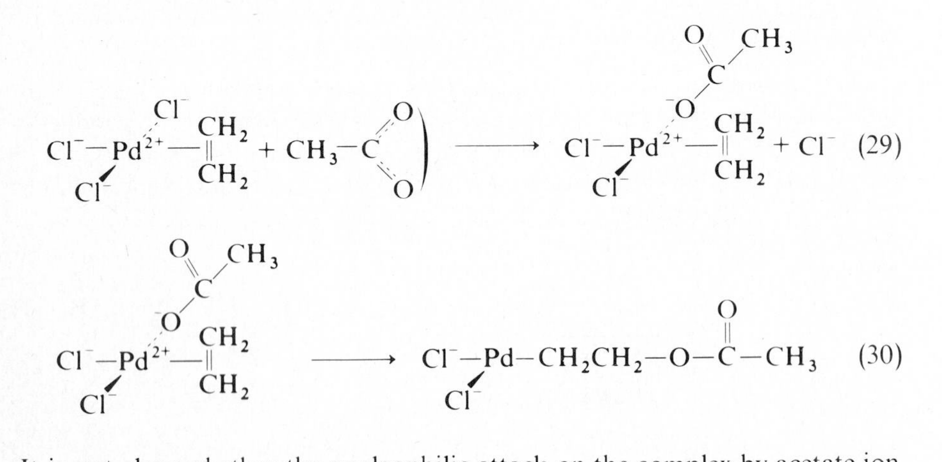

It is not clear whether the nucleophilic attack on the complex by acetate ion to give the product of step (29) actually takes place; alternatively, the attacking species might be acetic acid, and the product might be a complex which, on loss of a proton, would give the complex shown. The existence of the σ complex shown in step (30) has not been demonstrated, and the mechanism of its supposed decomposition to vinyl acetate and other reaction products is unknown [11].

Process Design

Processing schemes are similar to those discussed for the Wacker process. For example, in the single-reactor process of Fig. 2-16, oxygen, ethylene, and acetic acid are fed to a bubble-column reactor containing the palladium and copper salts. Reactor temperature is about 100 to 130°C, and pressure is about 30 atm [19], although pressures as low as atmospheric may be used. Gas is continuously recirculated at a high rate through the reactor to maintain efficient gas-liquid contacting; a continuous bleed stream prevents the buildup of gaseous impurities. The overhead vapor-product stream from the reactor is rich in vinyl acetate and acetaldehyde and also contains acetic acid, water, and other components. This stream is treated in several purification steps. The heavier components, including water, acetic acid, and the dissolved salts, are recirculated to the reactor. Purified acetaldehyde product may be oxidized separately to acetic acid to be recycled to the reactor; the process can be run economically with all the acetic acid produced internally from ethylene.

The aqueous solution in the reactor contains about 3×10^{-4} M Pd^{2+} and about 0.08 M Cu^{2+}, concentrations much lower than those used in the Wacker process. Higher palladium concentrations are avoided since they cause butene formation to be a significant side reaction [19]. The concentration of Cl^- ion is regulated to maintain catalyst efficiency, presumably to ensure that palladium is predominantly in the form of a soluble chloride in the products. Sodium acetate may be included in the reactant solution, since at low concentrations acetate ion increases the rate of vinyl acetate formation [11]. The water concentration is optimized, since excess water leads to high rates of acetaldehyde formation and

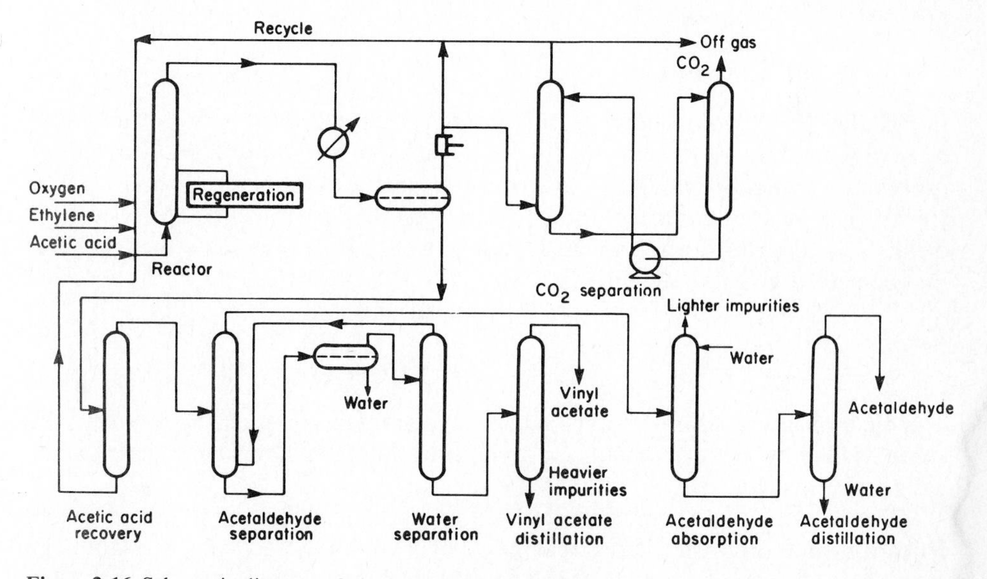

Figure 2-16 Schematic diagram of vinyl acetate process [19].

low water concentrations prevent sufficient solubility of the palladium and copper salts.

The commercial process is reported by Krekeler and Krönig [19] to give a yield of vinyl acetate plus acetaldehyde of 90 percent of stoichiometric based on ethylene and a yield of at least 95 percent based on acetic acid. The conversion of acetic acid per pass may be 30 percent, with the ethylene conversion dictated by the explosive limits of ethylene-oxygen mixtures. For example, at 130°C and 30 atm, the upper ignition limit is at an ethylene-to-oxygen ratio of 94.5 : 5.5 (Fig. 2-15). Therefore, with a 50 percent conversion of oxygen and a feed mixture of 5 percent oxygen in ethylene, there is only a 5 percent conversion of ethylene.

The available data give only very imprecise indications of reaction rates at industrial processing conditions. For the conditions cited, the average residence time of liquid per pass is of the order of an hour, and the average residence time of gas is of the order of a few seconds.

The corrosion problems in the vinyl acetate process are the same as those in the Wacker process, and materials contacting the palladium-copper salt solution may be of titanium. The trade-offs between one- and two-reactor processes are similar in the vinyl acetate and Wacker processes.

A fixed-bed process involving supported palladium metal catalyst is now predominantly applied in vinyl acetate production [19]; since the acetaldehyde formation rate is very low, this process is preferable when acetic acid is available economically from a separate source, such as a process for hydrolysis of poly(vinyl acetate) to give poly(vinyl alcohol) and acetic acid.

THE OXO PROCESS: HYDROFORMYLATION OF OLEFINS

Reactions and Catalysts

In 1938 Roelen of Ruhrchemie AG in Germany discovered the hydroformylation, or oxo, reactions, which account for the greatest chemical production based on processes involving catalysis by transition-metal complexes, about 3.5×10^9 kg/year. The hydroformylation (oxonation) reactions lead to conversion of olefins to straight- and branched-chain aldehydes, which can then be hydrogenated to give oxo alcohols:

$$RCH{=}CH_2 + CO + H_2 \longrightarrow RCH_2CH_2CHO + R\overset{\overset{\displaystyle CHO}{|}}{C}HCH_3 \qquad (31)$$

The alcohols are used as solvents and in the manufacture of plasticizers, and the straight-chain C_{12} to C_{15} alcohols especially are sulfonated on a large scale to give detergents.

The oxo reaction is carried out in a liquid organic phase containing catalyst, reactants, and products; if the reactant olefin is of low molecular weight, e.g., propylene, an inert diluent is used. The most widely used industrial catalyst is a cobalt carbonyl, $Co_2(CO)_8$, the structures of which are shown on page 125; during reaction the catalyst is predominantly present as the hydridocarbonyl, $HCo(CO)_4$. Newer catalysts incorporate phosphine ligands, as in the Shell catalyst, $HCo(CO)_3PBu_3$, and rhodium complexes such as $Rh(CO)Cl(PPh_3)_2$ have found application in the newest processes for propylene hydroformylation.

Reaction Kinetics

Kinetics data obtained from measurements of absorption of reactant diisobutylene into solutions of the cobalt carbonyl complex, $HCo(CO)_4$, are summarized by the equation [24]

$$r = \frac{kC_C C_{RCH=CH_2} P_{H_2}}{P_{CO} + KP_{H_2}} \tag{32}$$

The second term in the denominator is negligible in the pressure range usually applied [25], which implies almost pressure-independent kinetics in the absence of mass-transfer influence for 1 : 1 H_2/CO ratios. Equation (32) is based on data for reaction of diisobutylene at 150°C, $P_{CO} = 20$ to 150 atm, and $P_{H_2} = 50$ to 275 atm [24], but it is not entirely clear whether the data were obtained free of mass-transfer influence.

Even though the rate is almost independent of pressure, high pressures are used in the oxo process, since a high liquid-phase CO concentration is required to

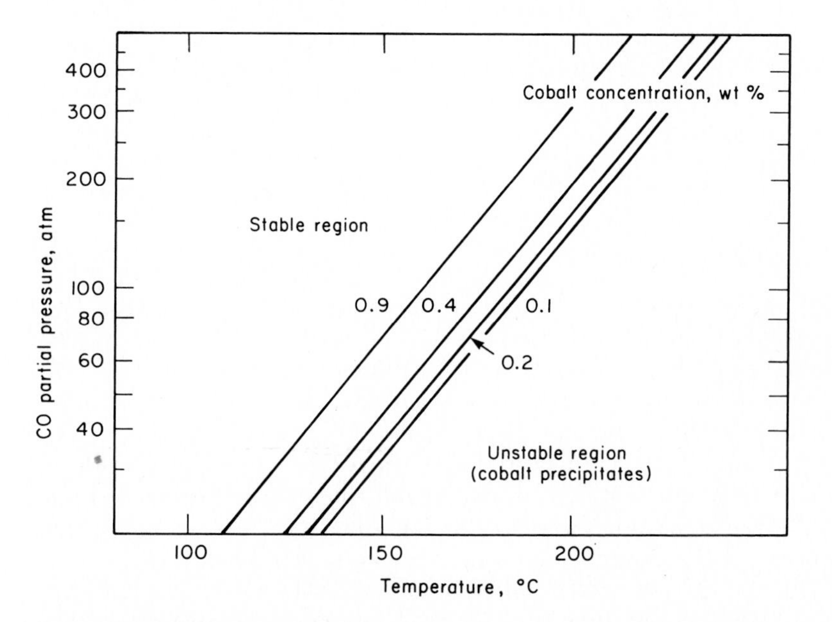

Figure 2-17 Stability of cobalt catalyst $[Co_2(CO)_8 + HCo(CO)_4]$ as a function of temperature and equilibrium carbon monoxide partial pressure [25].

prevent decomposition of the cobalt carbonyl complex into inactive species like $Co_4(CO)_{12}$ or possibly metallic cobalt. The critical concentration of CO is determined by the equilibrium partial pressure, which is shown as a function of temperature in Fig. 2-17.

Newer catalysts like the Shell catalyst, and even more the rhodium complexes, are resistant to decarbonylation at much lower CO partial pressures, and consequently these catalysts are applied at lower pressures. For example, the Shell process may be operated at 100 atm and a process with a rhodium catalyst at only 15 atm, whereas a pressure of about 200 to 300 atm is required for the Ruhrchemie process using $HCo(CO)_4$ catalyst [25, 26].

The Shell catalyst has only about 20 to 50 percent of the activity of the conventional catalyst; the much more expensive rhodium catalysts, on the other hand, are as much as 1000 times as active as the more conventional catalyst [25, 27]. The kinetics of propylene hydroformylation catalyzed by $HRh(CO)(PPh_3)_2$ in the presence of excess PPh_3 is described by [28]

$$r = kC_C C_{CH_3CH=CH_2} P_{H_2} \tag{33}$$

This equation is based on data obtained in the temperature range 79 to 107°C and in the pressure range 6 to 12 atm. The lack of dependence of rate on CO partial pressure is not general, and the data should be extrapolated with caution.

The oxo reaction applies to olefins of any structure, and a discussion of the relative reactivities and product distributions is given by Falbe [25]. Table 2-4 is a summary of reactivities of various olefins in the presence of $HCo(CO)_4$. The data show a variation by a factor of 50, the terminal straight-chain olefins being the most reactive. Internal olefins are less reactive, and branching in the olefin structure reduces reactivity. The reactivity patterns are similar when rhodium complexes are used as catalysts [25, 27]. The data of Table 2-4 may be used with Eq. (32) to provide preliminary reactor-design estimates, as illustrated in Example 2-2.

Product Distribution

The distribution of hydroformylation products has been regulated by temperature, pressure, solvent polarity, catalyst ligand basicity, and the steric hindrance offered by ligands. Yields of 80 percent aldehyde are typically obtained with the cobalt catalyst, and the ratio of straight- to branched-chain aldehydes formed from straight-chain terminal olefins is about 3 : 1 or 4 : 1. Yield of straight-chain aldehydes, normally the desired products, is favored by lower temperatures and higher CO partial pressures [25]. Bulky ligands (like the PR_3 groups in the Shell catalyst) improve selectivity for straight-chain products, at least in part by steric hindrance [30]. For example, the ratio of straight-chain to branched-chain product may be 88 : 12 in the Shell process compared with 80 : 20 in the Ruhrchemie process [25]. This ratio may approach 30 : 1 when rhodium-complex catalysts are used in the presence of excess PPh_3 [26].

Table 2-4 Hydroformylation of olefins at 110°C [29]
Reaction conditions: 0.5 mol of olefin; 65 mL of methylcyclohexane solvent; 2.8 g (8.2 mmol) of dicobaltoctacarbonyl; $CO/H_2 = 1:1$; initial pressure at room temperature = 233 atm

Class of olefin	10^5 × reaction rate constant, s^{-1} [a]
Straight-chain, terminal:	
Pentene-1	114
Hexene-1	110
Heptene-1	111
Octene-1	109
Decene-1	107
Tetradecene-1	105
Internal:	
Pentene-2	35.5
Hexene-2	30.2
Heptene-2	32.2
Heptene-3	33.3
Octene-2	31.3
Branched, terminal:	
4-Methylpentene-1	107
2-Methylpentene-1	12.2
2,4,4-Trimethylpentene-1	7.98
2,3-Trimethylbutene-1	7.10
Camphene	3.7[a]
Internal:	
4-Methylpentene-2	27.0
2-Methylpentene-2	8.12
2,4,4-Trimethylpentene-2	3.82
2,3-Dimethylbutene-2	2.25
2,6-Dimethylheptene-3	10.4
Cyclic:	
Cyclopentene	37.3
Cyclohexene	9.70
Cycloheptene	42.8
Cyclooctene	18.0[a]
4-Methylcyclohexene-1	7.8

[a] Except for camphene and cyclooctene, the values were determined in duplicate. The error, determined by a statistical analysis of 55 experiments, was ±1.5%.

Source: Reprinted with permission from *Journal of the American Chemical Society*. Copyright by the American Chemical Society.

Many products are formed besides aldehydes; the aldehydes may be hydrogenated directly to the corresponding oxo alcohols, and even paraffins are formed. Hydrogenation activity of the Shell cobalt carbonyl catalyst, in contrast to the conventional catalyst, is so high that alcohols are obtained in high yield in a one-step synthesis. Rhodium carbonyl catalysts, on the other hand, have much less hydrogenation activity, and aldehydes are produced in high yield [27]. Carboxylic acids and ketones are side products of the oxo reaction, and higher-molecular-weight products are formed in esterification reactions and in condensation reactions involving alcohols and aldehydes. Migration of double bonds in olefins also occurs, especially at low CO partial pressures, greatly increasing the number of products. Product-distribution data are collected in the reviews of Falbe [25] and Paulik [27].

Reaction Mechanism

The mechanism of hydroformylation catalyzed by complexes of rhodium has been reviewed by Wilkinson and coworkers [2, 31], and their interpretation is summarized in Fig. 2-18. The figure illustrates an associative mechanism, involving reaction of an olefin with a five-coordinate complex, and a dissociative mechanism, involving reaction of an olefin with a four-coordinate complex.

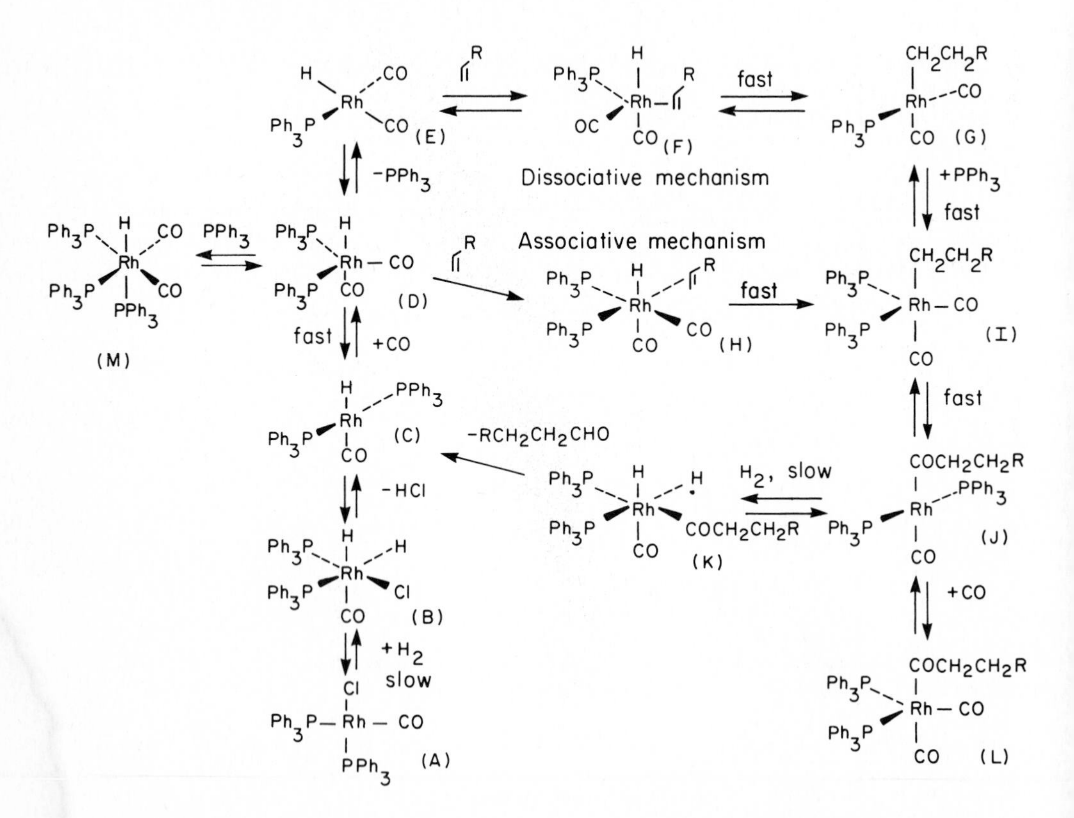

Figure 2-18 Mechanism of the oxo reaction catalyzed by Rh-phosphine complexes. For simplicity, only the mechanism for linear aldehyde formation is shown. (Adapted from ref. [2] with permission.)

The relative importance of each mechanism is dependent on equilibria between species D and E shown in Fig. 2-18. In the presence of excess PPh_3, the equilibrium is shifted toward complex D over complex E, and the associative mechanism predominates. This mechanism leads to high selectivities for linear over branched aldehydes because of the steric hindrance offered by the PPh_3 groups. As the concentration of PPh_3 is increased, the reaction rate passes through a maximum [32].

The complexes formed by addition of olefin (F and H) rapidly rearrange to alkyl complexes (G and I), and the CO-insertion step follows. The resulting square-planar complex (J) then experiences slow oxidative addition of H_2, forming a coordinatively saturated Rh(III) complex (K). The final steps are another H transfer to the carbon atom of the acyl group in K and a reductive elimination, leading to the loss of aldehyde and regeneration of the square-planar Rh(I) complex (C). An excess of CO over H_2 probably leads to formation of complex L, which accounts for the occasionally observed inhibition of hydroformylation by CO.

The mechanism of hydroformylation catalyzed by cobalt carbonyl complexes is less well understood than the reaction involving rhodium, but it is clear that the mechanisms are similar for the two metals. The literature of the hydroformylation mechanism with cobalt has been reviewed by Orchin and Rupilius [30], who suggested a mechanism, shown in Fig. 2-19, similar to that originally formulated

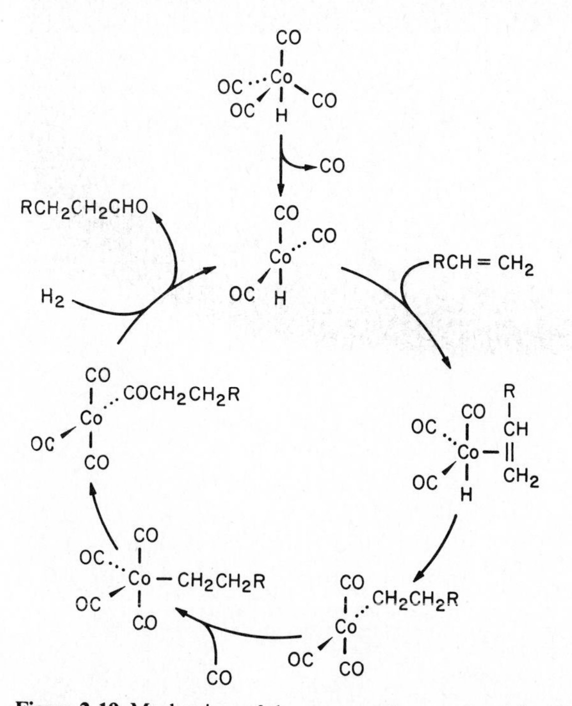

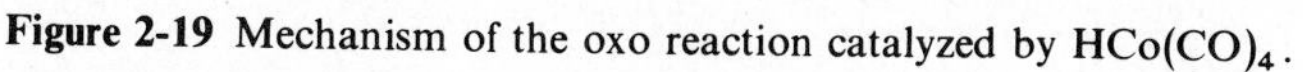
Figure 2-19 Mechanism of the oxo reaction catalyzed by $HCo(CO)_4$.

by Heck and Breslow [33]. Although many details are not understood, the important result is the occurrence of the CO-insertion reaction and the rapid interconversion of the σ- and π-bonded complexes.

Process Design

Lemke [34] discussed several designs for the oxo process, in particular that represented schematically in Fig. 2-20, which is referred to as the *Kuhlmann process.* Reaction in the presence of the unmodified cobalt carbonyl catalyst, $HCo(CO)_4$, takes place in a single gas-liquid reactor at 110 to 180°C and 200 to 250 atm; CO and H_2 are fed to the reactor at a molar ratio of 1 : 1 to 1 : 1.3 [25]. With liquid residence times in the reactor of 1 or 2 h, the catalyst concentration may be 0.001 to 0.01 g of cobalt per gram of olefin; the most common reactor is a stirred tank constructed of stainless steel, but narrow tubular reactors designed for piston flow are also applied [25]. Since the reactions are exothermic (−28 to −35 kcal/mol, depending on the olefin structure), the reactor design requires efficient heat removal. Heat may be removed by coolant fluid in internal coils or by boiling water surrounding the reactor. Efficient reactor design also requires good gas-liquid contacting, since otherwise the transfer of reactant from the gas to the liquid phase may affect the rate; the mass-transfer effect is discussed in Example 2-2. Conversion in the reactor may be high, for example, 90 to 95 percent. Yields of aldehyde may also be high, about 85 to 90 percent of the olefin converted.

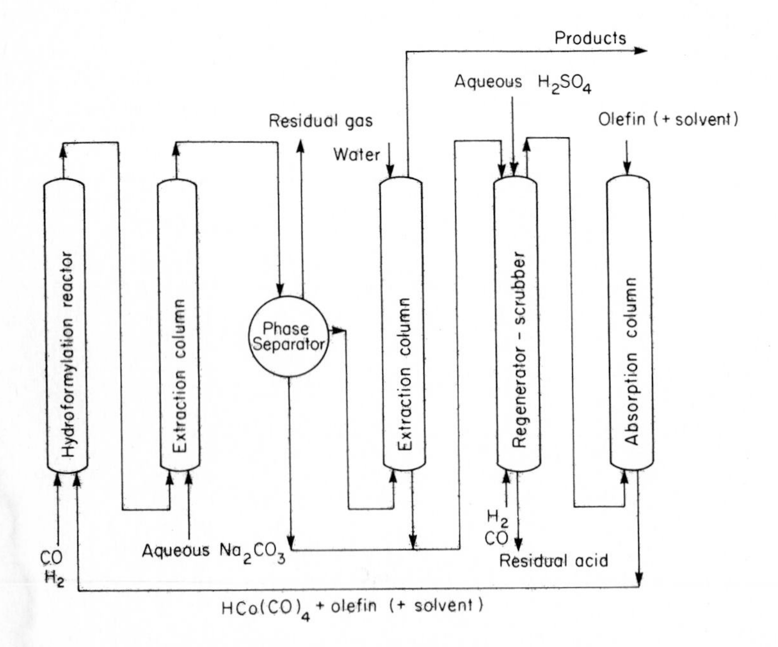

Figure 2-20 Schematic diagram of Kuhlmann oxo process [34].

In the Kuhlmann process, the reactor-product stream flows to a column where the catalytically active species, $HCo(CO)_4$, is extracted from the organic phase into an aqueous Na_2CO_3 solution; the following reaction takes place

$$2HCo(CO)_4 + Na_2CO_3 \longrightarrow 2NaCo(CO)_4 + CO_2 + H_2O \qquad (34)$$

The next vessel in the system is a phase separator, in which residual gases are removed. The remaining catalyst is removed from the organic product in a second extraction column, and the aqueous phases from the two extraction columns flow to another column, where they are regenerated with acid and stripped:

$$2NaCo(CO)_4 + H_2SO_4 \longrightarrow 2HCo(CO)_4 + Na_2SO_4 \qquad (35)$$

The catalytically active cobalt hydridocarbonyl formed in this reaction is now volatile and sparingly soluble in the aqueous acid. It is carried overhead in a stream of CO and H_2, flowing then to an absorption column, where it is dissolved in a stream of olefin, which may be mixed with diluent. The catalyst, now regenerated and mixed with reactant, is recycled to the reactor. The products flowing from the second extraction column are subjected to further purification steps.

Lemke [34] and Falbe [25] have discussed several of the numerous other processing schemes, which differ most significantly from the Kuhlmann process in the method of removal of catalyst from products. An advantage of the newer catalysts is their stability at high enough temperatures to allow their separation from products by distillation. Some earlier processes involved decomposition of the cobalt carbonyl to form metallic cobalt, which has the disadvantage of forming adhesive deposits; the deposits can seriously affect heat transfer from the reactor and may even lead to unstable operation. (Instability phenomena are considered in Chap. 4.)

Processes based on catalysis by rhodium complexes have been reviewed by Cornils et al. [35], who concluded that the economics are about the same as for the conventional processes with cobalt catalysts.

METHANOL CARBONYLATION

Reaction and Catalyst

A process for the production of acetic acid from methanol yielding 135 million kilograms per year with more than 99 percent selectivity has been reported by Paulik and Roth and their coworkers, and about five plants are now in operation [36–36*c*]. The reaction

$$CH_3OH + CO \longrightarrow CH_3C\begin{matrix}\nearrow O \\ \searrow OH\end{matrix} \qquad (36)$$

takes place in solution in the presence of CH_3I promoter (which may be formed from the reaction of HI with methanol), and it is catalyzed by a complex of rhodium, which is formed from any of a number of complexes, including that previously mentioned for the oxo process, $RhCl(CO)(PPh_3)_2$. The carbonylation process takes place under conditions as mild as atmospheric pressure and 150°C, but the commercial conditions have been suggested to be 15 atm and 175°C [37].

The rhodium catalyst is a significant improvement on the previously applied soluble cobalt complexes, which require pressures of 475 to 600 atm and temperatures of 210 to 250°C while giving yields of only about 85 percent based on methanol [37].

Reaction Kinetics

Roth et al. [36*b*] reported kinetics of the following form for the methanol carbonylation reaction

$$r = kC_{CH_3I}C_{Rh\ complex} \tag{37}$$

The rate constant is $3.5 \times 10^6 e^{-14.7/RT}$ L/mol · s, where the activation energy is given in kilocalories per mole [38].

Product Distribution

The selectivity of the process based on methanol is reported to exceed 99 percent [36–36*c*]. The side products include dimethyl ether (which is ultimately converted into methanol and then product) and methyl acetate; the product distribution is determined by the acid-to-ester ratio in the reactant solution, as follows [36]:

Alcohol-to-ester molar ratio	Major product
0–2	Acetic acid
2–10	Methyl acetate
10–∞:	
Methanol conversion $\lesssim 90\%$	Methyl acetate
Methanol conversion $\gtrsim 90\%$	Acetic acid

Source: Reprinted from [36] with permission of the authors.

The water content is also important in determining product distribution. Other side products are methane, hydrogen, and CO_2.

Reaction Mechanism

The reaction mechanism suggested by Roth et al. and Forster [36*b*, 36*c*] (Fig. 2-21) has many characteristics in common with the hydroformylation mechanism shown in Fig. 2-18. The first step in the sequence is the oxidative addition of CH_3I promoter to the catalytically active form of the Rh(I) complex; the ligands of this anionic complex are CO and I^- [36*a*]. The oxidative addition

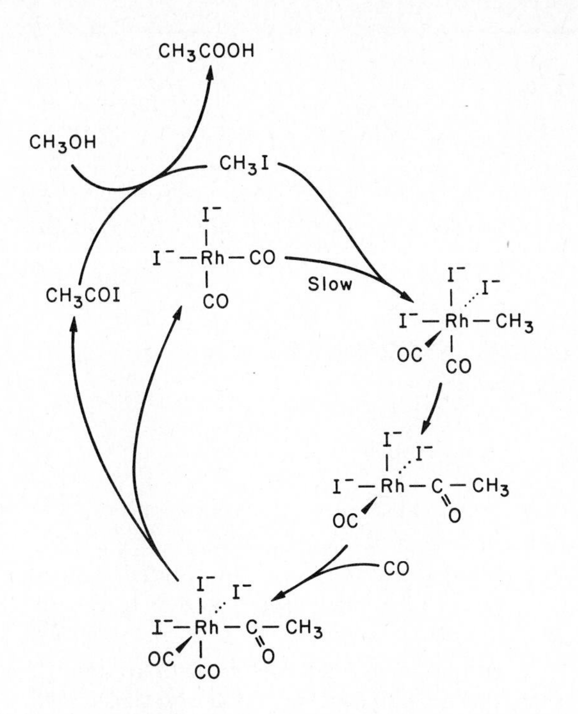

Figure 2-21 Mechanism of methanol carbonylation [36c].

step is inferred from the kinetics to be rate-determining. The subsequent steps include bonding of CO cis to the methyl group, the familiar insertion of the carbonyl group, and a poorly characterized reductive elimination step or series of steps to yield the product and regenerate the catalyst and promoter. The catalytic cycle shown in Fig. 2-21 rests on a firmer foundation than any other presented in this book, since the key intermediates have been identified spectroscopically [36c].

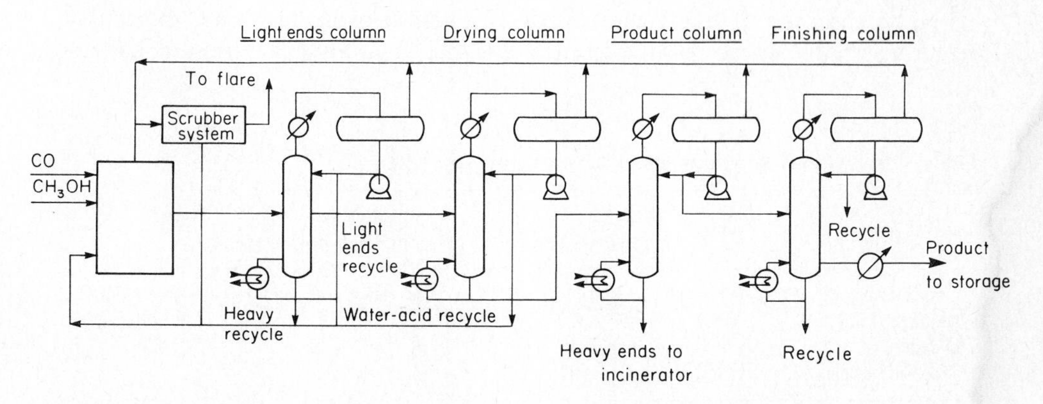

Figure 2-22 Process flow diagram of methanol carbonylation process [36, 37].

Process Design

A process flow diagram is shown in Fig. 2-22. The gas-liquid reactor might be a stirred tank, a bubble column, or a trickle bed. Details of the purification system are lacking, but the corrosive nature of the iodide-containing catalyst-promoter solutions and the high cost of rhodium suggest that the devices must be constructed of expensive noncorrosive materials and must operate with a high efficiency of catalyst recovery.

THE ZIEGLER-NATTA PROCESS: STEREOSPECIFIC POLYMERIZATION OF α-OLEFINS

Reactions and Catalysts

The discovery of catalytic reactions of α-olefins to give stereoregular polymers is among the most significant in the science and practice of catalysis. Ziegler discovered the polymerization catalysts, and Natta recognized the unique stereoregularity of the polymers. For their pioneering work, Ziegler and Natta were awarded the Nobel prize in chemistry in 1963.

Polymerization of propylene, for example, takes place in solutions containing complexes of zirconium [39]. The stereospecific reaction takes place on surfaces of solid compounds of transition metals, and the following discussion of polymerization is focused on industrial applications with solid catalysts prepared from α-$TiCl_3$ and metal alkyls like $Al(C_2H_5)_2Cl$.

The early technology of olefin polymerization involved free-radical reactions at pressures as high as 2000 atm. These reactions are essentially the reverse of thermal cracking, which was discussed in Chap. 1. Such processes are used on a large scale to give low-density polyethylene, a solid having highly branched chains and a low degree of crystallinity. High-density polyethylene formed in the presence of a Ziegler catalyst is characterized by much less chain branching, a high degree of crystallinity, and a melting point about 20°C higher than that of low-density polyethylene (Table 2-5).

Natta discovered that when propylene or an α-olefin of higher molecular weight was polymerized in the presence of a Ziegler catalyst, the product had an

Table 2-5 Typical physical properties of polyethylene

	Ziegler-Natta linear polyethylene	Free-radical branched polyethylene
Crystallinity, %	90	50–60
Melting point, °C	135	115
Density, g/cm^3	0.95–0.97	0.91–0.94
Modulus of elasticity, dyn/cm^2	7×10^{11}	1.5×10^{11}

unusually high melting point. X-ray diffraction showed that the polypropylene was highly crystalline, and the following stereoregular structure, referred to as *isotactic polypropylene*, was inferred:

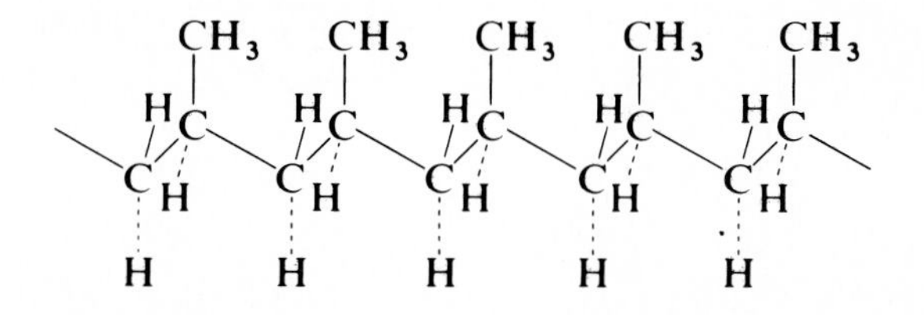

Actually the polymer chain is not the linear (stretched) form shown above but a helix with all the CH_3 groups pointing toward the outside. The helical structure of this isotactic polymer is depicted in Fig. 2-23.

Polymers with this regular structure have more desirable physical properties than polymers with irregular structures. For example, isotactic polypropylene has greater mechanical strength and a higher melting point (about 175°C compared with about 35°C) than the polymer with a random orientation of substituent

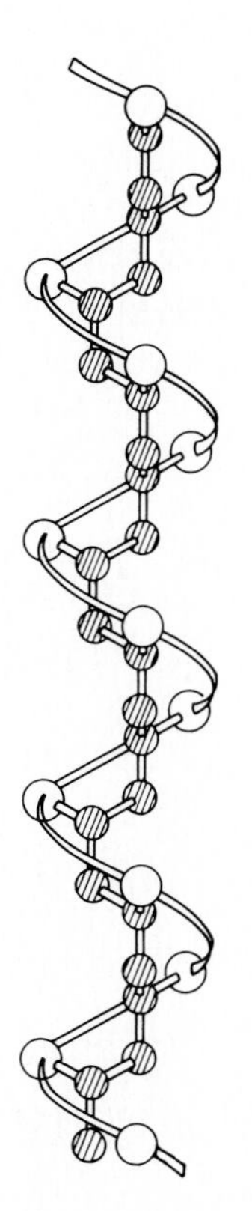

Figure 2-23 Helical structure of isotactic polypropylene chain. The methyl groups define a helix, as indicated by the ribbon.

methyl groups, *atactic polypropylene*, which is produced in the presence of non-stereospecific catalysts. The following structure represents the stretched form of this amorphous atactic polypropylene:

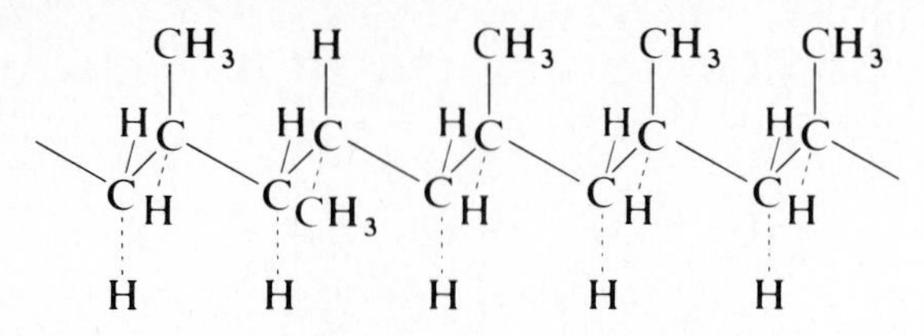

A polymer in which the methyl groups are arranged alternately in the in and out positions is called *syndiotactic*. This polymer has been synthesized, but it has not found practical application.

Polymerization is not confined to α-olefins; monomers such as butadiene and isoprene can be polymerized to give rubberlike substances; polyisoprene is similar to natural rubber, a stereoregular *cis*-1,4 polymer. Rubberlike substances can also be produced by copolymerization, for example of C_2H_4 and C_3H_6, in the presence of Ziegler-Natta catalysts. The Ziegler-Natta systems can be used for the production of polymers with widely different physical properties.

Reaction Kinetics

Ziegler-Natta polymerizations are carried out in gas-liquid-solid reactors, and rates of reaction depend on many variables, which are identified in the following paragraphs and illustrated for the most part by the batch-reactor data of Berger and coworkers [40, 41]. There are conflicting literature reports about the kinetics [42], and the results cited here are the ones which appear to have the greatest

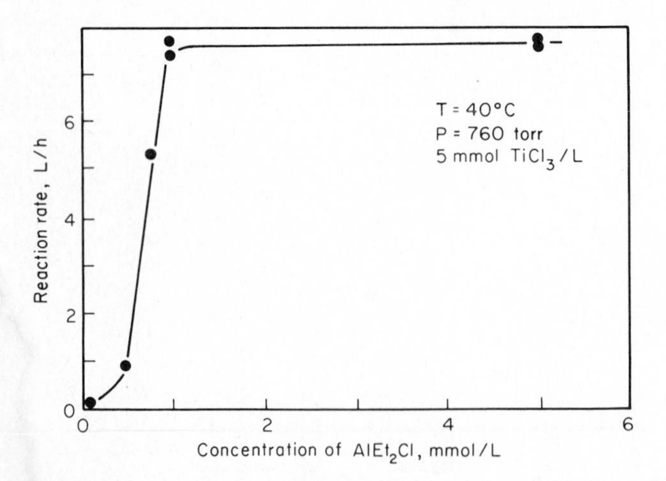

Figure 2-24 Effect of aluminum alkyl concentration on rate of ethylene polymerization [41]. (Reprinted with permission of Hüthig and Wepf Verlag.)

internal consistency and are least likely to have been misinterpreted because of mass-transfer influence.

The effect of concentration of aluminum alkyl, $Al(C_2H_5)_2Cl$, is demonstrated in Fig. 2-24. The rate of ethylene polymerization was found to increase to a maximum with increasing $Al(C_2H_5)_2Cl$ concentration and then remain constant. Evidently 1 mol of the solid $TiCl_3$ catalyst was saturated with about 0.2 mol of the aluminum alkyl. An excess of the latter is normally used as a scavenger of poisons like water.

Rates of polymerization are sensitive to the nature as well as the amount of the metal alkyl. The data of Fig. 2-25 show differing stabilities of catalysts prepared from two different added aluminum alkyl compounds. The effects of these additives are not well understood, and their choice is an important part of the industrial art; often a combination of several metal alkyls is used. Under properly chosen conditions, catalyst activity is maintained, and a batch reactor can be run until a further increase in viscosity of the reactants prevents adequate stirring.

The data of Fig. 2-26 illustrate a nearly constant rate of polymerization of ethylene until a 17 percent slurry is formed. More typically, however, a decreasing rate is observed [43]. The catalyst surface evidently remains accessible to reactant even though the catalyst particle is being buried in growing chains of attached

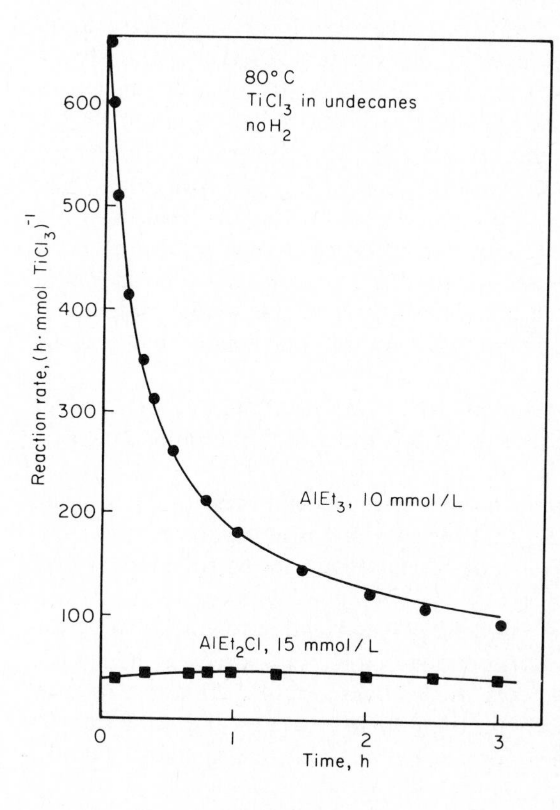

Figure 2-25 Effect of aluminum alkyl composition on catalyst stability. The monomer is presumed to have been ethylene [40]. (Reprinted with permission from *Advances in Catalysis*. Copyright by Academic Press.)

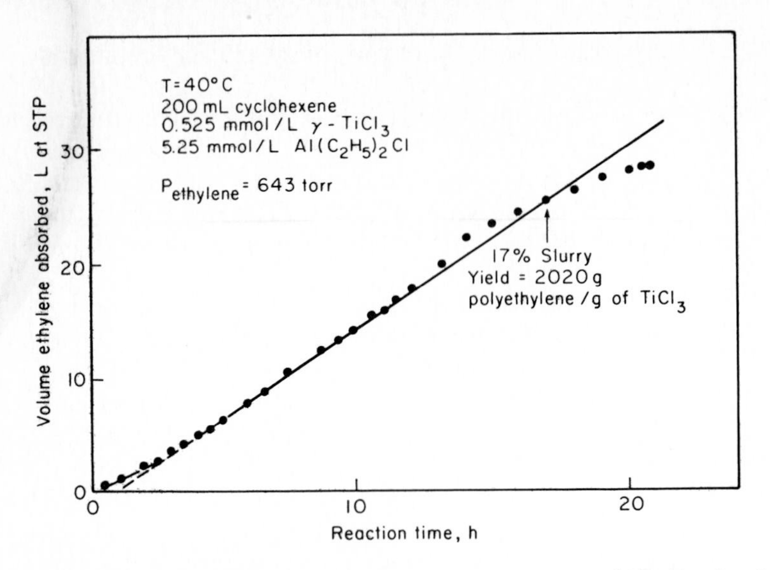

Figure 2-26 Polymerization of ethylene in a batch reactor [41]. (Reprinted with permission of Hüthig and Wepf Verlag.)

polymer. A porous catalyst particle breaks up under the forces of the expanding polymer mass, and the growing polymer particle contains islands of catalyst surrounded by a sea of polymer [44, 45] (Fig. 2-27). Especially under conditions of high yield of polymer (perhaps exceeding 250 kg of polymer per gram of $TiCl_3$), the diffusion resistance offered by the mass of growing polymer chains has a significant effect on the rate of polymerization; the rate may decrease by as much as several orders of magnitude just during the first minutes of polymerization and then decrease slowly [45, 46]. The molecular-weight distribution also changes significantly as polymerization proceeds, since the diffusion resistance increases and the number of catalytic sites decreases; high-molecular-weight polymer is formed initially, then lower-molecular-weight material, and finally intermediate product [46].

Under conditions of slow reaction when external-phase mass-transfer processes do not affect the rate, the reaction is first order in the amount of catalyst (Fig. 2-28).

Reaction is also found to be first order in olefin partial pressure under such conditions (Fig. 2-29), and, assuming that Henry's law is applicable in this pressure range, the result indicates a first-order dependence on the concentration of dissolved monomer.

The temperature dependence of ethylene polymerization rate is illustrated in Fig. 2-30. The data show an Arrhenius dependence at temperatures less than about 50°C, with an apparent activation energy of 13.5 kcal/mol [41]. At higher temperatures, either the solubility of ethylene in the solvent is so low that mass transfer from the gas to the liquid determines the rate, or the catalyst is unstable, with the

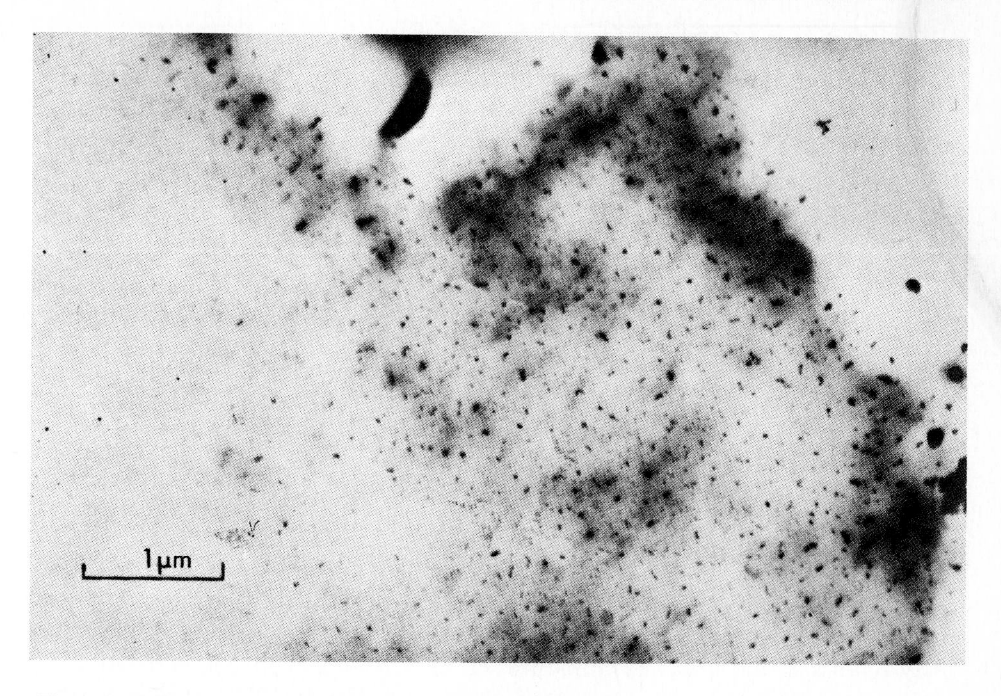

Figure 2-27 Electron micrograph of polypropylene after 6 min of reaction. $TiCl_3$ particles with dimensions of 100 to 1000 Å are distributed throughout the polymer mass [45].

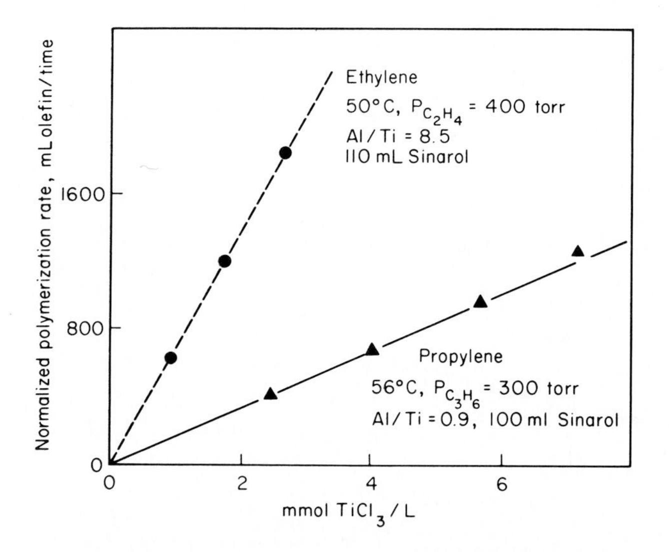

Figure 2-28 First-order dependence of polymerization rate on catalyst loading [40]. (Reprinted with permission from *Advances in Catalysis*. Copyright by Academic Press.)

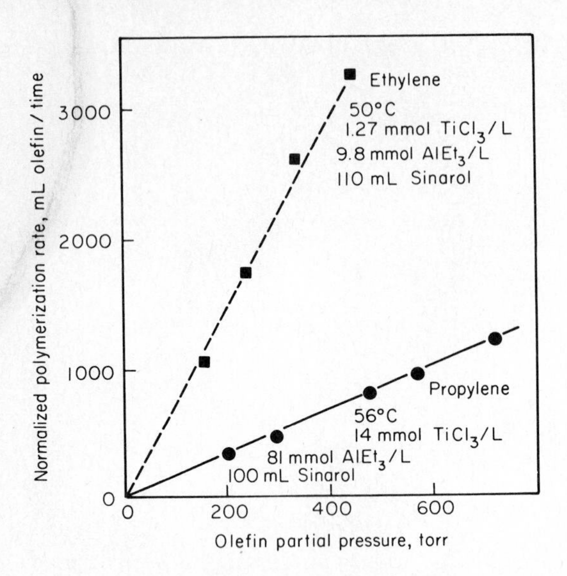

Figure 2-29 Dependence of polymerization rate on olefin partial pressure [40]. (Reprinted with permission from *Advances in Catalysis*. Copyright by Academic Press.)

data indicative of the deactivation. In a stirred reactor with a well-designed gas-liquid contacting device, the liquid-phase mass-transfer influence is probably avoided even at the higher temperatures of industrial reaction. The calculation of mass-transfer influence is illustrated in Example 2-1, page 170.

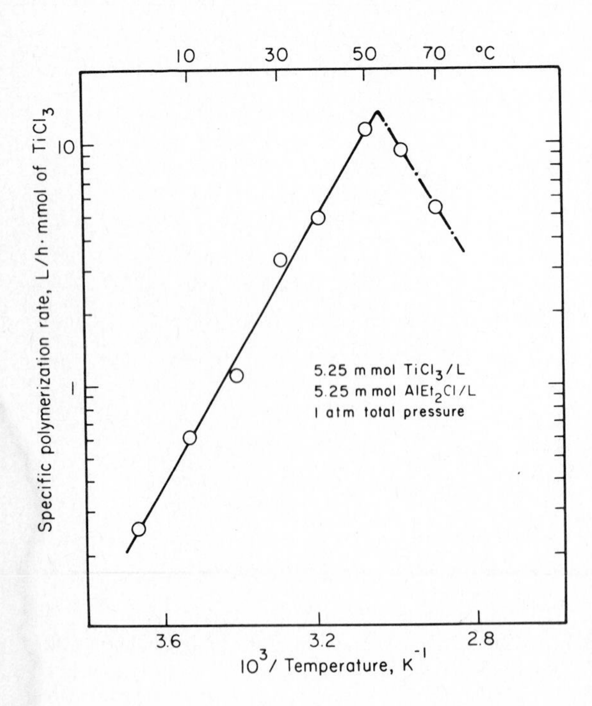

Figure 2-30 Temperature dependence of ethylene polymerization rate. At temperatures exceeding 50°C, the rate data were influenced by gas-liquid mass transfer or by catalyst deactivation [41]. (Reprinted with permission of Hüthig and Wepf Verlag.)

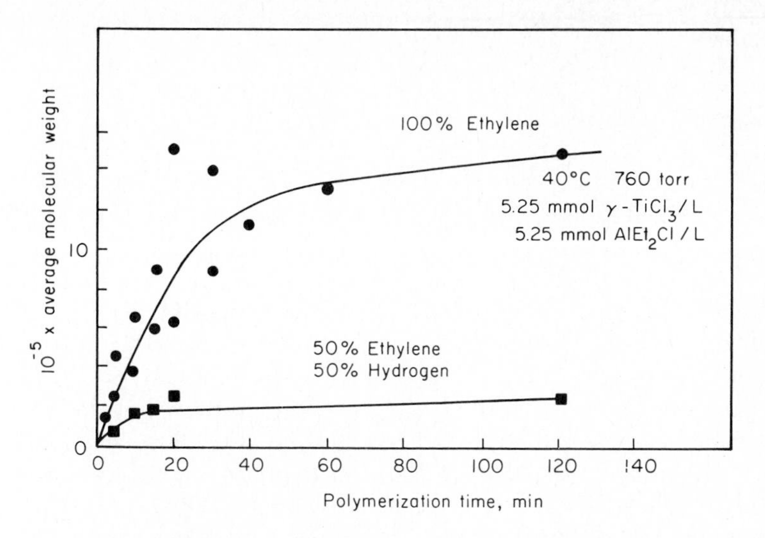

Figure 2-31 Effects of reaction time and H_2 partial pressure on average molecular weight of polyethylene [40]. (Reprinted with permission from *Advances in Catalysis*. Copyright by Academic Press.)

Some average molecular weights for polyethylene are shown in Fig. 2-31. Average molecular weights of 10^6 are achieved in the order of 20 min. The data show the effect of added hydrogen in reducing the average molecular weight of polyethylene product (Fig. 2-31). Hydrogen competes with monomer for catalytic sites and reacts with attached polymer chains, as will be discussed later. Other chain-transfer agents such as metal alkyls, hydrogen halides, or hydrogenated hydrocarbons may also be used [47].

Product Distribution

Atactic polymers, which may be formed in yields of 5 to 10 percent, are the only significant side products in Ziegler-Natta processes. The selectivity for production of isotactic polypropylene is sensitive to catalyst composition, and this composition in practice represents a trade-off between catalyst activity, selectivity, and stability. Since atactic polypropylene (unlike isotactic polypropylene) is soluble in hydrocarbons, it is easily separated from the desired product. A disadvantage of high yields of atactic polymer is that its accumulation in solution in the reactor leads to difficulty in stirring and reduced rates of heat and mass transport.

The distribution of polymer molecular weights is sometimes sharp when reaction is catalyzed by soluble transition-metal complexes. The distribution would be expected to be sharp if the solid-catalyzed reaction were unaffected by intraparticle diffusion and took place on uniformly active catalytic sites. The broad distributions of molecular weight generally observed provide evidence that the catalytic sites are nonuniform in activity and/or accessibility.

Reaction Mechanism

The structures of isotactic and syndiotactic polymers themselves provide strong evidence of the Ziegler-Natta reaction mechanism, for they are unmistakable evidence of the occurrence of stereospecific reactions. The polymers have asymmetric C atoms, and if certain monomers are used, the polymers are optically active. Catalysts for polyolefin formation are not only Ti compounds or their combinations with Al alkyls; other transition-metal ions and metal alkyls and different combinations show similar activities. Formation of crystalline polymers and in particular stereoregular polymers is more restricted: the stereospecificity of a catalyst is associated with details of its crystal structure. For example, α-$TiCl_3$ and γ-$TiCl_3$ produce polypropylene with a high degree of stereoregularity, but β-$TiCl_3$, also an active polymerization catalyst, shows less tendency to catalyze the formation of isotactic polymers, although the polyethylene formed in its presence is highly crystalline. The tendency of a catalyst to produce highly crystalline polymers (straight chains) is evidently not the same as the tendency to produce isotactic polymers. A practical example of this distinction is provided by a catalyst of an entirely different composition, the Phillips Cr_2O_3/SiO_2–Al_2O_3 catalyst. Ethylene is polymerized to high-quality crystalline polyethylene, but propylene forms a relatively low-molecular-weight polymer which is a tacky semisolid.

The catalyst systems most commonly used consist of a slurry of $TiCl_3$ powder in a solution of, for example, $Al(C_2H_5)_2Cl$, $Al(C_2H_5)_3$, or the dimer $Al_2(C_2H_5)_6$ in *n*-hexane or a similar solvent or in the liquid monomer. More recently developed industrial catalysts are supported on porous solids prepared, for example, from Mg(OH)Cl [48]. Compared with the unsupported catalysts, the new catalysts have increased activity (by a factor of perhaps 100) and give much higher yields of polymer, up to 250 kg of polymer per gram of catalyst. It is often unnecessary to remove the small amount of catalyst from the polymer product.

The interaction of solid $TiCl_3$ and the aluminum alkyl leads to an exchange of ligands:

$$Al(C_2H_5)_3 + TiCl_3 \longrightarrow Al(C_2H_5)_2Cl + TiCl_2(C_2H_5) \qquad (38)$$

This exchange is not a stoichiometric process, and only a fraction of the Cl, that present on the surface of the solid, is exchanged for alkyl. The demonstration of this assertion was given by Natta and Pasquon [50],† who studied the reaction of radioactive $Al(^{14}C_2H_5)_3$. When this compound in solution was contacted with $TiCl_3$, the solid became radioactive, retaining the ^{14}C even after repeated washings with solvent. The number of C_2H_5 groups found to be present on the $TiCl_3$ was orders of magnitude less than the number corresponding to stoichiometric exchange with the bulk solid. Corresponding results were obtained when the solid was prepared from radioactive Cl.

† The collected papers of Natta and his coworkers are available in English [49]; a summary of the work is referred to in the following paragraphs [50].

Natta and Pasquon used the radioactive $TiCl_3$ containing $^{14}C_2H_5$ to polymerize propylene. They found that every polymer molecule contained one $^{14}C_2H_5$ group. This observation leads to the following important conclusions:

1. Since the $^{14}C_2H_5$ is present on the surface, the polymerization site is inferred to be a surface site containing one C_2H_5 group.
2. Once the C_2H_5 group becomes incorporated in the polymer, it can no longer be directly bonded to the surface, although the polymer still is. Therefore, the monomer molecules must be inserted between the surface site and the C_2H_5 group. Consequently the polymerization reaction is an insertion reaction.

Another essential detail of the mechanism is the chain-termination reaction, i.e., the reaction which stops the growth of the polymer. In homogeneous free-radical polymerization, the termination reaction usually involves two free-radical chains, but Natta and Pasquon showed that the chain terminations in stereospecific polymerization are of different nature. One process involves a chain-transfer reaction:

$$Al_2(C_2H_5)_6 \longrightarrow 2Al(C_2H_5)_3 \tag{39}$$

$$Al(C_2H_5)_3 + \underline{*P_n} \longrightarrow Al(C_2H_5)_2P_n + \underline{*C_2H_5} \tag{40}$$

where $\underline{*P_n}$ is a polymer chain bonded to the surface and $\underline{*C_2H_5}$ is an alkyl group bonded to the surface. This alkyl exchange reaction is similar to that giving rise to the Cl^-–C_2H_5 exchange. A consequence of this chain-termination process (which explains the industrial use of metal alkyls for molecular-weight control) is that an increase in the metal alkyl concentration leads to a decrease in number average degree of polymerization (the average number of monomer units per product polymer molecule). The evidence that the chain-transfer process involves a dissociation of the dimeric Al alkyl is given by the experimental result that the number average degree of polymerization is proportional to the square root of the concentration of the Al alkyl in the solution [50].

At low concentrations of Al alkyl, another chain-termination reaction predominates:

$$\underline{*P_n} + C_3H_6 \longrightarrow \underline{*C_3H_7} + C_nH_{2n} \tag{41}$$

A similar sequence involves H_2, which explains its use in the industrial process:

$$\underline{*P_n} + H_2 \longrightarrow \underline{*H} + C_nH_{2n+2} \tag{42}$$

$$\underline{*H} + C_3H_6 \longrightarrow \underline{*C_3H_7} \tag{43}$$

The first and third type of chain transfer can ultimately give rise to saturated polymers, because in the further processing to give products (which is needed if yields are low or products must be very pure), the aluminum alkyl can be eliminated from the surface by treatment with alcohol or water:

$$>Al{-}P_n + H_2O \longrightarrow AlOH + C_nH_{2n+2} \tag{44}$$

The second chain-termination reaction produces polymers containing a double bond, as has been confirmed by infrared studies showing the presence of terminal double bonds. Moreover, when high aluminum alkyl concentrations are applied, i.e., when the chain termination involves reactions (39) and (40), it is evident that when radioactive $^{14}C_2H_5$ groups are contained in the metal alkyl, the product should contain $Al—C_nH_{2n+1}—^{14}C_2H_5$. The expected number of Al atoms per radioactive $^{14}C_2H_5$ group was found experimentally [50].

The foregoing results formed the basis of Cossee's interpretation of the stereospecific polymerization as a cis-insertion reaction taking place on $TiCl_3$ surface sites formed from $Al(C_2H_5)_3$ [51]. Cossee started from the knowledge that Ti^{3+} is octahedrally surrounded by Cl^- ions in solid $TiCl_3$. Since the catalytic site is on the surface, it was assumed to be coordinatively unsaturated. Cossee represented it as five-coordinate:

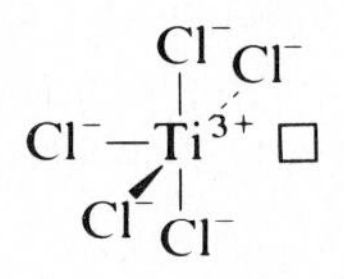

Ligand exchange with $Al(C_2H_5)_3$ can take place via a bridged structure

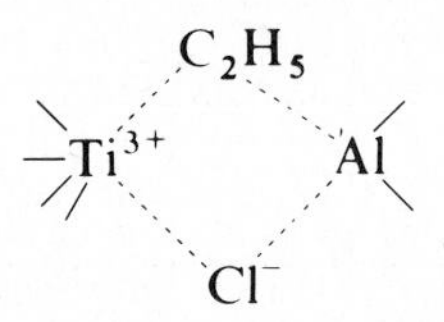

which breaks up to give the coordinatively unsaturated surface site

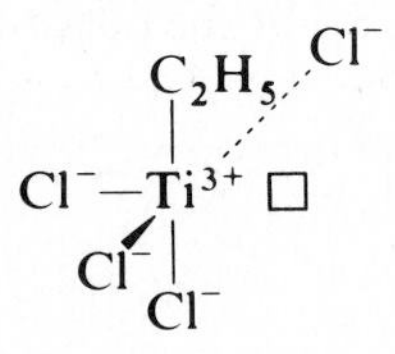

Cossee assumed that this surface site was catalytically active,† capable of adding a C_2H_4 ligand in the manner discussed previously:

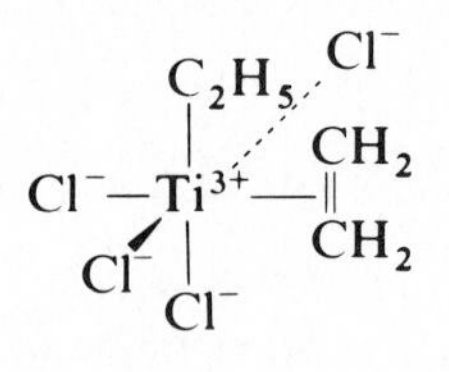

† There is considerable but inconclusive evidence, reviewed by Boor [42], that the catalytic site includes Al. The nature of the mechanism need not be revised significantly to accommodate the different catalytic site.

Cis insertion now leads to the formation of a σ-bonded alkyl incorporating the new monomer unit:

$$\underset{\text{(C}_2\text{H}_5\text{)}}{\text{—Ti}}\cdots\begin{matrix}\text{CH}_2\\ \|\\ \text{CH}_2\end{matrix} \longrightarrow \square\text{—Ti}^{3+}\text{—CH}_2\text{CH}_2\text{CH}_2\text{CH}_3 \qquad (45)$$

Again, the site contains an alkyl group adjacent to an empty site. The reaction can therefore be repeated ad infinitum to form a longer and longer polymer chain attached to the surface site.

The influence of H_2 in molecular-weight control points to the following reaction, which accounts for chain termination:

$$\square\text{—Ti—C}_n\text{H}_{2n+1} + \text{H}_2 \longrightarrow \square\text{—Ti—H} + \text{C}_n\text{H}_{2n+2} \qquad (46)$$

The Cossee model therefore accounts for most of the mechanistic details of Ziegler-Natta polymerization. Several difficulties remain, however:

1. There are two possible positions for the methyl group in propylene bonded to the surface site:

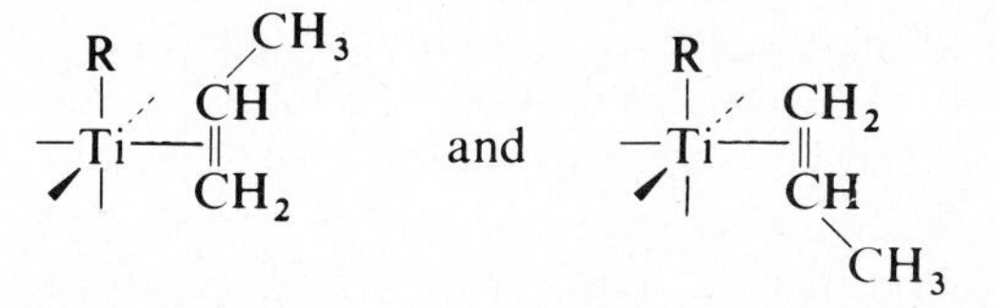

   If these two structures were equivalent, the arrangement of the methyl groups in the resulting polypropylene would be random; i.e., the polymer would be atactic not isotactic.
2. If only one of these two structures could form, the polypropylene would be stereoregular but syndiotactic instead of isotactic, since the two positions from which insertion can occur

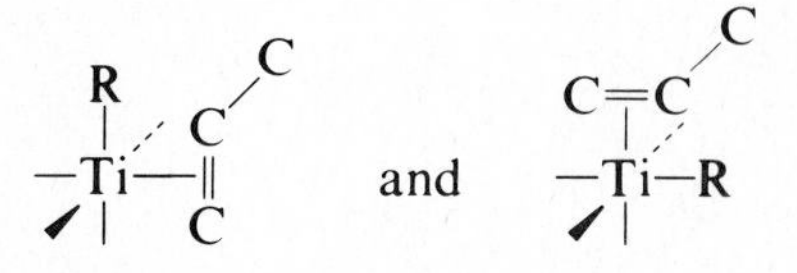

   lead to an in-out sequence.
3. The theory does not account for the differences in selectivity between the various crystal forms of $TiCl_3$, since they all have Ti^{3+} in an octahedral surrounding.

These difficulties have been resolved by Arlman and Cossee [52], who developed models of surface sites accounting for the subtle differences in structures of the various forms of $TiCl_3$. The differences are reflected in the surface-site symmetry.

All the solid structures of $TiCl_3$ are based on close packings of Cl^- ions. They differ in the arrangements of the cations and occasionally in the type of close packing, which may be cubic or hexagonal. A close packing of Cl^- ions has one octahedral interstice per Cl^- ion; since there can be only one Ti^{3+} ion for every three Cl^- ions, only one of the three octahedral interstices is occupied.

The structure of α-$TiCl_3$ is a layer structure, as shown in Fig. 2-32. Each layer of Cl^- ions is close-packed. The octahedral interstices are occupied in the following sequence: filled-empty-filled-empty, etc. Therefore, two layers of Cl^- ions are held together by the positive ions, but there are no cations in the interstices between the next layers. In each filled layer, one of three octahedral sites remains empty. Figure 2-32 shows one sandwich consisting of two anion layers with two-thirds of the octahedral sites filled by Ti^{3+} ions. The α-$TiCl_3$ crystal consists of a stacking of such layers, preserving hexagonal close packing.

In the related structure of γ-$TiCl_3$, the stacking instead leads to a cubic close packing. The difference between these two packings is discussed in detail in Chap. 3 for the example of the aluminas. The octahedral c_3 axes are oriented at a fixed angle with respect to the layer, as shown in Fig. 2-33.

To preserve stoichiometry, there must be empty anion sites (anion vacancies) at the crystal surfaces. Arlman and Cossee [52] used estimates of the Coulomb energies required to remove a Cl ion from crystallographically different sites to infer that the anion vacancies are exclusively situated at the edges of the layers. Figure 2-34 shows a possible structure of an α-$TiCl_3$ crystal viewed at the edge. Each Ti^{3+} ion in the surface layer is only five-coordinate, corresponding to Cossee's assumption; three Cl^- ions are buried in the crystal, and the other two are attached either to one or two Ti^{3+} ions.

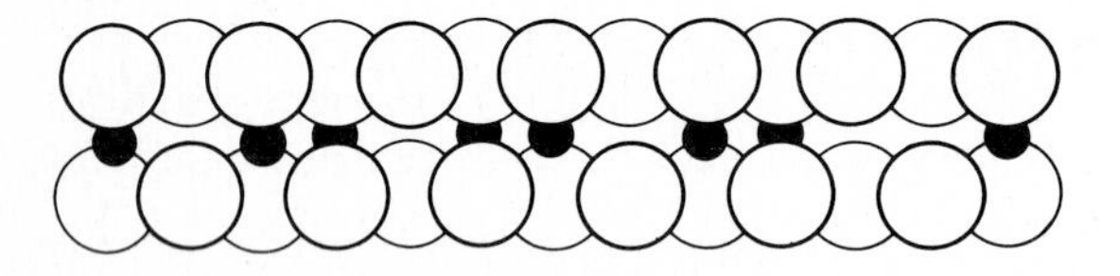

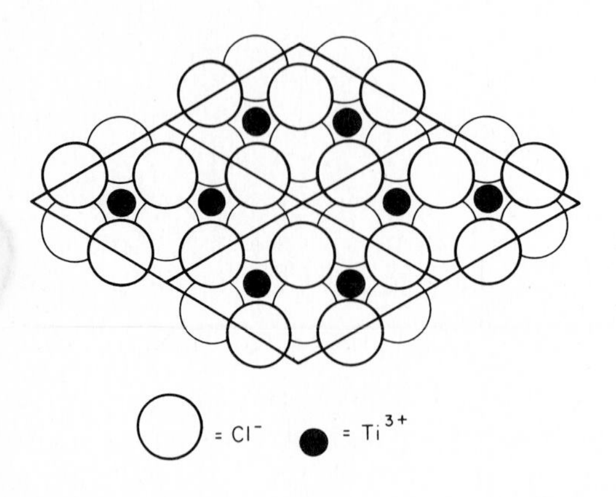

Figure 2-32 Arrangement of ions in the layer structure of α-$TiCl_3$.

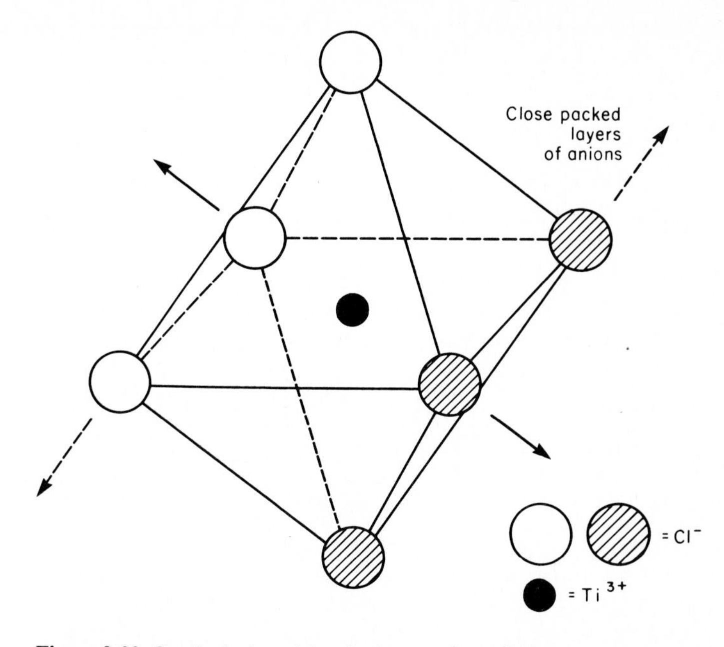

Figure 2-33 Octahedral position in layers; the solid arrows represent the direction of the c_3 axis perpendicular to the layers. The octahedra are oriented with the c_4 axis at a fixed angle with respect to this direction.

Figure 2-34 One of the four possible structures of the $(10\bar{1}0)$ face of α-$TiCl_3$. The black balls represent Ti^{3+}, the white balls Cl^-. (Reprinted from Ref. [9] by courtesy of Marcel Dekker, Inc.)

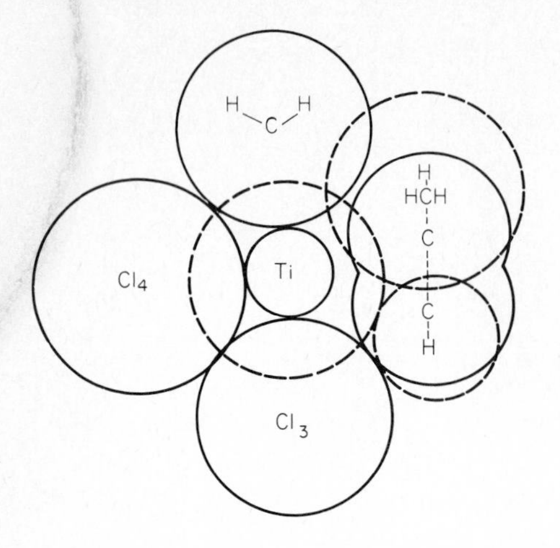

Figure 2-35 Diagram of the complex of C_3H_6 with the active center. Radii r have been drawn to scale: $r_{CH_2} = 2.0$ Å; $r_{Cl^-} = 1.8$ Å; $r_{Ti^{3+}} = 0.8$ Å; r_C (in C_2H_4) = 1.7Å. (Reprinted from ref. [9] by courtesy of Marcel Dekker, Inc.)

If sites on such a surface are to be the catalytic sites, the geometry must allow for an alkyl group (the growing polymer chain) to replace one of the Cl^- ions. Figure 2-35 demonstrates that the geometric requirement is met and further that the attached alkyl group and monomer ligand can occupy positions close enough to each other for the insertion reaction to appear feasible. The symmetry of a catalytic site at the edge of the crystal layer is shown in Fig. 2-36. The two octahedral coordination sites (I and II) are supposed to be either empty or filled with an alkyl group. The Cossee-Arlman model states that because of a difference in steric restrictions, one of the two positions has a stronger affinity for the alkyl group than the other. If the group is transferred from II to I, e.g., as a consequence of the insertion reaction, it tends to return rapidly to II before undergoing another insertion. A more complete picture of the differences in the two positions is given in Fig. 2-37.

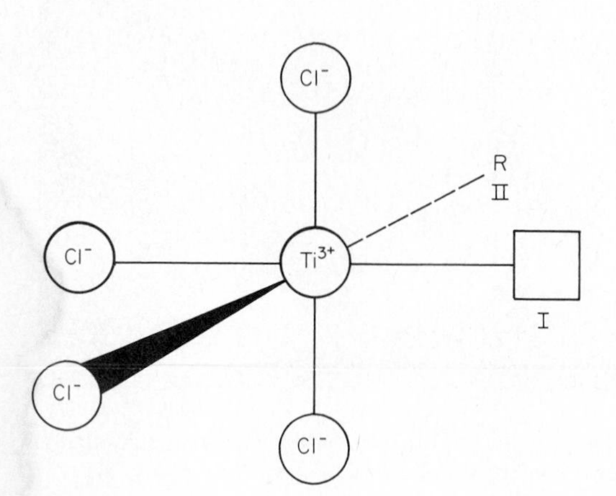

Figure 2-36 Cossee-Arlman site at the edge of a layer of α-$TiCl_3$. R is an alkyl group (the growing polymer chain) and □ is a vacant coordination site.

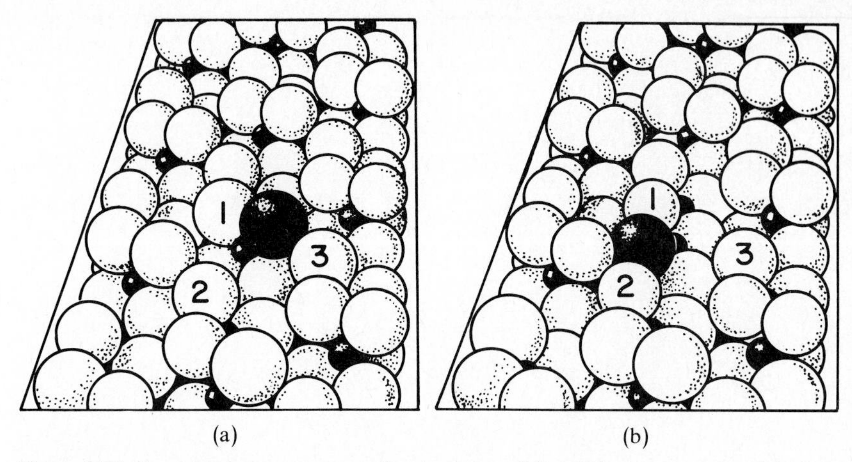

Figure 2-37 Cossee-Arlman structure of active center (a) and the same center with the vacancy and alkyl group interchanged (b). The small black spheres represent Ti^{3+} ions, the large black spheres alkyl groups, and the white spheres Cl^- ions. Structure (b) is converted to structure (a) by migration of the alkyl group, the driving force being the greater steric crowding of the alkyl group in (b). Complexing of propylene with structure (a) can take place in only one sterically preferred configuration, with the CH_2 end downward and the CH_3 group on the side of the Cl^- ion marked 1. (Reproduced from ref. [9] by courtesy of Marcel Dekker, Inc.)

A further assumption in the development of the Cossee-Arlman model is that now there is only one way the propylene can be bonded to site I, which is the only position allowed since the alkyl group rapidly shifts to site II. The assumption is that the CH_2 end of propylene points downward and the CH_3 group is on the side of the Cl^- ion marked 1 (Fig. 2-37).

With the incorporation of this assumption, the model defines a unique template for the "foot" of the growing polymer chain, and the first two difficulties mentioned on page 161 are resolved. We emphasize that this intricate template model finds its origin in the incomplete occupation of the octahedral interstices and in the oblique position of the fourfold axis of the octahedron with respect to the layer direction and therefore also the layer edge (Fig. 2-33).

The mechanism of isotactic polymer formation can therefore be summarized as follows:

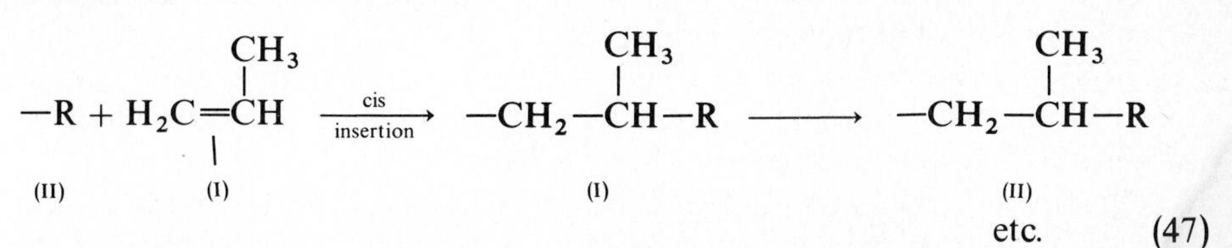

(47)

The β-$TiCl_3$ structure is different from the α-$TiCl_3$ structure since there is a different distribution of cations in the octahedral interstices. The Ti^{3+} ions form a linear arrangement and are surrounded by parallel rows with empty sites. The structure can be visualized as rows of $Ti^{3+}(Cl^-)_3$ packed as fibers (Fig. 2-38). To preserve stoichiometry, the ends of the rows, i.e., the surface sites, must be Ti^{3+}

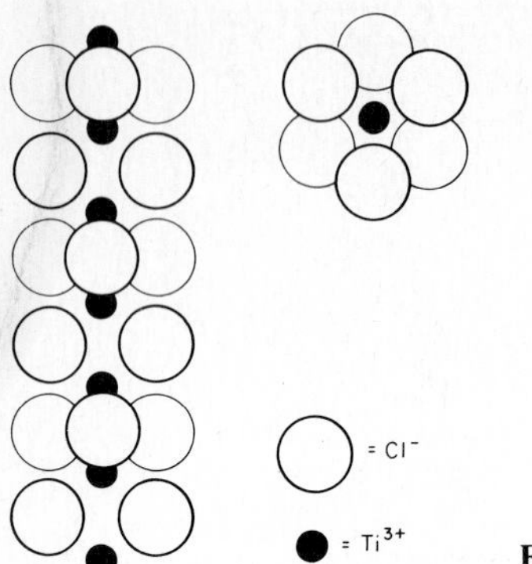

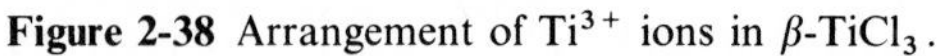

Figure 2-38 Arrangement of Ti^{3+} ions in β-$TiCl_3$.

ions with either one or two vacant coordination sites. The sites are strictly equivalent, so that there is no driving force for a change in the alkyl position. The site symmetries are complicated (Fig. 2-39), and although the possibility of stereospecific polymerization does not seem to be necessarily excluded, the differences in site symmetry are great enough to suggest that polymerization would occur randomly, which is in agreement with experimental results.

A convincing demonstration of the correctness of the hypothesis of polymer growth from sites on the edges of the layers of α-$TiCl_3$ was given by Rodriguez and Gabant [53]. The electron micrograph of Fig. 2-40 shows polymer chains as dots localized on a spiral which is identified as a crystal growth spiral of α-$TiCl_3$. There is accordingly hardly any doubt that the Cossee-Arlman model is essentially correct, although it may be incorrect in some details, e.g., whether the catalytic site incorporates Al.

Process Design

Processes for carrying out the stereospecific polymerization of α-olefins have been reviewed by Compostella [54] and Valvassori et al. [47]:

1. *Suspension polymerization*, the most commonly applied process, is to be discussed in detail in the next paragraphs.
2. *Bulk polymerization* takes place homogeneously in the liquid monomer, from which insoluble polymer is precipitated. The process may require high pressures to maintain monomer in the liquid state.
3. *Solution polymerization* is like bulk polymerization except that a solvent is present. The process is not used much since atactic polymer is formed in excessive amounts.
4. *Fluidized-bed polymerization* takes place as gaseous olefin flows upward through a bed of fluidized catalyst particles. The catalyst may be removed, separated from polymer particles, and recycled to the reactor, or it may remain

Figure 2-39 The $(3\bar{1}\bar{1}2)$ face of β-$TiCl_3$. The Cl^- ions attached to only one Ti^{3+} ion are dark. (Reprinted from ref. [9] by courtesy of Marcel Dekker, Inc.)

incorporated in the polymer. The process is continuous and affords good temperature control since the bed contents are well mixed. Reactants are efficiently transported to catalyst surfaces since there is no liquid phase to offer resistance; the solvent-recovery costs of other processes are avoided. The fluidized-bed process is applied industrially in polyethylene manufacture with a chromium-containing catalyst rather than a Ziegler catalyst [55].

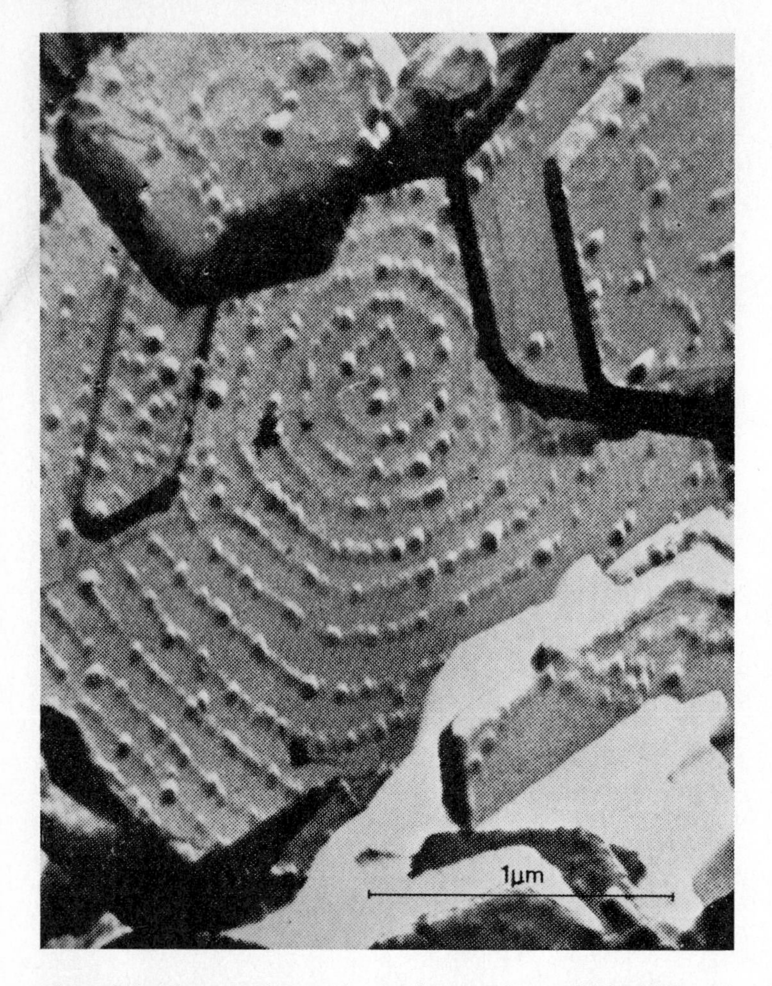

Figure 2-40 Electron micrograph of a hexagonal crystal of α-$TiCl_3$. The dots are believed to represent polypropylene chains growing on sites located along a crystal growth spiral [53].

Ziegler catalysts are used in industrial processes for production of polyethylene and polypropylene. They have recently been applied in production of poly(1-butene) [56] and poly(4-methyl-1-pentene) [54, 57].

A schematic representation of a suspension-polymerization process for polypropylene manufacture is given in Fig. 2-41. The process can be operated continuously with one or a series of stirred-tank reactors, or it can be operated batchwise with single reactors. Ethylene is converted at temperatures of 50 to 80°C and pressures of 1 to 10 atm. Propylene is converted at temperatures of 75 to 125°C and pressures of about 10 atm. The stirred reactor contains liquid monomer and sometimes solvent, e.g., hexane, with suspended polymer and catalyst,

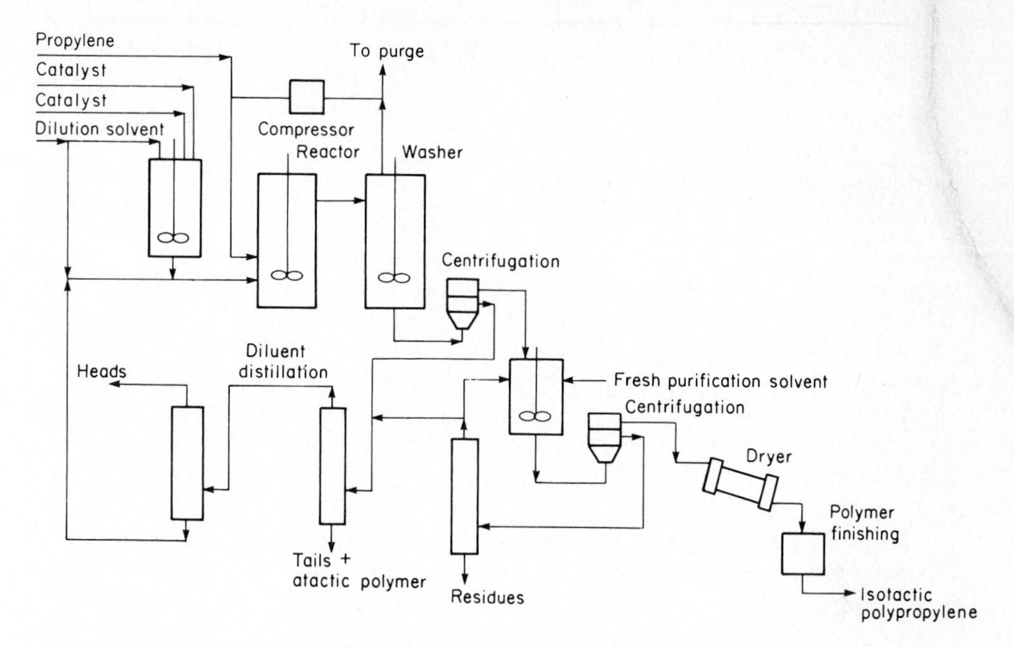

Figure 2-41 Schematic diagram of Ziegler-Natta polypropylene process. (Reprinted with permission from ref. [47]. Copyright by Ernest Benn Limited.)

e.g., α-$TiCl_3$ combined with an aluminum alkyl such as $Al(C_2H_5)_2Cl$. Purified monomer may be added as liquid or gas to the reactor. Hydrogen is usually present to exert control over the molecular weight of the polymer. Oxygen-containing compounds such as water and carbon dioxide are carefully excluded because they poison the catalyst by combining with it in competition with reactants.

The polymerization reactions are strongly exothermic

$$C_2H_4 \longrightarrow \frac{1}{n}(C_2H_4)_n \qquad \Delta H^\circ = -25.9 \text{ kcal/mol [58]} \tag{48}$$

$$C_3H_6 \longrightarrow \frac{1}{n}(C_3H_6)_n \qquad \Delta H^\circ = -24.9 \text{ kcal/mol [47]} \tag{49}$$

and temperature is controlled by the rate of heat removal by cooling water in a jacket surrounding the reactor; supplementary heat transfer may result from evaporation of monomer or solvent, which is condensed and recycled to the reactor. The heat-transfer requirements are difficult to predict, since viscosities of reactants are variable and surfaces become fouled with polymer.

PROCESS ENGINEERING AND MULTIPHASE REACTORS

MASS-TRANSFER INFLUENCE

The subject of mass-transfer influence on reaction rates in gas-liquid reactors is considered by example in the following paragraphs. The purpose of the examples is to indicate briefly the methods of reactor design and to direct the reader toward the well-developed literature of this subject. The quantitative estimates are intended to be representative of industrial practice, but they should be accepted with caution because of the paucity of data on which they are based.

The subject of mass-transfer influence is conveniently introduced with the example of Ziegler-Natta polymerization of ethylene.

Example 2-1: Design of a suspension polymerization reactor for production of polyethylene Provide a preliminary design of a reactor to produce batches of 1000 kg of polyethylene.

SOLUTION A preliminary design is based on kinetics data given in this chapter and on an estimate of the mass-transfer influence. For simplicity, the heat-transfer requirements are ignored. Initially, temperature and pressure are chosen to be values representative of industrial practice, 80°C and 8 atm. From the data of Fig. 2-26, which suggest that the suspension should not exceed about 20 weight percent solid, and from the assumption of a rather low yield of 1 kg of polymer per gram of catalyst, the amount of catalyst to be charged to the reactor is calculated:

$$\text{Catalyst charge} = \frac{10^6 \text{ g polymer}}{10^3 \text{ g polymer/g catalyst}}$$

$$= 1000 \text{ g} = 6400 \text{ mmol } TiCl_3$$

$$\text{Volume of product} \approx \frac{10^6 \text{ g}}{10^6 \text{ g/m}^3} = 1 \text{ m}^3$$

If the solvent in the suspension-polymerization reactor is chosen to be cyclohexane (sp gr ≈ 0.8), then (with the volume of solution defined as V_s)

$$\frac{20 \text{ g product}}{80 \text{ g solution}} \approx \frac{10^6 \text{ g product}}{(8 \times 10^5 \text{ g soln./m}^3 \text{ soln.})(V_s \text{ m}^3 \text{ soln.})}$$

or $V_s \approx 5 \text{ m}^3$

The catalyst loading is therefore

$$\frac{6400 \text{ mmol}}{5 \times 10^3 \text{ L}} = 1.3 \text{ mmol/L}$$

A reactor volume of 8 m^3 would amply allow for the solution, product, and gas holdup. The reactor is therefore chosen to be a stirred tank with an inside radius of 1 m and a height of 2.5 m.

An estimate can now be made of the time for reaction based on the kinetics data. It is first assumed that mass-transfer resistance offered by the liquid phase is negligible (the assumption is to be tested later). The rate, which may remain nearly constant during operation, is estimated by extrapolation of the Arrhenius plot of Fig. 2-30. At 80°C and an ethylene partial pressure of 1 atm, the rate is

$$r \approx 3.8 \times 10^5 \text{ L } C_2H_4/\text{h} \approx 17 \text{ mol } C_2H_4/\text{h}$$

We assume that the rate is proportional to ethylene partial pressure and, for a modern supported catalyst is 100 times that of an unsupported catalyst.

$$r \approx 6 \times 17 \times 10^2 = 1.0 \times 10^4 \text{ mol/h}$$

(assuming the partial pressure of ethylene is 6 atm). Since the total conversion is $(10^6 \text{ g})/(28 \text{ g/mol}) = 3.6 \times 10^4$ mol of ethylene, the reaction time for one batch is approximately

$$\frac{3.6 \times 10^4 \text{ mol}}{1.0 \times 10^4 \text{ mol/h}} = 3.6 \text{ h}$$

The assumption of negligible liquid-phase mass-transfer resistance is now to be tested. Calculation of the mass-transfer resistance requires characterization of the hydrodynamics of the reaction mixture. It is assumed that an ethylene-hydrogen mixture is sparged into the tank at a mole ratio of 3 : 1, with the ethylene flow rate equal to 3 times the rate at which ethylene is consumed by reaction:

$$q_{C_2H_4} = 3.0 \times 10^4 \text{ mol/h} \qquad q_{H_2} = 1.0 \times 10^4 \text{ mol/h}$$

$$q_{tot} = 4.0 \times 10^4 \text{ mol/h} = 146 \text{ m}^3/\text{h}$$

(assuming the gas is ideal at 8 atm and 80°C).

For a rough estimate it is assumed that the interfacial area a_g is 1 cm^2 per cubic centimeter of slurry, which is typical of gas well dispersed in stirred liquid [59]. Since the gas-phase resistance to mass transfer can be neglected in all cases except for very fast reaction, the remaining problem is to estimate the liquid-phase mass-transfer coefficient k_L. The value of k_L may be affected by chemical reaction in the liquid, and according to the procedure of Astarita [60, 61], k_L is estimated for the appropriate reaction regime, which is defined by the following time parameters.

The time required for reaction t_R is $C^*_{C_2H_4}/r$, where $C^*_{C_2H_4}$ is the concentration of ethylene in solution in equilibrium with the gas phase, approximately equal to 5×10^2 mol/m^3 [62];

$$t_R \approx \frac{5 \times 10^2 \text{ mol/m}^3}{(1.0 \times 10^4 \text{ mol/h})/(5 \text{ m}^3)} = 0.25 \text{ h} = 900 \text{ s}$$

The time characteristic of diffusion t_D is defined from the penetration theory as $D/(k_L^0)^2$, where k_L^0 is the physical absorption coefficient, i.e., the liquid-phase mass-transfer coefficient in the absence of chemical reaction. The value of k_L^0 can be estimated from the following standard correlation [61, 63]:

$$k_L^0 \frac{d_B}{D} = 2 + a\left[\mathrm{Re}^{0.484}\mathrm{Sc}^{0.339}\left(\frac{d_B g^{1/3}}{D^{2/3}}\right)^{0.072}\right]^b \tag{50}$$

For swarms of bubbles, $a = 0.0187$ and $b = 1.61$ [62]. Approximate values of D and d_B are 10^{-5} cm²/s and 0.2 cm, respectively [64]. The Reynolds number is calculated approximately as

$$\mathrm{Re} = \frac{d_B u_s \rho}{\mu} \approx \frac{(0.2)(40)(0.8)}{4 \times 10^{-3}} = 1.6 \times 10^3$$

where u_s, the slip velocity of a gas bubble, has been calculated from the superficial velocity and gas holdup:

$$v_s \approx \frac{146\ \mathrm{m^3/h}}{\pi\ \mathrm{m^2}} = 46\ \mathrm{m/h} = 1.3\ \mathrm{cm/s}$$

$$H \approx \frac{a_g d_b}{6} = \frac{(1)(0.2)}{6} = 0.03\ \mathrm{cm^3\ gas/cm^3\ slurry}$$

$$u_s \approx \frac{1.3\ \mathrm{cm/s}}{0.03\ \mathrm{cm^3/cm^3}} = 40\ \mathrm{cm/s}$$

The Schmidt number Sc is

$$\mathrm{Sc} = \frac{\mu}{\rho D} \approx \frac{4 \times 10^{-3}}{0.8 \times 10^{-5}} = 500$$

The remaining dimensionless group is

$$\frac{d_B g^{1/3}}{D^{2/3}} \approx \frac{0.2 \times 980^{1/3}}{(10^{-5})^{2/3}} = 4.3 \times 10^3$$

Therefore

$$k_L^0 \frac{d_B}{D} \approx 2 + 0.0187(535^{1.61}) = 463$$

$$k_L^0 \approx \frac{10^{-5}}{0.2}(463) = 2.3 \times 10^{-2}\ \mathrm{cm/s}$$

The time characteristic of diffusion is

$$t_D \approx \frac{10^{-5}\ \mathrm{cm^2/s}}{(2.3 \times 10^{-2})^2\ \mathrm{cm^2/s^2}} = 1.9 \times 10^{-2}\ \mathrm{s}$$

Since $t_D \ll t_R$, the *slow reaction regime* prevails [60, 61]. If the reaction is sufficiently slow within this regime, the *kinetic subregime* prevails and the mass-transfer influence is negligible. The further criterion for this subregime is [60, 61]

$$\frac{1}{a_g} r \ll k_L^0 C^*_{C_2H_4}$$

The left-hand side of the inequality is approximately

$$\frac{1}{1\ \text{cm}^{-1}} \left[\frac{1.0 \times 10^4\ \text{mol/h}}{5\ \text{m}^3} \frac{1}{(3.6 \times 10^3\ \text{s/h})(10^6\ \text{cm}^3/\text{m}^3)} \right]$$

$$= 5 \times 10^{-7}\ \text{mol/cm}^2 \cdot \text{s}$$

The right-hand side of the inequality is approximately

$$(2.3 \times 10^{-2}\ \text{cm/s})[(5 \times 10^2\ \text{mol/m}^3)(10^{-6}\ \text{m}^3/\text{cm}^3)]$$

$$= 1.1 \times 10^{-5}\ \text{mol/cm}^2 \cdot \text{s}$$

Therefore the inequality is satisfied, and the reaction takes place in the kinetic subregime; this conclusion confirms the earlier assumption that mass transfer has no influence on the rate.

To be complete, the design calculation should also include an estimate of the heat-transfer requirement. The heat-transfer requirement may be the primary consideration in the choice of reaction temperature.

The results of the example suggest that mass-transfer limitations (aside from intraparticle diffusion) do not affect rates of Ziegler-Natta polymerization in industrial practice, since with olefins of higher molecular weight than ethylene, solubilities are higher and reaction rates lower. This conclusion is to be contrasted with results from laboratory reactors, which suggest a significant liquid-phase mass-transfer resistance (Fig. 2-30).

The presence of a mass-transfer influence can be demonstrated by an increase in rate with increased stirring speed and by an inverse dependence of rate on catalyst loading, as described by the following equation, derived by Satterfield [65]:

$$\frac{C^*}{r} = \frac{d_B}{6k_L H} + \frac{\rho_p d_p}{6m}\left(\frac{1}{k_c} + \frac{1}{k}\right) \tag{51}$$

The laboratory data of Berger et al. [40] (Fig. 2-42) conform to this equation.

Example 2-2: Design of an oxo reactor Provide a preliminary design of a reactor for hydroformylation of 5×10^6 kg/year of diisobutylene.

SOLUTION The design is based on kinetics data of Martin [Eq. (32)], the criterion of minimum CO partial pressure for stability of the cobalt hydridocarbonyl catalyst (Fig. 2-17), and an estimate of mass-transfer influence.

A stirred-tank reactor is chosen, and the molar ratio of CO to H_2 in the feed is set equal to the stoichiometric value of 1. Temperature is set at 150°C and pressure at 200 atm, both representative values for which kinetics data are available [24]. The cobalt concentration is chosen to be 0.5 weight percent.

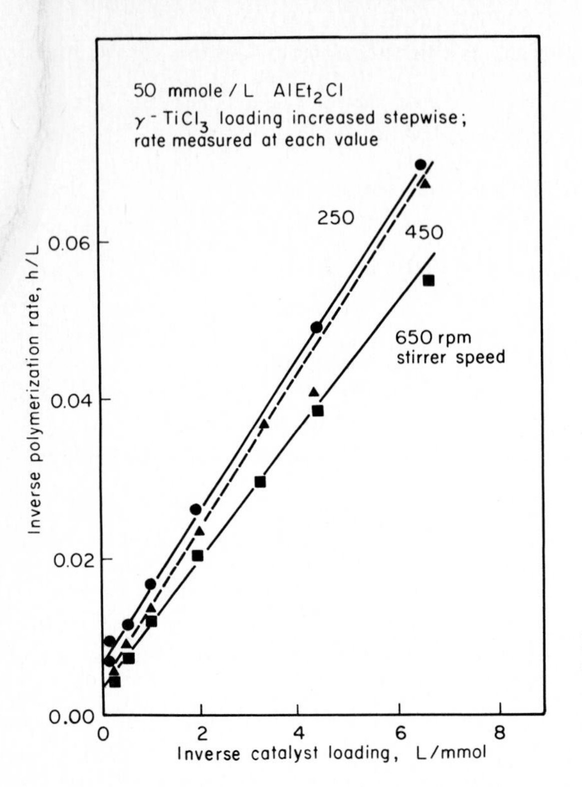

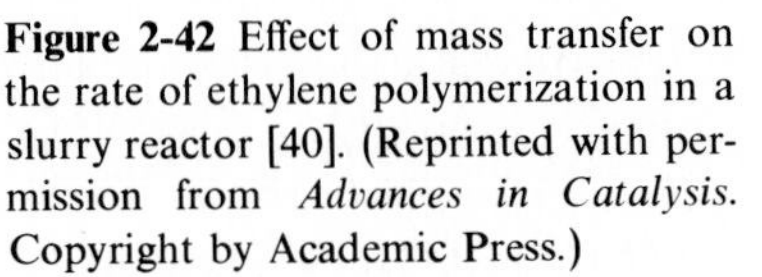
Figure 2-42 Effect of mass transfer on the rate of ethylene polymerization in a slurry reactor [40]. (Reprinted with permission from *Advances in Catalysis*. Copyright by Academic Press.)

The minimum CO partial pressure is determined from Fig. 2-17 to be 50 atm, which appears to be safely less than the design value of about 100 atm. If there is much mass-transfer resistance in the liquid phase, however, the concentration of dissolved CO may be below the critical value, leading to precipitation of cobalt. This situation could prevail only if there were very fast reaction or very high conversion of CO. The possibility is to be tested by methods used in Example 2-1.

The reaction rate equation given by Martin simplifies to

$$r = 0.19 C_{\text{Co complex}} C_{C_8H_{16}}$$

where the concentrations are in moles per liter and r has units of moles per liter per second.

This equation indicates that rate is independent of CO and H_2 partial pressures, hence independent of the liquid-phase concentrations of these reactants. Rate is therefore independent of mass-transfer influence, provided that the rate equation remains valid, i.e., provided that there is sufficient CO in solution to maintain catalyst stability.

If the conversion of diisobutylene is assumed to be 90 percent, the steady-state value of the reaction rate is

$$r \approx (0.019)(0.075)(0.6) = 9 \times 10^{-4} \text{ mol/L} \cdot \text{s}$$

Since the contents of the well-stirred reactor may be assumed to be nearly perfectly mixed, the outlet liquid concentration is the same as the liquid concentration at any position in the reactor. The equation of conservation of mass of diisobutylene can then be written as follows, provided the density change of the liquid on reaction is negligible. The equation determines V_S, the volume of solution in the reactor

$$q(C_{C_8H_{16}})_{in} - q(C_{C_8H_{16}})_{out} - rV_S = 0$$

The feed rate of diisobutylene is 5×10^6 kg/year = 1.1 mol/s. If the density of liquid is estimated as 0.7 g/mL, then q is equal to 0.18 L/s:

$$(0.18)(6.2 - 0.62) = 9 \times 10^{-4} V_S$$

or

$$V_S = 1.1 \times 10^3 \text{ L} = 1.1 \text{ m}^3$$

A reactor volume of 1.3 m^3 would easily accommodate the liquid plus gas holdup. Consequently, a pressure vessel with a radius of 0.5 m and a height of 1.7 m is selected.

The average residence time of liquid in the reactor is approximately

$$\frac{1.1 \times 10^3 \text{ L}}{0.18 \text{ L/s}} \approx 6.1 \times 10^3 \text{ s} = 1.7 \text{ h}$$

Rough estimates of mass-transfer influence are now to be made to determine whether the catalyst is stable. Since the methods of estimation are like those of Example 2-1, calculations are omitted and only rough estimates are used.

Typical values of the important parameters, as exemplified by the calculations of Example 2-1, are $k_L^0 \approx 0.02$ cm/s and $t_D \approx 0.01$ s [61]. Since the reaction is zero order in CO,

$$t_R \approx \frac{\frac{1}{2}C_{C_8H_{16}}}{r} = \frac{0.31}{9 \times 10^{-4}} \qquad t_R \approx 10^2 \text{ s}$$

An estimate can now be made to determine whether the concentration of CO in the bulk liquid is close to the saturation value. The estimate is based on the following mass-balance equation for CO, where k_L has been set equal to k_L^0, since the slow reaction regime prevails:

$$k_L^0(C_{CO}^* - C_{CO})a_g V_S = rV_S$$

$$0.02(C_{CO}^* - C_{CO})(1) = 9 \times 10^{-4}$$

$$C_{CO}^* - C_{CO} = \frac{9 \times 10^{-4}}{2 \times 10^{-2}} \approx 0.05\ M$$

The value of C_{CO}^* can be determined from an estimated Henry's law constant; it is roughly a few tenths of a mole per liter. The result therefore suggests that the concentration of CO in the bulk liquid is a large fraction of the saturation value, and so the catalyst is expected to be stable.

This result indicates that the design is adequate and suggests that a pressure nearer to 100 atm might be preferable. It is clear, however, that if conversion of CO were too close to 100 percent, the interfacial area would be small and the mass-transfer limitation could be significant. Further, olefins such as propylene evidently require higher pressures than diisobutylene because they are more reactive (Table 2-4) and, with higher vapor pressures, have lower concentrations in the reactor liquid (which is predominantly product).

The design of tubular gas-liquid flow reactors (like those used in the Wacker and oxo processes) is complicated by the dependence of reactor performance not only on intrinsic reaction kinetics but also on the complex fluid mechanics, which influences the rate of transfer of reactants and products between the two phases. Design methods are available for certain flow regimes, but they are only approximate and not well tested by comparison with experimental results. Design first requires estimation of the regime of gas-liquid flow, which may involve, for example, bubbles of gas in continuous liquid, slugs of gas separating liquid volumes, strata of gas and liquid, drops of liquid carried in gas, etc. Flow-distribution patterns of each phase are then estimated, and design calculations are based on rates of interphase mass transfer and intrinsic kinetics of the liquid-phase reaction.

Russell and coworkers [66, 67] have presented the appropriate mass-balance equations, i.e., design equations, for many cases and indicated the available parameter-estimation procedures. The design calculations are often quite complex, and industrial reactors may often have been designed by scale-up procedures more primitive than those cited.

HOMOGENEOUS CATALYSIS

Among the large-scale catalytic processes based on solution reactions are those involving transition-metal-complex catalysis, acid-base catalysis, and enzyme catalysis. The contents of this chapter suggest the following summary of the practical advantages of homogeneous catalysis:

1. Compared with solid surfaces, soluble catalysts are easily characterized, e.g., by spectroscopy, and knowledge of reaction mechanisms can be extended to prediction of performance of families of catalysts and reactants.
2. The highly dispersed dissolved catalysts are effectively utilized since all the molecules are usually accessible to reactants.

The recurrent patterns in process design, however, suggest the following general disadvantages of homogeneous catalysis:

1. There is difficulty and expense involved in separating catalysts from reaction products.
2. Corrosion and catalyst losses often result from catalyst recycling.
3. Gas-phase reactants may be inefficiently contacted with catalysts because of mass-transfer resistance in the liquid phase.

MULTIPHASE REACTORS AND CATALYST DESIGN

Most catalytic processes involve catalysis by solids; a great advantage of these processes is the ease of separation of fluid-phase products from catalysts. But solid catalysts are inefficient, exposing only small fractions of their atoms on accessible surfaces, and since surfaces are generally difficult to characterize, the development of new catalysts results far less from scientific prediction than from trial-and-error experimentation. Because of these disadvantages of typical solid catalysts, there is growing interest in design methods, summarized in the following paragraphs, for isolating a catalyst in a separate phase immobilized in a reactor while at the same time eliminating the disadvantages of surface catalysis by effecting chemistry nearly the same as that of homogeneous catalysis.

Catalyst Solutions in the Pores of Solids

There are a number of examples of catalysts prepared from solutions of acids and transition-metal complexes contained within the pores of inorganic solids. Phosphoric acid on kieselguhr is used industrially as a catalyst for aromatic alkylation and olefin polymerization. Transition-metal complexes in porous solids have been reported to catalyze hydroformylation of propylene to butyraldehyde [68, 69]. The catalyst was contained in chloroform solution within the narrow pores of alumina or carbon, where vapor pressures of the solvents were considerably reduced. As much as 1 g of rhodium complex $RhCl(CO)(PPh_3)_2$ was contained in 8 g of activated carbon. Catalyst activity in a fixed-bed reactor was maintained for more than 300 h in operation, with the catalyst reported to be stable at temperatures up to 335°C.

Related examples involve catalysis by solutions on solid surfaces. SO_2 oxidation is believed to be catalyzed by a liquid film on vanadium pentoxide [70], and early oxo catalysts were metallic cobalt, the surfaces of which were carbonylated by reactant CO. (These catalysts were impractical because the volatile cobalt complex was not confined to the reactor.) Stable methanol carbonylation catalysts have been prepared by impregnating porous carbon with Rh complexes [71].

Catalysis by Matrix-bound Complexes

Some advantages of both homogeneous and heterogeneous catalysis can be realized in applications of insoluble organic polymers. Industrially applied polymeric catalysts include ion exchangers containing easily accessible sulfonic acid groups, and similar polymeric catalysts have been prepared containing transition-metal complexes [72]. Polymers containing Co and Rh complexes, e.g.,

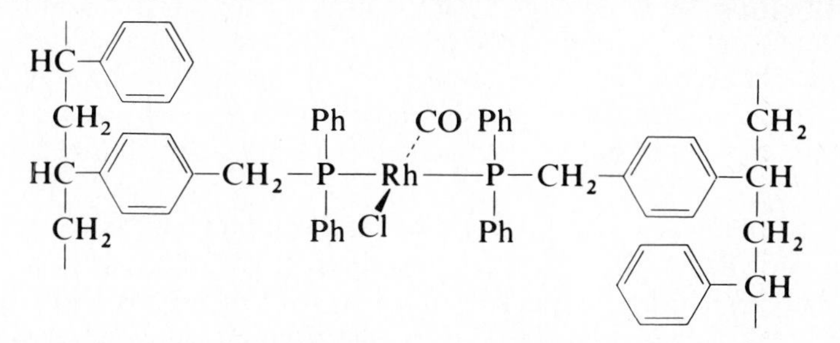

have been used for hydroformylation [73, 74], and the Rh-complex-containing polymer has been used for methanol carbonylation [75].

Haag and Whitehurst [76] reported a catalyst incorporating Pd

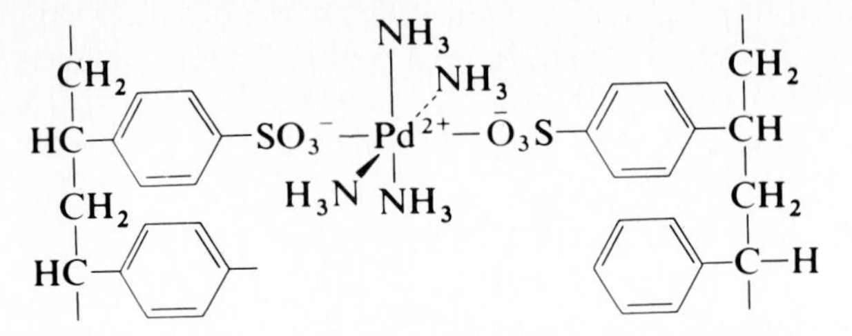

prepared from a macroporous polystyrenesulfonate resin, which was used to catalyze carbonylation of allyl chloride at 100°C and 69 atm:

$$CO + H_2C{=}CHCH_2Cl \longrightarrow H_2C{=}CHCH_2C({=}O)Cl \qquad (52)$$

The catalytic activity per Pd ion was found to be the same for the complex in the resin as for the analogous complex (the salt of *p*-toluenesulfonic acid) in solution. This result demonstrates the uniform accessibility of the matrix-bound groups and the occurrence, in at least one example, of the same chemistry in the solution and in the resin matrix. But the comparison between soluble and attached catalytic groups is often more complicated; for example, the reduction of interactions among catalytic groups held apart from one another by attachment to a support may allow catalysis to take place when it cannot in the presence of freely mobile soluble complexes which bond preferentially to one another. One can design the flexibility and swellability of a polymeric support to optimize both the interactions between catalytic groups and the local concentrations of reactants. One can further influence the catalytic properties by building into the polymer noncatalytic functional groups which may play a role analogous to that of a solvent affecting the distribution of components between two liquid phases.

The predictability of properties of these polymeric catalysts points to many potentially useful possibilities for extending the chemistry of homogeneous catalysis to solid organic phases; the ease of interpreting polymeric reaction mechanisms with the aid of spectroscopic methods may approach the ease of interpretation of mechanisms of solution reactions [75]. Organic polymers containing acidic and basic functional groups and transition metals can perhaps one day be synthesized to give useful imitations of enzymes. One important disadvantage seems to have hindered the application of the polymeric catalysts in such processes as hydroformylation and carbonylation: the reactions are exothermic, and the temperatures in the catalyst particles can rise so much that their stability is inadequate.

The class of matrix-bound transition-metal-complex catalysts is not restricted to organic polymers. Silica surfaces have been used for attachment of Ziegler catalysts [48] and of Rh complexes, which were evaluated as hydroformylation catalysts [77]. It is clear that the class of matrix-bound complex catalysts can be extended to many other solids [78].

The solid catalysts just described are especially important because they have for the most part been designed from a priori understanding of reaction mechanism and not discovered by trial-and-error experimentation. They may foreshadow the preparation of complex surfaces of inorganic solids designed for particular catalytic reactions. The solid Wacker catalyst reported by Evnin, Rabo, and Kasai [20] is one of the clearest available indications of the possibilities of such a design. The Wacker catalyst was required to contain a surface-bound Pd^{2+} complex, since this metal is uniquely suitable to the Wacker reaction. It was necessary to combine the Pd with a solid which, like Cu^{2+} in solution, was able to reoxidize Pd^0 to Pd^{2+} rapidly. The solid was chosen to be supported V_2O_5, which could be reoxidized by O_2 via electron transfer at the Pd site. The resulting catalyst was so active for ethylene oxidation that the reaction rate was apparently determined by the rate of gas-phase mass transfer to the catalyst surface. An explanation of the catalyst functions is given in more detail in Chap. 4, which is concerned with partial-oxidation reactions catalyzed by solid oxides.

NOTATION

a_g	gas-liquid interfacial area per unit reactor volume, l^{-1}
C	concentration, catalyst
C^*	saturation concentration in liquid, mol/l^3
D	diffusion coefficient, l^2/t
d_B	average bubble diameter, l
d_p	average particle diameter, l
E	energy
g	gravitational acceleration, l/t^2
H	gas holdup (volume fraction of gas in reactor)
H_0	hamiltonian operator
K	concentration equilibrium constant, variable dimensions
k	reaction rate constant, variable dimensions
k_c	mass-transfer coefficient (liquid to solid), l/t
k_L	mass-transfer coefficient (gas to liquid), l/t
k_L^0	mass-transfer coefficient (gas to liquid) in the absence of reaction, l/t
m	catalyst loading, m/m
P	pressure or partial pressure, F/l^2
q	flow rate, l^3/t or mol/t
r	reaction rate, mol/l^3t
T	temperature, °C or K
t_D	time characteristic of diffusion, t
t_R	time characteristic of reaction, t
u_s	slip velocity (velocity of bubbles relative to liquid), l/t
V_S	volume of solution, l^3
v_s	superficial velocity (volumetric flow rate per cross-sectional area of reactor), l/t

Dimensional

E = energy
F = force
l = length

m = mass
t = time

Dimensionless numbers

Re Reynolds number
Sc Schmidt number

Greek

β exchange integral
ΔH° standard enthalpy change on reaction, E/mol
μ viscosity, m/lt; ionic strength, mol/l^3
ρ density of liquid, m/l^3
ρ_p density of particle, m/l^3
ϕ_L ligand orbital
ϕ_M metal orbital
$\int d\tau$ integral over all space

REFERENCES

1. Halpern, J., *Adv. Chem. Ser.*, **70,** 1 (1968).
2. Cotton, F. A., and G. Wilkinson, "Advanced Inorganic Chemistry." Wiley-Interscience, New York, 1972.
3. Fukui, K., *Bull. Chem. Soc. Jap.*, **39,** 498 (1968); K. Fukui, and H. Fujimoto, ibid. **41,** 1989 (1968).
4. Bader, R. F. W., *Can. J. Chem.*, **40,** 1164 (1962).
5. Salem, L., and J. S. Wright, *J. Am. Chem. Soc.*, **91,** 5947 (1969).
6. Pearson, R. G., *Theor. Chim. Acta.*, **16,** 107 (1970); *Acct. Chem. Res.*, **4,** 152 (1971).
7. Rösch, N., R. P. Messmer, and K. H. Johnson, *J. Am. Chem. Soc.*, **96,** 3855 (1974).
8. O'Connor, C., and G. Wilkinson, *J. Chem. Soc.*, **A1968,** 2665.
9. Cossee, P., in A. D. Ketley (ed.), "The Stereochemistry of Macromolecules," vol. 1, pp. 145–175, Marcel Dekker, New York, 1967.
10. Smidt, J., W. Hafner, R. Jira, J. Sedlmeier, R. Seiber, and H. Kojer, *Angew. Chem.*, **71,** 176 (1959).
11. Stern, E. W., *Catal. Rev.*, **1,** 73 (1967).
11*a.* Jira, R., and W. Freiesleben, in E. I. Becker and M. Tsutsui (eds.), "Organometallic Reactions," vol. 3, pp. 1–190, Wiley-Interscience, New York, 1972.
12. Henry, P. M., *J. Am. Chem. Soc.*, **86,** 3246 (1964).
13. Moiseev, I. I., O. G. Levanda, and M. N. Vargaftik, *J. Am. Chem. Soc.*, **96,** 1003 (1974).
14. Szonyi, G., *Adv. Chem. Ser.*, **70,** 53 (1968).
15. Hartley, F. R., *Chem. Rev.*, **69,** 799 (1969).
16. Moiseev, I. I., *Am. Chem. Soc. Div. Petrol. Chem. Prepr.* **14,** B49 (1969).
17. Smidt, J., *Chem. Ind.* (*Lond.*), **54,** July 13, 1962.
18. Smidt, J., W. Hafner, and R. Jira, Canadian Patent 625,435 (1961).
19. Krekeler, H., and W. Krönig, *7th World Pet. Cong. Proc.*, vol. 5, pp. 41–48, Elsevier, Barking, Essex, England, 1967.
20. Evnin, A. B., J. A. Rabo, and P. H. Kasai, *J. Catal.*, **30,** 109 (1973).
21. Moiseev, I. I., M. N. Vargaftik, and Y. K. Syrkin, *Dokl. Akad. Nauk SSSR*, **130,** 801 (1960); **133,** 377 (1960).
22. Zachry, J. B., *Ann. N.Y. Acad. Sci.*, **125,** 154 (1965).
23. Stern, E. W., in G. N. Schrauzer, (ed.), "Transition Metals in Homogeneous Catalysis," pp. 93–146, Marcel Dekker, New York, 1971.
24. Martin, A. R., *Chem. Ind.* (*Lond.*), Dec. 11, 1954, p. 1536.

25. Falbe, J., "Carbon Monoxide in Organic Synthesis," (trans. C. R. Adams), chap. 1. Springer-Verlag, New York, 1970.
26. Pruett, R. L., and J. A. Smith, U.S. Patent 3,527,809 (1970).
27. Paulik, F. E., *Catal. Rev.*, **6,** 49 (1972).
28. Olivier, K. L., and F. B. Booth, *Prepr. Am. Chem. Soc. Div. Petrol. Chem. Gen. Pap.*, **14**(3), A7 (1969).
29. Wender, I., S. Metlin, S. Ergun, H. W. Sternberg, and H. Greenfield, *J. Am. Chem. Soc.*, **78,** 5401 (1956).
30. Orchin, M., and W. Rupilius, *Catal. Rev.*, **6,** 85 (1972).
31. Evans, D., J. A. Osborn, and G. Wilkinson, *J. Chem. Soc.* **A1968,** 3133.
32. Olivier, K. L., and F. B. Booth, *Hydrocarbon Process.*, **49**(4), 112 (1970).
33. Heck, R. F., and D. S. Breslow, *J. Am. Chem. Soc.* **83,** 4023 (1961).
34. Lemke, H., *Hydrocarbon Process.*, **45**(2), 148 (1966).
35. Cornils, B., R. Payer, and K. C. Traenckner, *Hydrocarbon Process.*, **54**(6), 83 (1975).
36. Paulik, F. E., A. Hershman, W. R. Knox, and J. F. Roth, U.S. Patent 3,769,329 (1973).
36*a*. Paulik, F. E., and J. F. Roth, *Chem. Commun.*, **1968,** 1578.
36*b*. Roth, J. F., J. H. Craddock, A. Hershman, and F. E. Paulik, *Chemtech.*, October 1971, p. 600.
36*c*. Forster, D., *J. Am. Chem. Soc.*, **98,** 846 (1976).
37. Lowry, R. P., and A. Aguilo, *Hydrocarbon Process.*, **53**(11), 103 (1974).
38. Hjortkjaer, J., and V. W. Jensen, *Ind. Eng. Chem. Prod. Res. Dev.*, **15,** 46 (1976).
39. Ballard, D. G. H., *Adv. Catal.*, **23,** 263 (1973).
40. Berger, M. N., G. Boocock, and R. N. Haward, *Adv. Catal.*, **19,** 211 (1969).
41. Berger, M. N., and B. M. Grieveson, *Makromol. Chem.*, **83,** 80 (1965).
42. Boor, J., Jr., *Macromol. Rev.*, **2,** 115 (1967).
43. Buls, V. W., and T. L. Higgins, *J. Polym. Sci. Polym. Chem.*, **11,** 925 (1973).
44. Hock, C. W., *J. Polym. Sci. Part A-1*, **4,** 3055 (1966).
45. Buls, V. W. and T. L. Higgins, *J. Polym. Sci. Part A-1*, **8,** 1037 (1970).
46. Crabtree, J. R., F. N. Grimsby, A. J. Nummelin, and J. M. Sketchley, *J. Appl. Polym. Sci.*, **17,** 959 (1973).
47. Valvassori, A., P. Longi, and P. Parrini, in E. G. Hancock (ed.), "Propylene and Its Industrial Derivatives," pp. 155–213, Wiley, New York, 1973.
48. Chien, J. C. W., and J. T. T. Hsieh, *J. Polym. Sci. Polym. Chem.* **14,** 1915 (1976).
49. Natta, G., and F. Danusso (eds.), "Stereoregular Polymers and Stereospecific Polymerizations," vols. 1 and 2, Pergamon, Oxford, 1967.
50. Natta, G., and I. Pasquon, *Adv. Catal.*, **11,** 1 (1959).
51. Cossee, P., *J. Catal.*, **3,** 80 (1964).
52. Arlman, E. J., and P. Cossee, *J. Catal.*, **3,** 99 (1964).
53. Rodriguez, L. A. M., and J. A. Gabant, *J. Polym. Sci., Part C*(4), 125 (1963).
54. Compostella, M., in A. D. Ketley (ed.), "The Stereochemistry of Macromolecules," vol. 1, pp. 309–387, Marcel Dekker, New York, 1967.
55. Rasmussen, D. M., *Chem. Eng.*, Sept. 18, 1972, p. 104.
56. Anonymous, *Hydrocarbon Process.*, **52**(10), 129 (1973).
57. Boor, J., Jr., *Ind. Eng. Chem. Prod. Res. Dev.*, **9,** 437 (1970).
58. Parks, G. S., and H. P. Mosher, *J. Polym. Sci. Part A*, **1,** 1979 (1963).
59. Calderbank, P. H., in V. W. Uhl and J. B. Gray (eds.), "Mixing: Theory and Practice," vol. 2, pp. 2–114, Academic, New York, 1967.
60. Astarita, G., "Mass Transfer with Chemical Reaction," Elsevier, Amsterdam, 1967.
61. Schaftlein, R. W., and T. W. F. Russell, *Ind. Eng. Chem.*, **60**(5), 13 (1968).
62. Goldman, K., in S. A. Miller (ed.), "Ethylene and Its Industrial Derivatives," p. 218, Ernest Benn, London, 1969.
63. Hughmark, G. A., *Ind. Eng. Chem. Process Des. Dev.*, **6**(2), 218 (1967).
64. Calderbank, P. H., *Trans. Inst. Chem. Eng.*, **36,** 443 (1958).
65. Satterfield, C. N., "Mass Transfer in Heterogeneous Catalysis," MIT Press, Cambridge, Mass., 1970.

66. Cichy, P. T., J. S. Ultman, and T. W. F. Russell, *Ind. Eng. Chem.*, **61**(8), 6 (1969).
67. Cichy, P. T., and T. W. F. Russell, *Ind. Eng. Chem.*, **61**(8), 15 (1969).
68. Robinson, K. K., F. E. Paulik, A. Hershman, and J. F. Roth, *J. Catal.*, **15,** 245 (1969).
69. Rony, P. R., *J. Catal.*, **14,** 142 (1969).
70. Mars, P., and J. G. H. Maessen, *J. Catal.*, **10,** 1 (1968).
71. Robinson, K. K., A. Hershman, J. H. Craddock, and J. F. Roth, *J. Catal.*, **27,** 389 (1972).
72. Bailar, J. C., Jr., *Catal. Rev. Sci. Eng.*, **10,** 17 (1974).
73. Ragg, P. L., German Patent 1,937,225; 1,937,232; 2,000,829 (1970).
74. Haag, W. O., and D. D. Whitehurst, *Proc. 5th Int. Cong. Catal.*, p. 465, North-Holland, Amsterdam, 1973.
75. Jarrell, M. S., and B. C. Gates, *J. Catal.*, **40,** 255 (1975).
76. Haag, W. O., and D. D. Whitehurst, paper presented at *Meet. Catal. Soc., Houston, Tex., 1971;* Belgian Patent 721,686 (1969).
77. Allum, K. G., R. D. Hancock, S. McKenzie, and R. C. Pitkethly, *Proc. 5th Int. Cong. Catal.*, p. 477, North-Holland, Amsterdam, 1973.
78. Burwell, R. L., *Chemtech*, June 1974, p. 370; L. L. Murrell, in J. J. Burton and R. L. Garten (eds.), "Advanced Materials in Catalysis," Academic, New York, 1977.

PROBLEMS

2-1 Design a one-reactor Wacker process. Use the kinetics data of Moiseev, and estimate the appropriate mass- and heat-transfer coefficients. Assume a reaction temperature of 100°C and a pressure of about 3 atm. It has been suggested that heat evolution and reaction temperature can be well controlled by the pressure at which water is allowed to vaporize. Consider this suggestion in the design. Provide detailed calculations for the reactor but not the remainder of the process.

2-2 Design a tubular fixed-bed flow reactor for oxidation of ethylene to acetaldehyde. Base the estimate on the result of Evnin et al. [20] indicating that gas-solid mass transfer is rate-determining. Consult Satterfield [65] for prediction of mass-transfer rates. Assume a feed temperature of 140°C and a pressure of 1 atm.

2-3 Design an oxo reactor like that of Example 2-2 but use propylene rather than diisobutylene as the feedstock.

2-4 Design a backmix reactor for methanol carbonylation at 175°C and 15 atm assuming an acetic acid production rate of 13.5 million kg per year. For what range of catalyst concentration does mass transfer have a significant influence on rate at these conditions?

2-5 Design a suspension-polymerization reactor for polyethylene production. Use the appropriate estimates of Example 2-1 but assume that an improved catalyst is available giving 100 times the yield of that referred to in the example.

2-6 In the oxo process, the aldol condensation reaction is usually an undesired side reaction. In a process called the aldox process, however, this reaction forms part of a reaction network by which propylene, CO, and H_2 are converted into 2-ethylhexanol, an important plasticizer alcohol.

(*a*) Deduce the reaction network and predict the side reactions.

(*b*) The reactions are catalyzed homogeneously in the industrial process; choose the catalysts.

(*c*) Design a solid polymer catalyst for all the reactions of the aldox sequence.

2-7 Roth et al. [36*b*] found that all the following complexes of Rh gave the same rates of methyl iodide-promoted methanol carbonylation under comparable conditions: $RhCl_3$, $Rh(CO)_2Cl_2$, $Ph_4AsRh(CO)_2Cl_2$, $Ph_4AsRh(CO)_2I_2$, and $RhCl(CO)(PPh_3)_2$. Explain this result.

2-8 The hydrogenation of olefins is catalyzed by Rh complexes. Use the information in this chapter to predict the mechanism of the reaction and compare your suggestion with Wilkinson's interpretation [2, chap. 24].

2-9 Rhodium-complex catalysts are also active for olefin isomerization. Predict the mechanism and compare your suggestion with Wilkinson's [2, chap. 24].

2-10 What products would you expect from an ethanol carbonylation process with a Rh-complex catalyst?

2-11 Suggest a mechanism for ketone formation when ethylene is replaced by propylene in the Wacker process.

2-12 Use the laboratory data of Fig. 2-42 and the appropriate predictive methods given by Satterfield [65] to estimate the parameter values of Eq. (51). Which parameter values are significantly different from those expected for a large-scale design?

2-13 Test the hypothesis that significant temperature gradients exist in a polymeric catalyst bead containing Rh-complex groups used for methanol carbonylation. Use the results of Ref. 75 as a basis for the estimate.

2-14 Recent patent literature suggests that rhodium carbonyl clusters such as $[Rh_{12}(CO)_{30}]^{2-}$ are active catalysts for the direct conversion of synthesis gas ($CO + H_2$) into ethylene glycol and methanol. Consult the most recent literature and report on the reaction chemistry and potential processes.

2-15 Compounds such as pentachlorothiophenol, known as *pseudohalides*, function as promoters for the Rh-complex-catalyzed carbonylation of methanol. Suggest a mechanism for the promoter action of pentachlorothiophenol. What advantages might it offer over hydrogen iodide?

2-16 One of the side products in the industrial methanol-carbonylation process is CO_2. Suggest a mechanism for formation of this product.

2-17 Consider the polymeric catalyst for methanol carbonylation described in Ref. 75 in the light of the more recent mechanistic information given in Ref. 36*c*. Do you expect the polymeric catalyst to be stable? What would you expect the causes of deactivation to be?

2-18 When $(RuCl_4)^{3-}$ is contacted with acetylene and water, a catalytic cycle takes place and the acetylene is hydrated to give acetaldehyde. Predict the catalytic cycle.

2-19 Ziegler catalysts [α-$TiCl_3$ solid promoted by a compound like $Al(C_2H_5)_3$] are selective for the production of isotactic polypropylene, but they give about a 5 percent yield of atactic polypropylene. Some compounds called *stereoregulators*, such as cyclic triolefins and diamines, modify the Ziegler catalyst and increase the fraction of the polypropylene product which is isotactic. They also decrease the activity of the catalyst, i.e., the rate at which isotactic polypropylene is formed. Suggest how the stereoregulators work in terms of explicit chemical structures.

2-20 The methanol homologation reaction catalyzed by complexes of Co or Rh may be useful in the conversion of CO and hydrogen (synthesis gas) into chemicals. What chemicals might be produced? Consult the recent literature and summarize what is known about the catalytic chemistry and potential processes.

2-21 Refer to Prob. 2-15 and suggest the design of a solid for methanol carbonylation which incorporates both promoter and catalyst groups.

2-22 Molecular-orbital calculations reported for the proposed intermediate in the Wacker reaction have led to the suggestion that the mechanism presented in this chapter is not quite correct [D. R. Armstrong, R. Fortune, and P. G. Perkins, *J. Catal.*, **45,** 339 (1976)]. Consult the reference and prepare a summary of the new evidence and the mechanism suggested by the authors.

2-23 Rösch and Johnson [*J. Mol. Catal.*, **1,** 410 (1976)] used the self-consistent-field Xα scattered-wave method to characterize the bonding in a model structure representative of the surface complex involving ethylene bonded to the Ziegler catalyst. Consult the reference and prepare a summary of the results. Compare the bonding of ethylene in the Ti complex with that of ethylene in Zeise's salt (Fig. 2-9).

2-24 Organozirconium complexes in solution are often poor polymerization catalysts since they form intermediates during the catalytic cycle which are lost through reactions like oligomerization of the metal centers. The loss can be prevented by attaching the Zr complexes to a silica surface holding them apart from each other. Active, stable catalysts prepared by attachment of Zr complexes to surfaces have found industrial application. Consult Ballard [39] as an initial reference, and prepare a summary of the catalytic chemistry and the design of the solid polymerization catalysts.

CHAPTER

THREE

REFORMING

INTRODUCTION

Catalytic reforming is one of the basic petroleum refining processes, yielding more than 3 million barrels per day in the United States. In reforming, gasoline-range molecules, including those formed from larger hydrocarbons by cracking, are reconstructed, or *reformed*, without changing their carbon numbers [1–5]. The reactions, which include isomerization, hydrogenation, dehydrocyclization, and dehydrogenation, among others, lead to a marked improvement in fuel quality as measured by the octane number. Catalysts for reforming reactions are small crystallites of Pt or Pt alloys supported on porous, promoted alumina; the catalysts are said to be bifunctional, since both the metal and oxide components play active roles.

This chapter proceeds with an introductory description of the reforming reactions, catalysts, and industrial processes and follows with detailed accounts of the catalytic chemistry and process engineering. The account of the chemistry is built upon a foundation including the theory of catalysis by metals, and this subject is therefore developed in detail.

REACTIONS

The low-octane feedstocks to be reformed contain large quantities of straight-chain paraffins and relatively small quantities of branched-chain paraffins, naphthenes, olefins, and aromatics. Table 3-1 gives the octane numbers of pure

Table 3-1 Octane numbers of pure hydrocarbons [1]

Hydrocarbon	Blending research octane number (clear)
Paraffins:	
n-Butane	113
n-Pentane	62
n-Hexane	19
n-Heptane	0
n-Octane	−19
2-Methylhexane	41
2,2-Dimethylpentane	89
2,2,3-Trimethylbutane	113
Naphthenes (cycloparaffins):	
Methylcyclopentane	107
1,1-Dimethylcyclopentane	96
Cyclohexane	110
Methylcyclohexane	104
Ethylcyclohexane	43
Aromatics:	
Benzene	99
Toluene	124
1,3-Dimethylbenzene	145
Isopropylbenzene	132
1,3,5-Trimethylbenzene	171

Source: From "Catalysis," vol. VI, P. H. Emmett (ed.). Copyright 1958 by Litton Educational Publishing Company. Reprinted with permission of Van Nostrand Reinhold Co.

hydrocarbons and indicates the types of hydrocarbons desired for high-octane gasoline. The reactions used to achieve increased octane number include the following:

Paraffins. Typical virgin petroleum naphthas and many refinery charge stocks contain 15 to 75 percent straight-chain paraffins with a typical research octane number less than 50. The paraffins can be isomerized to branched-chain molecules, e.g.,

$$n\text{-}C_7H_{16} \longrightarrow H_3CCH_2CH_2C(CH_3)_2CH_3 \qquad (1)$$

They can also undergo dehydrocyclization to give cycloparaffins:

$$n\text{-}C_7H_{16} \longrightarrow \text{(methylcyclohexane)}\ C_6H_{11}CH_3 + H_2 \qquad (2)$$

or

$$n\text{-}C_7H_{16} \longrightarrow \text{(cyclopentane)}\text{-}CH_2CH_3 + H_2 \tag{3}$$

Paraffins can also undergo hydrocracking:

$$n\text{-}C_9H_{20} + H_2 \longrightarrow n\text{-}C_5H_{12} + n\text{-}C_4H_{10} \tag{4}$$

Olefins. Olefins are usually not present in straight-run naphthas but may be present in other refinery charge stocks. They are undesirable in high concentrations and are usually hydrogenated. They can also be hydroisomerized to give isoparaffins:

$$\text{Heptene-1} + H_2 \longrightarrow H_3CCH_2CH_2CH_2\underset{}{C}(CH_3)HCH_3 \tag{5}$$

They can also undergo cyclization.

Naphthenes. Typical feeds contain from 18 to 50 percent naphthenes as cyclopentanes and cyclohexanes. Often there are more cyclopentanes than cyclohexanes. The cyclohexanes can be dehydrogenated to give aromatics, e.g.,

$$\text{(cyclohexane)}\text{-}CH_3 \longrightarrow \text{(benzene)}\text{-}CH_3 + 3H_2 \tag{6}$$

Cyclopentanes can be hydroisomerized to give cyclohexanes with subsequent dehydrogenation to give aromatics:

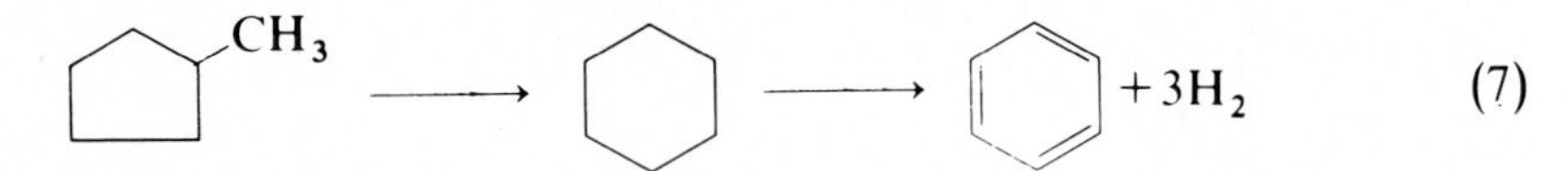

(7)

Unsaturated cyclic hydrocarbons. Some of these are present in charge stocks, and others are generated in the reactor. Reaction of substituted aromatics can occur as follows:

$$\text{(o-xylene)} + \text{(benzene)} \longrightarrow 2\,\text{(benzene)}\text{-}CH_3 \tag{8}$$

Aromatics can also experience hydrodealkylation:

$$\text{(benzene with } CH_3, C_2H_5) + H_2 \longrightarrow \text{(benzene)}\text{-}CH_3 + C_2H_6 \tag{9}$$

Species containing S or N can undergo hydrodesulfurization and hydrodenitrogenation, respectively, as discussed in Chap. 5:

$$\text{(thiophene)}\text{-}CH_3 + 4H_2 \longrightarrow C_5H_{12} + H_2S \tag{10}$$

Table 3-2 Thermodynamic data for typical reforming reactions [3]

Reaction	$K_p{}^a$ at 500°C, P_i in atm	ΔH_r, kcal/mol of hydrocarbon
Cyclohexane $\rightleftharpoons$ benzene + $3H_2$	6×10^5	52.8
Methylcyclopentane $\rightleftharpoons$ cyclohexane	0.086	−3.8
n-Hexane $\rightleftharpoons$ benzene + $4H_2$	0.78×10^5	63.6
n-Hexane $\rightleftharpoons$ 2-methylpentane	1.1	−1.4
n-Hexane $\rightleftharpoons$ 1-hexene + H_2	0.037	31.0

[a] For the reaction

$$(HC)_1 \rightleftharpoons (HC)_2 + nH_2$$

the equilibrium constant is defined as

$$K_p = \frac{P_{(HC)_2} P^n_{H_2}}{P_{(HC)_1}}$$

Source: Reprinted with permission from *Advances in Chemical Engineering.* Copyright by Academic Press.

THERMODYNAMICS

Thermodynamic data for typical reforming reactions at 500°C are given in Table 3-2. At equilibrium, cyclohexanes are substantially converted into aromatics at low hydrogen partial pressures, and cyclopentanes are substantially favored over cyclohexanes. At equilibrium only very small concentrations of olefins can exist with paraffins. The major reforming reactions are endothermic.

The equilibrium between cyclohexane, benzene, and hydrogen as a function of temperature and pressure is illustrated in Fig. 3-1, which shows how operating

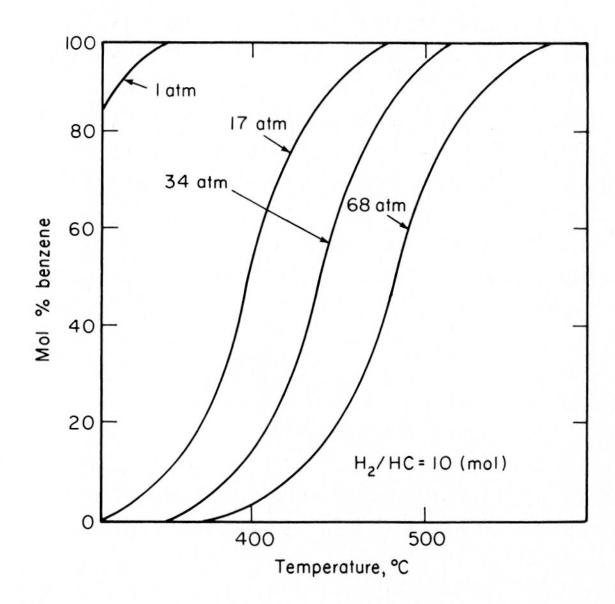

Figure 3-1 Equilibrium distributions of cyclohexane, benzene, and hydrogen [1]. (From "Catalysis," vol. VI, P. H. Emmett (ed.). Copyright 1958 by Litton Educational Publishing Company. Reprinted by permission of Van Nostrand Reinhold Company.

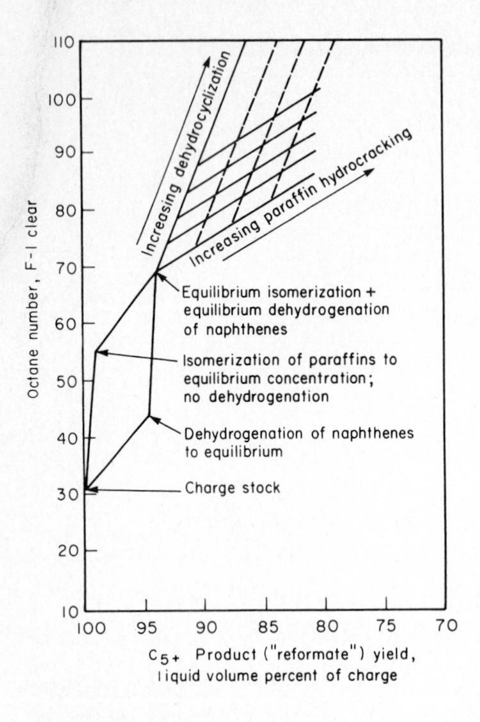

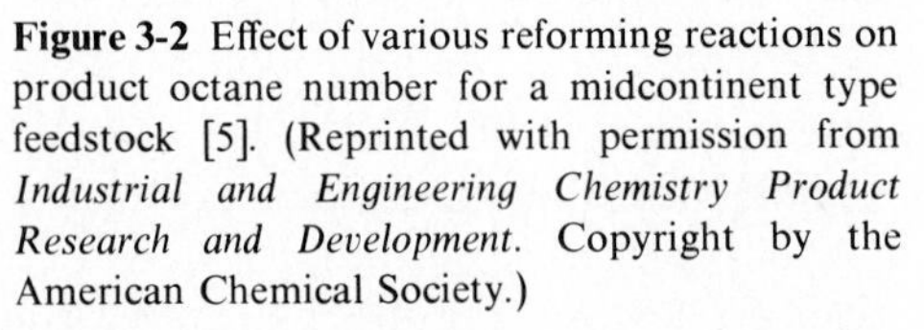
Figure 3-2 Effect of various reforming reactions on product octane number for a midcontinent type feedstock [5]. (Reprinted with permission from *Industrial and Engineering Chemistry Product Research and Development*. Copyright by the American Chemical Society.)

temperature and pressure affect conversion to aromatics. Since typical operating conditions include temperatures from 455 to 510°C and pressures from 6.5 to 50 atm, only partial conversion to aromatics is achievable; consequently recent practice has been to operate at temperatures in the upper end of this range and to use pressures as low as 10 atm [4]. At these conditions there is almost complete conversion of naphthenes into aromatics at equilibrium.

Figure 3-2 shows how the various reactions lead to improvement in octane number for a typical midcontinent feedstock. The feedstock was assumed to contain equimolar amounts of C_7, C_8, C_9, and C_{10} paraffins and naphthenes, each having a characteristic isomeric distribution for this stock; it had a volumetric composition of 49 percent paraffins, 44 percent naphthenes, and 7 percent aromatics and an octane number of 31. Equilibrium isomerization of paraffins and equilibrium dehydrogenation of naphthenes under medium-pressure reforming conditions each provide significant increases in octane number, but dehydrocyclization of paraffins and some hydrocracking are essential to achieve the maximum octane increase. Design of reforming catalysts has been directed primarily to the last two reactions.

KINETICS

To understand the behavior of reforming reactors, we must understand the reaction kinetics and the influence of heat effects. A qualitative summary of the rate

Table 3-3 Rate behavior and heat effects of important reforming reactions[a]

Reaction type	Relative rate[b]	Effect of increase in total pressure	Heat effect
Hydrocracking	Slowest	Increases rate	Quite exothermic
Dehydrocyclization	Slow	None to small decrease in rate	Endothermic
Isomerization of paraffins	Rapid	Decreases rate	Mildly exothermic
Naphthene isomerization	Rapid	Decreases rate	Mildly exothermic
Paraffin dehydrogenation	Quite rapid	Decreases conversion	Endothermic
Naphthene dehydrogenation	Very rapid	Decreases conversion	Very endothermic

[a] Partly from Krane et al. [6].
[b] Relative rates are for a modern bifunctional catalyst.

behavior of the important reaction classes catalyzed by a modern bifunctional catalyst is given in Table 3-3 with statements of the heat effects. This information largely determines what is needed for the reactor design. Naphthene and paraffin dehydrogenation reactions are so rapid that they are essentially in equilibrium, and rates need not be considered explicitly. Similarly, the equilibrium between *n*-paraffins and isoparaffins is usually closely approached. In contrast, the rates of cyclization and hydrocracking are typically low and in need of explicit consideration. Hydrogen partial pressure is an important variable since it strongly affects the conversion to aromatics and the rate of hydrocracking. Since reforming reactions which produce hydrogen, especially aromatization, predominate over those which consume hydrogen, the process is a net hydrogen producer.

CATALYSTS

Reforming reactions typically proceed through a number of elementary steps. For example, a straight-chain paraffin is converted into an isoparaffin by first being converted into an olefin, which is isomerized to an isoolefin and then converted into an isoparaffin. Correspondingly, the catalyst has two functions, a hydrogenation-dehydrogenation function for the paraffin-olefin conversions and an isomerization function, which is associated with the catalyst acidity, as described in Chap. 1. The catalysts used until the early 1950s were chromium oxide or molybdenum oxide supported on alumina, which incorporated both the catalytic functions on the surface of the metal oxide. More recently developed reforming catalysts have crystallites of a metal such as Pt on an acidic support such as alumina, and the two functions are present in separate phases. The metal (Pt, Pt and Re, or a noble-metal-containing trimetallic alloy) provides the hydrogenation-dehydrogenation activity, and the promoted acidic alumina provides the isomerization activity. The hydrogenation-dehydrogenation activity of the supported metal and the isomerization activity of the alumina are much greater than the respective activities of the early-generation metal oxides.

Catalyst deactivation by coke formation involves both the metal and support but primarily the support. Coking results from secondary reactions of the hydrocarbons, particularly olefins. Catalyst deactivation is rapid when the hydrogen partial pressure is low and when the temperature is high. Deactivated catalyst is regenerated in place by slowly burning off the coke.

OPERATING CONDITIONS

In processes operated at the relatively high pressures of 34 to 50 atm, the high partial pressure of hydrogen results in high rates of hydrocracking, limited degrees of conversion to aromatics (Table 3-3), low rates of catalyst deactivation, and low yields of hydrogen. In contrast, low-pressure operation (8.5 to 20.5 atm) gives high conversions to aromatics, low hydrocracking conversions, and high hydrogen yields; however, it also leads to rapid deactivation of the catalyst by coke formation.

Reaction temperatures are chosen to balance the advantage of increased catalytic activity and the disadvantage of increased deactivation rate as temperature is increased. The values range from 460 to 525°C and are usually between 482 and 500°C. Low-pressure processes are operated at slightly higher temperatures than the others to optimize conversion to high-octane-number products. As catalysts lose activity in operation, reactor temperature is gradually increased to maintain a constant octane number in the product reformate.

Space velocity ranges from 0.9 to 5 vol of liquid feed per volume of catalyst per hour, with 1 to 2 most common. The choice represents a compromise between allowable hydrocracking and desirable dehydrocyclization. Aromatization and isomerization are not affected by changes in space velocity, since these reactions approach equilibrium even at high space velocities.

Hydrogen-to-hydrocarbon-feed mole ratios vary from 3 to 10, and ratios of 5 to 8 are preferred. A high hydrogen-to-hydrocarbon ratio adversely affects aromatization, increases hydrocracking, and reduces catalyst deactivation rates. The value chosen is a trade-off and is usually governed on the lower bound by the desired amount of hydrocracking and the maximum acceptable deactivation. High ratios require high hydrogen recycle rates and correspondingly high operating costs.

Straight-run naphthas generally constitute the major feedstocks to reformers. Stocks which have an appreciable concentration of unsaturated hydrocarbons must usually be hydrotreated before reforming in order to prevent undue hydrogen consumption in the reformer and excessive catalyst deactivation. The concentrations of impurities in the feed that can act as catalyst poisons must also be rigorously controlled. Sulfur poisons the metal function of the catalyst and must be maintained at concentrations less than about 1 ppm in the feed contacting the newer catalysts. Fuels with higher sulfur content must be catalytically hydrodesulfurized before being reformed (Chap. 5). Organic nitrogen compounds are converted into ammonia and poison the acid function of the reforming catalyst. Their concentrations must also be kept below about 2 ppm; nitrogen compounds are usually removed in a catalytic hydrotreating step, which simultaneously leads to hydrosulfurization. Water- and chlorine-containing compounds are undesirable feed components because they alter the acidity of the support and thus upset the balance between the reactions occurring. Since these compounds are not easily eliminated, their concentrations are carefully controlled to maintain the proper catalyst acidity. Metals such as As, Pb, and Cu must be kept at extremely low concentrations, since they alloy with the Pt component or otherwise deactivate it. Severe poisoning by As has been reported with 30 parts per billion of As in the feed [5].

REACTOR DESIGN

The need for controlling reactor temperature is important in dictating the design of the reforming reactor system. The overall reforming process is endothermic, which implies that the temperature of the fixed-bed reactor decreases in the direction of flow. A decreased temperature corresponds to a decreased reaction rate and, for aromatization, a decreased equilibrium conversion as well. Furthermore, the most endothermic of the reactions (Table 3-3), naphthene dehydrogenation, occurs very rapidly. It follows that large amounts of energy must be added to the reaction mixture to maintain its temperature, and much energy must be added near the reactor entrance. Because of the engineering problems associated with the addition of large amounts of energy to large packed-bed reactor systems, the design involves dividing the total catalyst charge among several reactors, with interstage heating between reactors (Fig. 3-3).

Table 3-4 gives typical operating conditions for a three-reactor system. Because of the rapid cooling of the reactant stream resulting from the endothermic dehydrogenation reactions, the residence time and the amount of catalyst in the first (upstream) reactor are considerably less than in each of the other reactors (Table 3-4). Energy-balance considerations essentially govern the design of the

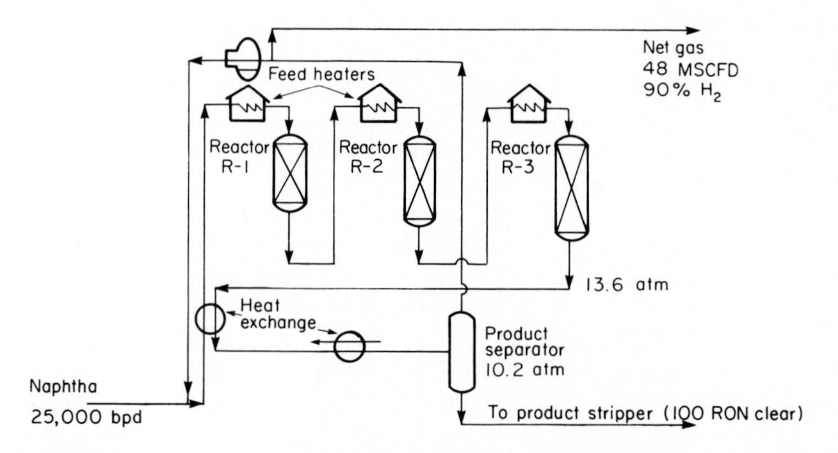

Figure 3-3 Process flow diagram for three-reactor reforming unit.

Table 3-4 Typical operating conditions for a three-reactor system[a]

	Reactor		
	1	2	3
Inlet temperature, °C	502	502	502
Exit temperature, °C	433	471	496
Temperature drop, °C	69	31	6
Octane number (F-1 clear)[b]	65.5	79.5	90.0
Octane-number increase	27.0	14.0	10.5
Principal reactions	Dehydrogenation, dehydro-isomerization	Dehydrogenation, dehydro-isomerization, hydrocracking, dehydrocyclization	Hydrocracking, dehydrocyclization
LHSV, h^{-1} per reactor	5.5	2.4	1.7
Percent of total catalyst charge	15	35	50

[a] Estimated from published information.
[b] Octane number of feed was 38.5.

first reactor. The amount of catalyst in the first reactor is usually about 15 to 20 percent of the total (Table 3-4). The octane-number increase in the upstream reactor is greatest because of aromatics formation and paraffin isomerization. The third (downstream) reactor is designed to optimize conversion involving the slower reactions, particularly dehydrocyclization, and it often contains half the catalyst in the system. The temperature change in this reactor is small because exothermic hydrocracking takes place. A controlled amount of hydrocracking is desired to increase the octane number by cracking long-chain paraffins and dealkylating aromatics with long side chains, but hydrocracking is expensive in terms of hydrogen consumption. This type of unit would run from 6 to 12 months between regenerations.

The recent development of much stabler reforming catalysts (supported Pt–Re and noble-metal-containing trimetallic catalysts) has increased interest in low-pressure (6.8 to 20.5 atm) reforming. Figure 3-4 shows a typical cyclic regenerative reforming system for low-pressure operation. There are three or four fixed-bed reactors, all the same size, and an additional "swing" reactor to replace one of the others when its catalyst is being regenerated. This scheme allows continuous operation of the process without shutdown for catalyst regeneration. The reactor which has been on stream the longest since regeneration is often used as the first (upstream) reactor, and the reactor containing freshly regenerated catalyst is used as the last reactor. Correspondingly, hydrocarbon with a low concentration of unconverted reactant contacts the most active catalyst. The design is suited to the low-pressure high-severity operation, with rapid catalyst deactivation, and catalyst regeneration may be as frequent as once a day to once a week.

Regeneration in general involves burning the coke off of the catalyst under

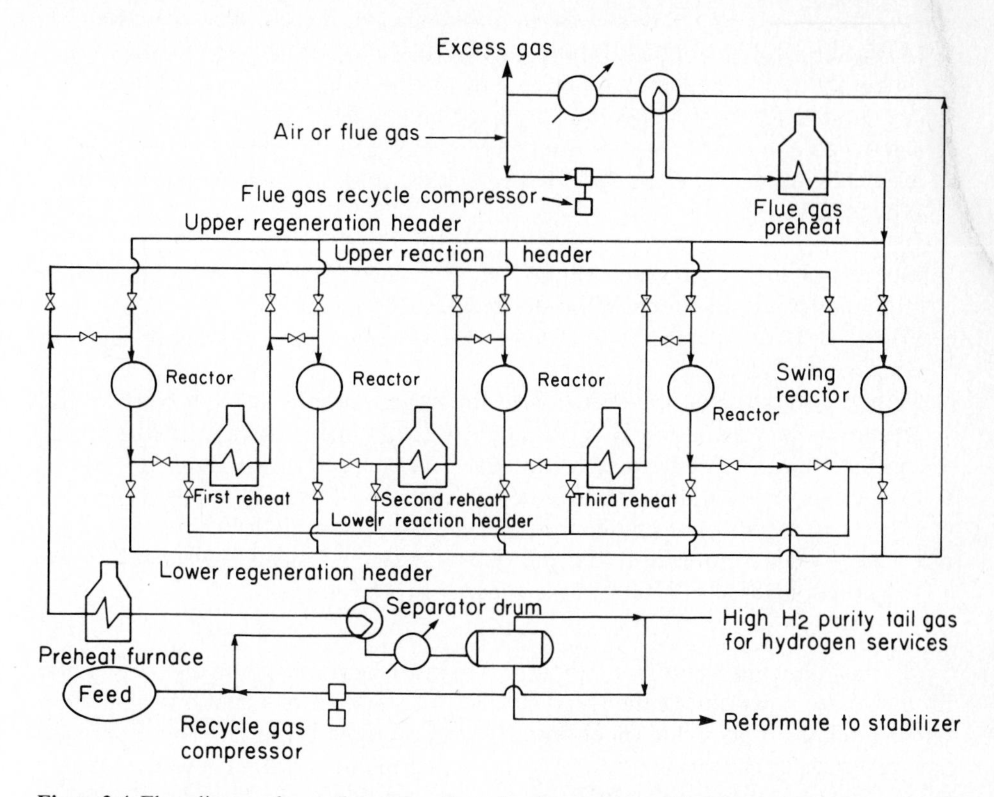

Figure 3-4 Flow diagram for cyclic regenerative reforming; upper and lower reaction headers involve several pipes so that any given reactor can be regenerated with the others arranged to give optimum operation.

carefully controlled conditions with a gas containing from 0.5 to 1 percent O_2; the maximum bed temperature does not exceed 450°C. Such careful procedures are required to prevent damage to the catalyst, particularly sintering and loss of surface area of the Pt component.

CATALYTIC CHEMISTRY

TRANSITION METALS AND THEIR CATALYTIC ACTIVITY

Introduction

Metals catalyze a broad range of reactions, of which hydrogenation and dehydrogenation are the most important examples. A theory of metal catalysis must account for the following general results and answer the following specific questions:

1. Many nonmetallic solids exhibit catalytic behavior, yet few can be found that effectively catalyze reactions catalyzed by metals. What properties of the metallic state are responsible for the unique behavior of metal catalysts?
2. Almost all metal catalysts are transition metals, and it is almost axiomatic that the behavior of metal catalysts is associated with the presence of *d* orbitals. Why is this so?
3. When transition metals are ordered according to their activity, each reaction usually has its own characteristic sequence of metals and there is not a common sequence for all reactions. What determines the sequences?
4. For a given set of reactants, different metals often have markedly different selectivities. Why?
5. Catalytic activity per unit surface area for some reactions depends on the metal crystal plane, on the metal crystallite size (if the metal is supported), on the support, and on preparation procedures. Why do these activity differences occur, and why are they reaction-dependent?
6. When two metals are combined to give an alloy, the catalytic behavior can vary strongly with composition and the behavior can be highly dependent on the reaction considered. What are the causes of these effects?

In the following sections, an attempt is made to summarize the theory of metal catalysis based on a simple theory of the metallic state, on observations of interactions at metal surfaces, and on observations of catalytic behavior. The discussion then proceeds to a consideration of alloys and concludes with a review of recent theories of the electronic properties of metal surfaces and their relations to catalytic behavior. These theories are much more sophisticated than models developed for nonmetallic catalysts; they hold promise of providing a quantitative base for understanding metal catalysis, although they are many years away from providing predictive capabilities. Our understanding of metal catalysis is increasing rapidly because of the recent development of electronic theories for metals and bonding on them and the development of sophisticated analytical techniques to study metal surfaces.

Bonding in Metals

An understanding of the catalytic properties of metal surfaces requires first an understanding of the bulk properties of metals and particularly the bonding in metal crystals. One might hope that a satisfactory model of bonding in the bulk would provide a basis for describing bonds between surface metal atoms and atoms, such as H, C, N, and O, interacting with the surface during catalysis. Because surface metal atoms are fixed in the metal lattice, they have often been treated as if they were bulk atoms, and interactions with adsorbates have been considered as secondary effects.

The approximation, however, fails to take account of the fact that surface atoms differ from bulk atoms because their environment is intermediate between those of free and bulk atoms. A surface atom has fewer nearest neighbors than a bulk atom but more than a free atom, and so it should have different electronic properties and different interactions with molecules which can be adsorbed. Since many catalysts consist of very small crystallites or clusters of metal atoms on an insulating, metal-oxide support, most of their metal atoms are exposed on a surface and a bulk, as it is usually understood, does not exist.

Bonding in organic compounds can be accounted for quantitatively because the number of atoms typically comprising in a molecule is small and because maximally only four atomic orbitals, one *s* and three *p* orbitals, are involved per atom. With transition metals, the situation is more complex because there are nine atomic orbitals which can become involved in bonding, one *s*, three *p*, and five *d* orbitals, and metal crystals usually contain a large number of atoms. This complex problem can be handled by the *band theory*, which adequately explains many of the important bulk properties of metals.

Consider a set of *N* atoms separated from each other by a distance of several hundred angstroms. Each atom can be described as a discrete entity with its own set of well-defined atomic orbitals given by the solution to the Schrödinger equation of quantum mechanics. If these atoms are brought together to form a crystal, their atomic orbitals overlap and interact, and the Schrödinger equation has to be solved for the many-atom system. If *N* is small and the atoms combine to form a small molecule, however, molecular-orbital theory can be used to provide a solution. For transition metals, if *N* is very small, a small cluster can be formed and molecular-orbital theory can provide a solution. If *N* is large, the problem becomes intractable but a simplification can be found by neglecting the surface and considering the crystal as a regular lattice of atoms.

The solution of the simplified problem indicates that all the *s*, *p*, and *d* orbitals have energies which fall within definite respective ranges and that these energy ranges span a few electron volts. Each orbital has its own energy level within this range, and since there are *N* orbitals, with *N* very large, the energy levels are so close together that they form an almost continuous band. The result is illustrated in Fig. 3-5 for *N* orbitals forming a so-called energy band. The energy levels are usually filled beginning with the lowest level, as spin-paired electrons are added until all electrons are accounted for. The Fermi level is defined as the highest

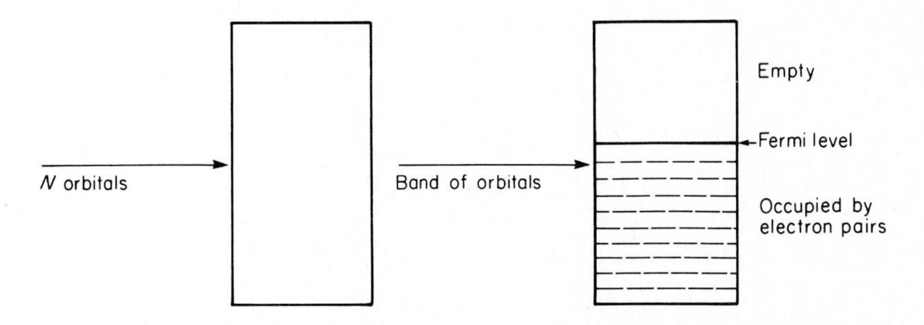

Figure 3-5 Combination of *N* orbitals to form an energy band and placement of spin-paired electrons in this band.

occupied molecular orbital (HOMO) or the energy level below which all single electron energy levels are filled and above which all energy levels are empty.

According to the band model, all electrons are completely delocalized over the crystal. The fact that the energy difference between the HOMO and LUMO is very small means that an electron can easily be promoted from a filled level below the Fermi level to an empty level above the Fermi level, and so conduction readily occurs even in a weak electric field. The band model explains many of the physical properties of metals, but the assumption of complete delocalization of electrons presents some difficulties which are discussed below.

The interactions between *s* orbitals are relatively strong, and the *s* band is usually quite broad, ~ 20 eV. Interactions between *p* orbitals and between *d* orbitals are usually weaker, and these bands are typically narrower, ~ 4 eV. Because relatively few electrons are involved in the *s* band, the energy levels are more widely spaced than in the *d* band, in which a large number of electrons are confined to a narrow band. The appreciable width of the *s* band means that it frequently overlaps the *d* band.

In a row of the periodic table of elements, there is an increase in ionization energies from left to right, which indicates that the Fermi level decreases in this direction. The relative positions of the *s*, *p*, and *d* bands also change. At the extreme left, the *s* band is the lowest, and the *d* band is the highest; at the extreme right, the *d* band is low and overlapped by the *s* band, and the *p* band is highest. Figure 3-6 shows the positions of the *d* band and *p* bands relative to the *s* band for three positions along a row in the periodic table.

The *s* and *d* bands are shown in Figure 3-7 in terms of the density of energy levels (density of states) for metals such as Ni and Cu. The *d* band overlaps the *s* band, is narrow, and has a higher density of states than the *s* band. In Figure 3-7*A* some energy levels remain unfilled in the *d* band. These empty levels are referred to as *holes* in the *d* band. In Figure 3-7*B*, the *d* band is completely filled, and the

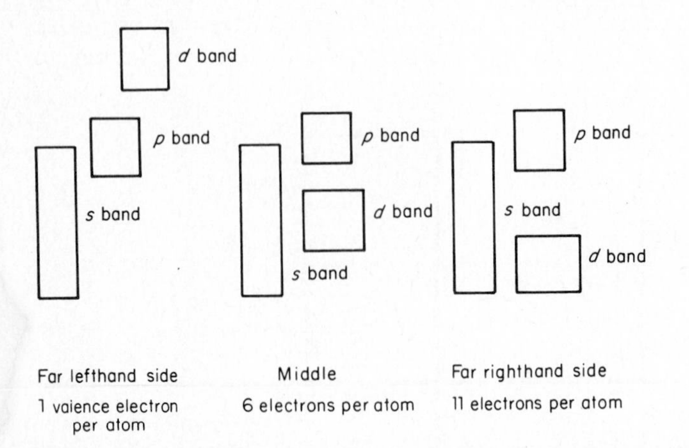

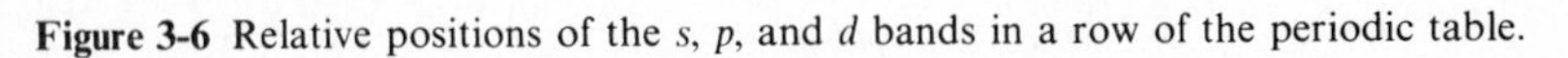
Figure 3-6 Relative positions of the *s*, *p*, and *d* bands in a row of the periodic table.

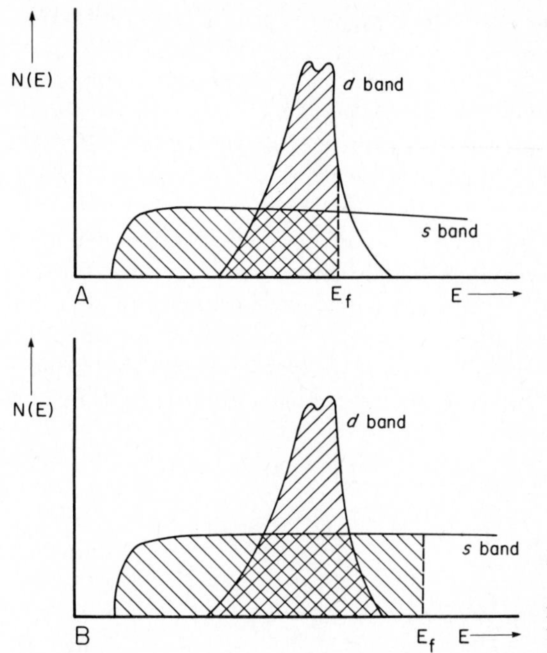

Figure 3-7 Representations of *s* and *d* bands showing two different degrees of filling; A might be considered typical of Ni, B of Cu.

highest occupied energy levels are of the *s* type. Modern analytical tools allow measurement of the density of states $N(E)$, so that the effects of system changes on electronic properties can be determined.

In the periodic table, the degree of occupancy of the bands by electrons varies within a row because of changes in the total number of electrons and in the relative positions of the bands. As one proceeds across a given row from left to right, one finds that electrons are added to the orbitals in the energy bands starting with the lowest one. The bands may be filled, for example, as electrons are placed in the orbitals, one per orbital with all spins parallel, until all the orbitals contain one electron, followed by addition of the remaining electrons, each with a spin antiparallel to that of the electron already present. Alternatively, the orbitals may be filled by addition of electron pairs to the lowest available orbitals, each pair having antiparallel spins. If the electrons are added by the second method, the order of filling the three bands for 1, 6, and 11 electrons is as indicated in Fig. 3-6. In this case, only the *s* band is occupied near the left-hand side of the row; the *s* and *d* bands are partially occupied near the center; and the *d* band is completely filled and the *s* band partially filled near the right.

When a band such as the *d* band is completely filled, it can no longer contribute to bonding. The maximum bonding should therefore exist when there are six electrons per atom, one each for the one *s* band and five *d*-bonding orbitals available. As more electrons are added, orbitals must become occupied by spin-paired electrons and become unavailable for bonding. Consequently, the maximum lattice binding energy and maximum heat of sublimation are expected when there are six electrons per atom. The expected behavior is generally confirmed by the data of Fig. 3-8, and the pattern is most closely followed in the last row of transition metals. In the first row, a minimum exists, which suggests a change in the order in which the orbitals are filled. A secondary maximum exists for two of the rows, and this is probably due to overlap of the *s* and *p* bands (Fig. 3-6), leading to an increase in the number of bonds.

The atoms in a metal phase are tightly packed, having cubic close packing, hexagonal close packing, or body-centered cubic packing. Each metal atom has 12 nearest neighbors, except in a body-centered cubic lattice, which means that it is not possible to have a fixed set of ordinary covalent bonds between all adjacent metal-atom pairs because there are neither sufficient atomic orbitals nor sufficient electrons. According to band theory, all valence electrons are distributed uniformly over the lattice in a band, but attempts have also been made to invoke

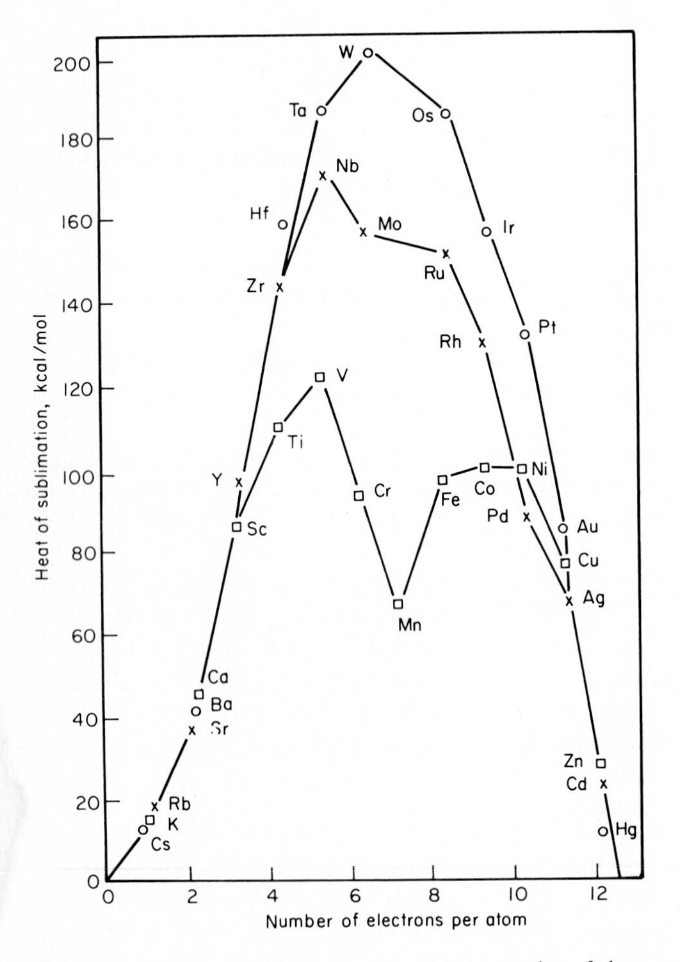

Figure 3-8 Correlation of bond strength with the number of electrons per metal atom.

resonance electron-pair bonding (valence-bond theory) involving all atom pairs. For W, for example, the six bonds per metal atom are presumed to resonate between the 12 nearest neighbors, each pair having one-half a bond. Therefore each bond of W, which has a high heat of sublimation, has an energy of only 32 kcal/mol; this is roughly half the C—C bond energy in diamond, in which each carbon atom has only four nearest neighbors.

Pauling [7] attempted to account for band overlap on the basis of resonance between different hybridization systems, each metal atom providing *dsp* hybrid orbitals which overlap to form metallic bonds. The arguments above suggest that Cu, which has a $3d^{10}4s^1$ electron configuration, would have only the one *s* electron involved in bond formation. Yet the sublimation energy of Cu is more than 4 times those of Cs and Rb, which have only one bonding electron each (Fig. 3-8). Empirical correlations [7] between the number of electrons engaged in bond formation and bond length suggest that 2.44 of the 5 *d* electrons of Cu are involved in bonding. If excited structures such as $(d^6)dd\ s\ pp$ or $(d^4)ddd\ s\ ppp$ contribute to bond formation (where the electrons in *d* orbitals enclosed in parentheses are assumed to remain in nonbonding atomic orbitals), the number of electrons in metal-metal bonding orbitals becomes greater and the bonds stronger. If hybridization between the excited states were to occur, bonding like that for Cu, for which the hybridization is about midway between the limiting cases, could explain the bond strengths observed.

Since one structure involves more *d*-orbital contribution to the *dsp* hybrids than the other, the actual resonating bond can be described by the contribution of *d* orbitals to the *dsp* hybrids or by its *percentage d character*. Pauling gave values of this percentage for each transition element, having estimated the values from the strength of the metal-metal bonds. Pauling's concept is important to the following discussion because it provides one of the parameters which has been used in correlating catalytic activities with physical properties of transition metals.

The partial contribution of *d* and *s* orbitals changes from left to right along a row of the periodic table; it first increases then decreases. The resonance theories, band theory, and Pauling's percentage *d* character represent different ways of accounting for this general phenomenon.

The 10-electron systems (Ni, Pd, and Pt) would appear to have their *d* bands completely filled. Because of electron overflow from the *d* to the *s* band (Fig. 3-9), however, the *d* band is not completely filled. One refers therefore to holes in the *d* band, i.e., the number of electrons per atom still missing from the *d* band. For ferromagnetic metals, holes in the *d* band can be measured quantitatively. Fe, Co, and Ni have permanent magnetic moments indicative of their electronic structures. Each atom has a magnetic moment equal to the vector sum of its electron spins. The vector sum of these magnetic moments of a crystal equals the ferromagnetic moment of the crystal; i.e., the magnetic moment equals the number of atoms in the crystal times the vector sum of electron spins per atom. The magnetic moment is measured by the magnetization per unit volume of a solid.

It is usually assumed that because of the broader width of the *s* band and smaller number of electrons involved, the energy levels are sufficiently spaced for

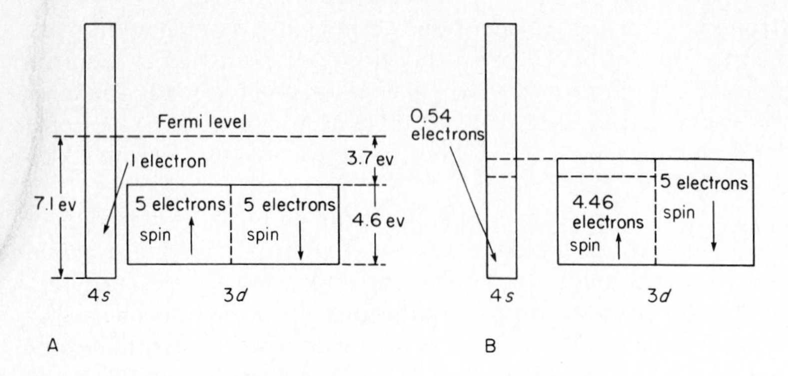

Figure 3-9 Schematic representation of electron structures of Cu (A) and Ni (B).

the electrons to be spin-paired. The closer energy spacing in the *d* band allows electrons to remain unpaired, and the magnetic moment arises from these unpaired electrons, as shown in Fig. 3-9 for Cu and Ni. In Cu, the *d* band is low enough to be completely filled, and Cu shows no magnetic moment. The measured magnetic moment of Ni is 0.6 per atom, and this result (when corrected for the magnetic-moment contribution due to orbital electronic motion) gives a deficiency of 0.54 electron per atom in the 3*d* band. To obtain the Ni electron number formally, we can remove one electron from Cu, but we now have on the average an overflow of 0.54 electron from the 3*d* band to the 4*s* band, resulting in a net magnetic moment of about 0.6. The agreement between the observed value and this calculated electron overflow supports the band model.

The changes observed in electronic properties from one element to the next in the periodic table are relatively large, but the band model suggests that the electronic properties of a given metal can be changed in a continuous manner by alloying since electrons can be added continuously to the metal band structure as the composition is changed. For example, addition of Cu to Ni should gradually increase the average number of electrons per Ni atom and gradually fill the vacancies in the *d* band. The changes should be evidenced by gradual changes in the ferromagnetic properties, indicating losses of holes in the *d* band. The effects of alloying are more complex than has been supposed above, however, and in fact there is no loss in the number of holes in the *d* band of Ni when Cu is alloyed with it. This subject is considered in detail below.

This completes our discussion of the simple band model and related considerations of bulk metals. We next consider surfaces, where the interactions essential to catalysis occur. The surface is a discontinuity not considered in the models of solids discussed above. The phenomena of chemisorption and bond formation between surface atoms and added atoms become the central issue of the following discussion.

Surface atoms represent a termination of the periodic bulk, having fewer nearest neighbors and existing in an asymmetric environment. The presence of the surface perturbs the periodic potential of the virtually infinite bulk of a solid and gives rise to solutions of the wave equation which do not exist for the infinite periodic lattice [8]. The use of appropriate boundary conditions leads to solutions that predict electronic states localized at the surface which have energy levels slightly above those of the bulk states. Therefore there is a dipole with its positively charged end at the surface of a metal.

Grimley [9] proposed that the surface orbitals are located near the top of the energy bands. Filling the bands then is accompanied by filling the surface orbitals, but because of their relatively higher energy levels, the surface orbitals are filled after the band is filled. From the point at which the surface orbitals begin to be filled with single electrons to the point at which they contain pairs of electrons, they should be similar to "dangling valencies" and should be quite reactive. When the surface orbitals become occupied by electron pairs at high degrees of band filling, they should become less reactive again. The strength of interaction of a singly occupied surface orbital with a molecular orbital of an approaching species depends on the relative energetical positions of the surface orbital and of the molecular orbital. Quantitative predictions cannot be made from this model, although it suggests that there should be a maximum in bond strength between the surface atoms and a given adsorbate and that there should be similar behavior from left to right in each row of the periodic table. A quantitative quantum-mechanical approach to these ideas is considered in a later section.

Bonding at Metal Surfaces

Energies of surface bonds have been determined from heats of adsorption of gases on surfaces of metals, but the data have been difficult to interpret because the heat of adsorption typically decreases with an increase in the amount of gas adsorbed. The variable heat of adsorption has been explained as an indication of surface heterogeneity and the corresponding differences in bond strengths for adsorption on different sites. It has also been proposed that repulsive interactions between adsorbed species reduce the observed heat of adsorption with increasing surface coverage and, alternatively, that adsorption on one site affects the bulk so that the heat of adsorption on neighboring sites is reduced. Recent studies with single crystals have shown that there are several distinct types of adsorption sites on each crystal plane, that these site types are filled one by one in order of decreasing heat of adsorption, and that the heat of adsorption is independent of coverage for each site type [10]. It is concluded that surface heterogeneity is the most probable explanation for decreasing heats of adsorption with increasing surface coverage.

If surface bond strengths are estimated from heats of adsorption at low surface coverages, the heat of adsorption decreases from left to right in the periodic table. The generalization is valid for a broad range of gases adsorbed on transition metals [11, 12] and is illustrated in Fig. 3-10 for oxygen adsorption on the transition metals of all three rows.

If surface bond strengths estimated from heats of adsorption at low coverages are compared with bond strengths of bulk compounds formed from the metal and the adsorbed species, a linear correlation is observed. This correlation is il-

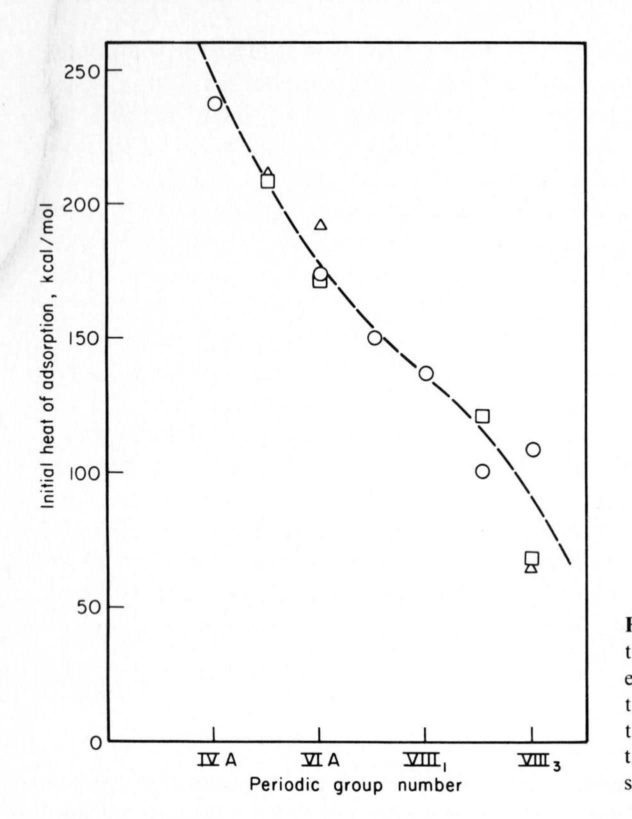

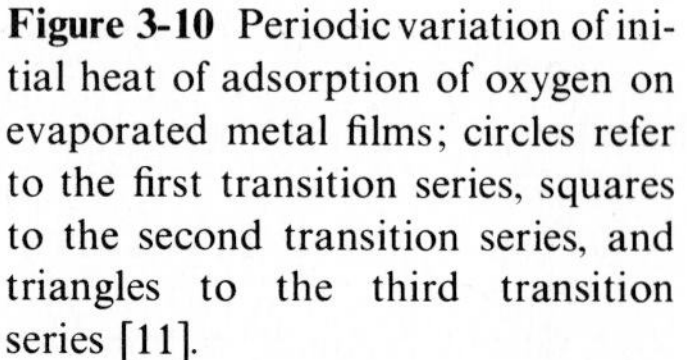
Figure 3-10 Periodic variation of initial heat of adsorption of oxygen on evaporated metal films; circles refer to the first transition series, squares to the second transition series, and triangles to the third transition series [11].

lustrated in Fig. 3-11 for O_2 adsorption on transition metals. The correlating parameter is the heat of formation of the "most stable" oxide of the metal per mole of oxygen [11]. A correlation of this type was proposed by Sachtler et al. [13]; it is referred to as a Sachtler-Fahrenfort correlation.

Tanaka and Tamaru [14] applied a slightly different correlation, plotting the heat of adsorption per mole of O_2 adsorbed vs. the heat of formation of the highest oxide per atom of metal in the oxide, where "highest" means highest valence state of the transition-metal ion. The upper line in Fig. 3-12 shows that this correlation appears to be better than that of Sachtler and Fahrenfort. These correlations suggest that the bond formed when oxygen adsorbs on the surface of a metal is similar to the bond formed in the bulk oxide and is suggestive of the formation of a *surface compound*, a surface metal oxide.

Figure 3-12 also shows that the heats of adsorption of C_2H_4, N_2, NH_3, and H_2 all follow a similar pattern. This result can be understood by analogy with the observation that heats of formation of compounds M_nL_m (where L is any ligand) are usually correlated linearly with the heats of formation of the respective oxides, as illustrated in Fig. 3-13 for nitrides. Thus if the heat of adsorption of a gas on a metal correlates with the heat of formation of its bulk compound, as is typically observed, then the heat of adsorption of that gas on a series of metals correlates with the heat of formation of metal oxides (Fig. 3-12).

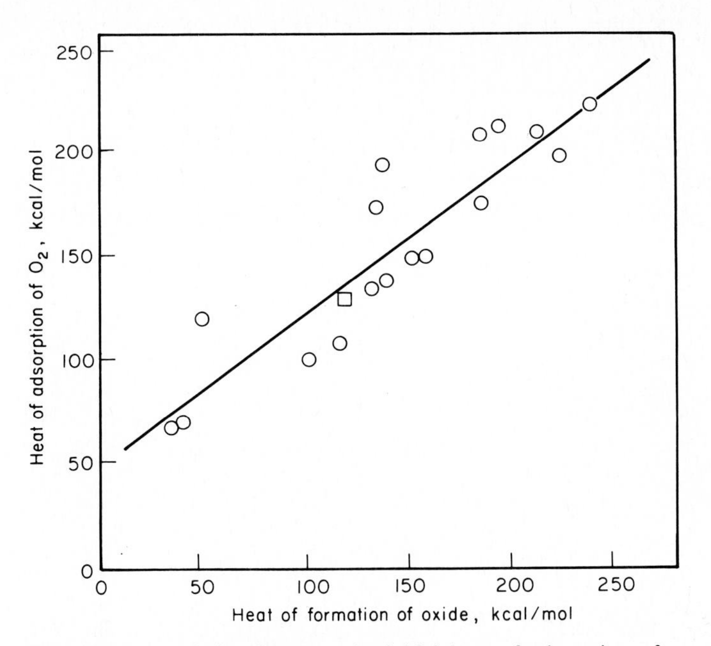

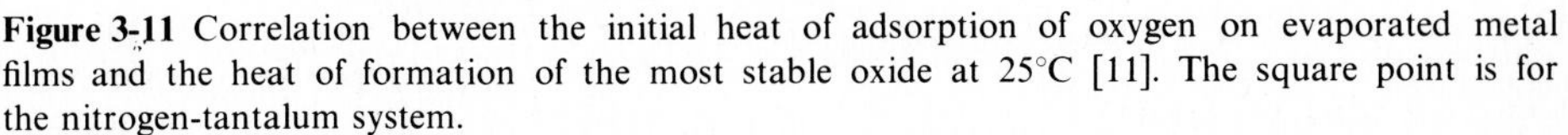

Figure 3-11 Correlation between the initial heat of adsorption of oxygen on evaporated metal films and the heat of formation of the most stable oxide at 25°C [11]. The square point is for the nitrogen-tantalum system.

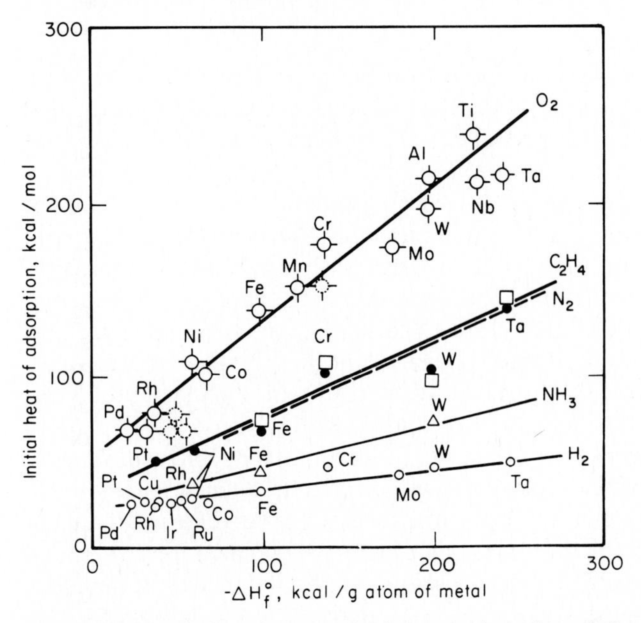

Figure 3-12 Correlations between the heat of adsorption of O_2, N_2, C_2H_4, NH_3, and H_2 on the respective reduced metal (as coverage approaches zero) and the heat of formation of the highest metal oxide per atom of metal [14]. (Reprinted with permission from *Journal of Catalysis*. Copyright by Academic Press.)

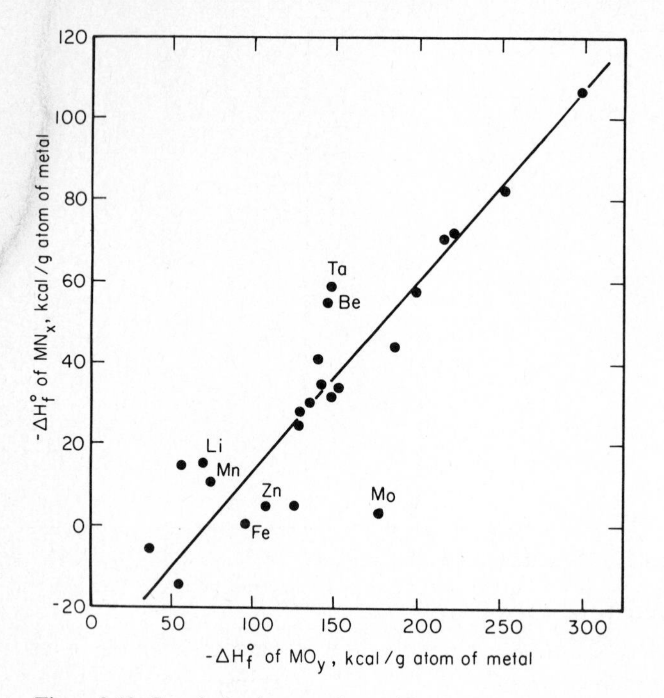

Figure 3-13 Correlation between heat of formation of the highest stable nitride and the highest stable oxide.

It follows from these results that bonds formed by adsorption of a gas on a metal surface are often similar to the bonds formed between the metal and the molecules of the gas in the formation of a three-dimensional bulk compound involving these two species; the adsorption process may be thought of as involving the formation of a two-dimensional surface compound. The idea of a surface compound has proved especially useful in understanding surface catalytic activity; it finds support in comparisons of metal surfaces with single-metal-atom complexes like those discussed in Chap. 2. Energies of metal-hydrogen bond formation resulting from adsorption on a clean metal surface or from addition to a metal complex are similar and only weakly dependent on the *d*-electronic properties of the metal [15]: the metal-hydrogen bond energy is about 65 kcal/mol for Ir, Rh, Ru, Pt, and Pd; 63.6 kcal/mol for Co; and 67 kcal/mol for Ni. It is 63 to 64 kcal/mol for some complexes of Ir and about 57 kcal/mol for some complexes of Co. Correspondingly, the metal-hydrogen vibrational frequencies are almost independent of whether the metal is part of a surface or part of a single-metal-atom complex [15]. Studies of combinations of CO with metals have provided confirmation of the foregoing idea, but they also have pointed to some differences, which are discussed below [16].

Surface species formed upon interaction of a gas with a metal surface frequently exhibit spectra similar to those of their bulk-phase counterparts

[17, 18]. For example, the adsorption of formic acid on metals leads to the formation of structures having infrared spectra like those of the appropriate metal formates [19]; the bonds between C and O are equivalent, as in

$$\left(HC\begin{smallmatrix}\cdot\cdot\cdot O\\ \cdot\cdot\cdot O\end{smallmatrix}\right)^-$$

Surface Ni atoms even appear to leave their original sites to form the two-dimensional surface compound [20].

Ferromagnetic-susceptibility measurements of supported Ni having crystallites less than 40 Å in size (so that a significantly large fraction of the Ni atoms are on the surfaces of the crystallites) show a decrease in ferromagnetic susceptibility occurring upon adsorption of a gas and directly related to the amount of gas adsorbed [21]. These results indicate that adsorption on a surface metal atom removes it completely from cooperative ferromagnetic interaction with the bulk and thus suggest that adsorption disengages the metal atom at least partly from the band structure of the bulk metal. In support of this suggestion is the observation that the conductivity of thin metal films ($\sim 10^3$ Å) decreases significantly upon chemisorption of gases, and the decrease is that predicted if it is assumed that chemisorption prevents the surface layer of metal atoms from contributing to the bulk electrical conductivity [20].

Secondary-ion emission spectroscopy (in which surface atoms or surface complexes of several atoms are sputtered off by bombardment of the surface with high-energy argon ions) also indicates that surface metal atoms are largely removed from the metal lattice upon chemisorption of a gas. Rates of secondary-ion (metal-ion) emission are quite low for clean metal surfaces but become several orders of magnitude higher when a gas is chemisorbed on the surface. It is inferred that chemisorption results in a significant decrease in the binding strength of surface metal atoms to the metal lattice. Analysis of surfaces, e.g., by low-energy electron diffraction (LEED) and field emission, has suggested that in many cases chemisorption causes a complete rearrangement of the surface layer (corrosive chemisorption) to accommodate surface-compound formation [20, 22].

All these results together suggest that chemisorption may often be considered as the formation of a surface compound; the bonding requirements of surface-compound formation result in a reduction in the strength of bonding of metal atoms to neighboring metal atoms and in a reduction in the contribution of the surface-metal-atom electrons to the band structure of the bulk metal, i.e., partial relocalization of electrons in surface-metal-atom bonding orbitals. The strength of bonding in the surface compound probably determines the extent of relocalization of electrons in the surface metal atom. For strong bonding, as in corrosive chemisorption, the localization of electrons in the surface metal atom is almost complete, and the surface metal atom probably contributes little to the band structure of the bulk metal; i.e., all electrons are localized in bonding orbitals to the added (adsorbate) atom(s) involved in surface-compound formation and to the metal atoms immediately associated with it in the bulk.

Reactions of Chemisorbed Species

Consider a simple reaction such as the decomposition of formic acid catalyzed by various metals [19]:

$$HCOOH \longrightarrow H_2 + CO_2 \tag{11}$$

Formic acid readily adsorbs on most metal surfaces, and, as noted previously, infrared evidence has shown that the adsorbed formic acid resembles a surface metal formate. The overall reaction therefore involves the formation of the surface intermediate followed by its decomposition into metal, CO_2, and H_2. Since it is a species like a metal formate which decomposes, one might expect the catalytic activity to be related to the stability of the metal formate. If the stability of the metal formate is low, as for silver and gold formates, the surface metal formate may not even be formed in appreciable concentration, corresponding to a low catalytic activity. On the other hand, if the stability of the surface metal formate is too high, decomposition is expected to be slow, corresponding to a low catalytic activity of the particular metal; the surface could be nearly covered with a stable metal-formate layer. It follows that the most active catalysts should be those having optimum bonding of the surface species and therefore an intermediate heat of formation or stability of the metal formate.

The foregoing argument is essentially a statement of the *principle of Sabatier*, dating from 1911 [23, 23a]; it has already appeared in a slightly different form in

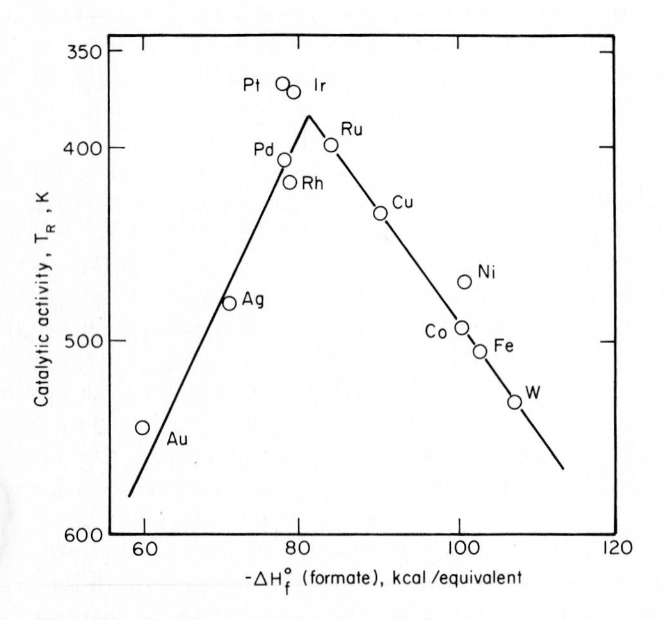

Figure 3-14 Catalytic activities of metals for formic acid decomposition as a function of heat of formation of metal formate. T_R is the activity given as the temperature required to achieve 50% conversion with all other conditions constant [23*a*].

Chap. 2. To check Sabatier's principle, catalytic activities of metals for formic acid decomposition are plotted as a function of the heat of formation of the respective metal formates (Fig. 3-14). The maximum in the curve (called a *Balandin volcano curve*) confirms the principle.

If the data of Fig. 3-14 are correlated with the heat of formation of the highest oxide per metal atom (a *Tanaka-Tamaru plot*), Fig. 3-15*A* is obtained; and if they

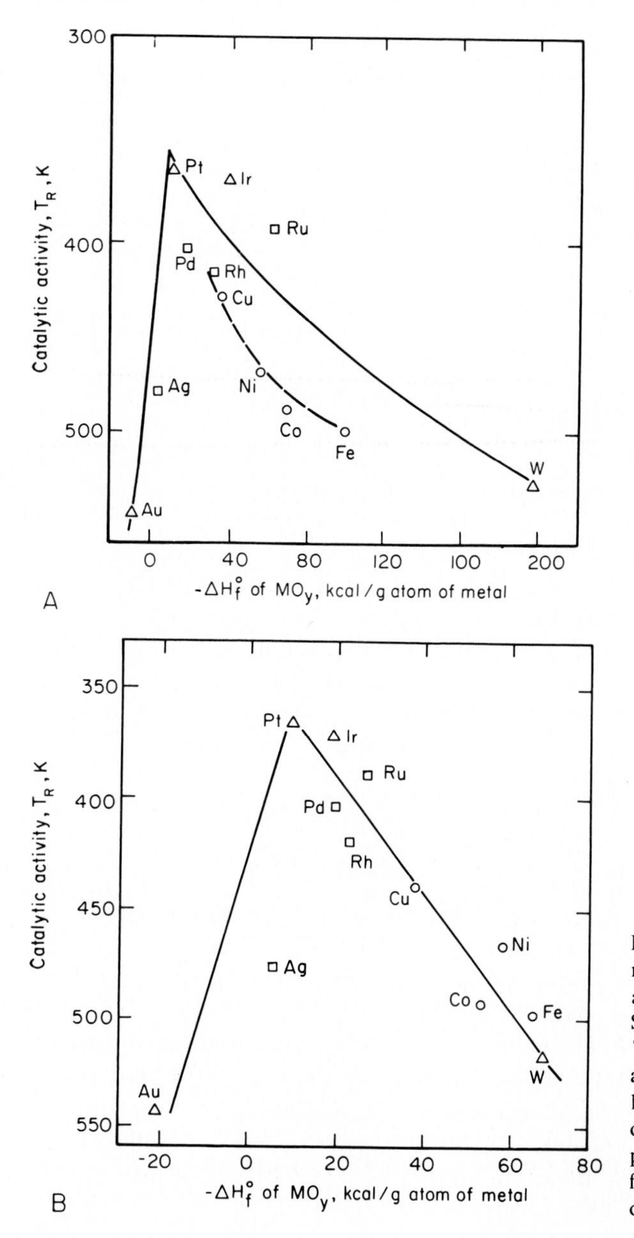

Figure 3-15 Catalytic activities of metals for formic acid decomposition according to the Tanaka-Tamaru and Sachtler-Fahrenfort correlations. The Tanaka-Tamaru parameter is defined as the heat of formation of the highest stable metal oxide per mole of metal. The Sachtler-Fahrenfort parameter is defined as the heat of formation of the most stable metal oxide per mole of oxygen.

are correlated with the heat of formation of the most stable oxide per mole of oxygen atoms (a *Sachtler-Fahrenfort plot*), Fig. 3-15*B* is obtained. The two are similar, and both provide confirmation of the Sabatier principle for formic acid decomposition.

The idea developed above is straightforward for the simple formic acid decomposition; the same rationale can also be applied to more complex reactions if it is assumed that any catalytic reaction proceeds on a surface by a series of elementary steps:

$$\begin{array}{l} R + M \longrightarrow R_1 \longrightarrow R_2 \longrightarrow R_3 \longrightarrow P + M \\ \qquad\qquad\quad \lfloor\!\longrightarrow R_4 \longrightarrow P_1 + M \end{array} \tag{12}$$

where R = reactant(s)
R_i = set of surface reaction intermediates
P = product(s)

The overall rate of reaction is determined by the rate of reaction of the most stable surface reaction intermediate [24]. The composition of this intermediate is usually unknown, but it can be postulated that the stabilities of possible intermediates are linearly related to the heats of formation of the corresponding bulk metal oxides, which are known for many examples. If the postulate is valid, one would expect (as suggested by Balandin [25, 26]) that for any reaction there should be a volcano curve generated by plots of observed relative activities of the metals vs. the heats of formation of the corresponding metal oxides. Our lack of knowledge of the nature of the most stable intermediate prevents an a priori prediction of the position of the top of the volcano curve.

To test the validity of Balandin's proposal, data for three additional reactions are plotted below. Since it is not clear whether the Tanaka-Tamaru or the Sachtler-Fahrenfort correlation is more appropriate, both are applied. Relative rates of three reactions,

$$C_2H_4 + H_2 \longrightarrow C_2H_6$$
$$2NH_3 \longrightarrow N_2 + 3H_2$$
$$C_2H_6 + H_2 \longrightarrow 2CH_4$$

are plotted in Fig. 3-16 according to the Tanaka-Tamaru correlation and in Fig. 3-17 according to the Sachtler-Fahrenfort correlation. The Tanaka-Tamaru correlations are more satisfactory, although they show a difference between first-row and second- and third-row transition metals. There is no such difference in the correlation of their heats of oxygen adsorption. However, the first-row transition metals are different in their heat-of-sublimation behavior (Fig. 3-8) and are ferromagnetic, so that perhaps the different catalytic behavior should not be considered so unexpected; it might be explained by assuming that (1) the thermodynamic stabilities are similar for all rows but (2) the reactions leading to and from the most stable intermediate (including adsorption and desorption steps) are slower

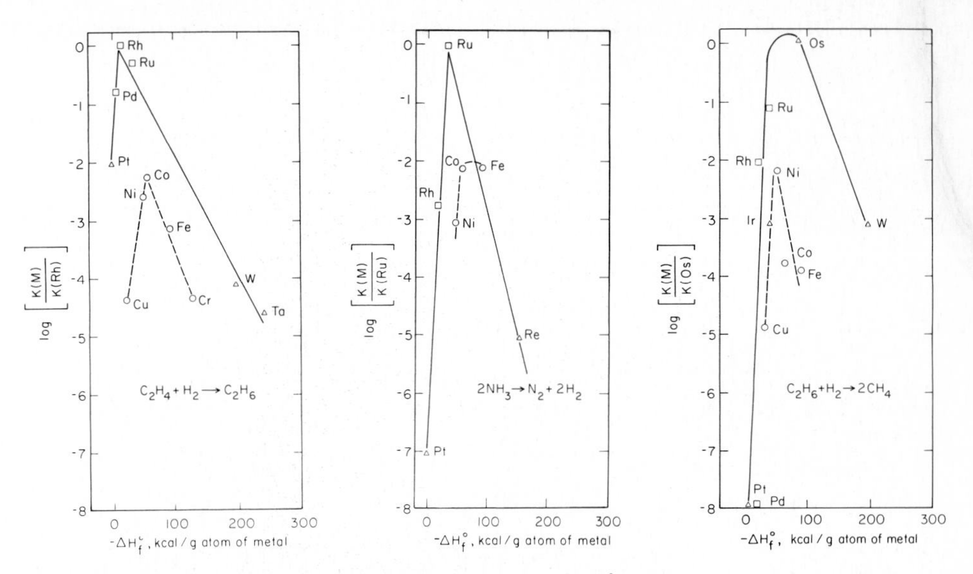

Figure 3-16 Tanaka-Tamaru correlation of catalytic activities of metals.

(or that side reactions play a different role for the first row). Plots according to the Sachtler-Fahrenfort correlation do not seem to discriminate between the rows in the periodic table, but they show some poor fits.

The empirical correlations are generally good, but they raise several points that need to be considered further. First, they are relations between thermodynamic quantities such as heats of adsorption and heats of formation; as such they are important but not necessarily relevant to kinetics, although the Polanyi relations between activation energies and heats of reaction suggest that they might be important [24]. The Polanyi relations hold for some homogeneous solution reactions, but they have not been demonstrated for activation energies and rates of reactions involving transition metals.

Second, in the discussion of formic acid decomposition it was suggested that on the right-hand side of the volcano curve the rate is low because the surface is covered by an increasingly stable compound and the rate-determining step is the decomposition of the stable surface compound. This is certainly the case, and the lower curve in Fig. 3-18 schematically illustrates the point: there is a low activation energy associated with the adsorption step, but the surface compound is very stable; its rate of decomposition controls the overall rate of reaction. On the left-hand side (the increasing portion) of the volcano curve, it was argued, the rate was low because of the low stability of the surface compound, which meant that there was little of it on the surface to decompose. It is implied by the Polanyi relation, however, that if the heat of formation of the adsorbed surface compound is too low, the activation barrier for its formation is high. The rate of formation of the surface compound is then the rate-determining step, and the low rate of formation of the surface compound accounts for the low surface concentration and the low rate of reaction.

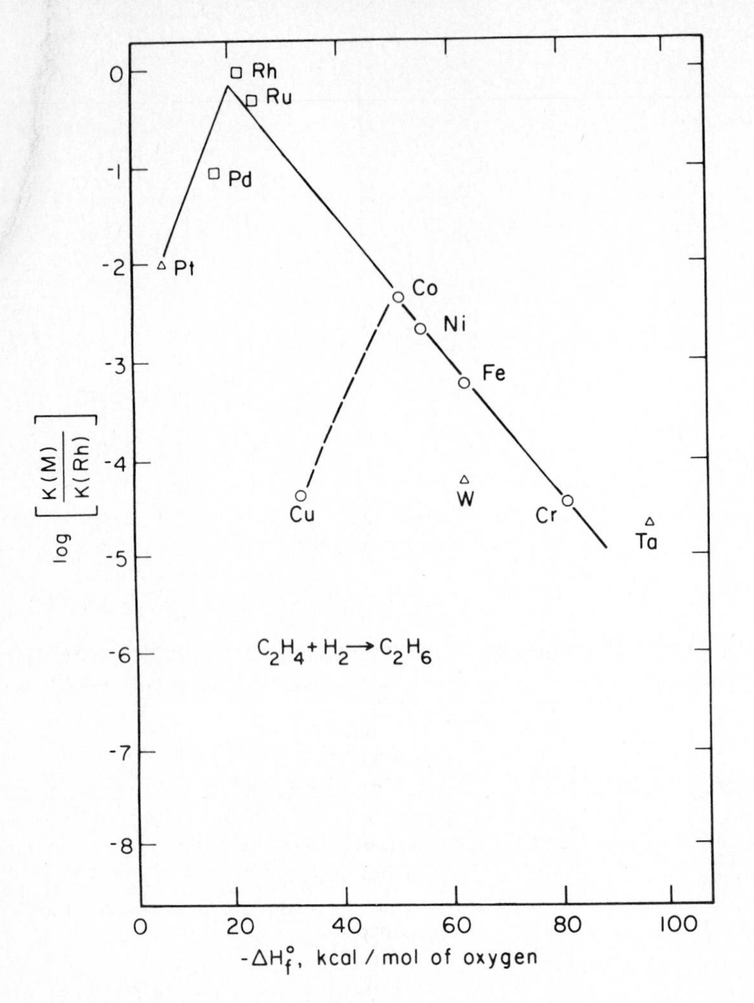

Figure 3-17 Sachtler-Fahrenfort correlation of catalytic activities of metals.

This concept is illustrated as follows. The best NH_3 synthesis catalysts are Ru and Fe, the same metals that are good NH_3 decomposition catalysts (Fig. 3-16). Tungsten is not very active for NH_3 decomposition because it binds nitrogen too strongly and forms a stable tungsten nitride surface phase. Platinum is not a good NH_3 decomposition catalyst because it does not bind and/or dissociate NH_3 well (Fig. 3-13). For NH_3 synthesis, it is necessary to break the strong N≡N bond (226 kcal/mol) to form NH_3. Tungsten chemisorbs nitrogen and effectively dissociates the N≡N bond. However, it is not a good NH_3 synthesis catalyst because the tungsten nitride formed is too stable and the nitrogen atoms are not easily hydrogenated off as NH_3. Both Fe and Ru chemisorb nitrogen and dissociate the N≡N bond; yet they form weakly stable metal nitrides (Fig. 3-13). They are good

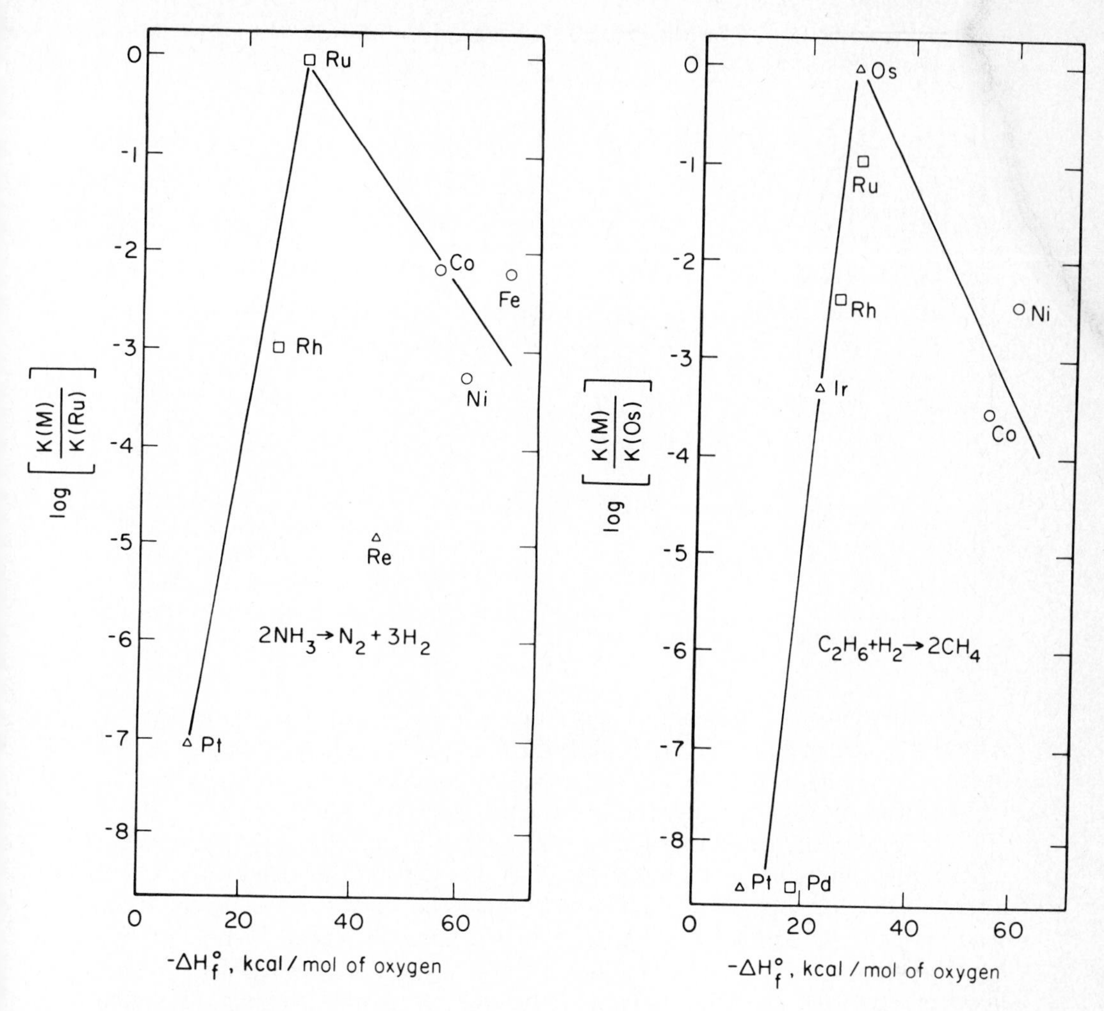

Figure 3-17 (Continued)

catalysts for both NH_3 decomposition and NH_3 synthesis. Platinum, on the extreme left-hand side, does not chemisorb nitrogen and does not dissociate the N≡N bond. Platinum is not a good NH_3 synthesis catalyst because of its inability to break the N≡N bond, which relates to the low stability of platinum nitride.

The third point relates to the complexity of some surface reactions as suggested by the indicated side reaction giving R_4 and product P_1 in Eq. (12). Such side reactions can lead to surface deactivation. For ethylene hydrogenation, for example (Fig. 3-16), there is usually some polymerization of ethylene to give higher-molecular-weight hydrocarbons which remain on the surface [27] and which come off only at higher temperatures and higher hydrogen partial pressures. Further dehydrogenation can also occur, leading to acetylenic and carbon

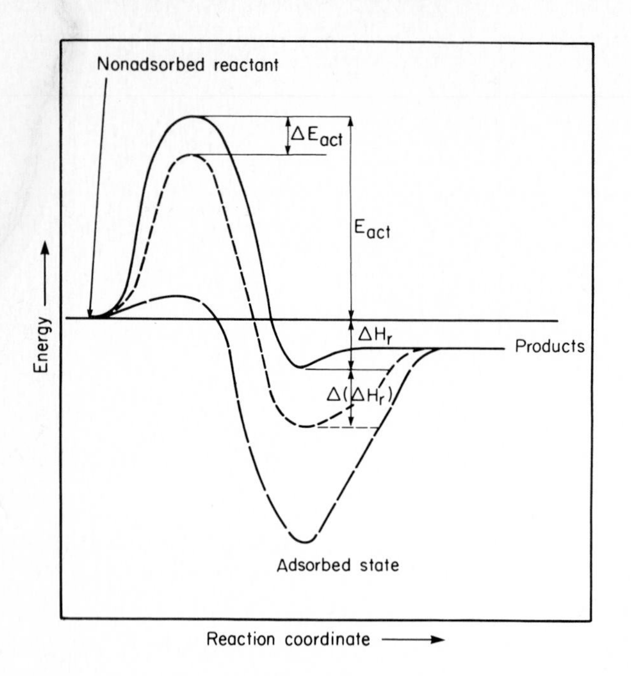

Figure 3-18 Polanyi relation showing potential relationship between heat of reaction and activation energy barrier for reaction, E_{act}; $E_{act} = -\alpha' \Delta(\Delta H_r)$.

residues on the surface. These residues block active sites and can alter the relative activities of the metals [28, 29]. Kouskova et al. [30] concluded that in ethylene hydrogenation Rh was most active because it was deactivated to a lesser extent by hydrocarbon residues than the less active metals, for example, Pt.

For formic acid decomposition on Ni, McCarty and Madix [31] observed in ultrahigh-vacuum-chamber studies that a clean Ni single crystal surface is very active for the decomposition of adsorbed formic acid, producing about equimolar quantities of CO and CO_2, whereas a precarbided Ni single crystal surface is much less active and produces predominantly CO_2. Decreases in activity and shifts in selectivity to CO_2 have been observed for freshly deposited Ni films [32] and for Ni wire [33], indicating that supported Ni [19] may have also experienced side reactions leaving carbon on the surface. The stability of the carbide may be the important parameter on the right-hand side of the volcano curve. This suggestion leads to the generalization that the reactions on metals involving reactants which may form stable compounds with the metal may actually involve reactions on the surface of the metal compound, e.g., metal oxides, metal nitrides, and metal carbides. Such surfaces are typically not as active as the clean metal surface.

As an alternative to the correlations between catalytic activity and thermodynamic properties of compounds related to surface intermediates, correlations between catalytic activity and percentage *d* character of the metal-metal bond have sometimes been found. Such a correlation is shown for ethane hydrogenolysis catalyzed by group VII and VIII transition metals (Fig. 3-19). In contrast to these data, the data for formic acid decomposition correlate poorly with percentage *d* character.

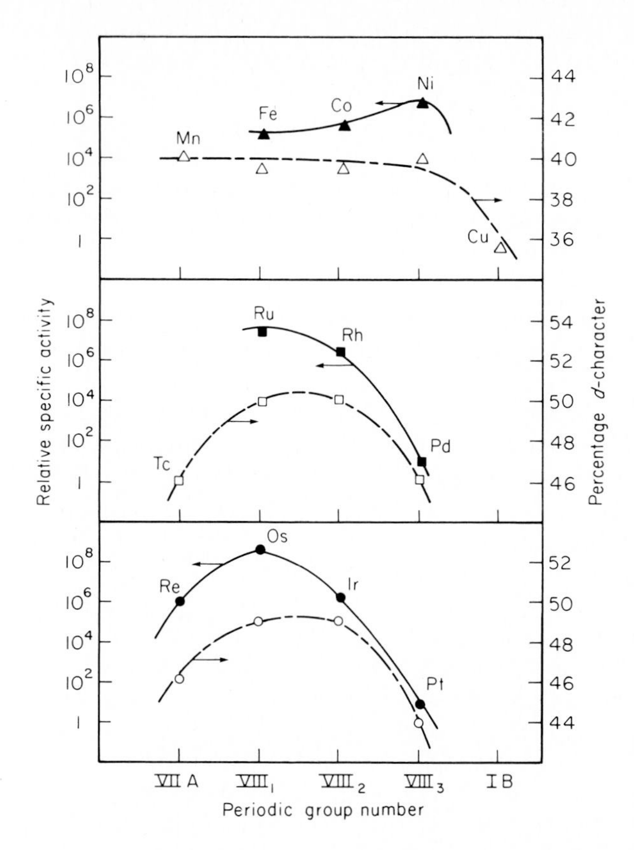

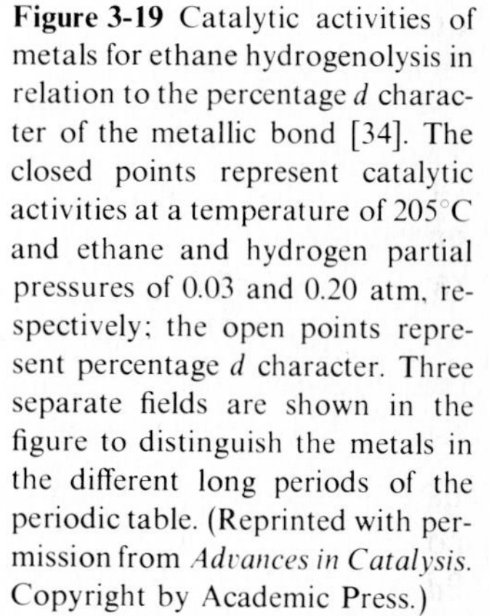
Figure 3-19 Catalytic activities of metals for ethane hydrogenolysis in relation to the percentage *d* character of the metallic bond [34]. The closed points represent catalytic activities at a temperature of 205°C and ethane and hydrogen partial pressures of 0.03 and 0.20 atm, respectively; the open points represent percentage *d* character. Three separate fields are shown in the figure to distinguish the metals in the different long periods of the periodic table. (Reprinted with permission from *Advances in Catalysis*. Copyright by Academic Press.)

The percentage *d* character is a fitting parameter from the valence-bond theory, a parameter derived from the bond length in a metal. When it provides a good correlation of catalytic activity, one concludes that there is a relation between the bulk-metal properties and the catalytic activity; but percentage *d* character is not always a good correlating parameter. Since the catalytic activity of a surface must depend on the formation of surface bonds with reactants, reaction intermediates, and products, and since the catalytic activity of the surface must depend on the rate of formation and breaking of these bonds, a fundamental correlation of catalytic activity must involve the strengths and reactivities of these bonds. A relationship between the rate of a surface chemical reaction and the stability of the surface compound or most stable intermediate is generally expected to exist. Such relationships provide the most soundly based correlations for catalytic activity.

Alloys

Binary metal systems have been studied for several decades in attempts to understand metal catalysis better. The major motivation for this work was the suggestion from band theory that varying the composition of an alloy would vary the electronic properties of the surface in a continuous manner and allow the effects of electronic properties on catalytic activity to be determined. Until the last decade progress has been slow because of a lack of understanding of the thermodynamics of alloys and because of insufficient attention to several other important complications. The most serious deterrent to progress was the lack of quantitative determinations of the compositions of the surfaces under study. Only recently have techniques of surface analysis been sufficiently refined to provide quantitative surface compositions. Coupled with the improvement in analytical capability has been major recent progress in the theoretical description of alloys, and consequently recent progress in understanding catalysis by alloys has been rapid and striking.

In the following section, the general properties of alloys are considered, along with their surface compositions and catalytic properties. The Cu–Ni system is used as the primary example because it is quantitatively well characterized and because it represents a unique model system. Copper has a $3d^{10}4s^1$ electronic configuration and has a very low catalytic activity for most reactions; Ni has a $3d^{9.4}4s^{0.6}$ electronic configuration in the metallic state ($3d^{10}4s^0$ as a gas-phase atom) and is several orders of magnitude more active than Cu for many reactions. If the rigid band model were to apply, alloying Ni with Cu would be expected to fill the d-band vacancies and lead to striking changes in the catalytic activity of the Ni. The expectation is not generally realized, as discussed below.

Phase behavior The bulk-phase properties of alloys are considered in terms of system thermodynamics, particularly the enthalpy of alloy formation. Alloy behavior can also be considered from the simplified point of view of the strength of pairwise bonds between metal atoms.

Mildly exothermic alloys, exemplified by Pd–Ag, are those for which the enthalpy of formation from the elements $\Delta H_f^\circ \leq 0$ and for which $|\Delta H_f^\circ|$ is small. In terms of pairwise bond energies involving atoms A and B, $(E_{AA} + E_{BB})/2 \approx E_{AB}$. For all temperatures, a single-phase solid solution exists for the whole composition range at equilibrium. There is no tendency for cluster formation, and atoms A and B are randomly dispersed; however, the surface layer, with a thickness of no more than several atoms, is enriched in the component having the lower surface free energy.

Endothermic alloys are characterized by values of $\Delta H_f^\circ > 0$ and $(E_{AA} + E_{BB})/2 > E_{AB}$. For temperatures $T > \Delta H_f^\circ/\Delta S_f^\circ$, the equilibrium alloy forms clusters of A atoms and clusters of B atoms within the bulk because of the greater strengths of A—A and B—B bonds relative to A—B bonds.

At temperatures $T < \Delta H_f^\circ/\Delta S_f^\circ$, a miscibility gap exists, and there are two phases of different composition at equilibrium. The phase diagram for the endothermic Cu–Ni alloy is shown in Fig. 3-20. For points beneath the miscibility envelope, which ranges from about 2 to about 80 atom percent Cu at 100°C, there are two phases with compositions given by the intersections of the envelope and a horizontal line drawn at the system temperature. The relative amounts of the phases are given by the lever rule. For the Cu–Ni system, the critical tempera-

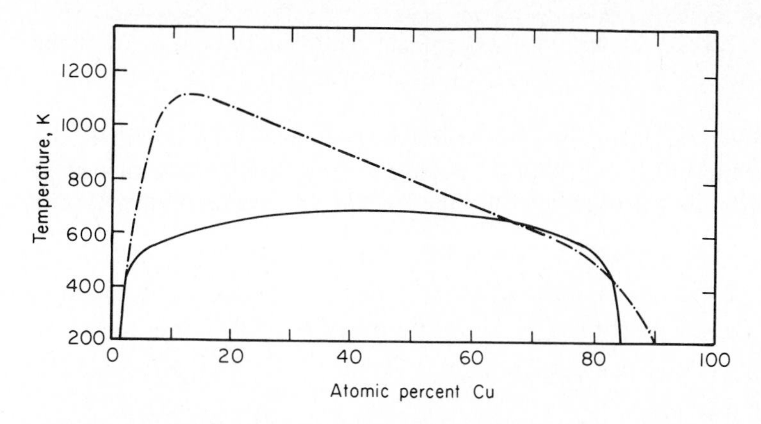

Figure 3-20 Phase diagram for Cu–Ni system as calculated from thermodynamic data from [35*a*] (dashed line) and [35*b*] (solid line) [35]. (Reprinted with permission from *Journal of Catalysis*. Copyright by Academic Press.)

ture above which only one phase can exist is about 320°C [35–35*b*], but the exact position and shape of the miscibility envelope are still somewhat in doubt. Because the critical temperature for the Cu–Ni system is low, long, careful annealing must be done to assure that equilibrium conditions are actually approached.

In the two-phase region, the phase rich in the component which has the lower surface free energy (or lower heat of sublimation) should form an outer layer around each crystallite, the core consisting of the other phase composition. It is well known that phase separation can occur within a crystallite, but there are no methods for predicting whether it will give a core of one composition and an outer layer of the other (the cherry model) or whether, alternatively, it will result in the formation of distinct crystallites of different composition (the marble model). Films of Cu–Ni about 200 Å thick, prepared by first evaporating Cu and then Ni, and vice versa, gave separation at about 200°C into a Cu-rich phase (80 atom percent Cu, 20 atom percent Ni) on the periphery and a Ni-rich phase (2 atom percent Cu, 98 atom percent Ni) in the core of the cherrylike crystallites [13, 35]. The composition of this Cu-rich phase was unchanged over a broad range of alloy compositions. The same type of behavior was observed for Ni–Au [36] and Pt–Au alloys [37].

When alloys are supported, however, as in reforming catalysts, the size of the metal crystallites, the nature of the support, the techniques of preparation, and the reduction procedure play decisive roles in determining whether phase separation occurs and how. When crystallites are small (< 100 Å), phase separation may not occur even when there is a miscibility gap for the bulk system [38, 39].

Highly exothermic or ordered alloys are characterized by values of $\Delta H_f^\circ \ll 0$ and $(E_{AA} + E_{BB})/2 \ll E_{AB}$. In these systems, dispersion in the form of clustering and/or phase separation does not occur, but because of the large free-energy reduction upon formation of A—B bonds, ordering and the formation of intermetallic compounds usually occur. An example is PtSn and Pt_3Sn in Pt–Sn alloys;

Ni–Al, Cu–Au, Cu–Pd, Cu–Pt, and Pt–Zn also form intermetallic compounds. The surface composition of the alloy depends on the crystal face.

Surface compositions An enrichment of the surface occurs in most alloys; the driving force is the minimization of total free energy. If we consider an ideal solid solution and regard only the outermost atomic layer as the surface, we find that at equilibrium [40],

$$\frac{X_B^s}{X_A^s} = \frac{X_B^b}{X_A^b} \exp\frac{(\gamma_A - \gamma_B)a}{RT} \tag{13}$$

where X = atomic fraction of A and B in an A–B alloy on the surface (s) and in the bulk (b)

γ = specific work required to form new surface (or surface tension or surface free energy of the pure element)

a = specific atomic area.

Equation (13) shows that the surface layer is enriched in the component which has the lower surface free energy (which corresponds to the lower heat of sublimation) and that small differences in the surface free energy cause large surface enrichments.

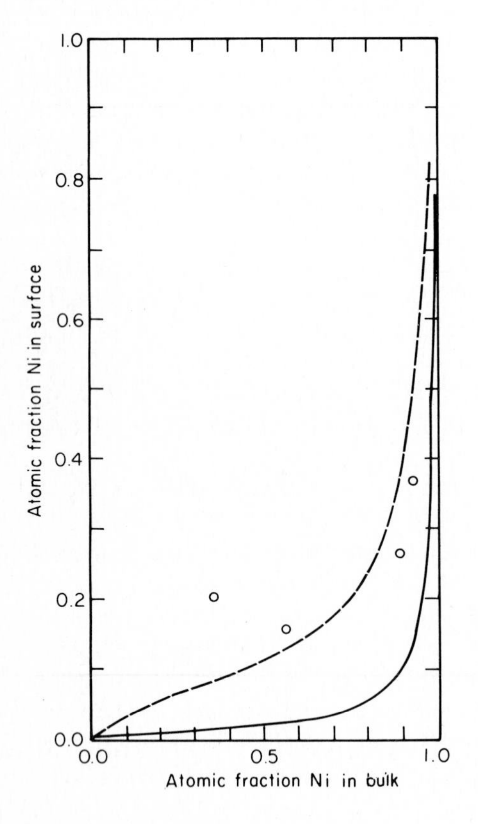

Figure 3-21 Predicted surface composition of Cu-Ni films; the calculations were done for a face-centered cubic [111] surface using effective metal interaction parameters of $\Delta H_{sub} = 10kT$ and $\Omega = 0.1kT$; solid line is for the free surface; dashed line is for the surface with hydrogen adsorbed on it with $\Delta H_{ads} = 5kT$ and $\Omega = 0.1kT$ [40]. Open circles are data from [41].

For a regular solution, the right-hand side of Eq. (13) is multiplied by [40]

$$\exp\left\{\frac{\Omega(l+m)}{RT}[(X_A^b)^2-(X_B^b)^2]+\frac{\Omega l}{RT}[(X_B^s)^2-(X_A^s)^2]\right\} \tag{14}$$

where $\Omega = H_{AB} - \dfrac{H_{AA}+H_{BB}}{2}$ = heat of formation of alloy [40]

l = fraction of nearest neighbors in same plane
m = fraction of nearest neighbors below plane

The predicted surface (first-layer) composition for a Cu–Ni alloy, which forms a nearly regular solution, is shown in Fig. 3-21. A reduction in the relative enrichment of the surface is predicted in the presence of hydrogen. The data points represent the fraction of strongly bound hydrogen determined experimentally by Sinfelt et al. [41]. Since hydrogen strongly chemisorbs on Ni but does not chemisorb on Cu, the fraction of strongly adsorbed hydrogen provides a measure of the fraction of the surface covered by Ni. The results show that the theory and experiment are in fair agreement. Figure 3-22 shows the effect of nonidealities on both surface composition and composition–depth profiles.

A summary of further important conclusions drawn from theoretical studies follows [40]:

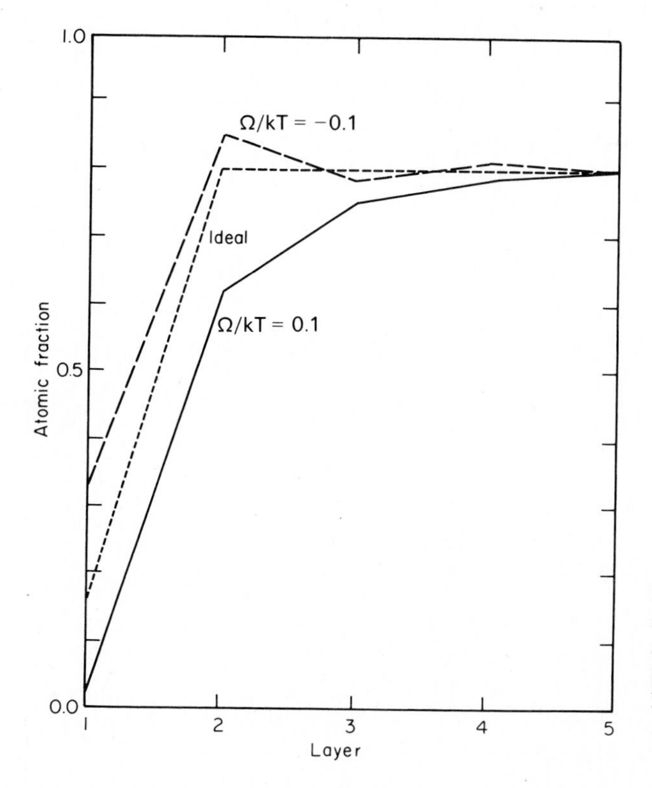

Figure 3-22 Theoretical composition profiles for segregation in a binary alloy face-centered cubic [111] surface; the difference in heats of vaporization is $\Delta H_{sub} = 10kT$ and the atomic fraction in the bulk is 0.8 (adapted from [40]).

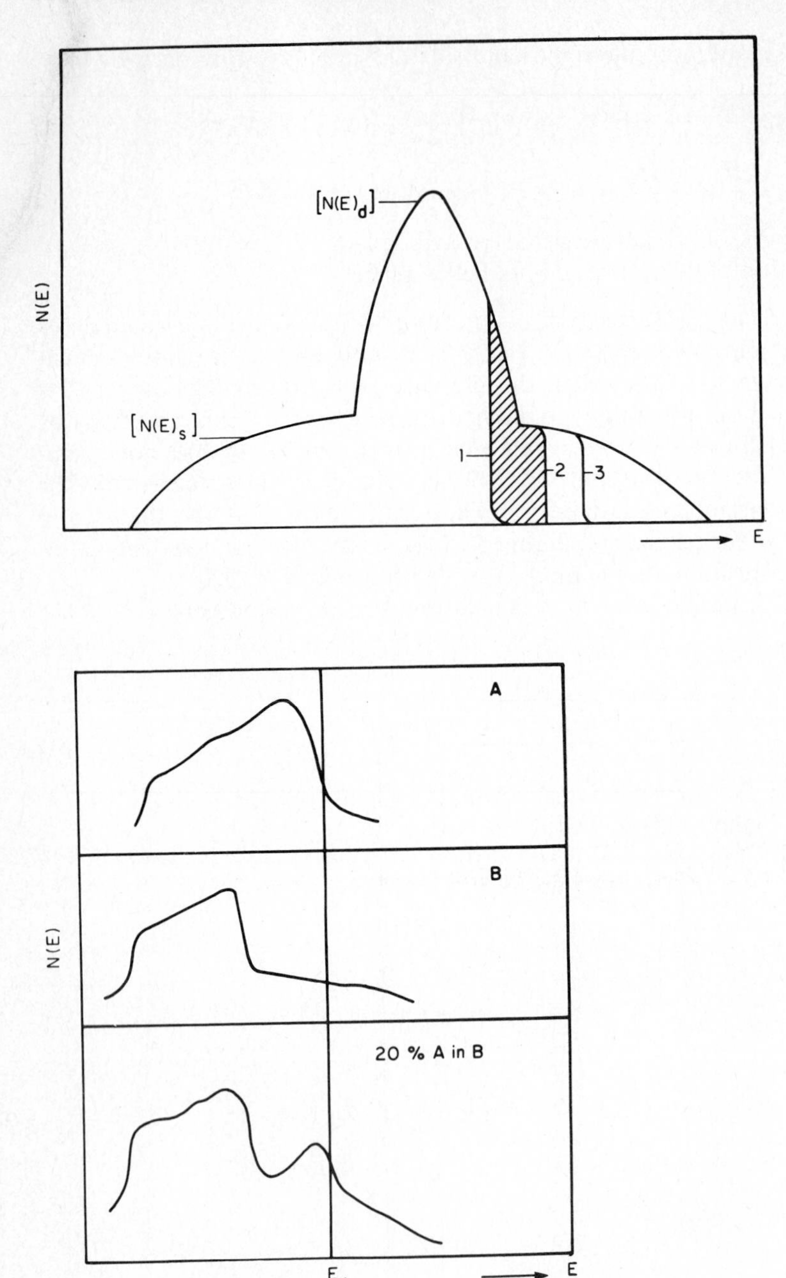

Figure 3-23 A. Comparison of predictions of rigid band theory and coherent potential theory with the measured density of states of Cu(0.6) – Ni(0.4) alloy [34]. B. Data for Cu–Ni alloy [46]. (Reprinted from *Catalysis Reviews* by courtesy of Marcel Dekker, Inc.)

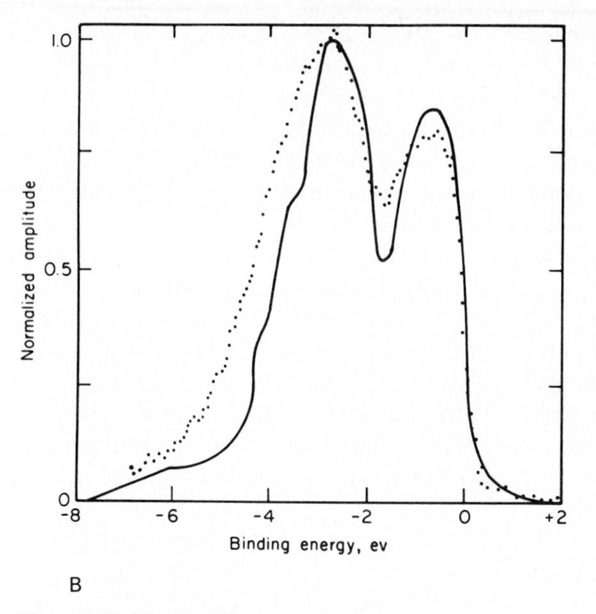

Figure 3-23 (*Continued*)

1. In the first layer, the enrichment in the element with the lower enthalpy per bond is determined by $\Delta H_{sub}/RT$ (where ΔH_{sub} = difference in the heats of sublimation of the two pure metals) and by the fraction of the nearest-neighbor atoms missing for atoms in the first layer.
2. Chemisorption of a compound leads to enrichment of the surface in the metal that bonds more strongly to the adsorbate; more aggressive adsorbates, those which form stronger bonds with the metal, lead to more substantial enrichment in the component that bonds more strongly with the adsorbate than less aggressive adsorbates.
3. If, due to surface relaxation, bonds are actually stronger at the surface than in the bulk, the enrichment is decreased. Changes also occur in the second layer.
4. The regular solution parameter Ω can strongly influence the enrichment in the first layer, enhancing it further if Ω is positive.
5. In regular solutions with negative values of Ω, the greater stability of A—B bonds causes a reduced enrichment of A in the first layer and an enrichment of B (depletion of A) in the second layer.
6. These conclusions must be modified in the case of small particles because segregation can alter the bulk composition and because a range of different crystal orientations is present [42], some of which show significantly different segregation properties.

The theoretical predictions of surface enrichment in alloys have largely been confirmed by determinations of surface compositions by Auger electron spectroscopy, work function measurements, and selective chemisorption.

Electronic structures The rigid-band model suggests that alloying Ni with Cu should fill the *d* holes, and since *d* electrons appear to be important in at least some catalytic reactions (Fig. 3-19), the catalytic activity of Ni might be expected to change greatly upon alloying. But recent measurements have shown that in Cu–Ni alloys, even those containing more than 60 atom percent Cu, the number of *d* holes per Ni atom remains constant at 0.5 ± 0.1 per atom [43–45]. Evidently the Cu electrons remain largely localized on the Cu atoms and the Ni *d* holes remain largely localized on the Ni atoms. Figure 3-23 shows the predicted density of states for Cu, Ni, and a 77% Cu–23% Ni alloy and compares the results with the measured density of states for a Cu–Ni alloy of the same composition [46]. The results show clearly that the electronic properties and thus the chemical character of Ni are not markedly affected by alloying with Cu, and they suggest, contrary to the first supposition, that the catalytic behavior of Ni atoms should not be greatly altered by alloying with Cu. This conclusion is consistent with the fact that Cu–Ni is an endothermic alloy in which there probably is clustering of Ni atoms and in which the electronic interaction of Ni and Cu are not great.

The result is not general to all alloys, however, for in Pd–Ag the *d* band is completely filled for alloys containing less than 35 percent Pd. The number of holes in the *d* band is reduced from 0.4 per Pd atom to less than 0.15 [43–45], and x-ray photoelectron spectroscopy clearly shows the filling of the *d* band of Pd upon alloying with Ag [46] (Fig. 3-24). These results are consistent with the fact that Pd and Ag form an exothermic alloy, and therefore the bonding interaction between the two different metal atoms is larger than in Cu–Ni [$E_{AB} > (E_{AA} + E_{BB})/2$]; consequently the electronic structure (and presumably the chemical character) of Pd is affected by alloying.

Catalytic activity We now proceed to consider the catalytic activity of alloys, not only because of the industrial importance of alloy catalysts for processes such as reforming but also because they provide much insight into the catalytic activity of unalloyed metals. We have seen that alloys, for example, Cu–Ni, can be formed in which the electronic structure of the catalytically active metal (Ni) is not appreciably altered by alloying; yet other alloys, for example, Pd–Ag, show pronounced variations in the electronic structure of the catalytically active metal (Pd). Therefore, these alloys provide the means of answering the decades-old question of the relative importance of geometric and electronic effects in metal catalysis.

We begin by considering the effects of alloying on hydrocarbon hydrogenation-dehydrogenation reactions and hydrogenolysis reactions, which are important in reforming. Hydrogenation-dehydrogenation reactions and H–D exchange reactions of hydrocarbons involve C—H bond breaking and occur readily at low temperatures. Hydrogenolysis involves C—C bond breaking and is a more difficult reaction, occurring only at relatively high temperatures. Correspondingly, the former reactions occur in the ligand spheres of single-metal-atom complexes, but there is no evidence for C—C bond breaking with such simple catalysts. A similar distinction between reactions involving C—H and C—C

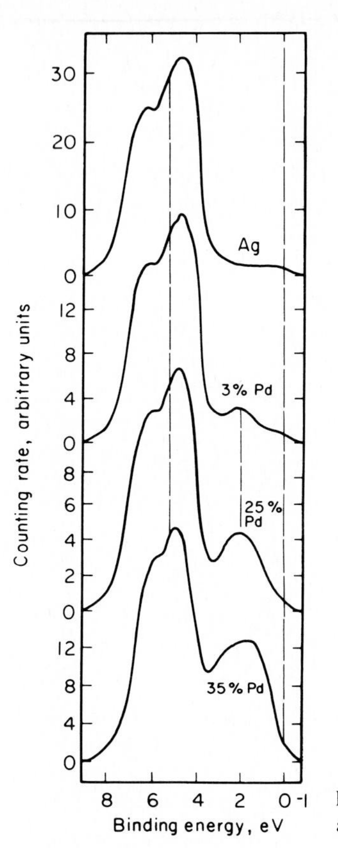

Figure 3-24 Photoemission spectra of density of states in Pd–Ag alloys [46].

bonds appears again in the consideration of structure sensitivity in reactions on surfaces of supported metals. The differences in these two reactions allow the effect of alloying to be tested by observation of the relative effect on each.

Some data showing the pattern of hydrogen adsorption on Cu–Ni alloys are given in Figs. 3-25 and 3-26. The adsorption isotherms show that a small amount of added Cu produces a marked decrease in the amount of hydrogen that is strongly adsorbed. This result is shown with particular clarity in Fig. 3-26, where the amount of hydrogen adsorbed is plotted as a function of the surface area measured by the physical adsorption of xenon. The fraction of the surface that adsorbs hydrogen decreases sharply with the first added Cu, which implies the enrichment of the surface by Cu in the composition range characterized by a single phase and by an insufficient fraction of the Cu-rich phase to cover the Ni-rich phase. For Cu concentrations exceeding about 15 percent, phase separation and complete encapsulation of the Ni-rich phase should occur (Fig. 3-20), and the composition of the outer, Cu-rich phase should not vary with added Cu; correspondingly, the surface composition should not vary. The chemisorption measurements confirm this behavior.

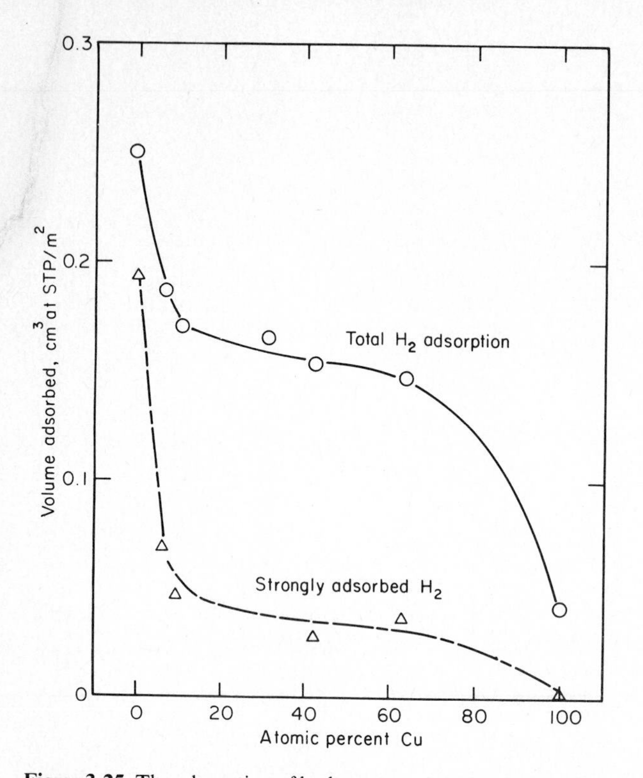

Figure 3-25 The adsorption of hydrogen on Cu–Ni alloy catalysts as a function of the Cu content. The circles represent the total amount of hydrogen adsorbed at room temperature at 1 atm pressure. The triangles represent the amount of strongly adsorbed hydrogen, i.e., the amount not removed by a 10-min evacuation at room temperature following the completion of the initial adsorption isotherm. The amount of strongly adsorbed hydrogen is determined as the difference between the initial isotherm and a subsequent isotherm obtained after a 10-min evacuation [41, 47]. (Reprinted with permission from *Journal of Catalysis.* Copyright by Academic Press.)

The pattern is confirmed by catalytic-activity measurements (Fig. 3-27) showing the effect of alloying Ni with Cu on the ethane hydrogenolysis and on the cyclohexane dehydrogenation reactions [47]. The ethane hydrogenolysis rate decreased by about four orders of magnitude upon addition of 20 percent Cu to Ni. Over the same range of alloy compositions, the cyclohexane dehydrogenation rate first increased somewhat and then became independent of alloy composition until the composition approached the limit of 100 percent Cu.

The same trends were observed for cyclopropane reactions [49] (giving propane and methane + ethane) catalyzed by Cu–Ni alloys (Fig. 3-28). Because of the strained character of the C—C bonds in cyclopropane, ring opening is considered to be more like double-bond hydrogenation than hydrogenolysis. For example, Pt readily catalyzes hydrogenation of cyclopropane, but there is almost no hydrogenolysis of the propane formed (see Fig. 3-16 for the relative activities of Pt and Ni for ethane hydrogenolysis) [50, 51]. Because of the ease of C—H bond-breaking reactions, alloying does not markedly alter their rates. For cyclohexane

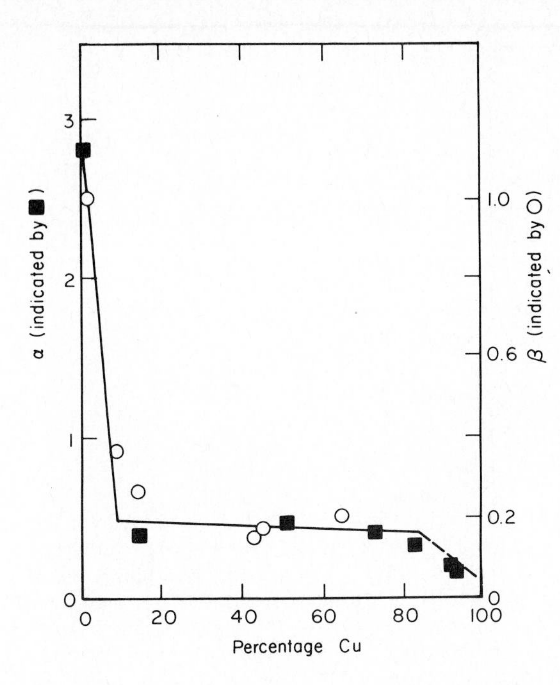

Figure 3-26 Hydrogen adsorption on Cu–Ni alloys: α = the number of hydrogen atoms adsorbed at 293 K per xenon atom adsorbed at 78 K on metal films [48]; β = irreversible hydrogen adsorption at 293 K per cm^2 of area, where $\beta = 1$ for pure Ni metal powders [41]. (Reprinted with permission from *Journal of Catalysis*. Copyright by Academic Press.)

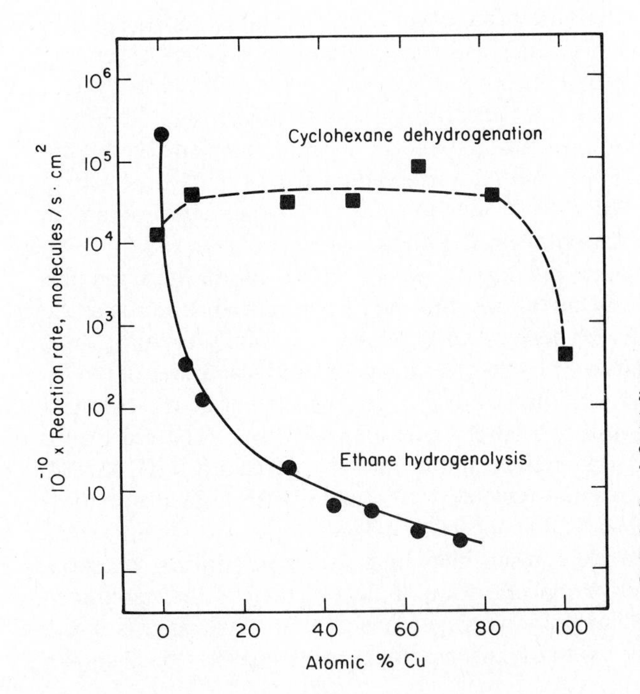

Figure 3-27 Activities of Cu–Ni alloy catalysts for the hydrogenolysis of ethane to methane and the dehydrogenation of cyclohexane. The activities refer to reaction rates at 316°C. Ethane hydrogenolysis activities were obtained at ethane and hydrogen partial pressures of 0.03 and 0.20 atm, respectively [47]. (Reprinted with permission from *Advances in Catalysis*. Copyright by Academic Press.)

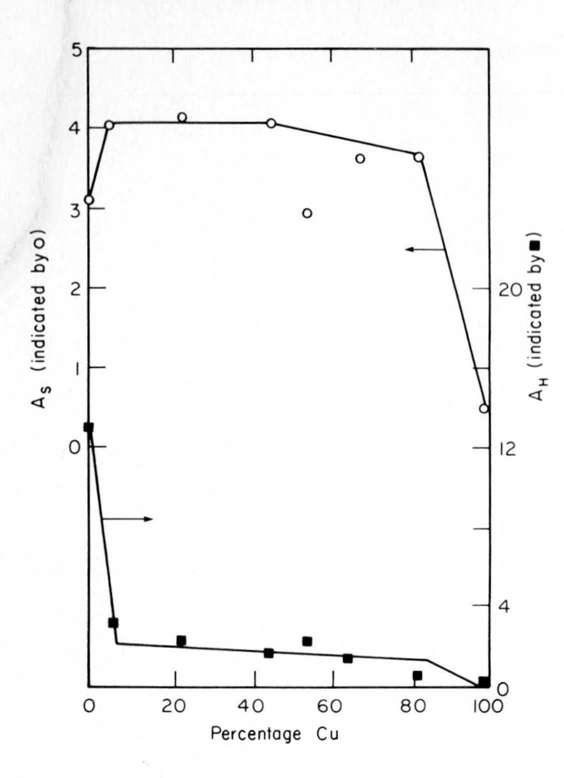

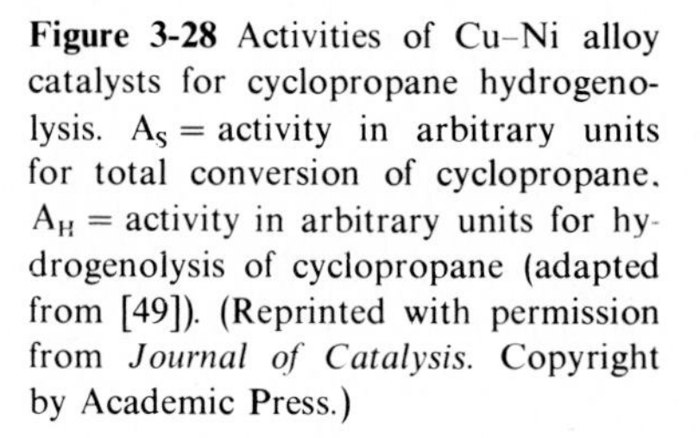
Figure 3-28 Activities of Cu–Ni alloy catalysts for cyclopropane hydrogenolysis. A_S = activity in arbitrary units for total conversion of cyclopropane. A_H = activity in arbitrary units for hydrogenolysis of cyclopropane (adapted from [49]). (Reprinted with permission from *Journal of Catalysis*. Copyright by Academic Press.)

dehydrogenation, the slightly increased rate observed upon initial addition of Cu to Ni is proposed to result from an increase in the rate of desorption of benzene due to a reduction in the strength of the π bond between benzene and Ni [47]. This result could be considered to be an electronic effect caused by Cu atoms surrounding the Ni atoms on the surface. Hydrogenation of acetylene on Au–Pd and Cu–Ni [52], hydrogenation of methyl acetylene on Cu–Ni [52], and hydrogenation of cyclopropane on Au–Pd [52] all show a lack of dependence of rate on alloy composition up to fairly high group IB metal concentrations.

The cyclopropane hydrogenation rate on Au–Pd alloys shows the same pattern as on Cu–Ni at low Au concentrations, but the rate begins to decrease at Au concentrations of about 40 to 50 percent and continues virtually to zero; these results suggest that catalytic activity cannot be related directly to filling of the Pd *d* band or to the number of *d* band holes but that this parameter is important. A more quantitative statement awaits further experimental results. The reason for the absence of a decrease in the hydrogenation activity of group VIII plus IB alloys as the surface is being rapidly depleted of the metal with higher activity is not clear. A possible explanation is that the olefin adsorbs aggressively and traps active metal atoms at the surface when they migrate there during reaction, and thus there is a higher effective concentration of the active metal than predicted from thermodynamics or measured by hydrogen adsorption. Hydrogen is a less aggressive chemisorbing species than olefin. A surface enrichment has been shown for Pt–Au [53]; on a Pt–Au alloy with a surface highly enriched in Au, chemisorption of CO resulted in marked surface enrichment in Pt, which binds more strongly to CO than Au. Upon evacuation of the CO, the surface returned to its original Au-rich composition.

In the case of C—C bond breaking, alloying markedly reduces the catalytic activity, as shown in Fig. 3-27. For hydrogenolysis to occur, the metal surface must induce dehydrogenation of the two carbon atoms involved and there must be at least a pair of adjacent metal-atom sites available to form bonds with the two carbon atoms. Sinfelt and coworkers [41] suggested that in ethane hydrogenolysis, dehydrogenation may be almost complete before C—C bond rupture occurs.

When Cu is alloyed with Ni, there is a decrease in the number of Ni surface sites, particularly site pairs, and there is a decrease in the strength of adsorption, as indicated by the large amount of weakly adsorbed hydrogen present on the Cu–Ni alloy surface (Figs. 3-25 and 3-26). The reduction in the number of site pairs with added Cu is a geometric effect; the decrease in the strength of adsorption is an electronic effect. Both geometric and electronic effects are important in catalysis.

Studies of Au–Pt alloys have shown that the geometric effect is important in catalysis [36, 54]. Pt catalyzes dehydrocyclization, isomerization, and hydrogenolysis of intermediate-chain-length *n*-paraffins. At low Pt content (1 to 12.5 percent Pt in Au), the Pt dissolves in the Au and should form a uniform dispersion throughout the Au (with the possible presence of clusters); but since Au has the lower surface free energy, the surface should be highly enriched in Au. It follows that in the range 1 to 4.8 percent Pt in Au, the surface should contain single Pt atoms or perhaps simple ensembles of at most several Pt atoms. Catalysts with 1 to 4.8 percent Pt in Au supported on silica gave almost exclusively isomerization products, and the alloys containing roughly 10 percent Pt gave both isomerization and dehydrocyclization products (Fig. 3-29). Pure Pt catalyzed isomerization, dehydrocyclization, and hydrogenolysis. The largest differences in activity occurred between 0 and 10 percent Pt and not between 10 and 100 percent, whereas magnetic measurements show that the largest variation in magnetic susceptibility occurs at the lowest Au concentrations; in the regions of highest dilution of Pt in Au, susceptibility varies only slightly [54*a*, 55]. Temperature-programmed desorption of hydrogen from Pt–Au alloys gives desorption peaks at the same temperatures as observed for pure Pt but with markedly reduced intensity, suggesting that the binding energy of hydrogen on Pt on the alloy surface is the same as that on pure Pt [56]. These results strongly suggest the importance of surface geometric effects in catalysis, since the critical region from a catalytic-activity (or selectivity) point of view is 1 to 14 percent Pt, and it is in this region that the surface composition (including ensembles, etc.) changes most markedly. There is little evidence of change in the electronic properties in this region, for if such changes had occurred, they should have been observed in the magnetic susceptibility and in the bonding energy of hydrogen with the Pt-alloy surface.

The conclusion is that hydrogenolysis requires large ensembles of Pt atoms, dehydrocyclization smaller ensembles, and isomerization the smallest ensembles. If the single-atom mechanism of isomerization proposed by McKervey et al. [57]

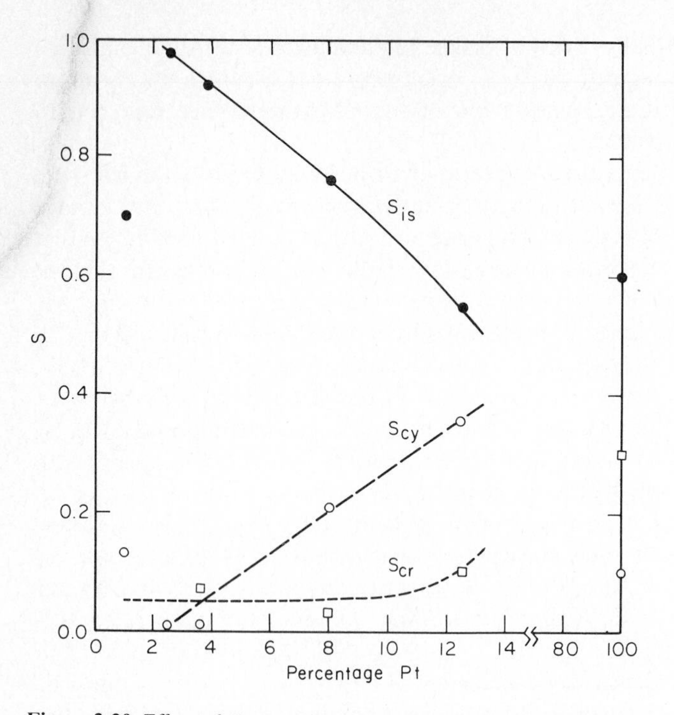

Figure 3-29 Effect of Pt–Au alloy composition on selectivity in *n*-hexane reactions at 360°C; S_{is}, S_{cy}, and S_{cr} are the fractions of the total conversion to isomerized products, cyclized products, and cracked (hydrogenolysis) products, respectively [54]. (Reprinted with permission from *Journal of Catalysis*. Copyright by Academic Press.)

is operative, as these results suggest, isomerization occurs on single Pt atoms dispersed in a sea of Au and behaving chemically as though they were on a pure Pt surface, whereas dehydrocyclization, which may require at least two metal atoms, is not possible. Hydrogenolysis is thus blocked by the geometric effect, by the presence of inactive atoms (Cu, Au) on the surface of the metal. Gray et al. [58] had already shown this blocking effect in 1960, giving evidence that carbon deposits on the surface of Pt markedly suppress hydrogenolysis.

Alloying may be important in affecting side reactions as well as desired reactions, and indeed the influence on side reactions is one of the major reasons for using alloy catalysts for reforming. For example, in benzene hydrogenation [59] catalyzed by Cu–Ni in the composition range for which there are no complications arising from phase separation, the alloy had a lower activity per unit surface area than pure Ni at temperatures between 20 and 250°C. At temperatures exceeding 220°C, however, the alloy appeared to be more active than Ni because side reactions producing CH_4, other nonring hydrocarbons, and carbon deposits occurred on Ni (but hardly on alloys) and caused deactivation of the Ni catalyst.

In summary, there has been much progress in understanding alloy catalysis during the last decade, and the insights gained have already improved our understanding of catalysis by single metals and presumably played a role in the development of modern reforming catalysts. Further progress would be stimulated by a better understanding of the bulk and surface properties of alloys. Some of the remaining important unanswered questions are the following:

1. Is it realistic to picture atoms such as C or N doubly bonded to a surface; is it realistic to picture a single C or N atom bonded with more than one metal atom? How is the formation of the surface bonds influenced by alloying?
2. If a certain molecule prefers a position in a surface site between several atoms for its adsorption on a pure metal, is the same kind of adsorption site also preferred on an alloy? Which sites on a surface and what size and geometry of metal ensembles are necessary to activate molecules for dissociation of C—H, C—C, and C≡O bonds, etc., and how does alloying influence them?
3. In many cases the role of a group IB metal in alloys with group VIII metals appears to be to prevent dissociation of C—H and C—C bonds. Can spectroscopic measurements of adsorbed species clarify this role?
4. Can improved methods of determination of surface compositions be developed for alloys, both before and after reaction (in vacuum) and during reaction under the influence of corrosive chemisorption of reactants?
5. How can clusters of atoms in alloys be detected, and how can their sizes be determined?
6. Since rates of some reactions depend on the size of small supported-metal crystallites, it is logical to expect new effects of alloying when the catalyst particles are very small. How does alloying in combination with variations in the crystallite size of a catalyst and interaction with the support open new ways of controlling the activities and selectivities of catalysts?

Theoretical Considerations

Catalysis, even more than the rest of chemistry, is an empirical science, and progress would be faster if there were a stronger basis of theoretical understanding. Quantum-chemical theories of catalysis by metals should provide estimates of the energies of adsorption of molecules on metal surfaces and of the activation energies of adsorption and of reactions of the adsorbed species. Although these goals are still far from being realized, the results currently available provide useful insight and are therefore summarized below.

Solid-state physics provides substantial theoretical information about bonds in metals, obtained by using the periodicity of the lattice to simplify solution of the wave equation [60]. Corresponding to the assumption of perfect periodicity, the theory deals with solids that extend to infinity in all directions. Bulk defects and surfaces give rise to difficulties, since they lead to disruption of the periodicity. In recent years progress has been made using models that assume infinite periodicity in two directions but only limited periodicity in the third. A pure metal then becomes a slab, infinite in two directions but finite in the third. For the adsorption of a compound on the metal, it is assumed that the layer of adsorbed species has a periodicity related to that of the solid in two directions; a full monolayer is a requirement. This model is referred to as the *periodic-surface-layer* or *solid-state model.*

Many practical catalysts are small clusters of atoms that are often better dealt with by the calculational methods used by chemists for organic molecules. These methods involve (1) calculation of the energy levels and bond strengths in a small cluster of metal atoms and (2) addition of other atoms to the outer metal atoms to form adsorption bonds. The method is referred to as the *cluster approximation method.*

Surfaces of vast expanse can also be handled by this method by isolating a small cluster of metal atoms at the surface and solving for the energy levels and bond strengths; an adsorbate may also be added. Then the cluster is embedded in the bulk metal, requiring that energy levels match.

In both methods, the complete solution of the many-electron Schrödinger equation leads to numerical computational programs that contain approximations. Those common in the chemist's method are known as MO-LCAO (molecular orbital as linear combination of atomic orbitals), VB (valence-bond), EH (extended Hückel), CNDO (complete neglect of differential overlap), and the Hartree-Fock method; the degree of empiricism decreases from beginning to end of this list, and the approximations become less drastic, so that the Hartree-Fock method is essentially *ab initio;* the calculations, however, are very time-consuming.

Since the repeating unit in an infinite crystal is the unit cell, all the difficulties encountered in the cluster method are also encountered with unit cells, which have even more atoms, orbitals, and electrons. Here, also, simplification is required, e.g., by use of approximate potentials, as shown by Anderson [61]. Calculations involving *d* orbitals are far more time-consuming and require more severe approximations than those for systems involving only *s* and *p* orbitals.

The self-consistent field Xα scattered-wave (SCF-Xα) method described by Slater and Johnson [62, 63] (and mentioned in Chap. 2) is particularly interesting. The approximation of an "average" potential (Xα) is intrinsically superior to those of earlier methods [63]. The SCF-Xα method requires considerable computing time but far less than the Hartree-Fock method. It may be the most suitable model for understanding surface catalysis, as it already appears to be for understanding defects in solids.

Slater and Johnson [62] and Schrieffer and Soven [64] have given surveys of the various methods. Comprehensive reviews have been written by Gadzuk [65], Messmer [66], Grimley [67, 68], and Fassaert [69]. Below is a short review of results following a historical approach.

A highly empirical model devised by Weinberg, Deans, and Merrill [70, 71], the crystal-field surface-orbital bond-energy bond-order (BEBO) model, combines features of the bond-energy bond-order relationships found in molecular spectroscopy with the concepts of crystal field theory. The model is based on the assumption that the surface bond is a highly localized covalent bond involving primarily *d* electrons of the solid and highly directional *d* orbitals (dangling bonds) having orientations consistent with those in the bulk. The bond energy of

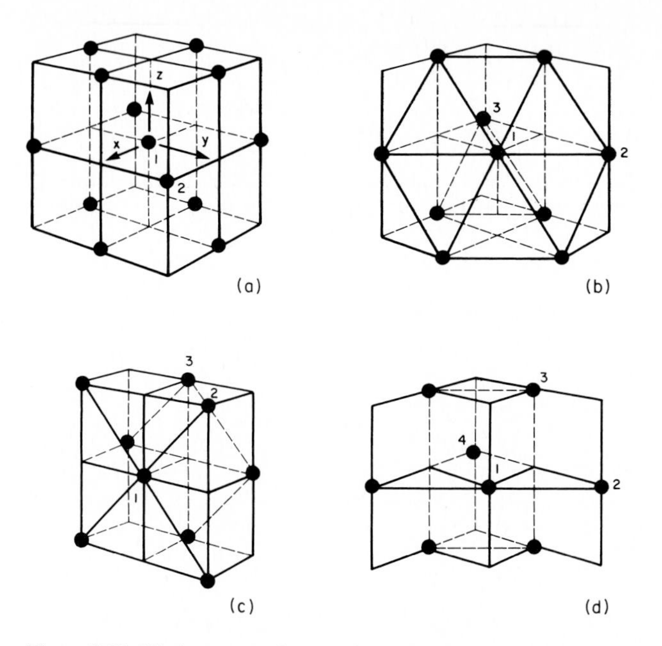

Figure 3-30 Ni clusters used as models in quantum mechanical cluster calculations [75].

a surface-adsorbate bond is assumed to vary with bond distance and bond order, as predicted by spectroscopic measurements of model compounds, and bond-energy assignments are obtained from the bond energies of bulk compounds. The model predicts bond strengths and allows prediction of activation energies of surface reactions and thereby allows prediction of probable and improbable reaction pathways. This model is highly empirical; it represents a quantification of the surface-compound model presented above.

The cluster calculations of several workers [69, 72–76] with the semiempirical methods (extended Hückel and CNDO) have provided some useful information about the bonding in metal clusters. Figure 3-30 shows the small clusters of Ni atoms used as models in some of the calculations. Binding energies calculated for the cluster models are low relative to those of bulk-metal crystals, as summarized in Table 3-5. This result is probably an artifact of inadequate parameterization and is partly the result of the smaller number of nearest neighbors per atom in the small clusters. Calculated orbital energies group into bands which are somewhat too narrow, again because of the small number of atoms and the small number of atomic orbitals. For the same reason, the Fermi levels are lower compared with the position of the ionization potential. The calculations for Ni confirm that *d* bands are narrow, *s* bands are broad (Table 3-5), and the *s* and *d* bands overlap; all these results are in agreement with observation.

Some impressive results were obtained by Slater and Johnson [62] using the SCF-Xα method for Cu and Ni clusters containing eight atoms each (Fig. 3-31). The Cu_8 cluster shows definite *s*- and *d*-band overlap, just as Ni_8 does. For the

Table 3-5 Results of quantum-mechanical calculations for Ni clusters[a] [75]

	Bulk cluster	Surface cluster		
		(111)	(100)	(110)
d bandwidth, eV	1.81	1.67	1.63	1.59
Fermi level, eV	−7.64	−7.72	−7.66	−7.69
Holes in *d* band	0.68	0.67	0.59	0.54
Total binding energy, eV	22.7	17.1	15.0	13.3
"Renormalized" cohesion energy, eV	3.8	4.3	4.5	4.7
Atomic charge $q(1)$	2.54	1.37	−0.12	−0.02
$q(1)$ surface − $q(1)$ bulk:				
$\Delta q(1)$	...	−1.17	−2.42	−2.52
$\Delta q(2)$	...	−0.01	−0.05	−0.04
$\Delta q(3)$	...	0.19	0.44	0.30
$\Delta q(4)$	...	...	...	0.35

[a] The numbering of the atoms is indicated in Fig. 3-30

latter case, the two spin directions give different band orientations, and so the cluster is paramagnetic. For Cu_8, the lower-lying orbitals are bonding, but those near the Fermi level are antibonding, with orbitals projecting into space. Messmer [66] has obtained similar results for Cu_8, Ni_8, and Pd_{13} clusters.

The general impression gained from these results is that bonding in metals such as Cu, Ag, and Au appears to be almost entirely due to *s* bonding (Fig. 3-32), and even for Ni and Pd the metal stability is largely due to *s*-orbital interactions. Since no calculations have been performed for clusters of metal atoms having electronic configurations other than those of Cu (Ag, Au) or Ni (Pd), calculated changes in bond strength with the number of electrons in the open shells are not available.

Van der Avoird and collaborators [69, 75–77] calculated strengths of surface-hydrogen bonds for the Ni-cluster-hydrogen system (Fig. 3-33). Bandwidths in the cluster, contrary to the results of Baetzold [73], were near those found in bulk metals. Bond strengths were similar to those in NiH_2, although smaller to an extent depending on the number of nearest neighbors. A surface Ni—H bond was weakest on the (111) plane (where each Ni atom has nine nearest neighbors), stronger on the (100) plane (with eight), and strongest on the (110) plane (with seven). The calculated bond strengths due to an *s*-orbital bond and to a *d*-orbital bond were almost equal.

A more extensive computation of the stagewise interaction of a pair of Ni atoms and a pair of Cu atoms with a H_2 molecule has also been performed under the assumption that there was only one metal-atom electron available in either an *s* or a *d* orbital [76, 77]. The system energy, computed as a function of the distance of the H_2 molecule from the pair of metal atoms, exhibited a high activation-energy barrier for bond formation in the case of *s* orbitals (Cu atoms), but it exhibited no activation-energy barrier in the case of *d* orbitals. The difference in

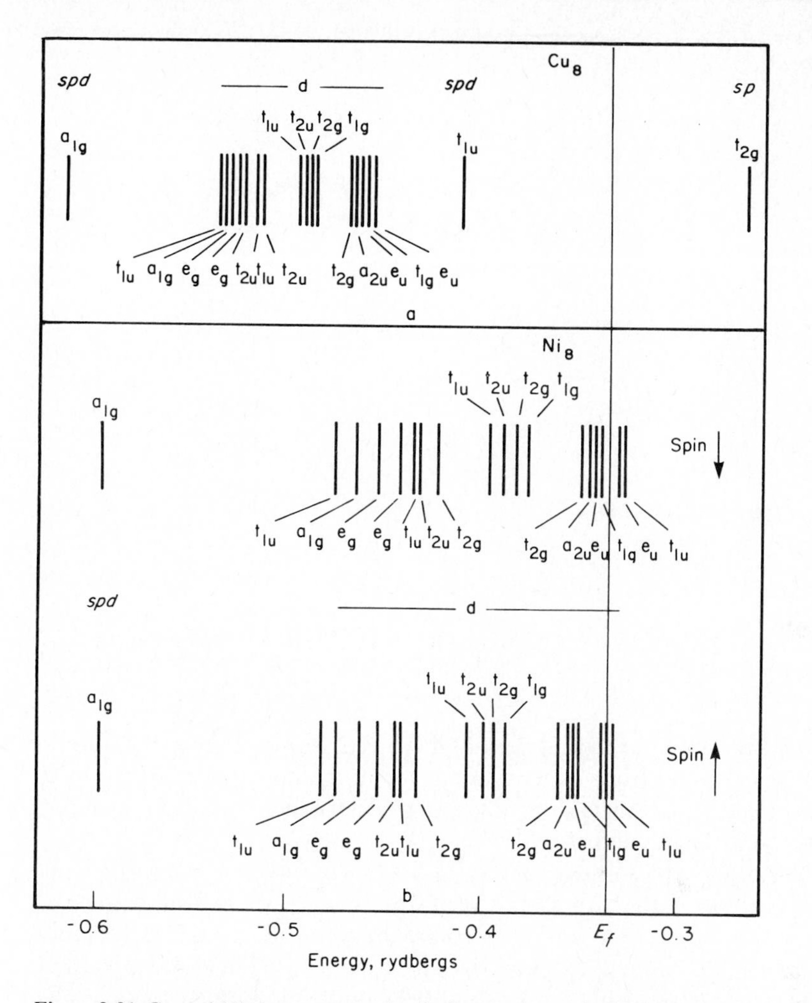

Figure 3-31 Cu and Ni cluster energy levels. Part a shows the SCF-Xα electronic energy levels for a cubic Cu_8 cluster; part b, the spin-polarized SCF-Xα electronic energy levels of a cubic Ni_8 cluster. For each cluster, the results are shown for a nearest-neighbor internuclear distance equal to that in the corresponding bulk crystal. The levels are labeled according to the irreducible representations of the cubic (O_h) symmetry group. The Fermi level E_f separates the occupied levels from the unoccupied ones [62]. (Copyright by the American Institute of Physics.)

interaction of Ni and Cu, for instance, with an adsorbing H_2 molecule appeared to be caused not by the strength of the bond formed (thermodynamics) but by the kinetics.

These results suggest that the rate of chemisorption of H_2 on Ni should be rapid because of the absence of an activation-energy barrier, whereas because of the high activation-energy barrier associated with H_2 interaction with a metal *s* bond, chemisorption of H_2 on Cu should be slow. This result is consistent with experiment: H_2 chemisorption on Ni is rapid at room temperature, whereas H_2

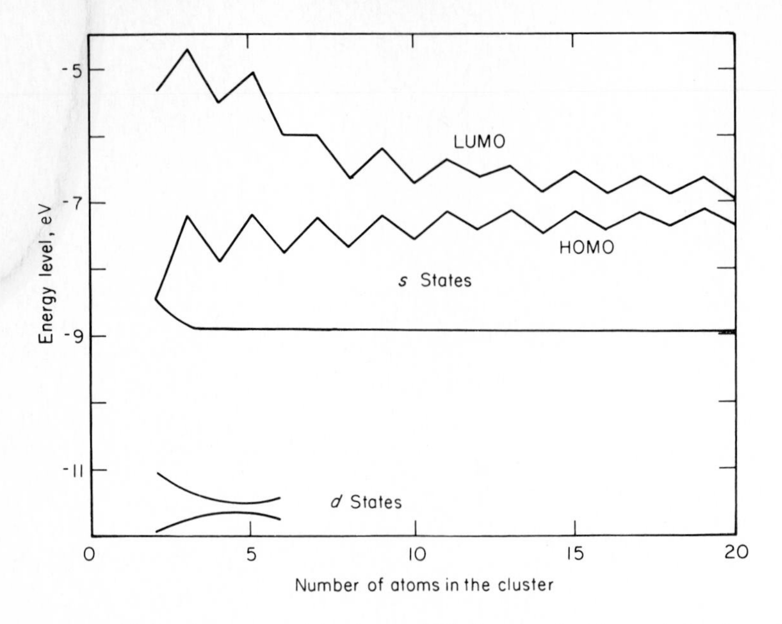

Figure 3-32 Band structure of Ag clusters [73]. (Reprinted with permission from *Journal of Catalysis*. Copyright by Academic Press.)

does not chemisorb on Cu except at very high temperatures or from the atomic state. Because of the similarities between chemisorption and surface migration in terms of bond breaking and formation, migration of adsorbed hydrogen would be expected to be rapid on Ni and slow on Cu.

The two-dimensionally infinite crystal with one (or two) surface(s) and the adsorption of H atoms on this surface has been treated using special mathematical techniques such as integral equations with boundary conditions and Green's functions (the resolvent technique). Tamm [78] and Shockley [79] had already found for one-dimensional models that "surface states" might be formed (outside the Bloch bands), such surface states being orbitals localized at the surface and rapidly dying out with distance into the bulk. At an early stage in the developments Grimley [80] suggested the possibility that adsorption could lead to localized bonding, and this idea is beginning to gain more favor.

The quantitative development of these earlier methods with three-dimensional geometry has been actively pursued in the last decades; it has led to a vast literature, much of which is clouded in an immense number of technical details that make it rather inaccessible to most workers in the field of catalysis. The various theoretical attacks may start from LCAO formulations with expansion of one-electron wave functions in localized basis orbitals or valence-bond methods constructing N-electron states from these localized orbitals. Pseudopotentials are introduced that lead to solution either by the resolvent or by the self-consistent-field scattered-wave method. Grimley and his collaborators have used Hartree-Fock type methods, and Schrieffer and Soven [64] have used many-

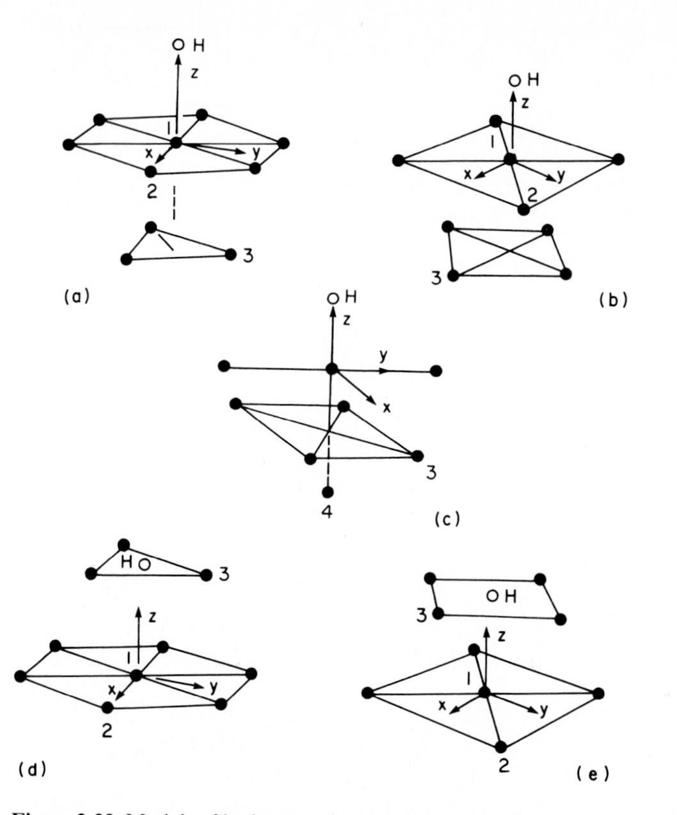

Figure 3-33 Models of hydrogen adsorption on Ni; (a) on the (111) surface, (b) on the (100) surface, (c) on the (110) surface, (d) in a (111) surface hole, and (e) in a (100) surface hole [75].

body approaches based on valence-bond (induced-covalent-bond) theory. To understand the outcome of all this work, it is advantageous to introduce the concept of a *surface compound* [65]. This concept has been further illustrated by the work of Paulson and Schrieffer [81] for the interaction of an H atom with the *s* orbital of the (100) surface of a cubic crystal ("cubium"). The calculations show that if the interaction between the adatom and the surface is weak, the states that arise are in the band and are strongly broadened; these states are referred to as *virtual states* (Fig. 3-34). If the interaction is strong, bonding and antibonding split-off states are formed below and above the band (Fig. 3-34); this situation looks like a surface compound which is not an integral part of the solid, confirming the suggestion that Grimley [67, 68, 82, 82*a*, 82*b*] had proposed earlier based on Hartree-Fock calculations. These results are supportive of the suggestions of a surface compound given above. The calculations of Schrieffer and Soven [64] also indicate that adsorption on one atom does not affect the electronic properties of its neighbors and that the hydrogen atom prefers sitting on top of a surface metal atom to being surrounded by two or four metal atoms.

One of the difficulties with the above calculational schemes is that they are

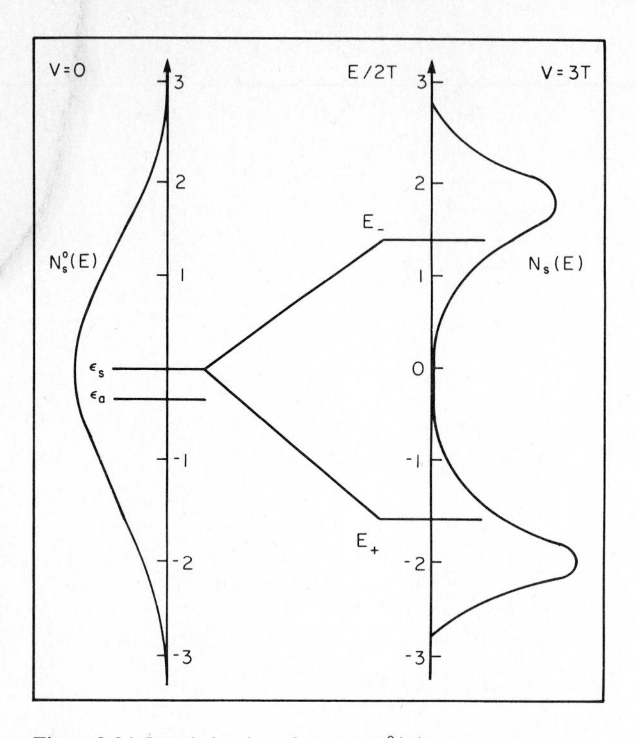

Figure 3-34 Local density of states $N_s^0(E)$ of a surface-atom orbital s for no interaction ($V = 0$) with an adsorbate level a (left) and $N_s(E)$ for an interaction $V = 3T$ (right), where T is an arbitrary unit of energy. For $V = 0$, atoms s and a form an isolated diatomic molecule, so that both the a and the s orbitals correspond to sharp levels, $e_s = 0$ and $e_a = -0.6T$. For $V \neq 0$, these orbitals hybridize to form the bonding (E_+) and antibonding (E_-) levels, leading to a splitting of $N_s(E)$. Curves are for the s atom of the (100) surface of a simple cubic s-band solid having a density of states $N_s^0(E)$ for $V = 0$, where the hopping matrix element between substrate is T. The two sharp molecular levels go over to two resonances, shifted somewhat from E_+ and E_-, illustrating both local (chemistry) and continuum (solid state) effects [64]. (Copyright by the American Institute of Physics.)

restricted to one electron per orbital and to an s orbital. Greater degrees of realism are virtually impossible unless one wants to accept the application of a less reliable semiempirical method such as the extended Hückel method. Fassaert, Verbeek, and van der Avoird [75, 69] changed the parameterization somewhat in a tight-binding extended Hückel scheme and then applied it to recalculate their former results and to study a finite periodic Ni crystal using five $3d$, one $4s$, and three $4p$ orbitals. The results agree closely with density-of-states calculations of the band structure of Ni in semi-infinite crystals and with experimental density-of-states information obtained from x-ray photoelectron, ultraviolet photoelectron, and electron-neutralization spectroscopy. It was concluded that both the $3d$ electrons and the conduction electrons take part in the chemisorptive bond with H. Stability of adsorption decreases in the order $(110) > (100) > (111)$ and top > bridge > centered. A similar calculation was made for a cluster. Adsorption is especially strong at the edges of a stepped surface.

The results emerging from the theoretical investigations, although difficult to assess and at first sight apparently not far-reaching, are nevertheless highly significant. The work has already produced some substantial contributions to our understanding of catalytic reactions. There is, first, the realization that adsorption on metal surfaces is most often a strongly localized phenomenon (localized chemical bonds are formed); consequently differences in adsorption energy are due to differences in the immediate surroundings and not to long-distance induction effects. This result alleviates the uncertainty which clouded our understanding of adsorption for a long time. Second, these results strongly support the surface-compound hypothesis, which has been in doubt for a considerable time. The H_2 adsorption calculation clearly shows that for some catalytic reactions, the rate-limiting step involves crossing the activation-energy barrier associated with the adsorption step, and the result supports our earlier conclusions about adsorption being the rate-limiting step on the left-hand side of the Balandin volcano curves (Fig. 3-14).

The theoretical results to date also provide insight into the role of *d* orbitals in transition-metal catalysis. The *s* band is broad and has a low density of states within a few eV above or below the Fermi level; this condition does not promote ready interaction with an adsorbing molecule. However, *d* orbitals are highly localized in space and in energy, resulting in easy, strong "mixing" between appropriate metal-surface orbitals and the orbitals of an adsorbing molecule. The strength of the bond formed and the reduction of energy barriers for reaction depend on the positions of the energy levels of the adsorbed species relative to the Fermi level. The lowering of activation energies for reaction requires that *d* orbitals facilitate electron transitions, e.g., by receiving electrons from orbitals that are being raised in energy and donating electrons to orbitals being lowered in energy. The result can be a ready occurrence of symmetry-forbidden reactions because the high density of states at the metal surface can lead to the formation of hybrid orbitals with the proper symmetry to promote reaction.

Current Research

Many questions about metal catalysis remain unanswered, but research is proceeding actively and can be expected to bring important progress in the next years. Some trends in current research are as follows:

1. The bonding of small groups of metal atoms on a surface is still not well described. Some information about supported-metal clusters might be obtained from studying well-defined cluster compounds, like those mentioned in Chap. 2, which can also be bonded to surfaces to make supported catalysts like those described in Chap. 2. These new catalysts may be expected to find important applications in the future.
2. It is necessary to know where adsorbed atoms are located at a metal surface and to know the geometry of molecular fragments and reaction intermediates adsorbed on metal surfaces. It is also of prime importance to know which

surface planes are present during catalysis, because this geometry governs the positions of the orbitals available from each surface atom, which then react with orbitals of adsorbing molecules. It also governs the relative orientation and distance with respect to each other of reactants involved in bimolecular surface reactions. Extensive research is currently concerned with the structures and reactivities of the various well-defined surface crystal planes and the defects, such as steps, which exist on them.

3. It is important to understand what influences the strength of surface bonds and contributes to the occurrence of an activation energy for their formation. The energy position of the surface *d* orbital relative to that of a molecular orbital of the molecule to be adsorbed is important; bond formation is favored by nearly equivalent energy levels. Also important is the number of electrons in the surface *d* orbital; partial filling is most desired. The latter consideration explains the slow reaction and high activation energy for H_2 chemisorption on Cu, Ag, and Au. The number of neighbors (coordination number) of an atom in the surface plane also influences the strength of bonds formed between surface atoms and H or other adatoms. The higher this number, the weaker the bond. A description of the bond formation and bond strength and a knowledge of the molecular orbitals involved in the bonding at the surface are essential to understanding and treating reactions on surfaces in a quantitative manner. A combination of theoretical calculations and detailed measurements using the most recently developed analytical tools will assuredly improve our understanding in these areas.

REFORMING CATALYSTS

We now turn to the platinum-on-alumina (Pt/Al_2O_3) reforming catalyst in detail, considering first the metal component and then the acidic support. Spurred mainly by the commercial importance of reforming, researchers in the preceding two decades have made considerable progress in understanding this catalyst.

The Metal

A summary of the dehydrogenation activity of several supported metals and metal oxides is given in Table 3-6. Although MoO_3–Al_2O_3 and Cr_2O_3–Al_2O_3 are among the most active oxides for dehydrogenation, they are much less active than the most active metal, which is Pt. Because of both its high activity and its unique selectivity characteristics, Pt has long been used as a reforming catalyst. Because Pt is expensive, and because its selectivity depends upon crystallite size, it is desirable to have it on the alumina in a highly dispersed form and to expose as large a fraction of the Pt atoms to reactants as possible. Therefore, Pt is used as very small crystallites on a porous support.

Table 3-6 Cyclohexane dehydrogenation activities of supported metal and metal-oxide catalysts [1]

Catalyst, wt%	Dehydrogenation activity, μmol C_6H_6/g catalyst · s[a]
34% Cr_2O_3 cogelled with Al_2O_3	0.5
10% MoO_3 coprecipitated with Al_2O_3	3
5% Ni on Al_2O_3 or SiO_2–Al_2O_3	13
5% Co on Al_2O_3	13
1% Pd on Al_2O_3	200
5% Ni on SiO_2	320
1% Rh on Al_2O_3	890
0.5% Pt on Al_2O_3 or SiO_2–Al_2O_3	1400–4000

[a] Differential flow reactor at 427°C, 6.8 atm, $H_2/HC = 6$ (mole ratio), activity determined after 30 min on stream, pretreated with H_2 at reaction conditions, LHSV varied to give differential operation.

Source: From "Catalysis," vol. VI, P. H. Emmett (ed). Copyright 1958 by Litton Educational Publishing Company. Reprinted with permission of the Van Nostrand Reinhold Company.

Preparation techniques Supported-metal catalysts are usually prepared by impregnation or ion exchange (ion adsorption) on a high-surface-area support such as silica or alumina [83–86]. The methods are discussed here for Pt/Al_2O_3 especially, but they also apply to many other catalysts.

In the first method, the porous support, Al_2O_3, is saturated with a solution of a salt such as H_2PtCl_6. If only a shell of metal around the outside of the catalyst particle is desired, as it may be when pore diffusion affects the catalyst performance, sufficient solution is used to coat only the outside of the porous particles. The particles are drained, dried [87, 88], and then calcined in air to convert the metal salt into a metal oxide. The oxide is reduced by hydrogen to give zero-valent metal. The ion-exchange technique, which is less commonly applied, involves introducing the metal onto the support by exchanging a cation complex such as $[Pt(NH_3)_4]^{2+}$ with acidic hydrogens on the surface [84, 86, 89–93]. The support is then washed with deionized water to remove all free salt, and the metal ions are left atomically dispersed on the surface. The catalyst may then be calcined and reduced.

Dispersion of metals in supported catalysts Catalysts prepared by these techniques have the metal dispersed on the surface of the support as small crystallites. These crystallites are usually 8 to 100 Å in diameter, depending on preparation; they are stable in comparison with other forms of dispersed Pt, for example, platinum black; and they are highly active. An examination of the effects of preparation conditions on crystallite size provides insight into the mechanism which produces the dispersion.

When there is no specific chemical interaction between the Pt salt and the support, as in impregnation of SiO_2 with H_2PtCl_6, the crystallite size is determined by the structure of the support. Figure 3-35 gives the Pt surface area as measured by hydrogen chemisorption (a method discussed later) for a high-surface-area silica (270 m^2/g); the samples were impregnated with H_2PtCl_6 solution to give a series of Pt loadings, varied by varying the solution concentration [92]. The surface area increased linearly with loading to about 5 weight percent Pt, and then it increased more gradually. The average Pt crystallite size remained constant at about 35 Å over the linear region of the figure, and it follows that the number of crystallites increased while their size remained unchanged; in contrast, for higher loadings, the number of crystallites remained constant and the size increased. The maximum number of crystallites was about the same as the estimated number of pores. When Al_2O_3 with a lower surface area and, correspondingly, fewer and larger pores was used instead of SiO_2, the resulting Pt crystallites were larger. These results indicate that the solvent was removed from the pores by evaporation until the solution became saturated with the Pt salt. Nucleation then occurred, and the crystallite size was determined by the amount of salt present in a pore, which was related to the pore size. When more concentrated impregnating solution was used, more pores were filled with solution when saturation was reached, and so more crystallites formed.

For catalysts prepared by the ion-exchange method, the metal surface area

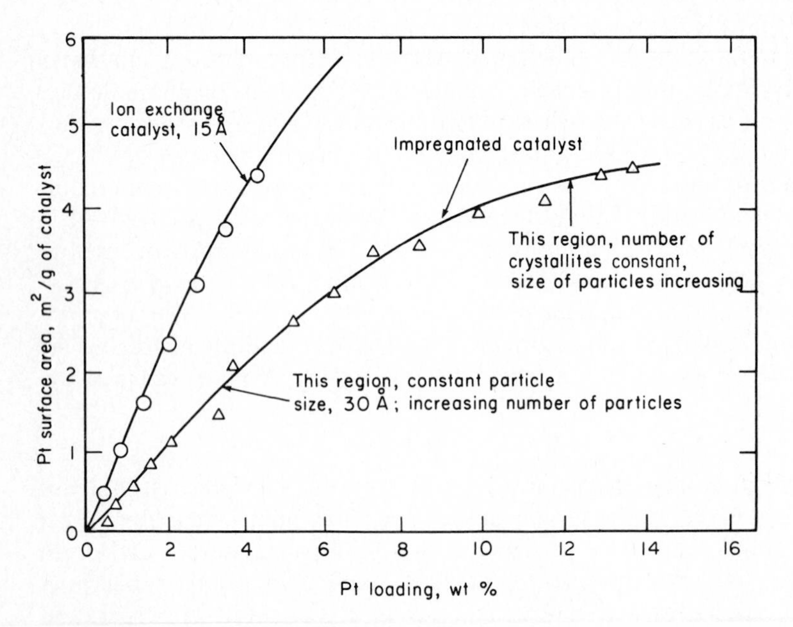

Figure 3-35 Effect of Pt loading and preparation technique on metal crystallite size. The metal loading was varied by varying concentration in solvent. The catalyst was dried at 120°C, then reduced (adapted from [92]). (Reprinted with permission from *Journal of Catalysis*. Copyright by Academic Press.)

was linearly dependent on metal loading up to about 7 weight percent Pt (Fig. 3-35). More and smaller crystallites resulted from the ion-exchange technique than from the impregnation technique. The average crystallite size was about 15 Å and was independent of loading and of the surface area and pore size of the SiO_2. After ion exchange the Pt was evenly distributed on the surface, and as reduction occurred, metal atoms within a certain volume migrated together to form each crystallite.

The size of the supported crystallites is also affected by the reduction and calcining conditions, as shown in Fig. 3-36. The catalyst prepared by ion exchange was more stable than that prepared by impregnation. The crystallite size growth for the impregnated catalyst was associated with the formation of volatile Pt chlorides formed from the chloride in the metal salt used in impregnation. The Pt salts apparently undergo vapor transport from small to large crystallites [92]. The catalyst prepared by ion exchange lacked chloride and showed much less growth in crystallite size. Reduction of the impregnated catalyst in hydrogen without precalcination resulted in little growth in crystallite size even at temperatures exceeding 500°C. Hydrogen reduction removed HCl, and subsequent heating in air resulted in little growth in crystallite size [92, 94].

Studies of the reduction and the calcination of a $[Pt(NH_3)_4]^{2+}$ ion-exchanged zeolite using infrared spectroscopy [93] showed that under hydrogen the complex began decomposing at about 125°C, with a resultant migration of the metal (probably as a zero-valent species) and loss of dispersion. Under air-calcination conditions the complex was stable up to a temperature of about 250°C, and the Pt

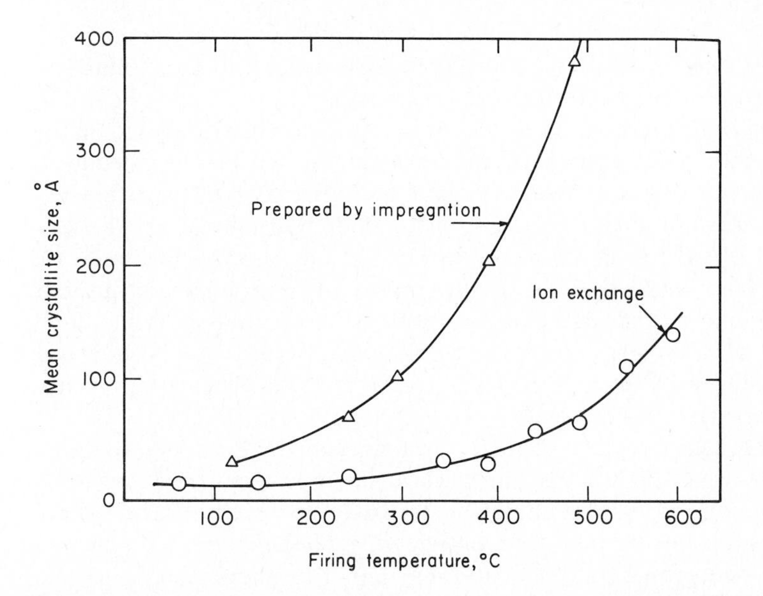

Figure 3-36 Pt crystallite growth due to firing of catalysts in air before the reduction stage [92]. (Reprinted with permission from *Journal of Catalysis*. Copyright by Academic Press.)

did not undergo migration because it remained bound by Coulomb forces to the exchange site as the Pt^{2+} ion. Reduction then gave a more highly dispersed catalyst (smaller Pt crystallites).

Heating in hydrogen or under vacuum typically results in little growth in crystallite size even at temperatures as high as 900°C [90, 95], but sintering is severe under such conditions when oxygen is present [96].

The dispersion of metal on a support can be determined by a number of techniques, each of which provides slightly different information. The techniques, reviewed by Adams et al. [97], include the following:

1. *Electron microscopy* is the only method of directly determining the individual sizes and shapes of metal crystallites. It can be used to measure crystallites from about 7 Å to sizes visible by light microscopy. For a statistically accurate size distribution or average size parameter to be determined, at least 1000 particles must often be counted. From this information, surface area and volume average particle diameters along with size distributions can be calculated [97, 98]. Figure 3-37 shows typical number and surface-area distributions for Pt/SiO_2 [98] and Pt/Al_2O_3 [99] obtained by electron microscopy. As illustrated, monodispersed systems cannot easily be produced, but size distributions can be kept quite narrow.

   Samples are prepared by crushing the catalyst particles and dispersing the powder on a collodion film or by embedding the catalyst particle in resin and microtoming sections from it. The major disadvantage of electron microscopy is that the sample volume studied is extremely small and unless careful sampling procedures are carried out to ensure that a representative sample is examined, the results can be misleading. A typical careful study involves the measurement of only about 10^{-17} g of Pt [97]. With catalysts containing less than 1 weight percent Pt, problems are encountered in even locating the metal.
2. *X-ray line broadening* is a method for determining crystallite sizes based on the fact that the width of an x-ray peak over that due to machine broadening is inversely proportional to the dimension of the crystallite giving rise to the x-ray reflection [97, 100, 101]. Line broadening is useful for crystallites in the range of about 40 to 1000 Å. The method gives a volume-averaged crystallize size for a sample about 10^{15} times larger than that examined in electron microscopy, but it provides no information about the crystallite size distribution. A platinum content as low as 0.5 weight percent can be measured by x-ray line broadening, but because of the small sizes of the Pt crystallites in reforming catalysts, they usually show no x-ray peaks at all.
3. *Small-angle x-ray scattering* uses the x-rays scattered by small metal particles; it provides a measure of the average particle size and a particle-size-distribution parameter [102, 103]. It is most applicable to particles in the 8 to 100 Å range and therefore supplements x-ray line broadening. The fine pore structures of most supports contribute to the scattering, and the interference must be masked by adsorption on the support of an organic compound with the appropriate electron density. Small-angle x-ray scattering has not been used much to

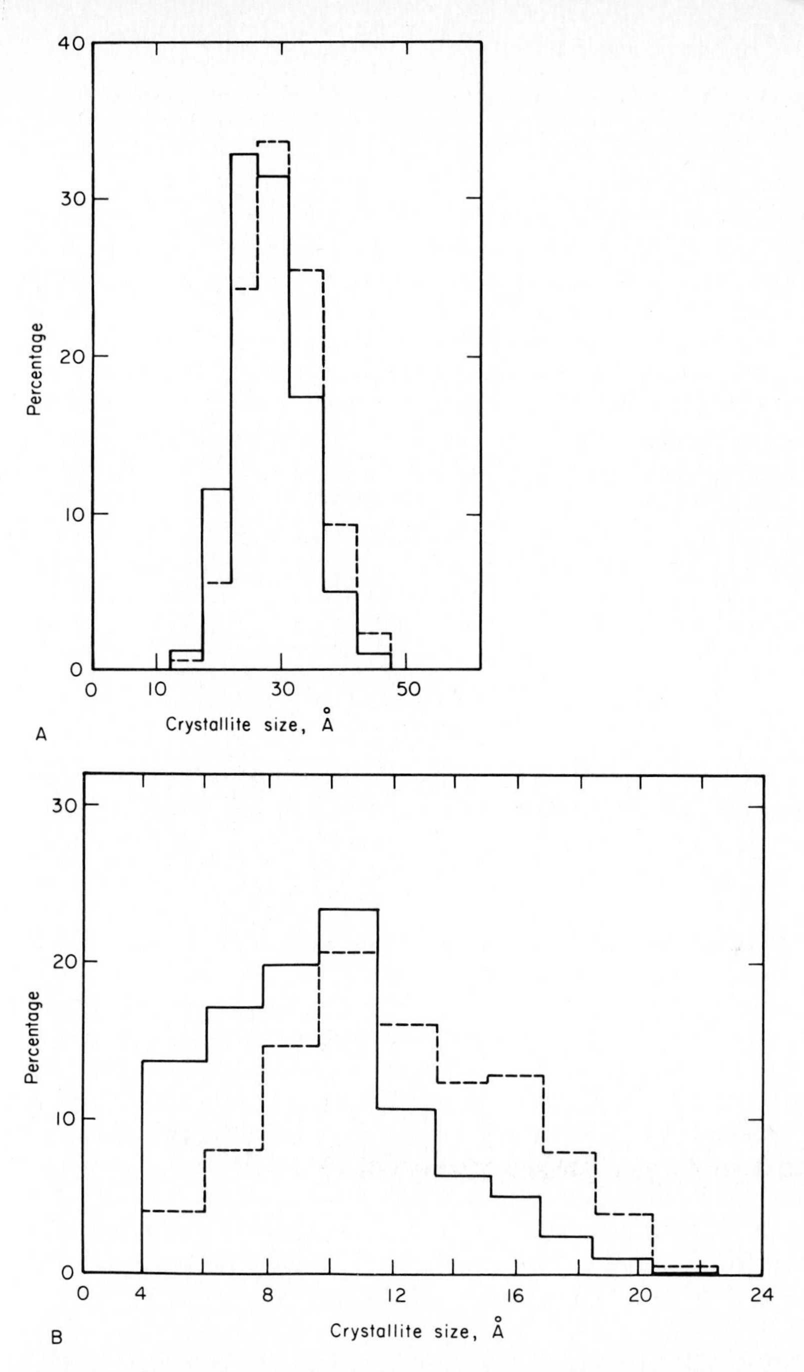

Figure 3-37 Pt crystallite size distribution determined by electron microscopy: (A) 2.5 wt % Pt on silica gel (adapted from [97]), (B) 2.83 wt % Pt on γ-alumina (adapted from [98]). Both catalysts were prepared by impregnation with chloroplatinic acid and hydrogen reduced; the full line refers to the number distribution, the dashed line to the surface area distribution. (A printed with permission from *Journal of Catalysis*. Copyright by Academic Press. B printed with permission from *Platinum Metals Review*. Copyright by Johnson Matthey and Company Limited.)

measure crystallite sizes, and the experimental techniques required have only recently been clarified [102]. However, it represents a powerful tool for studying reforming catalysts because of their high degrees of Pt dispersion.

4. *Gas chemisorption* is valuable for determining surface areas of metals on supports. Because the metal surface area is usually a small fraction of the total surface area of the metal and the support, it is not possible to measure the metal surface area by physical-adsorption techniques such as the Brunauer-Emmett-Teller (BET) method [104]. Selective chemisorption of a gas which does not chemisorb on the support but does chemisorb on the metal provides a means of measuring the metal surface area without interference from the support. Metal surface area is essential to interpretation of catalytic kinetics, since the proper way to express rates of metal-catalyzed reactions is per unit metal surface area. Rate per unit metal surface area, typically determined by hydrogen chemisorption, is called the *specific rate* or *specific catalytic activity*. The specific rate may also be stated in terms of the turnover number, the number of reaction events per surface metal atom per second. Chemisorption also allows metal crystallite size to be expressed in terms of dispersion, defined as

$$\text{Dispersion} = \frac{\text{g atoms hydrogen adsorbed}}{\text{total g atoms Pt present}}$$

A dispersion of 1.0 means that every metal atom present is on a surface of the very small crystallites; a dispersion of 0.1 means that 1 of every 10 metal atoms is present on a surface of the crystallites, which are about 100 Å in average diameter.

The chemisorption measurement gives the volume of gas adsorbed. If a gas-surface-metal-atom stoichiometry and an average surface-metal-atom spacing are assumed, the total metal surface area can be calculated. If a particle geometry is then assumed, e.g., a cube with five faces exposed, an average particle size can be calculated. The gas most appropriate for chemisorption is hydrogen; oxygen and carbon monoxide are also useful. The major problems associated with chemisorption measurements are the preparation of uncontaminated metal surfaces, determination of the presence or absence of adsorption on the support, and definition of the proper surface stoichiometry of the adsorption process.

Satisfactorily clean metal surfaces can be prepared by reduction with hydrogen at temperatures exceeding about 400°C for several hours, possibly preceded by an oxidation cycle, and then outgassing under high vacuum for several additional hours.

The surface adsorption stoichiometry is complicated. Spenadel and Boudart [96] showed that the surface area of platinum black calculated from hydrogen chemisorption assuming a surface Pt-atom-to-H-atom stoichiometry of 1 : 1 was the same as that determined from argon adsorption by the BET method. This 1 : 1 stoichiometry has also been observed for a series of supported Pt catalysts having crystallites with mean diameters from 42 to 143 Å [105]. It has been shown that for Pt crystallites in the 8 to 12 Å range, the H/Pt_s stoichiometry is apparently

1 : 1 and remains so with growth in crystallite size due to sintering or to recrystallization at high temperatures [91, 94, 99]. The Pt_s/O stoichiometry goes from 2 : 1 for very small crystallites (8 to 14 Å) to 1 : 1 after sintering of the crystallites to larger sizes or after high-temperature treatment without appreciable sintering [91, 94, 99, 106].

The oxygen chemisorption stoichiometry changes by a factor of about 2 upon heat treatment under hydrogen. This treatment prevents sintering of the Pt crystallites but allows them to reach a minimum free energy, e.g., by annealing out defects and recrystallization. This change in stoichiometry suggests the formation of more lower-index planes [91, 94, 99] or a reduction in the surface defects, but the reasons for the oxygen-chemisorption-stoichiometry change are still not clear.

Chemisorption of CO exhibits a Pt_s/CO stoichiometry of 1.15 : 1 on highly dispersed Pt [94, 96]. Upon sintering of the Pt to larger crystallite sizes, the Pt_s/CO stoichiometry increases to about 2 : 1 [94, 107]. The Pt_s/CO ratio in excess of 1 and the decrease in the ratio with heat treatment point to steric hindrance experienced by CO adsorbing on certain crystal planes. Electron micrographs suggest that large sintered crystallites are cuboctahedra bounded by (100) and (111) faces [94, 108–110], and calculations suggest that Pt_s/CO should be about 1 : 1 on the (100) plane. The limiting ratio on an infinite (111) plane should be 3 : 1 because of steric hindrance [111]. These considerations allow the prediction and the measurements to be reconciled.

Effects of crystallite size and support on catalytic activity The sites of the catalytic activity of supported metals have eluded precise definition for decades. Kobosev approached metal catalytic activity from the viewpoint of the theory of *active ensembles*. Taylor [112] proposed the concept of *active sites* for metals in 1925, and until recently little else that was definitive could be said. Studies with alloys (discussed above) have provided much information about the effect of ensemble size and the nature of the active sites on metals for a number of hydrocarbon reactions. In a similar sense, supported metals of different crystallite size can provide insight into the site requirements of metal-catalyzed reactions. If the crystallite size affects the reaction rate and/or the selectivity, observed crystallite size effects may contribute to our understanding of the following:

1. The nature of the active sites involved in the reaction, since changing the crystallite size changes the relative proportions of sites on the crystallite surface.
2. The effect of the support on the catalytic behavior of the metal, since as the metal crystallites become very small, support effects would be expected to become more important if they are present.
3. The effect of electronic factors in catalysis, since, as shown above, the electronic properties of very small crystallites are different from those of bulk metal.

The problem of sorting out the relative importance of these three factors when a crystallite size effect is observed still remains; understanding these effects is one of the fundamental challenges of catalysis by metals.

Boudart [113] and Taylor [112] proposed dividing metal-catalyzed reactions into two groups, *structure-sensitive* reactions and *structure-insensitive* reactions. Structure-sensitive (demanding) reactions are defined as reactions having rates sensitive to the details of the metal surface; they may consequently be dependent on the crystallite size. On the other hand, the rate of a structure-insensitive (facile) reaction is not affected by the details of structure, so that the specific catalytic activity (rate of reaction per unit surface area of metal or per surface metal atom) is not affected by changes in the mode of catalyst preparation or pretreatment or changes in the crystallite size or the support.

The structures of small metal crystallites (or clusters) unassociated with a support have been investigated theoretically by several authors [114–116], who concluded that icosahedral symmetry resulted in the most stable small crystallites; these are stabler than they would be with the crystallographic symmetry of the bulk metal. The effect of the support is proposed to be small [116], yet epitaxial growth of metals on such supports as mica is known [115, 117]. Baetzold [72, 73] concluded from molecular-orbital calculations that the metal-metal bond distances in small clusters are less than those in the bulk metal and that these bond distances depend on the support in contact with the cluster. Experimental evidence to test these predictions is meager, but metal-metal bond lengths have been shown to be less for smaller crystallites, approaching the values calculated theoretically [118–120].

A regularly faceted crystallite without bulk symmetry and without additional metal atoms on its surfaces provides a convenient example of the effect of crystallite size on the distribution of surface atoms and surface sites. Figure 3-38 represents a face-centered-cubic Pt crystallite and shows how the Pt atoms appear on the crystal faces for an octahedron and a cuboctahedron [121, 122]. Corner atoms have a coordination number of 4, edge atoms, 6 or 7, and atoms on the face of the plane, 8 or 9.

Table 3-7 gives the fraction of total crystallite atoms which are surface atoms, the total number of atoms in the crystal, and the average coordination number of surface atoms for an octahedral crystallite of increasing size [122]. As the crystallite size increases from 5.5 to 50 Å, the average coordination number increases from 4 to 8.64 (approaching the limit of 9.0), with the most pronounced change occurring in the smaller size range. This is the critical size range for catalytic activity if the activity is dependent on an average surface property (coordination number) or on the fraction of surface atoms which are on planes. However, if the activity is dependent on the relative number of corner and edge atoms, activity per unit metal surface area may continue to change over a much wider crystallite size range. For instance, the relative number of corner atoms continues to decrease as the size of the crystallite increases; if the activity is dependent on corner atoms as sites, the specific rate continues to decrease over a wide size range. This view of metal crystallite size effects is referred to as the *mitohedrical* (*face-edge*) *approach.*

The statistics of surface sites, defined as ensembles of two, three, four, or five atoms in a specific geometrical array, indicate that, with the exception of the five-atom site, the sites vary little in their relative proportion to each other as the

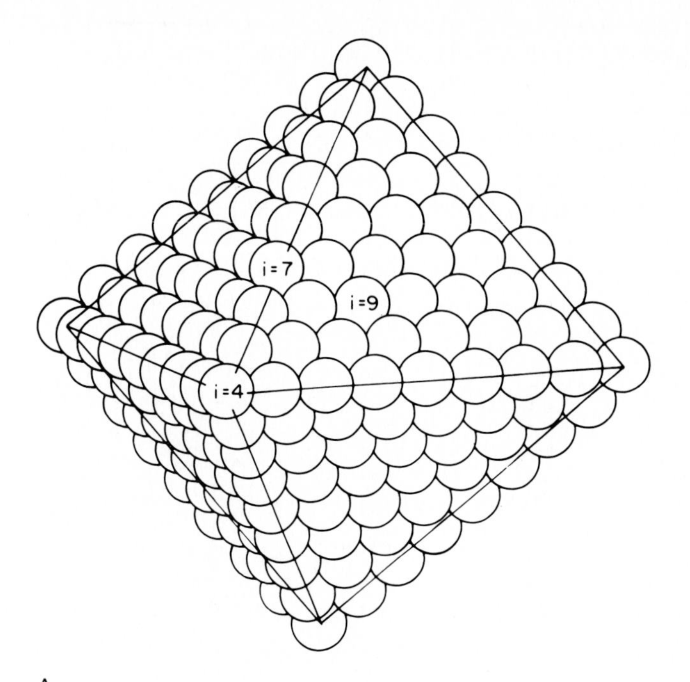

A

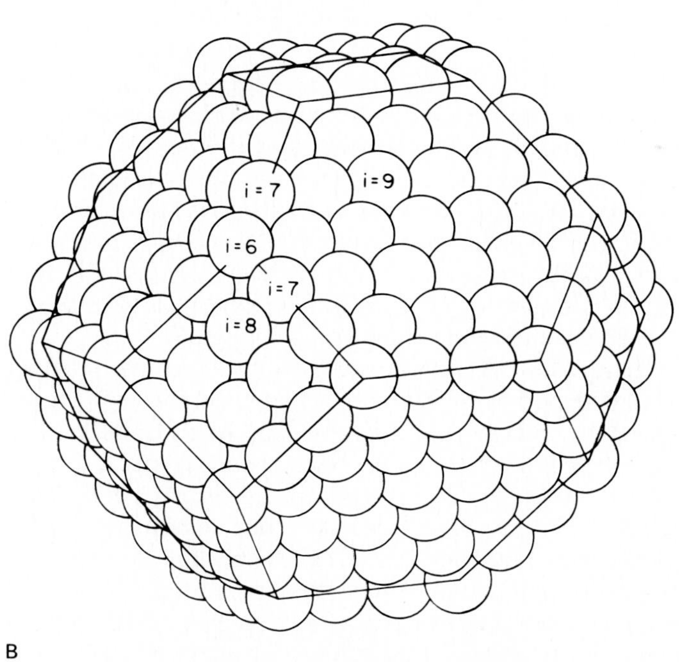

B

Figure 3-38 Surface atom arrangement for face-centered cubic crystallite: (A) octahedral and (B) cuboctahedral configuration [121]. Values of *i* specify number of nearest neighbors.

Table 3-7 Properties of platinum crystals of different sizes with regular faces [122]

Length of crystal edge		Fraction of atoms on surface	Total number of atoms in crystal	Average co-ordination number of surface atoms
Number of atoms	Å			
2	5.50	1	6	4.00
3	8.95	0.95	19	6.00
4	11.00	0.87	44	6.94
5	13.75	0.78	85	7.46
6	16.50	0.70	146	7.76
7	19.25	0.63	231	7.97
8	22.00	0.57	344	8.12
9	24.75	0.53	489	8.23
10	27.50	0.49	670	8.31
11	30.25	0.45	891	8.38
12	33.00	0.42	1156	8.44
13	35.75	0.39	1469	8.47
14	38.50	0.37	1834	8.53
15	41.25	0.35	2255	8.56
16	44.00	0.33	2736	8.59
17	46.75	0.31	3281	8.62
18	49.50	0.30	3894	8.64

crystallite size increases. The five-atom site B_5 is a terrace-edge ensemble which is not present on perfect crystallites but is present when there are incomplete layers. This site does not occur on metal crystallites smaller than 15 Å, and the relative surface concentration of these B_5 sites goes through a maximum for crystallites of 20 to 25 Å [123]. The surface density of these five-atom sites has been correlated with strong physical adsorption of molecular nitrogen on Ni and with changes in the infrared spectrum of adsorbed CO [124]. This same type of site is also present on stepped surfaces of single crystals, and their presence has been shown to impart high chemisorption capacity and catalytic activity [125].

It is also possible that crystallite size effects are indicative of energetic rather than geometric factors. Surface atoms of low coordination number adsorb adatoms more strongly than atoms of higher coordination number do; small crystallites ($\sim$ 10 Å) appear to melt at about one-half the bulk melting temperature [77]; and the atoms of the outer surface layer of very small crystallites may not be in the relative positions expected from the bulk structure; they appear to have a larger vibrational amplitude and have slightly shorter interatomic spacings [77].

Extremely small Pt clusters, estimated to have about six atoms each, have catalytic properties more like those of large particles of Ir [93], suggesting that they are electron-deficient. This behavior is suggested by the quantum-mechanical calculations discussed above, or alternatively, it may be induced by an interaction with the support.

Hydrocarbon reactions catalyzed by Pt which involve C—H bonds (hydrogenation-dehydrogenation) generally appear to be structure-insensitive [113]. For example, Boudart et al. [126] found that the specific catalytic activity for hydrogenation of cyclopropane was essentially independent of the degree of dispersion of Pt and independent of the support. The specific catalytic activity changed little as the catalyst changed from a highly dispersed 0.6 percent Pt/Al_2O_3 (the dispersion was 0.73) to Pt foil (the dispersion was 4×10^{-5}). Similar results were observed for cyclopropane hydrogenation by Poltorak et al. [127]. Poltorak and Boronin [122] demonstrated that the specific catalytic activities for hydrogenation of a number of compounds were independent of crystallite size of Pt between 10 and 70 Å. The specific catalytic activity for benzene hydrogenation has typically been observed to be independent of metal crystallite size [128–131], as has the specific catalytic activity for dehydrogenation of cyclohexane [132, 133] and methylcyclopentane [134]. Other reactions which appear to be structure-insensitive include isomerization of *n*-hexane to methylpentanes and aromatization to benzene [135, 136] and isomerization of neopentane [137].

This insensitivity to crystallite size and surface structure is consistent with the observation that hydrogenation-dehydrogenation reactions, including cyclopropane hydrogenation, are not affected by alloying an active metal with a much less active metal [36, 41]. These reactions are insensitive to both surface structure and surface composition. The structure insensitivity of reactions involving hydrogen is also consistent with the lack of dependence of hydrogen-surface Pt-atom stoichiometry on surface detail or crystallite size but is not required by it. This general insensitivity to structure, however, may not hold for dispersions which approach unity, i.e., for metals which approach being atomically dispersed. An apparent weak maximum in the specific rate of hydrogenation of benzene on Ni was observed for crystallites of about 12 Å, the specific rate decreasing markedly for smaller Ni crystallites [137*a*].

In contrast to reactions involving C—H bonds, reactions involving C—C bonds and certain other reactions, e.g., oxidation reactions, are structure-sensitive (demanding) [113]. The specific catalytic activity for neopentane hydrogenolysis to give isobutane and methane, which occurs in parallel with isomerization to isopentane, decreases by about two orders of magnitude with increasing crystallite size and with heat treatment to 900°C under gaseous hydrogen to prevent sintering [95]. Similarly, the specific rate of ethane hydrogenolysis increases markedly with decreasing crystallite size [138, 139]. The specific activities for isomerization of methylcyclopentane [140] and for dehydrocyclization of heptane [133] have also been shown to depend on metal crystallite size. Poltorak and Boronin [122], using the same series of Pt catalysis as in their hydrogenation studies, observed that the specific catalytic activity in oxidation reactions increased markedly as crystallite size increased and that the effect occurred entirely in the critical region between 8 and 50 Å. These results are in accord with the finding that the oxygen-to-surface Pt-atom stoichiometry is a function of crystallite size for small crystallites [91, 99, 106]. A similar large increase in specific catalytic activity with increasing crystallite size has been observed in ammonia oxidation on supported Pt [106, 141].

Structure-insensitive reactions are not influenced by the nature of the support; changing the support may well affect the specific catalytic activity for a structure-sensitive reaction. Highly dispersed Pt on SiO_2 was about 10 times more active (per unit of metal surface area) for ammonia oxidation than Pt with the same dispersion on Al_2O_3 [141]. This effect may be explained by the influence of the support on the surface detail of the metal crystallites or by induced electronic effects, which affect the activity. For metals such as Ni, which interact chemically with many supports, the situation is more complex and is not treated here.

Again consistency exists between the observed effect of crystallite size, alloy studies, and single-crystal studies. Hydrogenolysis reactions are sensitive to alloying, their rates decreasing markedly with decreasing concentration of active component on the surface [36, 41, 39, 47, 48]. Platinum single-crystal studies have shown orders-of-magnitude differences between the rates of *n*-heptane dehydrocyclization on low-index and stepped surfaces [125, 141*a*]. The stepped surfaces are most active, and activity correlates well with step density. Smaller metal crystallites show increased density of corners and edges, which are associated with steps on single crystals, and in C—C bond reactions the smaller crystallites have the higher specific catalytic activities.

Catalyst poisoning Catalyst poisoning typically results from strong chemisorption on the surface of the metal. The adsorbed poison either blocks the chemisorption of reactants or blocks the desired reactions. Catalyst poisoning is most appropriately considered from the surface-compound approach, discussed above. Thus those elements which form very stable compounds with the metal surface are typically severe poisons. Sulfur is the most prominent of these because of the great stability of metal sulfides, the equilibrium position of the reaction

$$H_2S + M \rightleftharpoons MS + H_2 \tag{15}$$

is far to the right for most metals. Generally molecules containing elements of group VB (N, P, As, Sb) and group VIB (O, S, Se, Te) may be strong catalyst poisons, depending on the electronic structures of the compounds containing them. If the element has unused bonding orbitals that can participate in the formation of a stable compound with the metal atoms at the surface, the substance acts as a poison. If the species containing the suspect atom cannot form a stable surface compound, poisoning does not occur. For example, sulfate ion is not a poison of some metals (V, Pt) because their sulfates are not stable; it is a poison of others, for example, Fe and Cu; H_2S almost always is a poison because of the great stability of almost all metal sulfides. Table 3-8 summarizes the poisoning characteristics of compounds of the above elements.

The first two elements of groups VB and VIB, nitrogen and oxygen, are weak poisons [142]. The crystallite size effect observed in oxidation reactions [106, 122, 141] may be an indication of differing susceptibilities of metal crystallites of different sizes to oxygen poisoning. The heavier elements of groups VB and VIB are much more troublesome in the practice of catalytic reforming, sulfur and arsenic [5] being of particular concern. Dilke et al. [143] have shown by magnetic-

Table 3-8 Poisons of metal catalysts: the influence of electronic structure on poisoning character

Element	Compounds which are poisons	Compounds which are not poisons
S, Se, Te	$\mathrm{H{:}\underset{\cdot\cdot}{\overset{\cdot\cdot}{S}}{:}H}$	
	$\left[\mathrm{O{:}\overset{O}{\underset{\cdot\cdot}{\ddot{S}}}{:}O}\right]^{2-}$ Sulfite	$\left[\mathrm{O{:}\overset{O}{\underset{O}{\ddot{\underset{\cdot\cdot}{S}}}}{:}O}\right]^{2-}$ Sulfate
	Also selenite, tellurite	Also selenate, tellurate
P, As, Sb	RSH; R_2S	RSO_3H; R_2SO_2
	$\mathrm{H{:}\overset{H}{\underset{\cdot\cdot}{\ddot{P}}}{:}H}$ Phosphine	$\left[\mathrm{O{:}\overset{O}{\underset{O}{\ddot{\underset{\cdot\cdot}{P}}}}{:}O}\right]^{3-}$ Phosphate
	Also arsine, stibine	Also arsenate
N	NH_3, pyridine, piperidine	NH_4^+, pyridinium ion piperidinium ion

susceptibility measurements that electrons from methyl sulfide enter the *d* band of the metal, forming a strong coordinative link upon adsorption. The process may be accompanied by a filling of the fractional deficiencies or holes in the *d* band of the metal due to *s*-band overlap, leading to even more pronounced deactivation than just blocking of sites. There is no quantitative information about this, but studies with alloys suggest that poisoning may be more of a geometric than electronic effect [36].

The state of an element entering the reaction zone is not nearly as important as the environment at the catalyst surface. Thus SO_2 seriously deactivates Pt under reducing conditions but is readily converted to SO_3 under oxidizing conditions, causing no deactivation. Similarly, arsenate ion fails to poison Pt under the strongly oxidizing conditions of hydrogen peroxide decomposition. Under the less severe oxidizing conditions of SO_2 oxidation, it severely poisons Pt, since it is converted into such species as arsenite, which forms a stable compound with Pt surface atoms.

The Alumina Support

Preparation Alumina is obtained by precipitation from aqueous solutions containing Al^{3+} ions. The first precipitate is a gellike substance giving a diffuse x-ray diagram. The slurry of the first precipitate is normally aged at temperatures between 40 and 80°C, and the details of the aging procedure are important in determining the properties of the final product. After aging, the precipitate is filtered, washed, and dried. The final operation in the preparation consists in heating at temperatures of up to 600°C.

During this operation, Al_2O_3 passes through various states of hydration, several distinct compounds being discernible:

$Al_2O_3 \cdot 3H_2O$, or $Al(OH)_3$, involves two different compounds, bayerite and gibbsite.

$Al_2O_3 \cdot H_2O$, or $AlO(OH)$, involves several different phases; only boehmite is of importance.

$Al_2O_3 \cdot nH_2O$ products are obtained upon heating $AlO(OH)$; n decreases with the severity of the calcination. Many compounds are discernible by x-ray analysis.

α-Al_2O_3 is the product of firing at 1200 to 1300°C; it has a low surface area.

Figure 3-39 shows how the various phases are formed from each other.

The initial precipitation leads first to a gel in which minute crystals of boehmite are present. If this is filtered without sufficient aging and then calcined, the calcination does not lead to any of the identifiable intermediate compounds with $0 < n < 0.6$ (Fig. 3-39). The material remains amorphous until after firing to temperatures greater than 1100°C. At higher temperatures, α-Al_2O_3 is formed. If the initial boehmite gel slurry is aged at 40°C, it is converted into bayerite, a form of $Al(OH)_3$. If this product is filtered, dried, and then calcined, a compound,

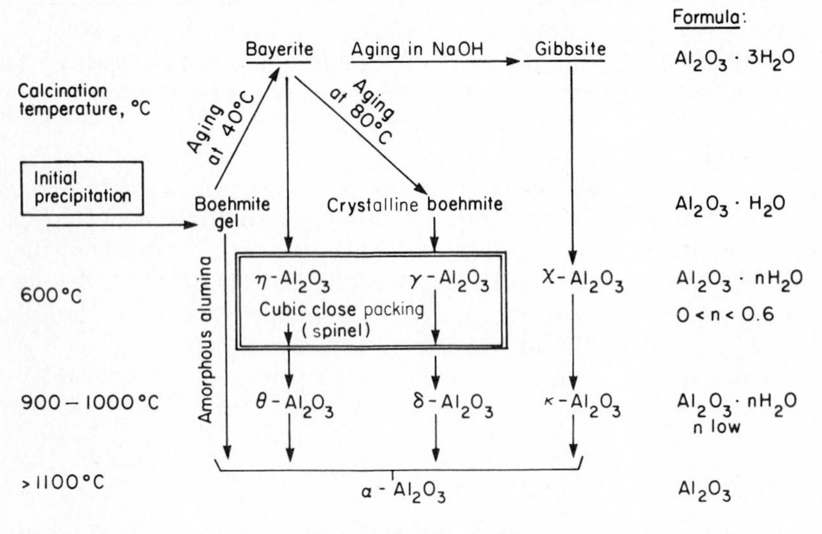

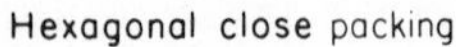

Figure 3-39 Schematic representation of formation of various Al_2O_3-hydrates; adapted from [144] and [145].

designated as η-Al_2O_3, is formed. Calcinating at yet higher temperatures produces another compound, θ-Al_2O_3, which at temperatures exceeding 1100°C is converted into α-Al_2O_3.

The bayerite may also be converted by further aging of the slurry at higher temperatures into another form, apparently a much more crystalline form of boehmite, referred to as *crystalline boehmite*. When this precipitate is heated after filtration and washing, it becomes another set of compounds, γ- and δ-Al_2O_3, which are very similar but not identical to η- and θ-Al_2O_3. Upon heating to temperatures higher than 1100°C, the compound γ-Al_2O_3 becomes α-Al_2O_3.

If the bayerite is aged in aqueous NaOH, gibbsite is formed. This is converted to χ- and κ-Al_2O_3 as its dehydrated forms. These materials are less important for our purposes. The important compounds for reforming catalysts are η- and γ-Al_2O_3, particularly the former. They represent supports with high surface area and high thermal stability, and the surface acidity can be controlled. One of their most important structural characteristics is cubic close packing of oxygen; they resemble spinels.

Oxides with cubic-close-packed structures Many oxide structures show close packing of their oxygen ions; there are two types, hexagonal and cubic. In the following description of close-packed structures, the oxygen anions are considered to be hard spheres. A single layer of close-packed spheres is considered first, and then the alumina structures are derived. The layer has a trigonal symmetry, with each oxygen ion located in the positions indicated as position 1 in Fig. 3-40. A similar layer is positioned above the first, with each sphere of the second layer just above a hole in the first layer, as indicated by position 2 in the figure. For the third layer, there are two possibilities:

1. Hexagonal close packing: the third layer is again placed in the position of the first layer, and the sequence of stacking is 1, 2, 1, 2, 1, 2, ...

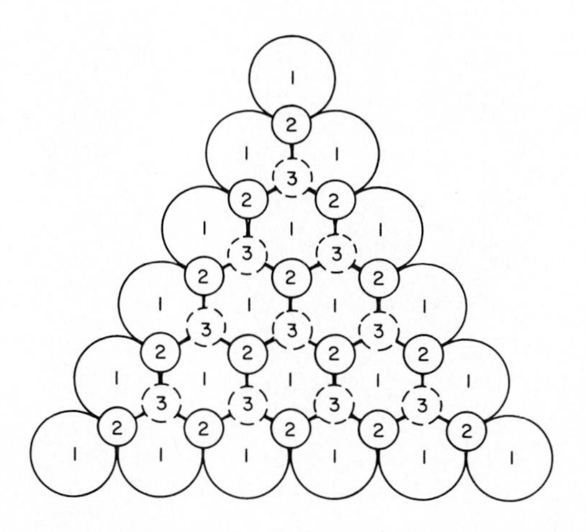

Figure 3-40 Single layer of close-packed spheres; 2 and 3 indicate sites of additional layers.

2. Cubic close packing: the third layer is placed above another set of holes in the first layer (position 3). This exhausts all the possibilities, and the fourth layer is again in the 1 position. The sequence is then 1, 2, 3, 1, 2, 3, ...

The hexagonal close packing applies to α-Al_2O_3; cubic close packing applies to η- and γ-Al_2O_3.

Positions of the Al^{3+} and H^+ ions in the close-packed anion structures The cations, in particular Al^{3+}, must be located in the interstices between the close-packed anion layers. The Al^{3+} ions are placed at positions between the two layers since these are the only locations offering sufficient room for them. One possibility is to place the Al^{3+} ions in the sites just above the triangularly formed holes shown in Fig. 3-40; the second oxygen layer in the 2 position is then placed over them. The Al^{3+} ion is then in an octahedral position, as shown in Fig. 3-41.

If we proceed with this method of stacking O^{2-}, Al^{3+}, O^{2-}, and Al^{3+} in hexagonal close packing, as in α-Al_2O_3, we find that there are just as many sites for the cations in the cation layer as sites for O^{2-} in the anion layer. With such an arrangement the system would fail to meet the requirement of electrical neutrality. To satisfy this requirement, one of every three cation sites must be vacant. The vacancies lead to several possibilities for symmetry of the Al^{3+} ions. The positions occupied in α-Al_2O_3 are the same as those for the starting compound, i.e., the $Al(OH)_3$ structures, and are shown in Fig. 3-42; γ-Cr_2O_3 has the same structure.

The description of α-Al_2O_3 is now complete. For $Al(OH)_3$, however, the positions of the protons are yet to be defined. First, two oxygen layers are stacked one on top of the other, "sandwiching" the Al^{3+} ions between them in the arrangement shown in Fig. 3-42. This layer is now electrically negative. But if a proton is attached to every O^{2-} at the outside, i.e., every O^{2-} is replaced by an OH^-, the layer is neutral and complete, as shown schematically in Fig. 3-43. The arrangement is an example of a layer structure; these structures occurred before in

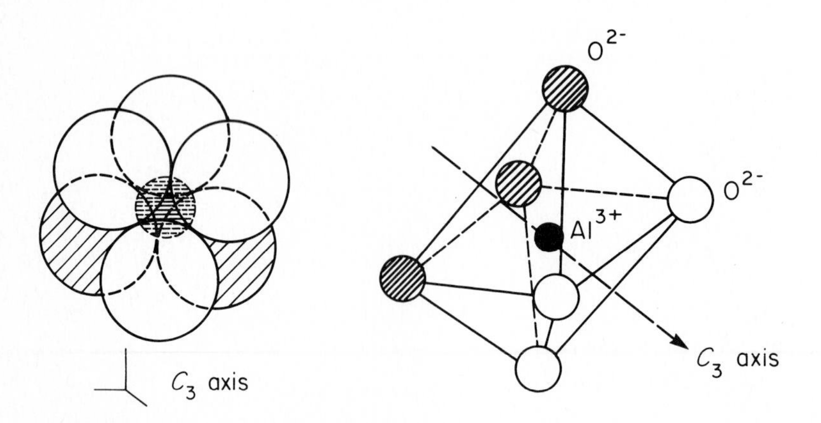

Figure 3-41 Octahedral position of Al^{3+} ion.

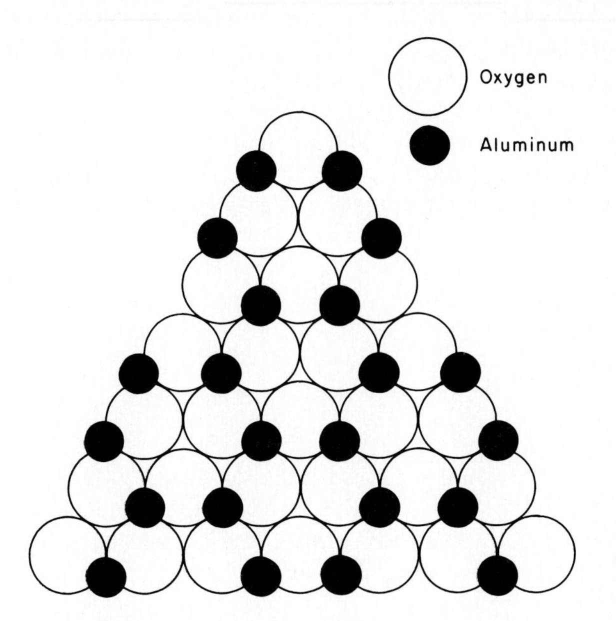

Figure 3-42 Positions of cations in α-Al_2O_3, α-Cr_2O_3, and aluminum trihydroxides.

the Ziegler-Natta catalysts of Chap. 2, and they appear again in the discussion of hydrodesulfurization catalysts in Chap. 5.

The two hydroxides differ in how the successive layers are packed. The two possible arrangements in terms of the O^{2-} position are

$$\underset{\text{Gibbsite}}{1, 2, 2, 1, 1, 2, \ldots} \quad \text{and} \quad \underset{\text{Bayerite}}{1, 2, 1, 2, 1, 2, \ldots}$$

The dehydration reaction which converts the hydroxides to the oxides can be summarized as

$$OH^- + OH^- \longrightarrow O^{2-} + \square + H_2O \tag{16}$$

where $\square$ represents an anion vacancy. The vacancy remains at the surface when two hydroxyls on the same surface interact with each other. If the hydroxyls from two different surface faces interact with each other, the O^{2-} of one and the anion vacancy of the other unite and the O^{2-} becomes part of the bulk structure as a larger unit is formed. The reaction results in the elimination of H_2O from the solid.

It might at first be expected that the similar orientation of the Al^{3+} ions in

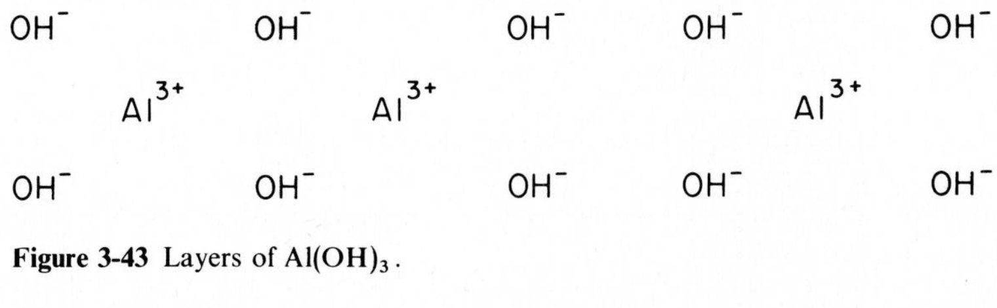

OH^- OH^- OH^- OH^- OH^-

Al^{3+} Al^{3+} Al^{3+}

OH^- OH^- OH^- OH^- OH^-

Figure 3-43 Layers of $Al(OH)_3$.

$Al(OH)_3$ and in α-Al_2O_3 would lead to a direct conversion of one into the other, but the actual conversion is far more complex.

The first reaction in the conversion to the catalytically important forms is the formation of AlO(OH) (boehmite). This reaction leads to a change in orientation of the Al^{3+} ions; they now occur in parallel rows, with rows of unoccupied octahedral sites between the Al^{3+}-ion rows. The orientation can be visualized as follows: First construct infinite chains in the following arrangement:

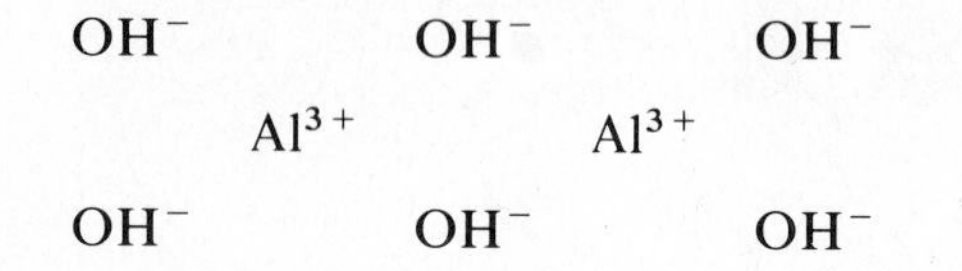

Then connect two of the chains as shown in Fig. 3-44. The chains are then connected together to form a three-dimensional structure as shown in Fig. 3-45. The structure involves a packing of double chains parallel to each other. The close-packed anion planes are now mixed O^{2-} and OH^-, and the Al^{3+} ions occur in rows parallel to the chain, every other row remaining vacant.

The intermediate aluminum oxides Upon further dehydration, two important changes occur: (1) the close packing changes from hexagonal to cubic, and (2) in the resultant cubic close packing, octahedral positions occur as shown in Fig. 3-45. However, the octahedra form interconnecting tetrahedra, and there is room for cations in the tetrahedral positions, provided that the cations are small, having radii less than 1.4 Å, the radius of O^{2-}; Al^{3+} ions have a radius of 0.5 Å

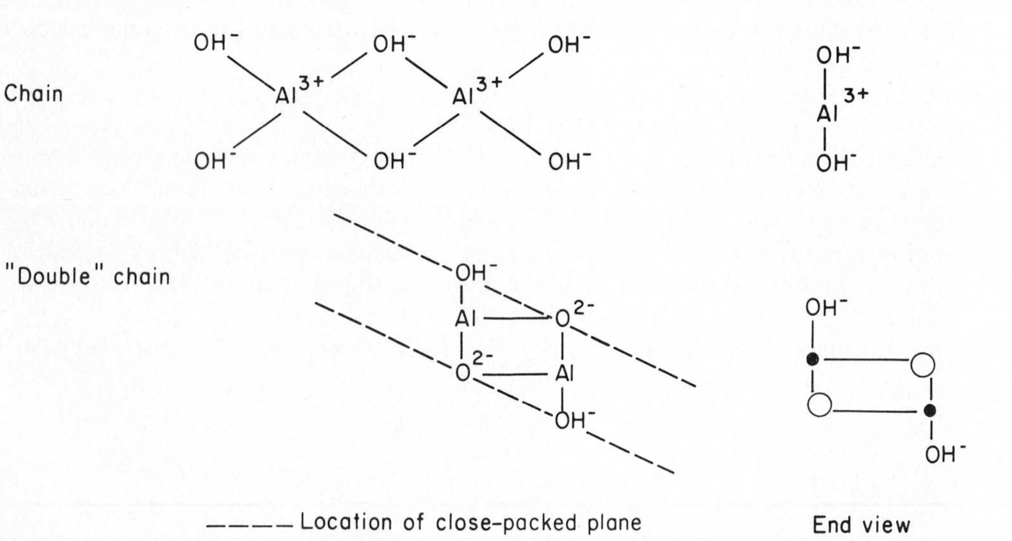

Figure 3-44 Boehmite crystal structure: relation between the double chain structural element in AlO(OH) and the single chain in $Al(OH)_3$.

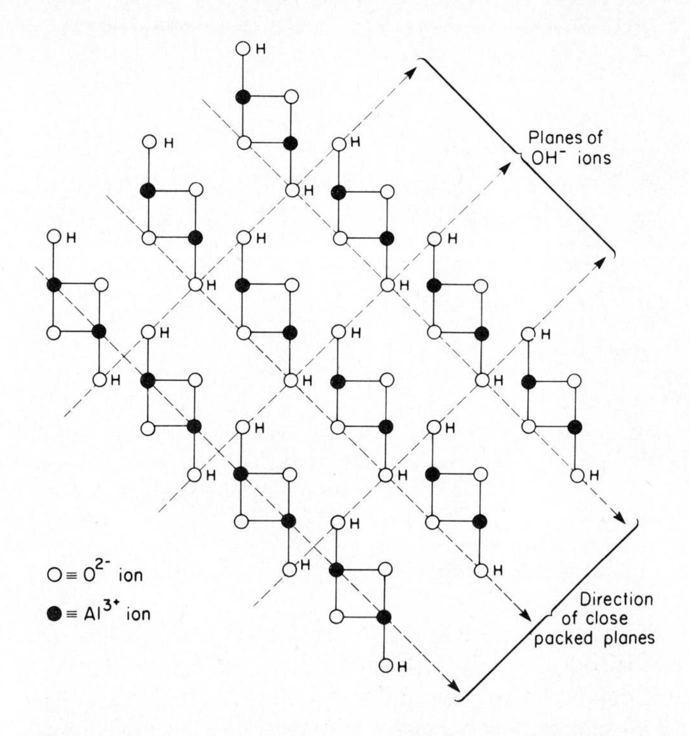

Figure 3-45 Boehmite crystal structure: packing of double chains in AlO(OH).

and can therefore fit into both octahedral and tetrahedral positions. In cubic close packing, appreciable numbers of the Al^{3+} ions are in the tetrahedral positions.

Aluminum oxides are members of a class of binary oxides in which the oxygen close packing is of the cubic type, with the cations situated such that one of the two cations prefers tetrahedral positions and the other prefers octahedral positions. These compounds are called *spinels* after the mineral spinel, which has the composition $MgAl_2O_4$. In spinel, Mg^{2+} occupies tetrahedral and Al^{3+} occupies octahedral positions. The structure of $MgAl_2O_4$ is shown in Fig. 3-46. The unit cell contains 32 O^{2-} ions, 16 octahedral sites, which are all occupied by Al^{3+} ions, and 64 tetrahedral sites, of which 8 are occupied by Mg^{2+} ions.

The x-ray diagrams of the intermediate aluminas are similar to those of the spinels. Because of this similarity and the presence of H in the structures of η- and γ-Al_2O_3, the formulas of the two are given formally as $(H_{1/2}Al_{1/2})Al_2O_4$ or $Al(H_{1/2}Al_{3/4})O_4$, in which some of the Al^{3+} ions occur in tetrahedral positions. In all probability, the protons are located not in tetrahedral sites, as suggested here, but at the surface in OH^- groups. One in every eight O^{2-} ions is therefore at the surface as OH^-, which means that the crystals are small and their surfaces consist largely of OH^- groups. These conclusions are in good agreement with the observations that the surface areas of η- and of γ-Al_2O_3 are high ($\sim$ 250 m^2/g) and that these structures contain a relatively large amount of "bonded water." Infrared

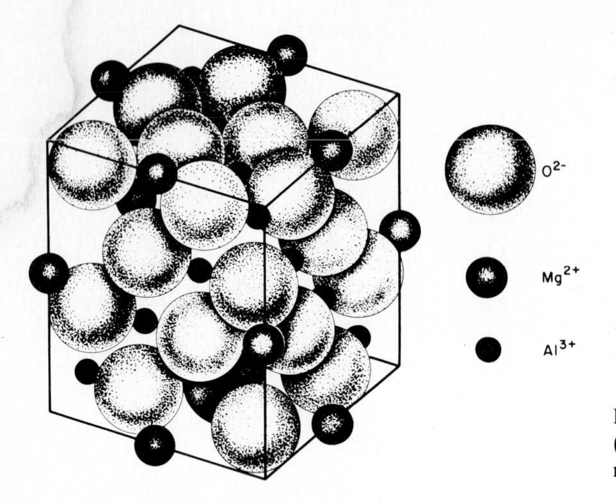

Figure 3-46 Unit cell of spinel ($MgAl_2O_4$) [60]. (Reprinted by permission of John Wiley & Sons, Inc.)

spectra indicate that the surfaces are almost completely covered with OH^- groups.

The several forms of the hydrated alumina, such as η-Al_2O_3, have different ratios of tetrahedral to octahedral Al^{3+} ions and different symmetries of occupation of Al^{3+} in the available octahedral and tetrahedral positions. Our knowledge is not sufficient to allow development of a conclusive model of the cation occupancies of the various forms. It is generally believed that η-Al_2O_3 contains relatively more tetrahedral Al^{3+} ions than γ-Al_2O_3 does.

Lippens [145] proposed the models of η- and γ-Al_2O_3 shown in Figs. 3-47 and 3-48, respectively. For η-Al_2O_3, we start with the close-packed oxygen layer shown in Fig. 3-40. This forms the (111) plane in cubic close packing. There are now two types of arrangements of the Al^{3+} ions. One (the B layer) has Al^{3+} ions located in octahedral positions only (Fig. 3-47). The other (the A layer) is obtained from this by transferring two-thirds of the cations from octahedral to tetrahedral positions. The resultant structure is a stacking of the layers, ABAB $\cdots$.

The γ-Al_2O_3 structure is more easily visualized by considering the (110) crystal plane which is oriented at an angle to the (111) plane. Again we have two types of layers, designated here as the C and D layers (Fig. 3-48). The D layer has only octahedrally located Al^{3+} ions, and the C layer has as many tetrahedral as octahedral sites. The packing of the structure is CDCD $\cdots$.

The important structural characteristic of alumina in catalysis is the surface, and since alumina occurs in the form of lamellae, it is most probable that only one type of surface plane is predominant. According to Lippens, this is the (111) plane for η-Al_2O_3 and the (110) plane for γ-Al_2O_3. These differences are important to the applicability of the alumina in catalytic processes; γ-Al_2O_3 appears better suited to hydrodesulfurization catalysts (Chap. 5), whereas η-Al_2O_3 appears better suited to reforming catalysts. Moreover, the η-Al_2O_3 is the more acidic of the two because of the higher density of Al^{3+} in tetrahedral sites on the surface.

If the dehydration is carried further by heating the solid to temperatures

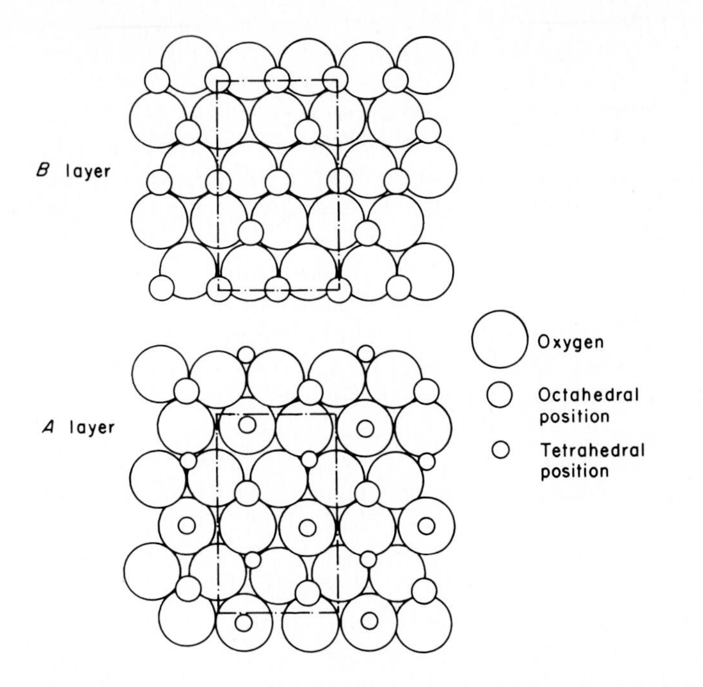

Figure 3-47 Structure of η-alumina showing tetrahedral and octahedral Al^{3+} ion positions [145].

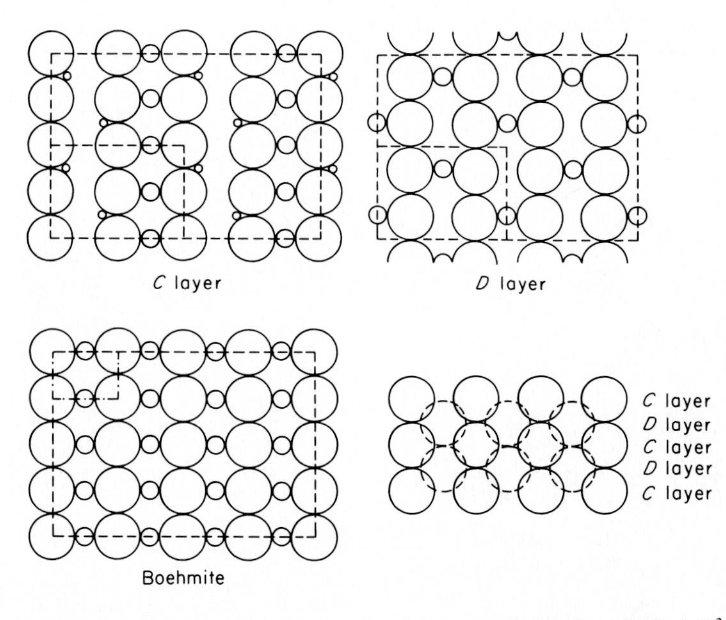

Figure 3-48 Structure of γ-alumina showing tetrahedral and octahedral Al^{3+} ion positions [145].

around 900°C, almost all the water is expelled. Since this dehydration results in a substantial loss in surface area, it appears to involve the formation of larger crystallites by interaction between crystallite surfaces. This result suggests that θ-Al_2O_3 and δ-Al_2O_3 are respectively similar to η-Al_2O_3 and γ-Al_2O_3, and the suggestion appears to be confirmed for δ-Al_2O_3 but not for θ-Al_2O_3, which has a structure similar to that of a boehmite, since the Al^{3+} ions occur in filled double rows separated by empty rows of octahedral positions (Fig. 3-49). This statement should apply only to some layers, because in others Al^{3+} ions should still exist in tetrahedral positions.

The surfaces of these dehydrated oxides are almost certainly devoid of protons and therefore should consist almost exclusively of O^{2-} anions and anion vacancies. In many respects their properties are quite different from those of other aluminas.

Acidity of aluminas Fully hydrated aluminas may have some surface Brønsted acidity since their surfaces contain OH^- groups. The surfaces of θ-Al_2O_3 and δ-Al_2O_3 contain Lewis acidity but probably do not contain Brønsted acidity. In the η- and γ-aluminas, both types of acidity may be present, depending on the degree of hydration. Peri [146, 147] has developed semiquantitative models of the surface of alumina as a function of its dehydration level.

In general the aluminum hydroxides, and therefore also the hydrated oxides, are not strongly acidic. They are more amphoteric in nature, as inferred from experience with the hydroxides in which all the Al^{3+} ions are octahedral. The OH^- groups attached to tetrahedral Al^{3+} ions may be more strongly acidic. The measurement of alumina surface acidity by the indicator methods described in

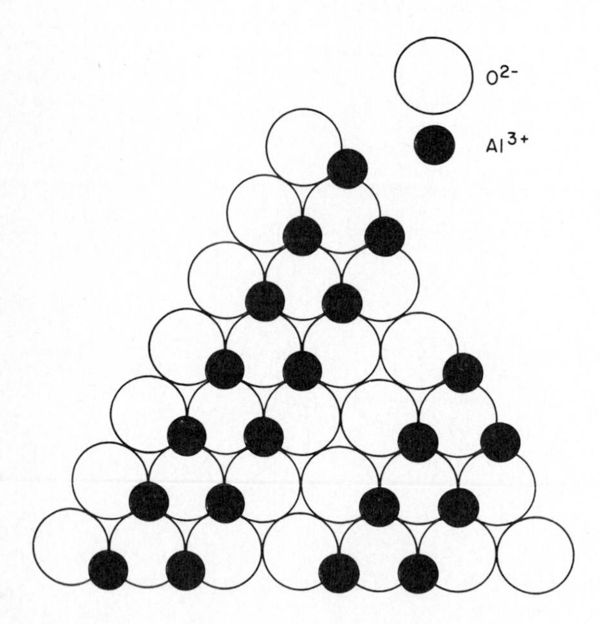

Figure 3-49 Structure and cation positions in θ-alumina.

Chap. 1 supports the hypothesis that certain groups on the surface of γ-Al_2O_3 and especially on η-Al_2O_3 are weakly acidic. The acidity of OH groups present on the surface can be markedly enhanced by the proximity of Cl^- ions: the development of surface acidity on alumina may be similar to that for silica-alumina (Chap. 1), for which the electronic asymmetry set up by the proximity of the different species draws electrons from the O—H bond, increasing the acidity of the group. Similarly, conversion of CH_3CO_2H into CCl_3CO_2H causes an increase in acidity. A proposed mechanism for such acidity enhancement is shown in Fig. 3-50.

Partial conversion of a fully hydrated alumina surface,

$$\begin{array}{cccccccccc} OH^- & & OH^- & & OH^- & OH^- & & OH^- & & OH^- \\ & Al^{3+} & & Al^{3+} & & & Al^{3+} & & Al^{3+} & \end{array}$$

to a surface consisting of Cl^- and OH^- groups (for example, by treating the surface with HCl)

$$\begin{array}{cccccccccc} Cl^- & & OH^- & & Cl^- & OH^- & & Cl^- & & OH^- \\ & Al^{3+} & & Al^{3+} & & & Al^{3+} & & Al^{3+} & \end{array}$$

enhances the acidity of the remaining OH groups markedly. The acidity of acetic acid increases progressively with addition of chlorine atoms to the carbon (Fig. 3-50). The sequence of acid strengths is acetic acid < monochloroacetic acid < dichloroacetic acid < trichloroacetic acid. In a similar way, it would

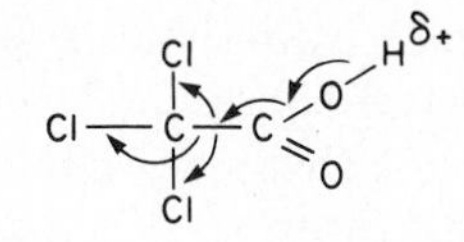

H H H$^{\delta+}$ H$^{\delta+}$ H H
O O O Cl O O O
Al Al Al

A

Electrons are pulled toward chlorine atoms and away from O-H bond, making the OH group more acidic.

ACID	pK_A
CH_3COOH	4.8
$ClCH_2COOH$	2.9
$Cl_2CHCOOH$	1.3
Cl_3CCOOH	0.7
$CH_3CH_2CH_2CHClCOOH$	2.9
$CH_3CH_2CHClCH_2COOH$	4.3
$CH_3CHClCH_2CH_2COOH$	4.8

B

Figure 3-50 Acidity enhancement in organic acids and in alumina: (*A*) mechanism of acidity enhancement; (*B*) effects of number and location of chlorine substituents on pK_a values of organic acids.

appear that the acidity of a group on the alumina surface can be progressively increased as more of the OH groups surrounding it are replaced by Cl^- ions. Thus both the amount of acidity and the acid-site strength can be controlled on alumina. This effect is probably stronger when the protons are already more readily donated as a result of being bonded via O^{2-} to tetrahedral Al^{3+}. In support of this suggestion, it is noted that $AlCl_3$ in the presence of water is an extremely strong acid (Chap. 1).

Alumina itself catalyzes easy isomerizations such as double-bond shifts in butenes; it does not readily catalyze skeletal isomerization because it is too weakly acidic. The increase in acidity induced by treatment of alumina with HCl or HF is sufficient to make it a highly active catalyst for skeletal isomerization, hydrocracking, and other strong-acid-catalyzed reactions which are desirable in reforming. It is especially important in the practice of reforming that the strength of the surface acid sites can be controlled by the extent of halogenation, and thus the relative rates of the several acid-catalyzed reforming reactions can be controlled.

REFORMING REACTIONS

The reactions related to catalytic reforming will now be discussed in some detail. Reforming involves reactions of hydrocarbons on the surface of the metal and on the promoted (acidic) alumina. The alumina-catalyzed reactions involve mainly the carbonium-ion chemistry of Chap. 1. In this case surface-catalyzed reaction mechanisms were inferred from a large body of well-established carbonium-ion chemistry for acid solutions, which provided an independent quantitative base. The reaction chemistry of soluble transition-metal-complex catalysts, discussed in Chap. 2, has been inferred from the large body of organometallic chemistry and in part from the application of a number of spectroscopic techniques to the reacting solutions. In the case of reactions on metal surfaces, there is no separate body of chemical information to draw on to infer the relevant surface-reaction mechanisms, and the direct application of spectroscopic techniques has not been as straightforwardly useful. Therefore, metal-surface-catalyzed hydrocarbon reaction mechanisms must be inferred directly from reaction studies. A particularly useful tool in this work has been H–D exchange studies.

Metal-catalyzed Reactions

Hydrogenation-dehydrogenation reactions Hydrogenation-dehydrogenation of hydrocarbons on metals is exemplified by the reaction

$$C_2H_4 + H_2 \rightleftharpoons C_2H_6 \qquad (17)$$

Typical hydrogenation-dehydrogenation processes involve olefins, paraffins, dienes, and aromatics. Hydrogenation reactions are usually first order in the hydrogen partial pressure and zero or negative order in olefin partial pressure. A simple mechanism for the reaction is the Bonhoeffer-Farkas mechanism [148], which involves chemisorption of both reactants followed by reaction. In this mechanism, a hydrogen molecule adsorbs on two surface sites (presumably single metal atoms), undergoing fission of the H—H bond (dissociative adsorption) [149]. The C_2H_4 molecule adsorbs associatively on two sites by forming σ bonds, one between each C atom and a metal atom [150]. Alternatively, it is sometimes assumed that the interaction of the olefin with the metal surface is essentially a π interaction [151, 152]:

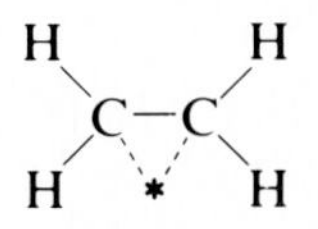

Demuth and Eastman [153] have shown by ultraviolet photoelectron spectroscopy that at low temperatures ($< -80°C$) ethylene adsorption on Ni involves a π-bonded species. Under their high-vacuum hydrogen-free conditions, π-adsorbed ethylene was converted into the σ-bonded form upon heating to room temperature. Except for aromatics, the π-bonded intermediate may under most conditions represent more of a transition state in adsorption than a stable species.

Early work by Turkevich and coworkers [154] showed that reaction (17) did not go by the simple addition of two hydrogen atoms to give ethane, since there was extensive H–D exchange between the hydrogens of C_2H_4 and D_2. This exchange indicates that the elementary steps are reversible and forms the basis of the following proposed reaction mechanism (asterisks represent surface sites):

$$H_2 + 2* \underset{k_{-1}}{\overset{k_1}{\rightleftharpoons}} 2\,\underset{*}{\overset{H}{|}} \tag{18}$$

$$2* + H_2C{-}CH_2 \underset{k_{-2}}{\overset{k_2}{\rightleftharpoons}} \begin{matrix} H & & & H \\ & C & - & C & \\ H & \underset{*}{|} & & \underset{*}{|} & H \end{matrix} \tag{19}$$

$$\underset{*}{\overset{H}{|}} + \begin{matrix} H & & & H \\ & C & - & C & \\ H & \underset{*}{|} & & \underset{*}{|} & H \end{matrix} \underset{k_{-3}}{\overset{k_3}{\rightleftharpoons}} \begin{matrix} H & & \\ & C{-}CH_3 & \\ H & \underset{*}{|} & \end{matrix} + 2* \tag{20}$$

$$\underset{*}{\overset{H}{|}} + \begin{matrix} H & & \\ & C{-}CH_3 & \\ H & \underset{*}{|} & \end{matrix} \underset{k_{-4}}{\overset{k_4}{\rightleftharpoons}} C_2H_6 + 2* \tag{21}$$

In the hydrogenation, hydrogen atoms on the surface add to the adsorbed olefin in two steps, leading first to an adsorbed intermediate, often referred to as an *adsorbed alkyl* [152, 155]. There is ample evidence for such a "half-hydrogenated" state from infrared spectra starting from the experiments of Eischens and Pliskin [156] and from extensive studies of the interactions of $C_2H_4 + D_2$ by Kemball [157–160] and others.

When H_2 is replaced by D_2, deuterated ethylenes along with deuterated ethanes are formed, indicating the reversibility of elementary reactions (19) and

(20). If reaction (20), for instance, occurs many times in both directions in combination with desorption of H_2 or HD and adsorption of D_2 [reaction (18)], the C_2H_5 radical finally becomes converted into C_2D_5 and may even leave the surface as C_2D_4 or C_2D_6. The degree to which C_2H_4 and C_2H_6 are deuterated is therefore related to the ratios of rates of the various elementary steps. Bond [152] tabulated the ratios of rate constants for ethylene hydrogenation and H–D exchange (Table 3-9). The metals tabulated show large differences in their behavior in ethylene hydrogenation; it is not yet possible to explain these differences quantitatively. The observed overall kinetics results from the strong aggressive adsorption of the olefin and the much weaker adsorption of the hydrogen; the olefin thus covers most of the surface and inhibits the reaction.

Since all reactions are assumed to be reversible, C_2H_6 or any paraffin should undergo exchange of its H atoms with D atoms. This exchange occurs readily only at somewhat higher temperatures because of the weaker adsorption of paraffins compared with olefins. Kemball [157–160] in particular has clarified the details of the exchange-reaction mechanism for stepwise exchange processes and has distinguished between stepwise and multiple exchange processes.

We turn now to another type of dehydrogenation, illustrated by the surface-reaction steps

$$\mathrm{H_2\overset{*}{C}{-}\overset{*}{C}H_2} \underset{k_{-5}}{\overset{k_5}{\rightleftharpoons}} \mathrm{H\overset{*}{C}{=}CH_2} + \mathrm{\overset{*}{H}} \qquad (22)$$

$$\mathrm{H\overset{*}{C}{=}CH_2} + 2* \underset{k_{-6}}{\overset{k_6}{\rightleftharpoons}} \mathrm{H\overset{*}{C}{=}\overset{*}{C}H} + \mathrm{\overset{*}{H}} \qquad (23)$$

Here there is further C—H bond scission, leading to what is essentially an adsorbed acetylene molecule. This reaction was first recognized by Beeck [28, 161] and shown to be the cause of a self-poisoning of the catalyst, particularly when the

Table 3-9 Ratios of first-order rate constants for ethylene hydrogenation and H–D exchange at 50°C [152]

Metal	k_3/k_{-2}	k_{-3}/k_4
Ru	0.22	2.1
Os	0.7	1.8
Rh	1.2	5.7
Pd	2.5	9
Pt	49	19
Ir	45	11.5

hydrogen partial pressure was relatively low. If the adsorption of olefin were to occur in this manner, the overall reaction would become essentially the following, which has been observed in the absence of adsorbed and molecular hydrogen:

$$\underset{\text{(Adsorbed)}}{2C_2H_4 \rightleftharpoons C_2H_6 + C_2H_2} \tag{24}$$

Direct information about the adsorption process can be obtained from ferromagnetic measurements, since Selwood [21] has shown that for ferromagnetic metals every surface atom engaged in an adsorptive interaction becomes "disconnected" from the cooperative ferromagnetic ordering, thereby leading to a decrease in the magnetic susceptibility of the metal. Knowledge of the amount adsorbed and the loss in ferromagnetism determines the number of sites occupied per molecule adsorbed. The numbers given for hydrocarbon adsorption correlate satisfactorily with the mechanism given; i.e., up to four metal atoms are removed per ethylene molecule adsorbed. The extent of dissociation upon adsorption follows the correlation given earlier for hydrogenation activity and the Tanaka-Tamaru parameter. Stronger adsorption results in more extensive dissociation; more surface area is blocked by acetylenic residues; and the observed hydrogenation activity is lower.

The possible involvement of the Eley-Rideal mechanism in hydrogenation must also be considered. In this mechanism, a nondissociated hydrogen molecule reacts with adsorbed ethylene, resulting in hydrogenation as follows:

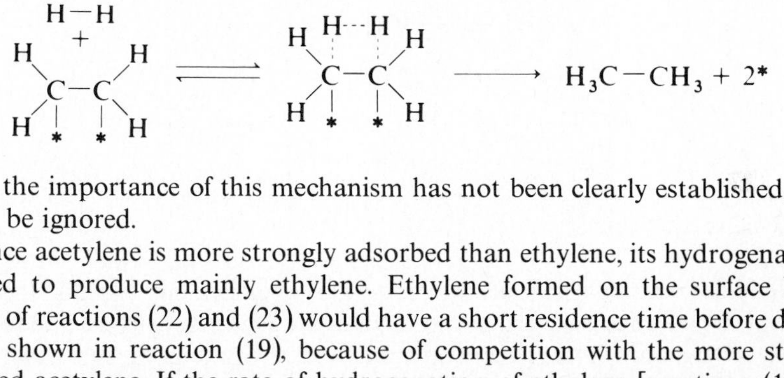

$$H{-}H + H_2\underset{*}{C}{-}\underset{*}{C}H_2 \rightleftharpoons H_2\underset{*}{C}{-}\underset{*}{C}H_2\cdots(H{\cdots}H) \longrightarrow H_3C{-}CH_3 + 2* \tag{25}$$

So far, the importance of this mechanism has not been clearly established, but it cannot be ignored.

Since acetylene is more strongly adsorbed than ethylene, its hydrogenation is expected to produce mainly ethylene. Ethylene formed on the surface by the reverse of reactions (22) and (23) would have a short residence time before desorbing, as shown in reaction (19), because of competition with the more strongly adsorbed acetylene. If the rate of hydrogenation of ethylene [reactions (19) and (21)] is low relative to the rate of the reverse of reactions (22) and (23), and if the surface concentration of ethylene is also low, predominantly ethylene will be produced. Because of the strong adsorption of acetylene relative to ethylene, the ethylene concentration on the surface remains low, and consequently little ethane is formed by ethylene hydrogenation as long as acetylene is present in the gas phase and, correspondingly, largely covering the surface. The selectivity for the reaction $C_2H_2 + H_2 \rightarrow C_2H_4$ vs. $C_2H_4 + H_2 \rightarrow C_2H_6$ depends on the rate of acetylene hydrogenation relative to the rate of ethylene hydrogenation and the relative concentrations of acetylene and ethylene on the surface. The selectivity therefore

Table 3-10 Selectivity of group VIII metals in acetylene hydrogenation [152]

Metal	Temp., °C	Selectivity, $(C_2H_4)/(C_2H_4 + C_2H_6)^a$
Pd	0	0.97
Pt	100	0.80
Rh	100	0.78
Ru	150	0.75
Os	150	0.60
Ir	150	0.55

[a] Fourfold excess of hydrogen and large excess of ethylene.

depends on the metal, on the relative gas-phase concentrations of the two species, and on the temperature. Some representative selectivity data are given in Table 3-10; high selectivity is characteristically observed.

Similar phenomena have been observed for butene hydrogenation; butadiene plays a role similar to that of C_2H_2 in C_2H_4 hydrogenation. Parenthetically, we note that during hydrogenation of butene, extensive double-bond isomerization occurs by the reaction sequence [162]

$$RCH_2{-}CH{=}CH_2 + 2* \rightleftharpoons RCH_2{-}\underset{*}{\underset{|}{CH}}{-}\underset{*}{\underset{|}{CH_2}} \underset{-H}{\overset{+H}{\rightleftharpoons}}$$

$$RCH_2{-}\underset{*}{\underset{|}{CH}}{-}CH_3 + * \underset{+H}{\overset{-H}{\rightleftharpoons}}$$

$$R\underset{*}{\underset{|}{CH}}{-}\underset{*}{\underset{|}{CH}}{-}CH_3 \rightleftharpoons RCH{=}CH{-}CH_3 + 2* \qquad (26)$$

where R is an alkyl group.

Aromatization reactions The aromatization of cyclohexane and alkylcyclohexanes to give aromatics is a specific type of dehydrogenation reaction and is important in catalytic reforming, occurring rapidly enough for equilibrium to be closely approached (Fig. 3-1 and Table 3-3). Aromatization occurs almost exclusively on the metal component, and with methylcyclohexane the dehydrogenation reaction is zero order in methylcyclohexane and in hydrogen at temperatures less than 372°C [149]. Under these conditions, the rate-limiting step appears to be desorption of toluene from the surface.

The mechanism of cyclohexane dehydrogenation involves the adsorption of cyclohexane, with either simultaneous or rapid subsequent dissociation of six hydrogen atoms. Dehydrogenation results in the formation of an aromatic structure bonded through π-electron interaction with metal d orbitals. A proposed stepwise mechanism in which all intermediate steps are rapid is [163]

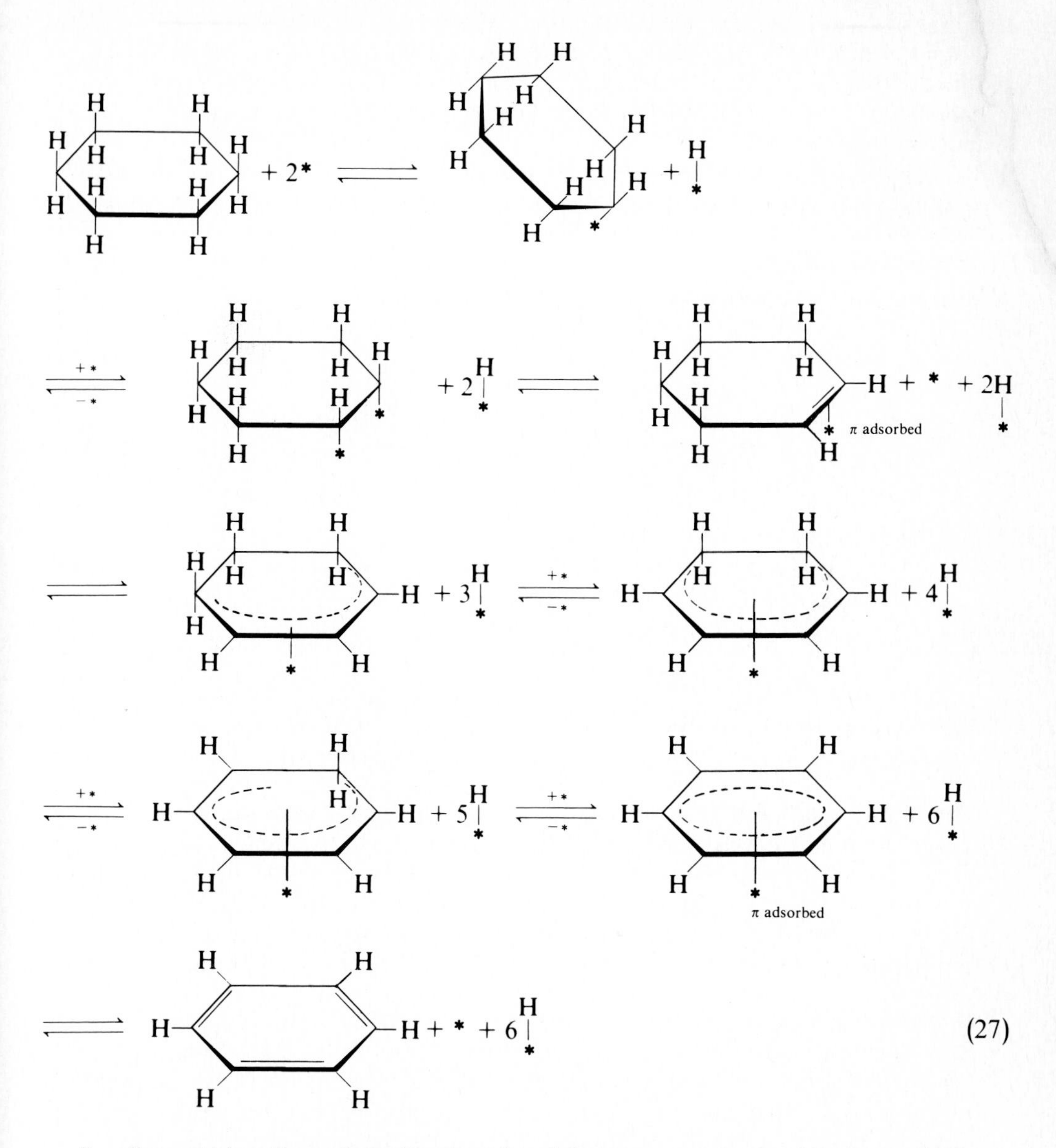

Small quantities of gas-phase dienes and cyclohexene were observed in the product; evidently the hydrogenation reaction occurs stepwise, and small amounts of the intermediates are desorbed [150, 155]. As long as all the intermediate steps are considered rapid and substantially complete in the absence of gas-phase hydrogen, the mechanism is consistent with magnetization studies at 150°C, which indicate that the adsorption of a cyclohexane molecule on Ni formed bonds involving 8 to 12 surface Ni atoms [21]. The above mechanism is also consistent with cycloparaffin H–D exchange studies, which have shown that exchange was almost exclusively multiple, giving predominantly d_6 and d_{12} for cyclohexane [151, 164]

Cyclopentane is not dehydrogenated to give dienes except at high temperatures, but it is readily hydrocracked to give pentane [165], indicating that the diene cannot easily desorb or is difficult to form because of symmetry problems with the metal surface.

Alkyl cyclopentanes and cyclohexanes with blocked positions (*gem*-dialkyl, spirane, or bridge) are not dehydrogenated to give dienes at temperatures at which cyclohexanes are dehydrogenated readily on Pt, but the blocked compounds are dehydrogenated at higher temperatures when intermolecular and intramolecular rearrangements begin to occur [166–169], for example,

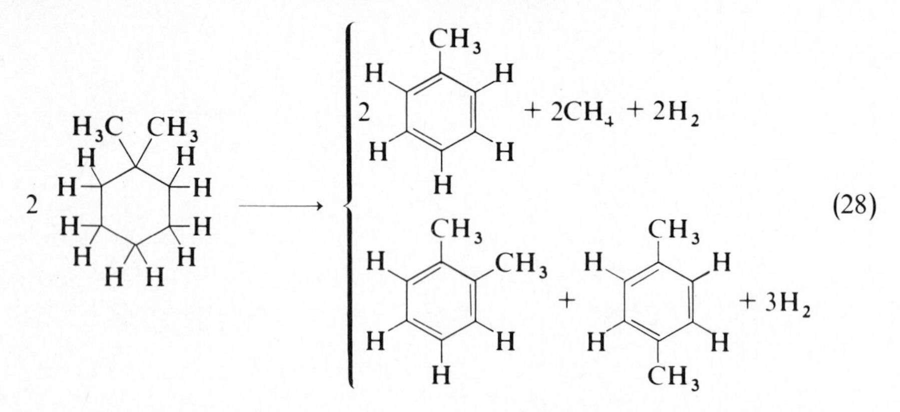

This result suggests that the molecule must lie flat on the metal surface for dehydrogenation to proceed beyond the initial stages of adsorption.

Isomerization, dehydrocyclization, and hydrogenolysis reactions So far only fission and reforming of C—H bonds have been considered, but, as shown above, these reactions can lead to double-bond isomerization [reaction (26)]. At higher temperatures, C—C bond formation and scission may also occur, leading to skeletal isomerization and hydrogenolysis. The temperatures required and the relative importance of the two reactions depend on the metal and the hydrocarbon.

Isomerization Isomerization also occurs on metal surfaces [170–173], by mechanisms that differ from those for isomerization catalyzed by acids (discussed in Chap. 1). When paraffin adsorption on a metal surface involves two adjacent carbon atoms, the paraffin can split off two hydrogen atoms and adsorb as an olefinic species. Desorption without readdition of the hydrogen atoms gives a gas-phase olefin:

$$H_3CCH_2CH_2CH_2CH_2CH_3 + 4Pt \rightleftharpoons$$

$$H_3C-\underset{\substack{|\\Pt}}{\overset{\substack{H\\|}}{C}}-\underset{\substack{|\\Pt}}{\overset{\substack{H\\|}}{C}}-CH_2CH_2CH_3 + 2\overset{\substack{H\\|}}{Pt} \rightleftharpoons$$

$$4Pt + H_3CCH{=}CHCH_2CH_2CH_3 + H_2 \qquad (29)$$

The adsorbed olefin can also undergo C—C bond rupture (hydrogenolysis), as discussed below.

If, on the other hand, the paraffin adsorbs through carbon atoms which are not adjacent, one alternative to desorption is the formation of a new C—C bond, leading to a five- or six-membered ring species. This reaction requires that the surface-bonded carbon atoms be separated by four or five carbon atoms and be bonded to adjacent metal atoms. This cyclic intermediate, which appears to be one route for skeletal isomerization of paraffins on metals, was discovered and characterized by Gault and coworkers [140, 170, 173–177] and Anderson and coworkers [171, 172, 178–180]. It is particularly evident on Pt and appears to involve the formation of intermediate five- and six-membered ring hydrocarbons and their subsequent opening at another C—C bond position. The reaction sequence in this isomerization is given as follows:

1. Two C—H bonds which are not contiguous are broken, and the hydrocarbon adsorbs on the surface with the C atoms bonded to neighboring metal sites. For example,

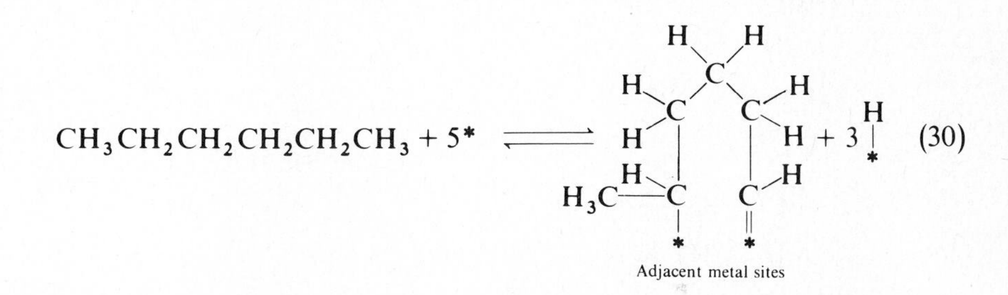

2. A C—C bond forms between the two C—∗ carbon atoms, resulting in the formation of a cyclopentane or a cyclohexane ring:

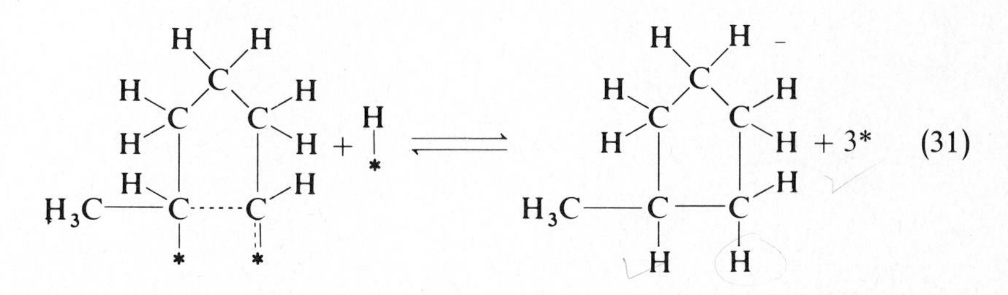

Desorption of a five-membered ring species (methylcyclopentane) or of six-membered ring species (cyclohexane) can occur.

3. The ring species can readsorb on the surface or change the carbon atoms which are actually bound to the surface without desorbing, and the C—C bond between the two adsorbed atoms (different from those adsorbed during ring closure) can be broken:

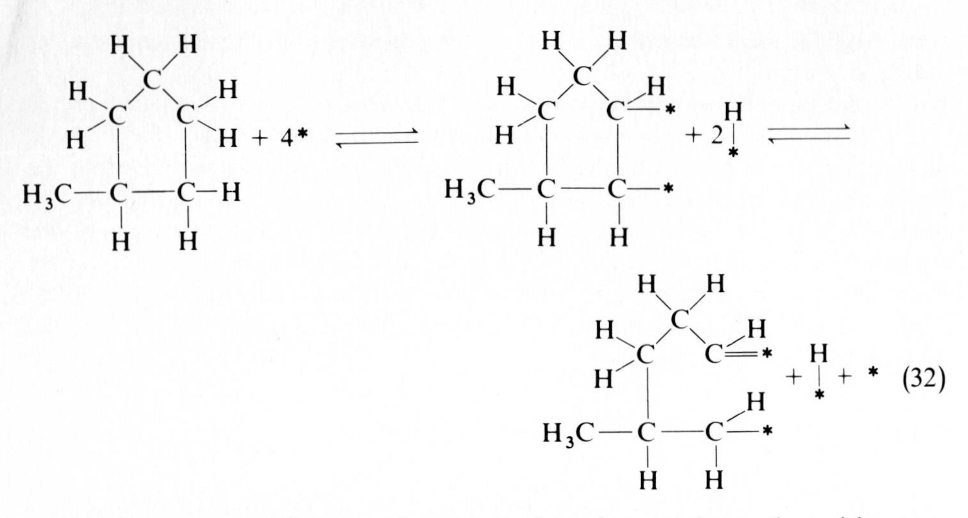

4. Hydrogen-atom addition to the adsorbed species and desorption without re-making the C—C bond leads ultimately to skeletal isomerization; i.e., the overall reaction is *n*-hexane → 2-methylpentane:

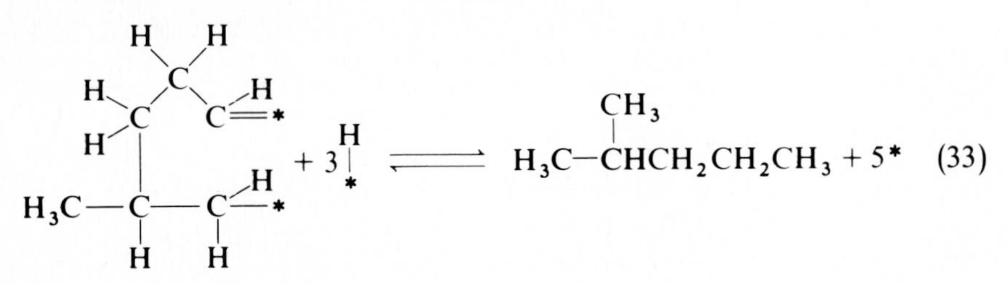

Figure 3-51 illustrates this reaction scheme for isomerization of C_6 paraffins on a metal surface. Six-membered-ring closure also occurs but does not lead to

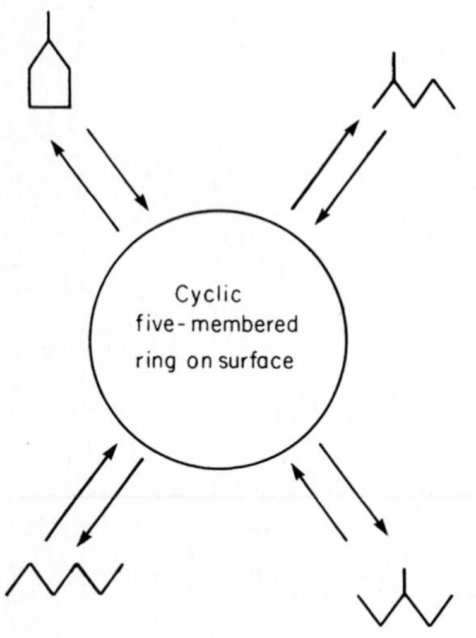

Figure 3-51 Reaction scheme for isomerization of C_6-paraffin on a metal surface.

isomerization, only cyclization. Once the six-membered ring is formed on the metal, dehydrogenation to an aromatic occurs very rapidly unless the hydrogen partial pressure is high enough to suppress the reaction.

The reaction network for isomerization and dehydrocyclization of *n*-hexane and 2-methylpentane catalyzed by Pt on nonacidic alumina is illustrated in Figs. 3-52 and 3-53, respectively. *n*-Hexane is readily converted into 2- and 3-methylpentane and methylcyclopentane. All three species show nonzero slopes at the origin, suggesting that they are primary products. Both five- and six-membered ring closure occur as parallel reactions. The following network, consistent with Fig. 3-52, is proposed for the Pt-catalyzed reactions of *n*-hexane under reforming conditions:

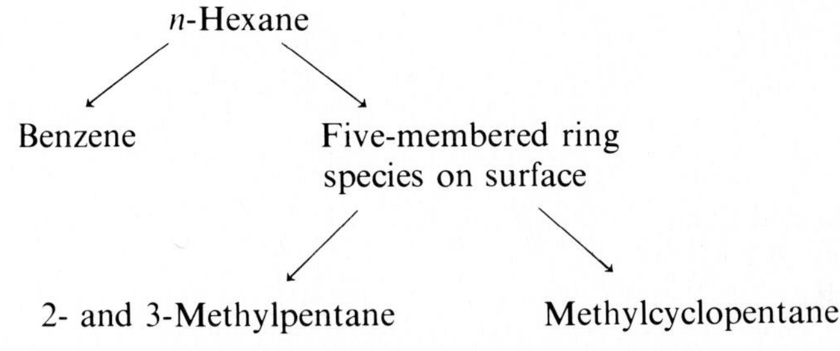

The reactions of branched paraffins containing only five carbon atoms in the chain involve the formation of five-membered ring species, which then give

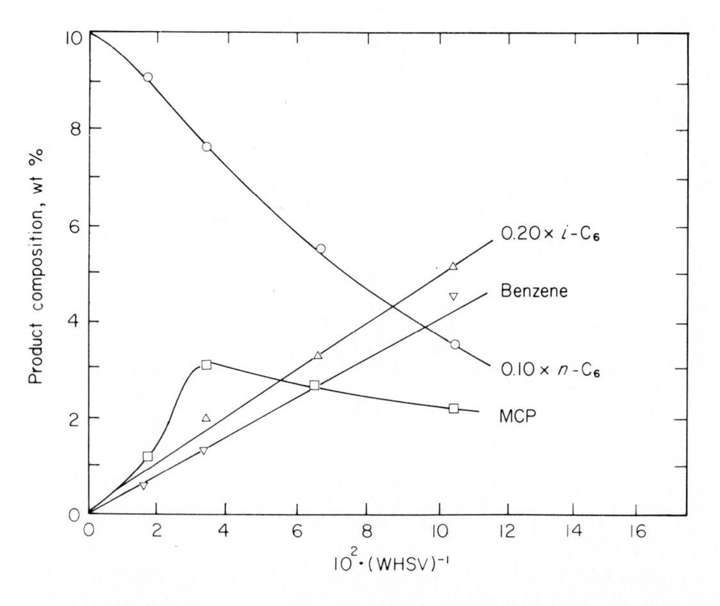

Figure 3-52 Reaction of *n*-hexane on Pt on nonacidic alumina. Reaction conditions: temperature-440°C; pressure = 9.5 atm; H_2: nC_6 molar ratio = 4; and catalyst = 0.5 wt % Pt on nonacidic alumina [136]. (Reprinted with permission from *Journal of Catalysis*. Copyright by Academic Press.)

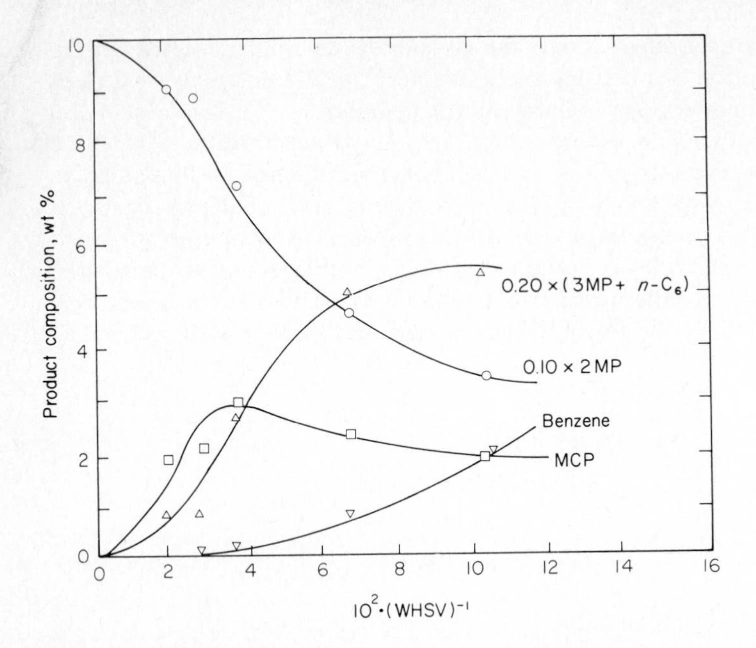

Figure 3-53 Product composition for reaction of 2-methylpentane on Pt on nonacidic alumina. Reaction conditions: temperature = 440°C; pressure = 9.5 atm; H_2: 2MP molar ratio = 4; and catalyst = 0.5 wt % Pt on nonacidic alumina [136]. (Reprinted with permission from *Journal of Catalysis*. Copyright by Academic Press.)

isomerized products, including *n*-hexane. The *n*-hexane can then react to give a six-membered ring, as indicated below (Fig. 3-53):

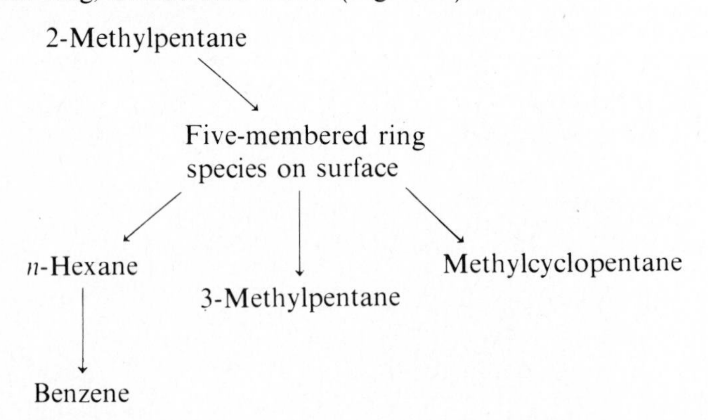

The benzene formation rate is zero near the origin, indicating that benzene is a secondary product.

Hydrogen atoms are readily available on the metal surface to hydrogenate any of the C—C bonds which might open to give an isomer from the cyclic C_5 species. Dimethylbutanes are not observed experimentally, in accordance with the proposed scheme. If an iC_6 paraffin is used as a feed, benzene appears as a secondary rather than a primary product because of the need to form *n*-hexane first.

The second metal-catalyzed paraffin-isomerization mechanism is a bond-shift mechanism. It has been attributed to the formation of $\alpha\alpha\gamma$-triadsorbed intermediates bonded to two contiguous metal atoms [181]. For neopentane, isomerization would involve (1) adsorption:

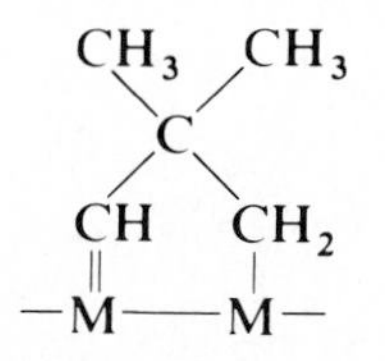

(2) formation of a bond between the carbon atoms bonded to the surface, and (3) breaking of one of the bonds in the short-lived cyclopropane ring to form 2-methylbutane. From H–D exchange studies during isomerization, Rooney and coworkers [57] proposed that covalently bonded alkyl groups on metal surfaces can rearrange much as they do in carbonium ions, via metallocarbonium ions. Formation of the proposed intermediate involves interaction of the $p\pi$ orbitals of the carbon with a metal $d\pi$ orbital (I); this bonding reduces the energy barrier for the methyl shift to give (II)

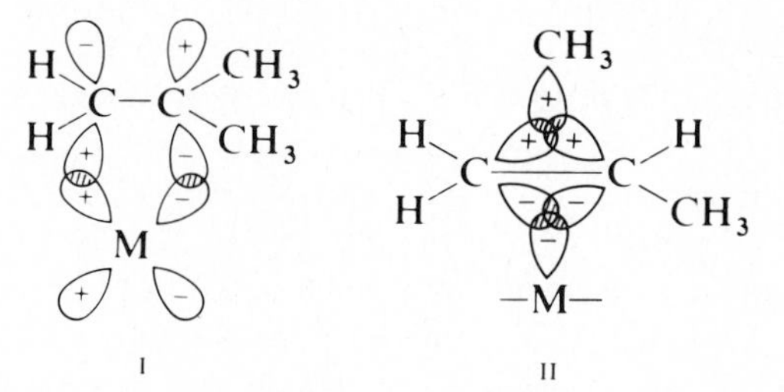

which involves three atomic orbitals and contains two electrons (see Chap. 1) and donates electrons to an empty σ orbital. This bonding in (II) is similar to that in olefin-metal complexes (as discussed in Chap. 2). Thus 5*d* transition metals, which form the strongest olefin-metal bonds, have the highest activities for 1,2 bond-shift reactions.

Tracer studies with ^{13}C have allowed the relative importance of bond shift and cyclic isomerization reactions to be determined as a function of catalyst properties. For 2-methylpentane-2-^{13}C, isomerization to give 3-methylpentane (Fig. 3-54) proceeds almost exclusively by the cyclic mechanism on highly dispersed Pt ($\sim$12 Å), whereas on large Pt crystallites ($\sim$180 Å), the bond-shift mechanism is predominant [174]. Further, the methylcyclopentane intermediate undergoes nonselective hydrogenolysis with almost equal probability of breaking each of the cyclic C—C bonds on highly dispersed Pt (Fig. 3-55*A*), whereas on large Pt crystallites the CH_2—CH_2 bonds are selectively ruptured (Fig. 3-55*B*) [182]. Evaluation of a series of different crystallite sizes showed that for crystallite sizes less than about 20 Å, cyclic isomerization and nonselective methylcyclo-

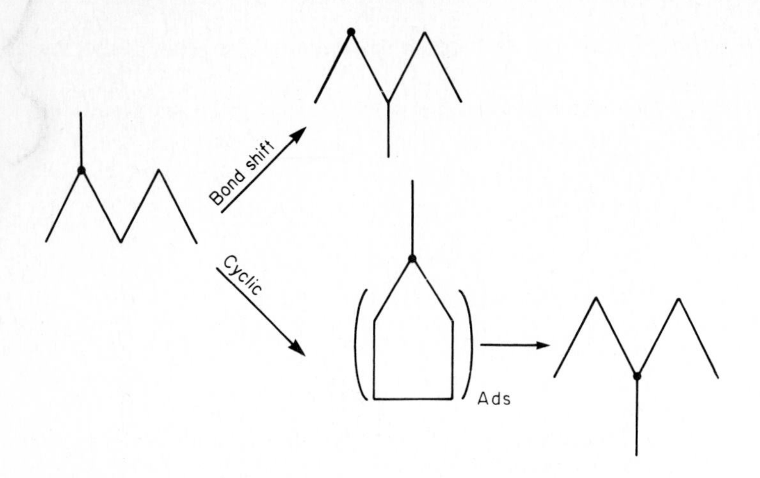

Figure 3-54 Bond shift and cyclic isomerization of 2-methylpentane-2-^{13}C.

pentane hydrogenolysis predominated, but for all larger sizes bond shift and selective hydrogenolysis dominated and did not vary with crystallite size. These results suggest (1) that the bond-shift mechanism is associated with B_5 sites, since they appear only on crystallites larger than about 20 Å and (2) that nonselective hydrogenolysis is associated with isolated edge atoms. This interpretation is not in accord with Rooney's proposal for the bond-shift mechanism [57].

Alloying Pt with Au, so that the surface becomes predominantly Au with small ensembles of Pt atoms and single Pt atoms, causes marked shifts in selectivity. For Pt–Au catalysts with small ensembles of Pt atoms (10 to 15 percent Pt in Au), the relative rates of isomerization, cyclization, and cracking are about the same as for large crystallites of pure Pt [54], but there is a change from the bond-shift to the cyclic isomerization mechanism and from selective to nonselective hydrogenolysis of methylcyclopentane [176]. These catalysts appear to be characteristic of highly dispersed supported Pt [174]. For very dilute (1 to 4 weight percent Pt) Pt–Au alloys, in which single Pt atoms may be present in a sea of Au, the reaction is isomerization with almost no cyclization or hydrogenolysis [54]. Under these conditions, bond shift via a single Pt atom appears to dominate, supporting the metallocarbonium-ion mechanism [57].

Again the effects of alloying are consistent with the effect of crystallite size, strongly indicating that structural factors predominate in explaining the selectivity properties of Pt in hydrocarbon reactions, thereby providing an additional route to an understanding of reaction mechanisms on metals. Studies with very dilute alloys can provide information that cannot be obtained with supported metals, since they allow the activities of metal atoms nearly isolated in the surface of another metal to be studied.

Dehydrocyclization Dehydrocyclization is similar to isomerization, since it can occur on the metal alone or via a reaction pathway involving dehydrogenation on

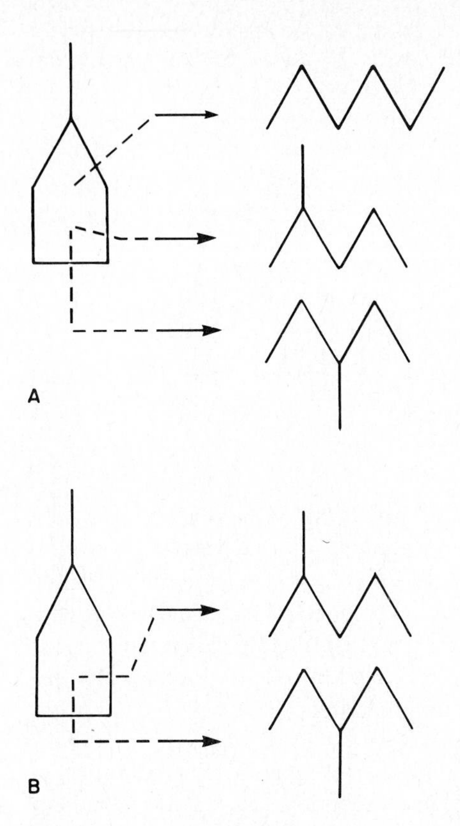

Figure 3-55 (*A*) Nonselective hydrogenolysis of cyclopentane and (*B*) selective hydrogenolysis of cyclopentane. The dashed arrows indicate which bonds break.

the metal and cyclization on the acidic support. The most probable mechanisms for dehydrocyclization reactions on metals are those discussed above in the consideration of isomerization on metals. For example [171],

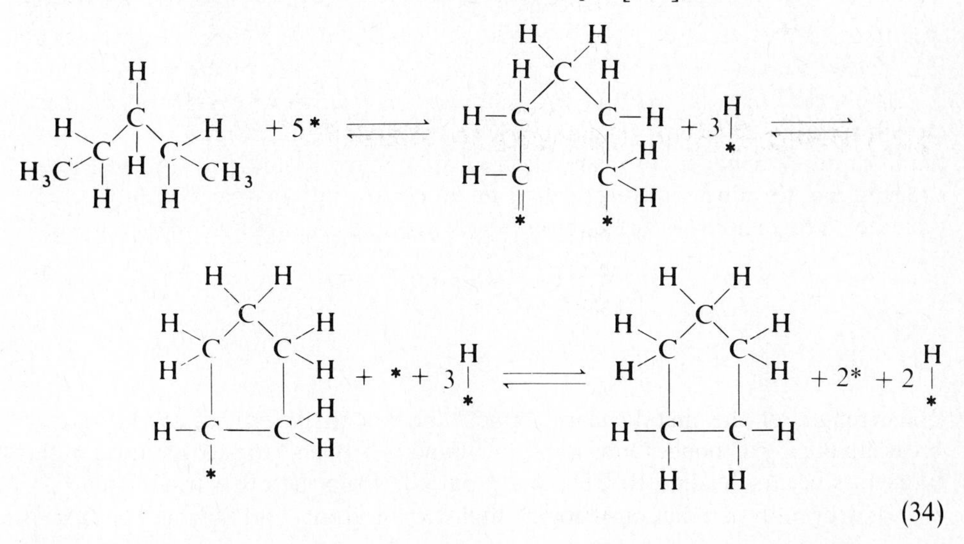

(34)

Alternatively, a mechanism which is consistent with the model for hydrogenolysis of cycloparaffins is [163]

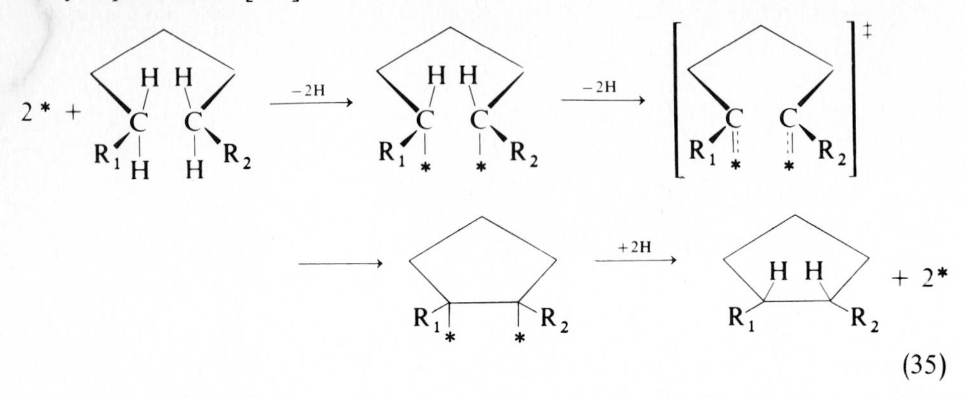

(35)

Mechanisms of adsorption and reaction involving only one metal atom [170, 171, 173] are open to serious question because of the results of studies [54] of the dilute Pt–Au alloy.

Direct six-membered ring closure also occurs in parallel with the above reactions on Pt [136, 183]. In general, direct cyclization on metals can readily occur between carbon atoms bonded to at least two hydrogen atoms, as long as they are separated sufficiently to give five- or six-membered rings.

Hydrogenolysis A common form of C—C bond scission occurs with formation of CH_4 along with smaller amounts of C_2H_6. In its simplest form, hydrogenolysis proceeds as follows on metal surfaces:

$$CH_3CH_2(CH_2)_nCH_3 + H_2 \longrightarrow CH_4 + CH_3(CH_2)_nCH_3 \qquad (36)$$

$$CH_3CH_2(CH_2)_{n-1}CH_3 + H_2 \longrightarrow CH_4 + CH_3(CH_2)_{n-1}CH_3 \qquad (37)$$

But direct formation of C_2H_2 also occurs; Ni gives almost exclusive formation of CH_4; and Pt favors fission of a C—C bond near the center of the molecule.

The mechanism of hydrogenolysis appears to involve adsorption of adjacent carbon atoms on adjacent metal sites with breaking of C—H bonds. For the C—C bond rupture to occur, the carbons appear to have to undergo further dehydrogenation, forming multiple carbon-metal bonds and leading to almost complete dehydrogenation of the carbon atoms in some cases [39, 41, 47]:

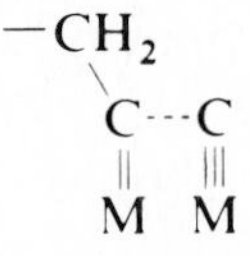

The strength of the metal-carbon bonds appears to be crucial to the rate of breaking the C—C bond. Once the C—C bond is broken, rehydrogenation of the fragments occurs, leading to CH_4 and a paraffin from the other fragment.

Hydrogenolysis reactions require high temperatures and strong bonding of reactants to the catalyst and are therefore difficult to accomplish; they are demanding compared with hydrogenation reactions, which occur much more readily. They appear to require at least a pair (and possibly a more complex ensemble) of metal atoms. Although the size of supported metal crystallites typically has little influence on the specific rates of hydrogenation reactions, it strongly influences hydrogenolysis rates; the smaller crystallites are the more active by as much as a hundredfold [113, 184]. This result is consistent with the requirement for strong carbon-metal bonding to break the C—C bond and the fact that bonding is stronger to metal atoms with a lower number of neighboring atoms [75], i.e., atoms on corners, edges, and higher index planes, which are relatively abundant on small metal crystallites.

Alloying provides information about the sizes of metal-atom ensembles required to catalyze various reactions, as discussed above. Alloying Ni with Cu [39, 41, 47, 49, 54] (or Pt with Au [185, 186]) leads to elimination of the hydrogenolysis activity, leaving only the slower isomerization (or isomerization and dehydrocyclization) activity. These effects are explained by the suggestion that the surface consists of small aggregates or perhaps even single atoms of Ni or Pt in a sea of inactive Cu or Au. Since hydrogenolysis is essentially eliminated but isomerization still occurs, it is concluded that hydrogenolysis requires more complex metal (Ni, Pt) sites than the isomerization reaction. Very dilute Pt–Au alloys having surfaces composed of almost pure Au catalyzed only isomerization, which suggests a single-atom site for this reaction [54, 57]. Dehydrocyclization appears to require larger site ensembles, and hydrogenolysis requires the largest site ensembles. These observations suggest that geometric effects are of primary importance and electronic effects of secondary importance.

However, since hydrogenolysis reactions require a site ensemble of two or more metal atoms, there must be a lower limit of crystallite size or cluster size below which the specific rate of hydrogenolysis decreases. In the limit of mononuclear metal complex catalysts, hydrogenolysis activity is absent. The ensembles of the most highly dispersed supported Pt catalysts may be small enough for isomerization to be favored over hydrogenolysis [170–173].

Alumina-catalyzed Reactions

Isomerization The equilibrium constants for hydrocarbon isomerizations are about 1.0 (Table 3-2), and these reactions form complex mixtures of hydrocarbons at equilibrium. Isomerization reactions on the alumina support proceed by carbonium-ion mechanisms, as illustrated in Chap. 1.

Skeletal isomerization of alkanes is much more difficult than isomerization of the corresponding alkenes, and strongly acidic catalysts [187, 188] or high temperatures are required to effect it. A difficult hydride abstraction or protonation by a highly acidic group is required for the alkane. Traces of olefins, which are readily protonated to form carbonium ions, markedly accelerate the isomerization of paraffins [189]. Intermolecular hydride transfer carried out by the carbonium ions formed initially from the olefin speeds the process; the pattern is familiar from Chap. 1:

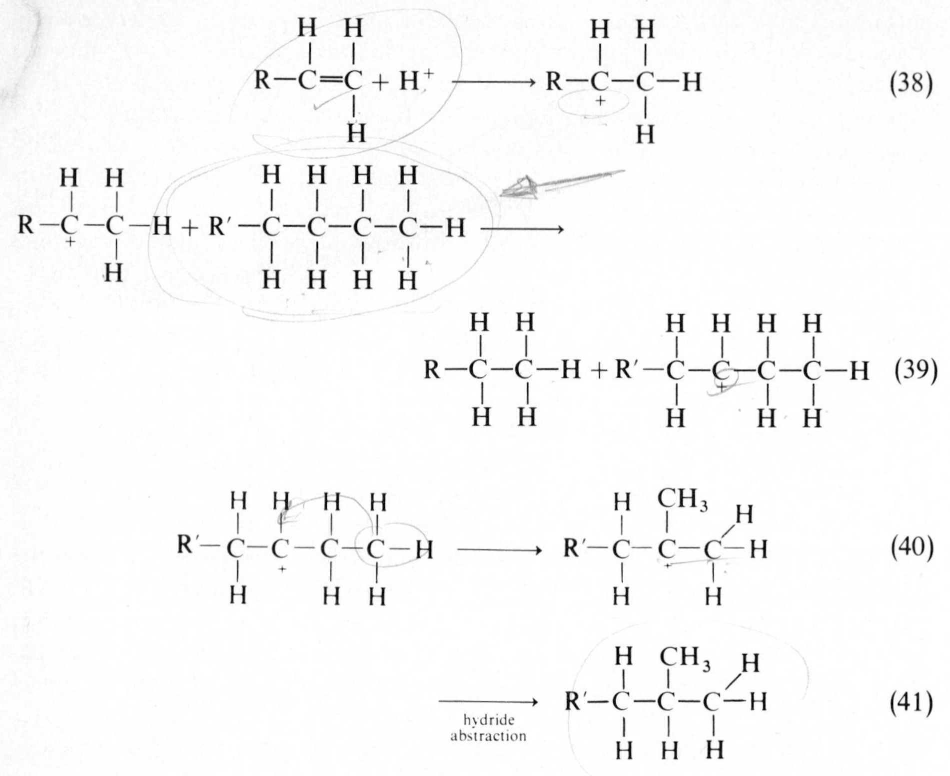

Therefore, for rapid isomerization of straight-chain paraffins, it is desirable first to dehydrogenate them to straight-chain olefins, to isomerize the olefins on the acid centers of the catalyst, and then to rehydrogenate the branched olefins to give isoparaffins. This reaction scheme is the heart of the catalytic reforming process; it allows application of catalysts with less strong, carefully controlled acidity which do not promote many undesirable reactions (including rapid hydrocracking and coke formation) as the strong acid sites of cracking catalysts do.

Olefin isomerization reactions occur readily on acid catalysts and become increasingly difficult in the following order: cis-trans, double-bond, and skeletal isomerization. Cis-trans and double-bond isomerization occur on weakly acidic catalysts, e.g., unpromoted alumina at low temperatures, whereas skeletal isomerization requires more strongly acidic catalysts, e.g., promoted alumina or silica-alumina and higher temperatures. Because of the relative ease of cis-trans and double-bond isomerization, these reactions usually approach equilibrium closely when skeletal isomerization is occurring, as in reforming. Even the rate of skeletal isomerization is rapid enough under reforming conditions for paraffin isomer distributions to be close to their equilibrium values (Table 3-3).

Olefins can be cyclized on acid centers by reaction mechanisms which have yet to be firmly established. However, the cyclization of olefins, e.g., of heptene-2 to give a 5-carbon ring structure, does not appear to occur via a normal carbonium-ion mechanism on promoted alumina, since this would involve a cyclopentyl carbonium ion, which could desorb as cyclopentene or abstract a hydride ion and desorb as cyclopentane; cyclopentenes are not observed under reforming conditions [190]. It is more probable that a concerted reaction occurs, involving double-bond protonation by a Brønsted acid site and simultaneous proton abstraction from a carbon atom along the chain. Ring closure could occur by reaction of the two polar centers, as follows:

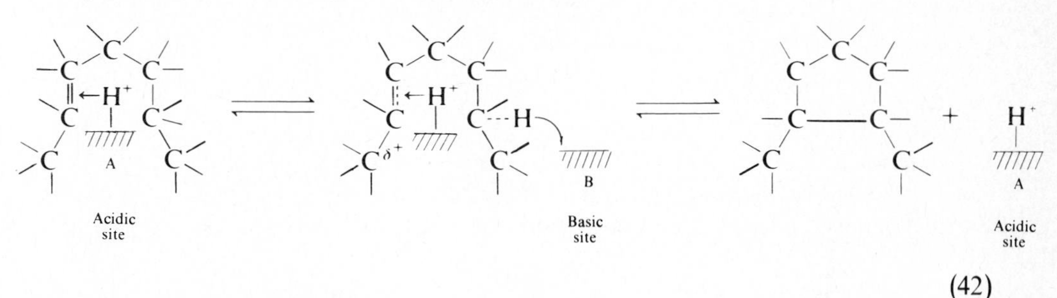

(42)

This reaction might be expected to occur on an adjacent pair of acid-base sites on the catalyst surface:

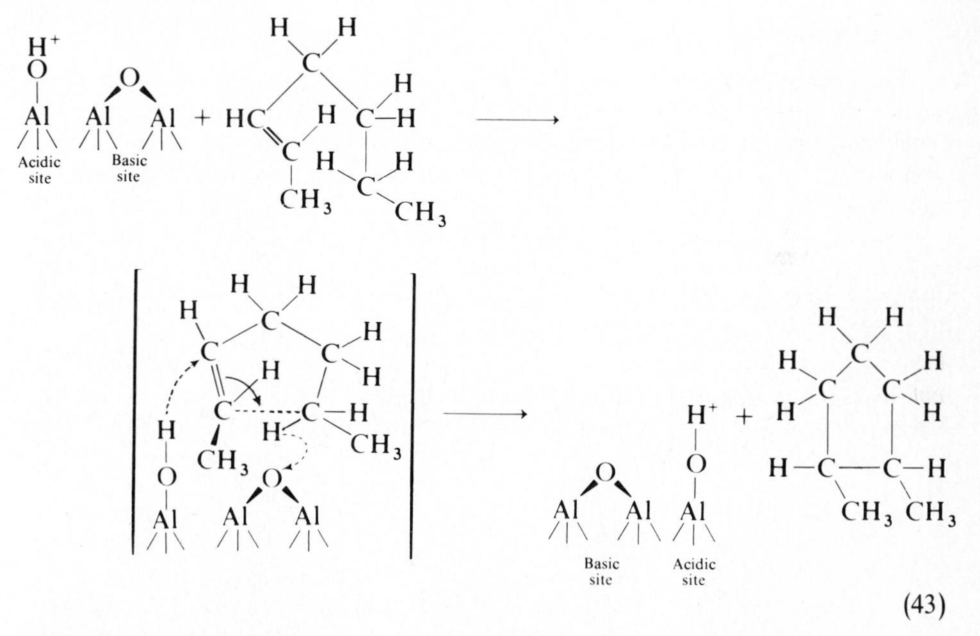

(43)

This concerted mechanism is consistent with the absence of cyclopentenes in the product when olefin is isomerized on promoted alumina. Ring closure of heptene-2 produces predominantly 1,2-dimethylcyclopentane with considerably smaller amounts of ethylcyclopentane and methylcyclohexane [190].

With hexene-1 on alumina, only five-membered ring closure occurs, indicating that the reaction proceeds via a mechanism similar to a carbonium-ion mechanism, since secondary carbonium ions are much more stable than primary carbonium ions. Metal-catalyzed cyclization reactions are significantly different, producing both five- and six-membered rings.

The acid-catalyzed ring opening of methylcyclopentane occurs directly by protonation of the ring structure with formation of an acyclic carbonium ion [191]:

$$\text{(methylcyclopentane, } CH_3\text{)} + {}^{+}H{-}\text{cat} \rightleftharpoons \text{(}CH_3\text{, }{}^{+}{-}\text{cat, } CH_3\text{)} \rightleftharpoons \text{(}CH_3\text{, } CH_3\text{)} + {}^{+}H{-}\text{cat} \qquad (44)$$

This reaction is like the reverse of (42). Hexenes are the primary products, and reaction is inhibited by methylcyclopentane, which indicates that cyclopentane is not the ring undergoing opening. A three-center intermediate is proposed:

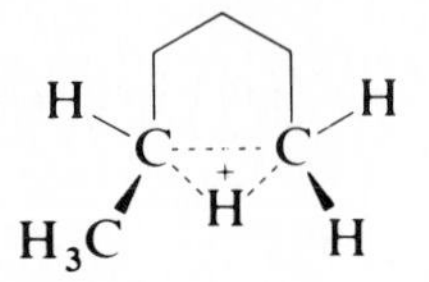

For dienes, the same type of concerted mechanism can be proposed to account for formation of methylcyclopentenes. For example, 1,5-hexadiene gives methylcyclopentane, methylcyclopentenes, and benzene. Methylcyclopentenes predominate, and the cyclization rates are roughly equal to those for hexene-1 cyclization to give methylcyclopentane. The reaction of dienes occurs by an internal (intramolecular) alkylation when the developing cation center is far enough away from the second double bond to produce a five- or higher-membered ring. This reaction mechanism is believed to prevail for 1,5-hexadiene cyclization [190]. Such an internal alkylation is illustrated for geraniolene in Fig. 3-56.

Acyclic conjugated trienes, e.g., hexatriene, cyclize readily without a catalyst at low temperatures; the reaction may be considered as an internal Diels-Alder reaction forming cyclohexadiene [136, 163]. This reaction occurs readily at low hydrogen partial pressures:

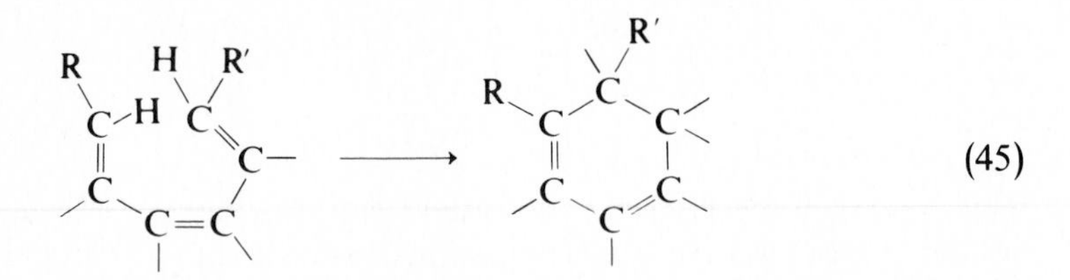

(45)

At higher hydrogen partial pressures, the equilibrium concentration of hexatrienes is so low that this mechanism becomes unimportant.

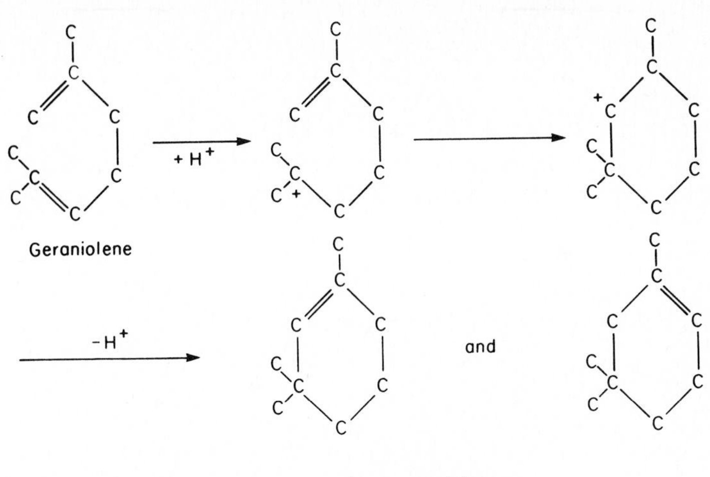

Figure 3-56 Internal alkylation of geraniolene involving a carbonium ion reaction in sulfuric acid solution [192].

Hydrocracking Hydrocracking is desired in measured amounts to split long-chain paraffins of low octane number to give shorter-chain, higher-octane species. Hydrocracking occurs on the acidic centers of the catalyst much in the same manner as discussed in Chap. 1; like isomerization, hydrocracking is facilitated by dehydrogenation of paraffins to olefins on the metal. The major characteristics of hydrocracking are that (1) all cracked species are saturated, (2) the process involves bifunctional catalysis, and (3) catalyst deactivation is much less rapid than in catalytic cracking. Saturated paraffins predominate among the cracked products because the olefins formed are hydrogenated on the metal component. The rate of hydrocracking increases rapidly with reactant molecular weight, making it easy to crack out undesired high-molecular-weight paraffins selectively. For example, nC_{16} paraffin cracks 3 times as fast as nC_{12} paraffin [193]. Hydrocracking results in indiscriminate cracking of both straight- and branched-chain paraffins.

Product distributions for *n*-heptane isomerization and hydrocracking catalyzed by Pt/Al_2O_3 at high and low conversions are shown in Fig. 3-57. The initial conversion data (Fig. 3-57*A*) show the role of each catalyst function in the reaction network. Isomerization (of olefins) gives about half of the total product (isoheptanes) and is attributed to the support (Al_2O_3). The metal is probably largely responsible for forming the naphthene and aromatic fractions and also the $C_6 + C_1$ fraction, which probably results from hydrogenolysis on the metal. The numbers in parentheses indicate that these are largely primary reaction products which are stable and do not undergo secondary reactions. The isoheptanes, on the other hand, are primary products which experience substantial conversions in secondary reactions. The secondary reactions are mainly hydrocracking of C_7 to give $C_2 + C_5$ and $C_3 + C_4$, which appear only at the higher conversions (Fig. 3-57*B*). Dehydrocyclization and aromatization also appear as secondary reactions.

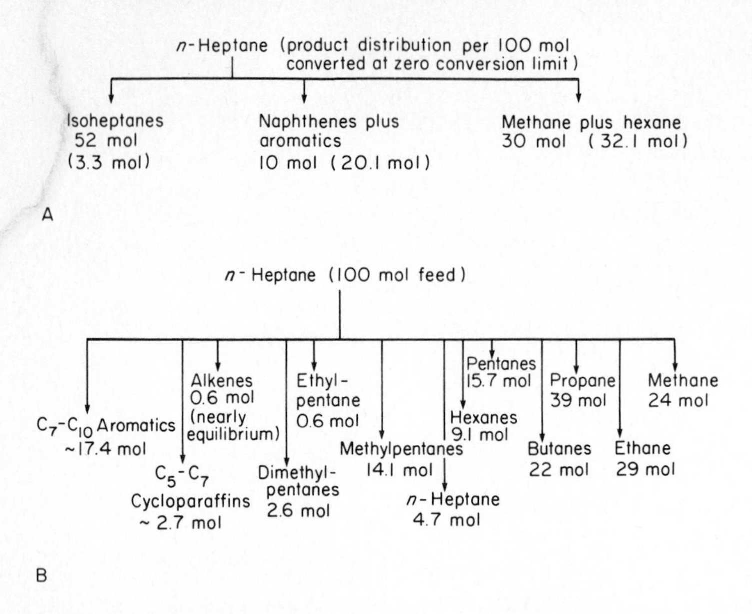

Figure 3-57 Product distribution for *n*-heptane isomerization and hydrocracking catalyzed by Pt/Al_2O_3: (*A*) per 100 moles of *n*-heptane converted in the limit as conversion approaches zero (numbers in parentheses represent conversion to these products at 95.3% conversion); and (*B*) per 100 moles of feed at 95.3% conversion of *n*-heptane (WHSV = 5.1). Reaction conditions: 0.6 wt % Pt on promoted alumina, 34 atm total pressure, 496°C, H_2: hydrocarbon molar ratio = 4.7 [194]. (Reprinted with permission from *Industrial and Engineering Chemistry*. Copyright by the American Chemical Society.)

Bifunctional Catalysis of Reforming Reactions

The reaction scheme Reforming reactions require two different functions: (1) a metal catalyzes dehydrogenation of paraffins into olefins and naphthenes into aromatics; it also catalyzes hydrogenation of isoolefins and contributes to dehydrocyclization and isomerization, and (2) an acid function provided by the support catalyzes isomerization, cyclization, and hydrocracking through carbonium-ion mechanisms. The two functions interact through the olefins, which are the key intermediates in the reaction network. The original statement of a reaction scheme for reforming, due to Mills et al. [195], is shown in Fig. 3-58.

Reactions drawn parallel to the abscissa in this figure occur on the acidic centers on the catalyst, and reactions drawn parallel to the ordinate occur on the hydrogenation-dehydrogenation centers. According to this scheme, the reactant (*n*-hexane) is first dehydrogenated on the metal to give straight-chain hexene. The hexene migrates to a neighboring acid center; there it is protonated to give a secondary carbonium ion, which can then isomerize and desorb as isohexene and migrate to the metal function, where it can be adsorbed and hydrogenated to give isohexane. Alternatively, the secondary carbonium ion can react to form methyl-

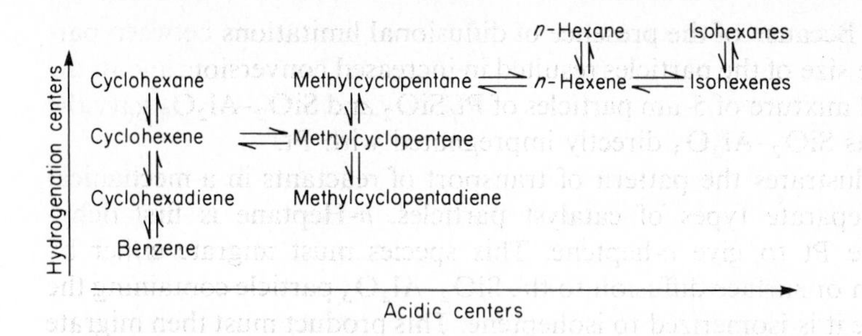

Figure 3-58 Reaction network for reforming of C_6 hydrocarbons according to Mills et al. [195]. (Reprinted with permission from *Industrial and Engineering Chemistry*. Copyright by the American Chemical Society.)

cyclopentane, which can react further to form cyclohexene and then benzene. Other routes to the aromatic also exist, and their importance is considered below.

The independent action of the metal and acidic centers of the reforming catalyst has been demonstrated by Weisz and Swegler in experiments involving *n*-heptane isomerization [193]. Figure 3-59 summarizes the results. The conversion was negligible when the catalyst was Pt on carbon or on SiO_2; when it was SiO_2–Al_2O_3, conversion was undetectible. In contrast, the conversion was appreciable when a 50-50 by volume mechanical mixture of 1000-μm particles of Pt/SiO_2 and SiO_2–Al_2O_3 was placed in the reactor. Similar behavior was observed when the catalyst was a mechanical mixture of 1000-μm particles of Pt/C

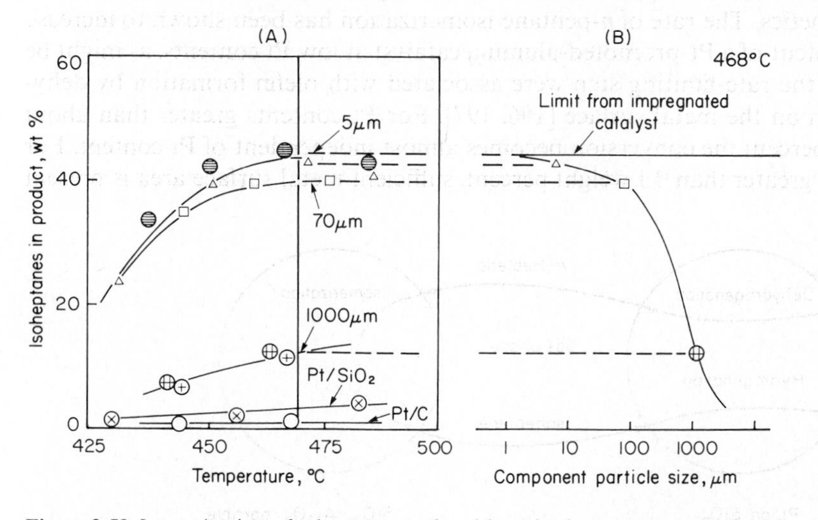

Figure 3-59 Isomerization of *n*-heptane catalyzed by mixed-component catalysts for varying sizes of the component particles: (A) conversion as a function of temperature; (B) conversion at 468°C as a function of component particle diameter. Reaction conditions: 2.5 atm *n*-heptane partial pressure, 20.0 atm H_2 partial pressure, residence time in catalyst bed 17 s, (○) Pt/C, (⊗) Pt/SiO_2, (⊕) SiO_2-Al_2O_3 with Pt/C (50-50 vol mix), (⊕) SiO_2-Al_2O_3 with Pt/SiO_2: (●) Pt/SiO_2-Al_2O_3 [193]. (Copyright by the American Association for the Advancement of Science.)

and SiO_2–Al_2O_3. Because of the presence of diffusional limitations between particles, reducing the size of the particles resulted in increased conversion, and in the limit a mechanical mixture of 5-μm particles of Pt/SiO_2 and SiO_2–Al_2O_3 gave the same conversion as SiO_2–Al_2O_3 directly impregnated with Pt.

Figure 3-60 illustrates the pattern of transport of reactants in a mechanical mixture of two separate types of catalyst particles. *n*-Heptane is first dehydrogenated on the Pt to give *n*-heptene. This species must migrate either by gas-phase diffusion or surface diffusion to the SiO_2–Al_2O_3 particle containing the acid centers, where it is isomerized to isoheptene. This product must then migrate back to a particle containing Pt, where it is hydrogenated to give isoheptane. Most transport probably occurs by gas-phase diffusion. The olefin intermediate is present in low concentration, restricted by equilibrium (Table 3-3). The two functions need not be in intimate contact; they need only be in the system so that the intermediate formed on one catalytic function can migrate to the other and allow the reaction sequence to continue.

The effect of particle size on the conversion at 468°C, illustrated in Fig. 3-59*B*, indicates a reduction of the intraparticle diffusional limitation with decreasing particle size. The presence of diffusional limitations can significantly affect the rate and selectivity of reforming in the presence of standard bifunctional catalysts. The influence of diffusion on multifunctional catalysis has been treated quantitatively by Weisz [2].

Isomerization Isomerization is now considered, in order to further establish the bifunctional nature of supported-metal reforming catalysts and to illustrate the reaction kinetics. The rate of *n*-pentane isomerization has been shown to increase with Pt content of a Pt-promoted-alumina catalyst at low Pt contents, as might be expected if the rate-limiting step were associated with olefin formation by dehydrogenation on the metal surface [196, 197]. For Pt contents greater than about 0.1 weight percent the conversion becomes almost independent of Pt content. For Pt contents greater than 0.1 weight percent, sufficient metal surface area is present

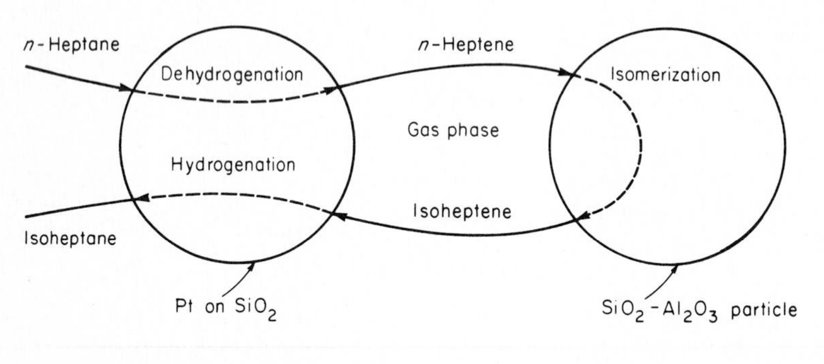

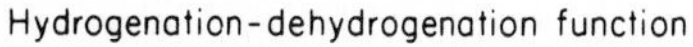

Figure 3-60 Transport pattern of reactants in *n*-heptane isomerization catalyzed by a mechanical mixture of particles, each containing only one catalytic function.

to provide an equilibrium concentration of olefin (Table 3-3), and then isomerization on the acid sites becomes rate-limiting. Since reforming catalysts contain 0.3 to 0.5 weight percent Pt, these data imply that near-equilibrium concentrations of olefins should be attained; correspondingly, equilibrium concentrations of olefins in products have been measured [198].

The overall reaction of n-pentane isomerization can be written as proceeding on the two catalyst functions via the following reactions:

$$nC_5 \xrightleftharpoons{\text{Pt}} nC_5\ (\text{olefin}) + H_2 \tag{46}$$

$$nC_5\ (\text{olefin}) \xrightarrow[\text{site}]{\text{acid}} iC_5\ (\text{olefin}) \tag{47}$$

$$H_2 + iC_5\ (\text{olefin}) \xrightleftharpoons{\text{Pt}} iC_5\ (\text{paraffin}) \tag{48}$$

For a typical reforming catalyst, reaction (47) is assumed to be slow, and reactions (46) and (48) are assumed to be close to equilibrium. If the rate of reaction is assumed to be proportional to the concentration of adsorbed olefin on acid sites,

$$r = k[nC_5\ (\text{olefin})] \tag{49}$$

and if the concentration of adsorbed olefin is assumed to be given by a Langmuir expression,

$$[nC_5\ (\text{olefin})] = \frac{KP_{nC_5\,(\text{olefin})}}{1 + KP_{nC_5\,(\text{olefin})}} \tag{50}$$

then the rate of reaction can be obtained in terms of the n-pentane and H_2 partial pressures by the application of the equilibrium relationship for reaction (46). This expression states that the isomerization rate is dependent only on the ratio of the n-pentane partial pressure to the H_2 partial pressure. The observed kinetics is consistent with this expression [3]. Figure 3-61 shows that the rate of n-pentane isomerization on a bifunctional reforming catalyst is the same as that measured for n-pentene skeletal isomerization on a Pt-free promoted alumina [196]. The addition of fluorine to a Pt/Al_2O_3 catalyst in incremental amounts up to 15 weight percent resulted in corresponding increases in the isomerization activity of the catalyst [199], indicating the importance of halogen-induced acidity. These results clearly show that isomerization on the acid centers of the catalyst is rate-limiting, and they provide additional evidence of the independent operation of the two catalytic centers on the surface.

Isomerization may also occur on the metal surface, as discussed above. The metal contribution is important but does not appear to be large. For n-heptane isomerization, the contribution appears to be of the order of 10 to 15 percent (Fig. 3-59 [193]); for n-pentane Fig. 3-61 suggests a 20 to 25 percent contribution due to the metal.

Dehydrocyclization Dehydrocyclization occurs by several distinct mechanisms on reforming catalysts, as discussed above. One involves bifunctional catalysis with dehydrogenation on the metal and cyclization on the acid centers, and another

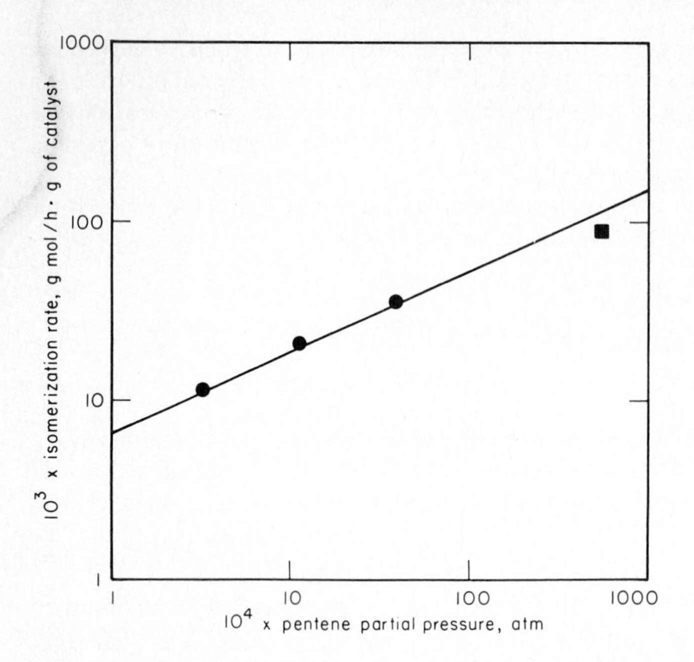

Figure 3-61 Dependence of *n*-pentane isomerization rate on *n*-pentene partial pressure at 482°C with a Pt/promoted-alumina catalyst (●) and comparison with rate of *n*-pentene isomerization on Pt-free promoted alumina at the same temperature (■) [196]. (Reprinted with permission from *Journal of Physical Chemistry*. Copyright by the American Chemical Society.)

involves monofunctional dehydrogenation and cyclization on the metal. Tables 3-11 and 3-12 provide data clarifying the importance of the two routes under reforming conditions [200]. The data of Table 3-11 show that dehydrocyclization occurs on the metal without need of acid centers and that the addition of 100 ppm of thiophene to the feed almost completely eliminates this cyclization but does not affect the dehydrogenation capability of the Pt. The concentration of 3-heptenes was unchanged and close to the calculated equilibrium value in these experiments. With 100 ppm of thiophene in the feed, essentially no cyclization of *n*-heptane occurred in the presence of Pt on carbon or Al_2O_3 separately (Table 3-12), but when the two components were mixed, dehydrocyclization occurred to an extent approximately equal to that observed with the Pt on carbon catalyst without thiophene in the feed (Tables 3-11 and 3-12). These data suggest that the mono- and bifunctional cyclization routes are of about equal importance.

Figure 3-62 shows how the dehydrocyclization rate depends on Pt surface area (Pt content). The heptene concentration in the experiments reported here was the equilibrium value for the catalyst with the smallest metal surface area. The extrapolation to zero therefore indicates the rate of cyclization occurring on the alumina alone (via heptenes); this value represents about 50 percent of the total rate of cyclization for a catalyst containing 0.4 weight percent Pt (12.5 μmol of H_2

Table 3-11 Effect of sulfur on production of ring compounds from *n*-heptane with Pt/C catalyst [200]

	Sulfur added to *n*-heptane, ppm	
	None	100
Percentage of *n*-heptane converted to[a]:		
Ethylcyclopentane	1.99	
1,1-Dimethylcyclopentane	0.01	
1,2-Dimethylcyclopentane	1.33	0.11
1,3-Dimethylcyclopentane	0.15	
Toluene	1.45	0.01
3-Heptenes	0.74	0.76

[a] Reaction conditions: 482°C, 12 atm H_2 partial pressure, 1.13 atm *n*-heptane partial pressure, *n*-heptane liquid flow rate = 16 cm^3/h, and Pt/C catalyst; catalyst weight = 0.25 g.

Source: Reprinted with permission from *Journal of Catalysis*. Copyright by Academic Press.

adsorbed per gram of catalyst, a result consistent with those of Silvestri et al. [200]). The aromatic products observed from the dehydrocyclization of C_8 and C_9 paraffins catalyzed by Pt on nonacidic alumina at atmospheric pressure, 482°C, and high hydrogen-to-hydrocarbon ratios are those predicted for direct cyclization to the six-membered ring [201].

Before five-membered rings can be converted into aromatics, they must

Table 3-12 Comparison of ring-compound production from *n*-heptane with and without mixing of catalyst components [200][a]

	Catalyst		
	0.25 g of Pt/C	0.50 g of Al_2O_3	0.75 g of mixed catalyst containing 33% Pt/C and 67% Al_2O_3
Percentage of *n*-heptane converted to:			
Ethylcyclopentane	...	0.02	0.28
1,1-Dimethylcyclopentane	...	...	0.09
1,2-Dimethylcyclopentane	0.11	0.06	1.03
1,3-Dimethylcyclopentane	...	0.07	1.04
Methylcyclohexane	...	...	0.12
Toluene	0.01	0.01	0.77
3-Heptenes	0.76	...	0.65

[a] Reaction conditions: 482°C, 12 atm H_2 partial pressure, 1.13 atm *n*-heptane partial pressure, and *n*-heptane liquid flow rate = 16 cm^3/h; sulfur added to *n*-heptane = 100 ppm.

Source: Reprinted with permission from *Journal of Catalysis*. Copyright by Academic Press.

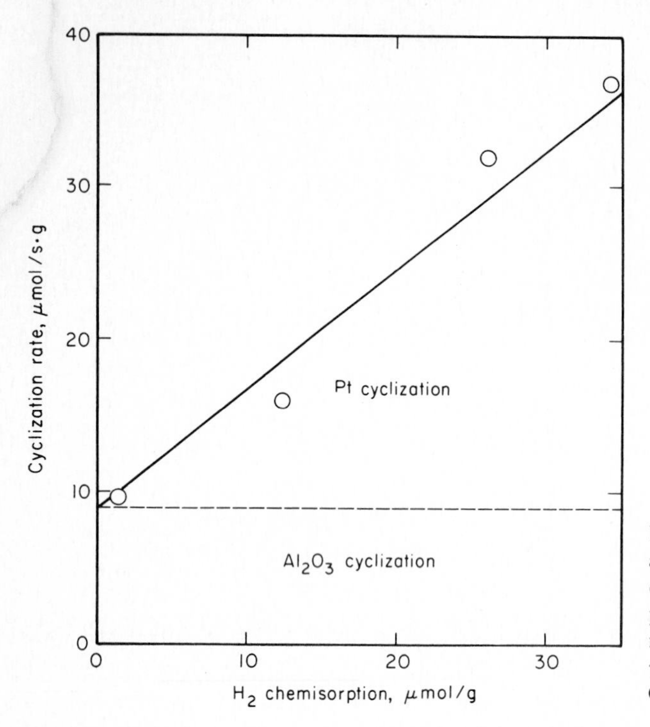

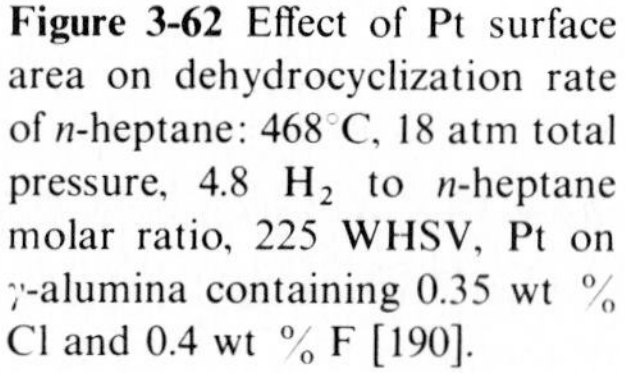

Figure 3-62 Effect of Pt surface area on dehydrocyclization rate of *n*-heptane: 468°C, 18 atm total pressure, 4.8 H_2 to *n*-heptane molar ratio, 225 WHSV, Pt on γ-alumina containing 0.35 wt % Cl and 0.4 wt % F [190].

undergo isomerization on the metal [136] or dehydrogenation to give cycloolefins and isomerization on the acid centers to give six-membered rings. The data of Table 3-13 confirm that (1) cycloolefin formation occurs on the Pt, (2) isomerization requires both the metal and the acid component, and (3) conversion to benzene results from the cooperative action of the two components [202].

The above information makes it necessary to alter the reaction scheme of Fig. 3-58 to include the additional reactions (particularly cyclization and isomeri-

Table 3-13 Conversion of methylcyclopentane catalyzed by acid, metal, and mixed catalysts [202][a]

	Liquid product analysis, mol %			
Catalyst	C	⇌ C	⇌ C	→
10 cm^3 of SiO_2–Al_2O_3	98	0	0	0.1
10 cm^3 of Pt/SiO_2	62	20	18	0.8
SiO_2–Al_2O_3 + Pt/SiO_2	65	14	10	10.0

[a] Reaction conditions: 500°C, 0.8 atm H_2 partial pressure, 0.2 atm methylcyclopentane partial pressure, 2.5 s residence time, and catalysts 0.3 wt % Pt/SiO_2 and SiO_2–Al_2O_3 with 420 m^2 per gram of surface area.

zation) which occur on the metal surface alone, without involving acid centers. Otherwise Fig. 3-58 states the dual-functional nature of reforming catalysts well.

Catalyst deactivation by carbonaceous residues Side reactions accompanying the reforming reactions lead to the formation of coke on the surface of the catalyst; the result is the deactivation common to all catalytic hydrocarbon processes. The reactions involve acid-catalyzed polymerization and cyclization of olefins to give higher-molecular-weight polynuclear compounds which undergo extensive dehydrogenation, aromatization, and further polymerization. These reactions were discussed in Chap. 1 in connection with catalytic cracking. They occur rapidly under reforming conditions at low hydrogen partial pressures, but their effect is markedly reduced by increased hydrogen partial pressure.

The dehydrogenation and polymerization which occur on the metal surface can be strongly retarded by increasing the hydrogen partial pressure, which shifts the hydrogenation-dehydrogenation equilibrium toward greater hydrogenation and causes increased hydrogenation of unsaturated species:

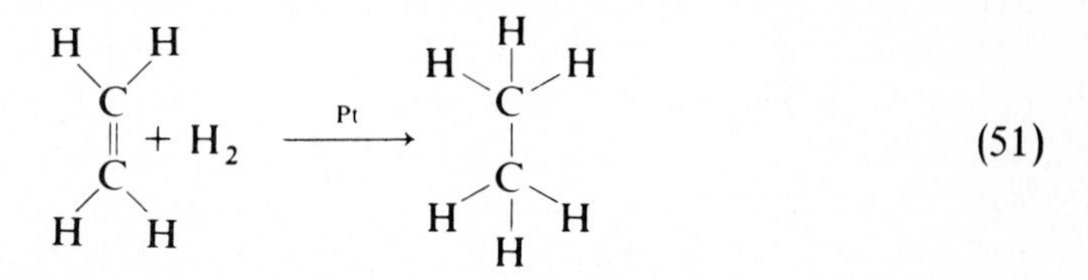

(51)

This effect is clearly illustrated by the hydrogen-partial-pressure dependence of the rate of dehydrocyclization of *n*-heptane catalyzed by a Pt/Al_2O_3 catalyst (Fig. 3-63). In the absence of H_2, no dehydrocyclization was detected because the

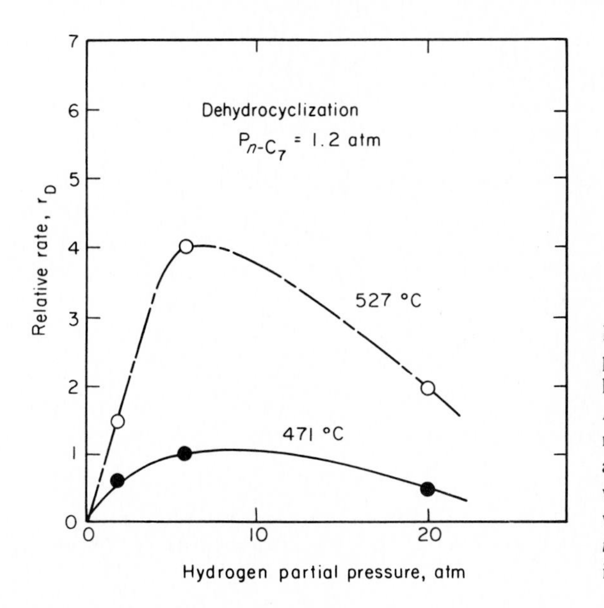

Figure 3-63 Effect of hydrogen partial pressure on the relative rate of dehydrocyclization of *n*-heptane on Pt/Al_2O_3; r_D is the rate relative to the rate of dehydrocyclization at 471°C and 5.8 atm hydrogen partial pressure with the same catalyst [203]. (Reprinted with permission from *Journal of Physical Chemistry*. Copyright by the American Chemical Society.)

metal surface was completely covered with coke. With increasing hydrogen partial pressure, the metal surface became partially cleaned of carbonaceous residues, and both dehydrogenation before ring closure on the alumina and dehydrocyclization on the metal occurred. At hydrogen partial pressures exceeding about 10 atm, removal of carbonaceous residues from the metal was not limiting, and further increases in hydrogen partial pressure restricted the extent of dehydrogenation of *n*-heptane to *n*-heptene and correspondingly the rate of dehydrocyclization.

A similar effect of hydrogen partial pressure has been observed in *n*-heptane isomerization to give methylheptanes [203] and in the isomerization-dehydroisomerization of methylcyclopentane to give cyclohexane and benzene [204] catalyzed by Pt/Al_2O_3.

In reforming with a paraffinic feed, poisoning begins on the metal with the formation of olefinic species and aromatics. These can slowly form coke on the metal, but they can also be transported by gas-phase diffusion and surface migration to the acid sites, where they slowly form more resistant coke [205], as shown schematically in Fig. 3-64. The long-term deactivation in reforming is probably due to this second type of coke formation. Just as in catalytic cracking, deactivation is more rapid with feeds containing intermediate-molecular-weight aromatics (C_8 to C_{10}), naphthenes, and paraffins than it is with lower-molecular-weight feeds. Most of the reactions are shown as reversible in Fig. 3-64 because a significant fraction of the catalyst activity could be regenerated by purging with nitrogen for several days [205]. This regeneration is probably the result of depolymerization and desorption. Regeneration with hydrogen was about twice as effective, but still about a week was required to regain most of the activity lost in a 1-week reforming run. More effective (but still slow) regeneration could be achieved by increasing the hydrogen partial pressure to more than 400 atm. This slow coke removal, particularly in the presence of hydrogen, represents

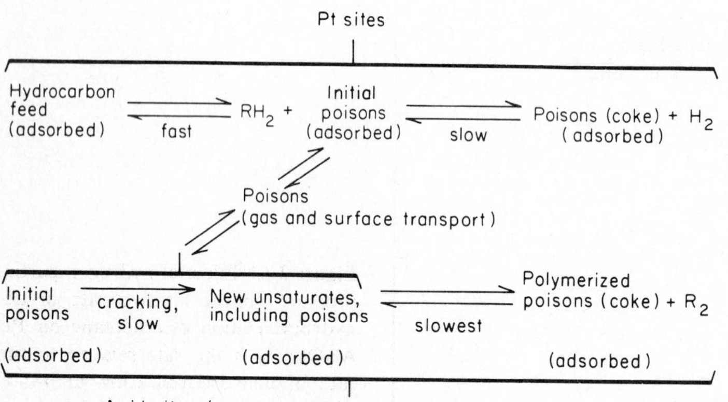

Figure 3-64 Schematic representation of deactivation processes involving coke formation on reforming catalysts [205]. (Reprinted with permission from *Industrial and Engineering Chemistry*. Copyright by the American Chemical Society.)

removal from the surface of the acidic support. Hydrogen also removes coke-forming residues on the acidic support during the reforming process.

The hydrogenation of carbonaceous residues on the acid support is responsible for keeping the acid centers clean. The metal plays a role in this cleaning, for the hydrogen must adsorb and dissociate into atomic species on the metal. The hydrogen atoms, probably with the help of hydrocarbon species bridging the two phases, migrate from the metal surface to the support and hydrogenate residues which are forming there. As the hydrogen partial pressure is increased, the quantity and reactivity of adsorbed hydrogen increase, leading to more effective hydrogenation of the residues on the oxide surface.

The transfer of dissociated hydrogen species from the metal to the support, called *spillover*, is important in other systems also [187, 188, 206–208]. Strongly acidic catalysts like hydrogen mordenite [188, 206, 207] and chlorided alumina [187] readily isomerize paraffins without metal present and under relatively mild conditions. Because of their high activity, however, these catalysts are rapidly deactivated via coke formation. A small amount of noble metal on the surface stabilizes the catalysts by dissociating hydrogen, which spills over onto the zeolite or alumina surface and both suppresses cracking (which leads to coke formation) and hydrogenates hydrocarbon residues. Spillover has been studied fundamentally by Boudart and coworkers [113, 209, 210] and by Neikam and Vannice [211, 212].

PROCESS ENGINEERING

CATALYST DESIGN AND OPERATION

Catalyst activity and selectivity patterns have been discussed in the preceding sections, and now we turn to a quantitative evaluation of the kinetics for process design. First, however, we reconsider catalyst aging, which plays a key role in the choice of processing conditions. Catalyst stability is determined by the operating temperature, pressure, and feed pretreatment to remove sulfur, nitrogen, and oxygen. Sulfur compounds poison the metal component, and nitrogen and oxygen compounds poison the acidic component. Coke deposition is reduced by the application of high hydrogen partial pressures. At low pressures, deactivation may be so rapid that catalyst life may be only a few days; at high pressures, catalyst life may approach 1 year.

Catalyst properties also affect deactivation rates. Platinum on promoted alumina has good stability. Platinum on silica-alumina, which has very strong acidity and catalyzes rapid hydrocracking, undergoes rapid initial deactivation (Fig. 3-65). Experiments mentioned earlier with a mixture of Pt/Al_2O_3 and Al_2O_3 (fluorided) catalysts show that the rapid deactivation is associated with the strongly acidic function. Separating the components results in more rapid deactivation because the deactivation-inhibiting Pt function is reduced (Fig. 3-65).

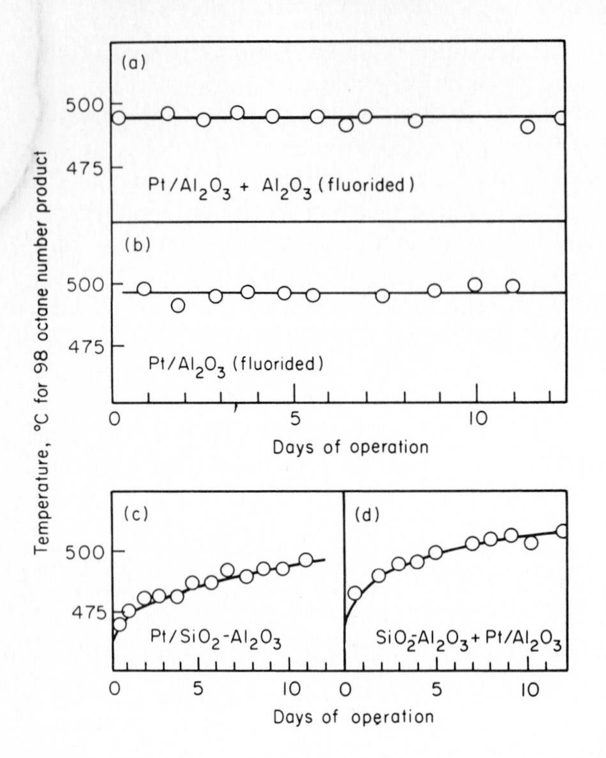

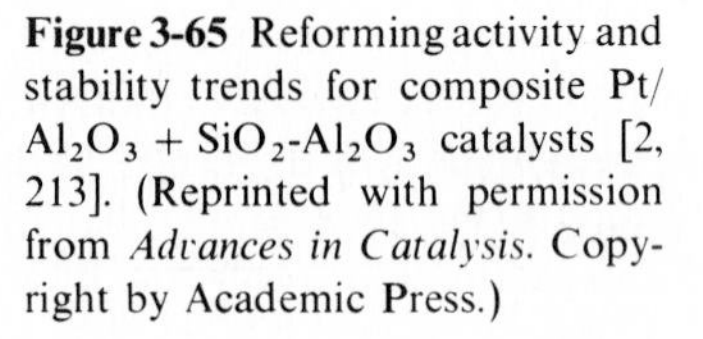
Figure 3-65 Reforming activity and stability trends for composite $Pt/Al_2O_3 + SiO_2$-Al_2O_3 catalysts [2, 213]. (Reprinted with permission from *Advances in Catalysis*. Copyright by Academic Press.)

Stability of industrial reforming catalysts has undergone marked improvement in recent years, first with the introduction of the Pt–Re/Al_2O_3 bimetallic catalyst [214, 215] and then with the introduction of multimetallic catalysts [216, 217]. Figure 3-66 compares the stabilities of Pt/Al_2O_3, Pt–Re/Al_2O_3, and multimetal/Al_2O_3 catalysts. The major advantage of Pt–Re/Al_2O_3 is its enhanced resistance to deactivation by coking, which allows relatively long runs at relatively low pressures of operation. The improved deactivation resistance of the Pt–Re/Al_2O_3 catalyst may be caused by a modification of the acidic support by the Re, at least a fraction of which is not reduced in the catalyst [218, 219]; Re also appears to stabilize the Pt component, so that the metal-surface-area loss during regeneration is less than in Pt/Al_2O_3 catalysts. The Pt–Re/Al_2O_3 catalyst has nearly the same initial activity as the Pt/Al_2O_3 catalyst. The multimetallic catalyst is even more stable because it accumulates coke at about half the rate characteristic of Pt–Re/Al_2O_3 [217]; it is also about 3 times more active in terms of octane number enhancement than the latter for highly paraffinic feeds (Fig. 3-66). The greater activity appears to result from increased metal activity, particularly for dehydrocyclization; this is the slowest reforming reaction and occurs predominantly on the metal function, giving marked octane number enhancement.

Since the two catalyst functions are carefully balanced, changes in activity of either markedly affect the product composition and quality. To compensate for changes in the metal function, changes in the strength and number of acid sites are controlled by the amount of Cl^- and/or F^- on the η-Al_2O_3. Since Cl^- is continually stripped from the surface as HCl by reaction with small amounts of water

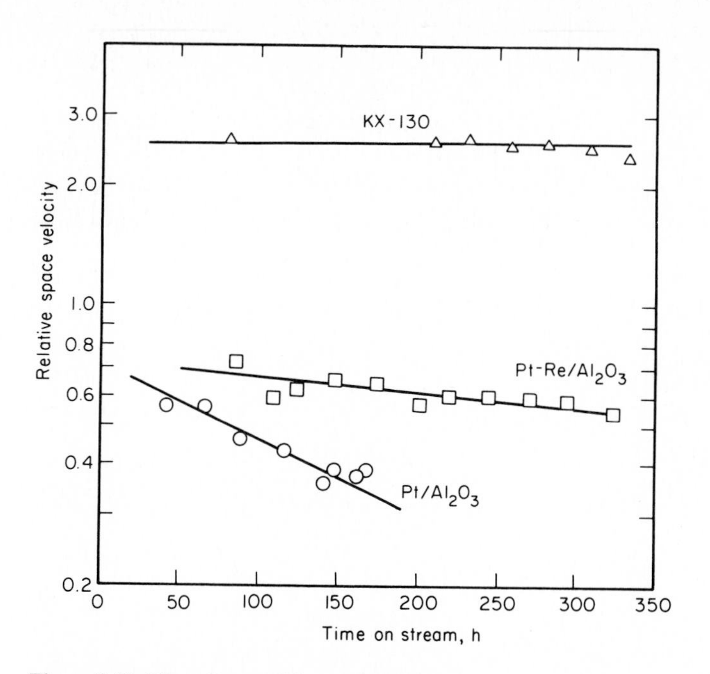

Figure 3-66 Pilot plant activity and stability data for Pt/Al_2O_3, Pt-Re/Al_2O_3, and multimetallic/Al_2O_3 (KX-130) for reforming of a paraffinic naphtha to give constant-octane product, Middle East feedstock (paraffinic naphtha), 9.2 atm total pressure, 499°C average temperature, product octane number set at 102.5 RON clear [217].

in the feed (or water produced from oxygen in the feed), the Cl content of the catalyst must be maintained by adding chlorinated organic compounds to the feed [220]; the concentrations are specified to fix the ratio of Cl to H_2O [221]. Figure 3-67 shows the effect of altered acidity on the octane number of the product reformate from pilot-plant operation. If the Cl content of the catalyst becomes too low, the acidity drops and the reactions which occur on the acid centers (isomerization, dehydrocyclization, and hydrocracking) slow down. The octane number of the reformate then drops (Fig. 67*B*). If excessive Cl (or too little water) is present, the acid strength of the catalyst increases; then the extent of hydrocracking increases relative to dehydrocyclization, and the octane number drops (Fig. 67*A*). Figure 3-67*C* shows how the effects are balanced with the proper ratio of Cl to water in the feed.

Regeneration of a deactivated catalyst involves burning off the accumulated coke (which may be as much as 20 weight percent of the catalyst [223]) under carefully controlled conditions. The procedure involves purging the system with nitrogen and cooling and then burning off the carbon with a gas stream containing from 0.5 to 1 percent oxygen. This operation is performed over a period of several days so that the bed temperature never exceeds 425°C. The temperature control is designed to prevent sintering of the Pt crystallites and subsequent loss of surface area, selectivity, and activity.

After the oxidation, the reactor is purged with nitrogen, and hydrogen is used to reduce the oxidized Pt to the metallic state. The reactor is then put back on stream with a special feedstock containing several hundred ppm of sulfur compounds. After several days, the feed is switched over to the refinery stock. This start-up procedure is similar to that used for a fresh catalyst, which, however, may first be sulfided and then brought on stream with a feed containing several hundred ppm of sulfur compounds. The procedure involves selective, controlled

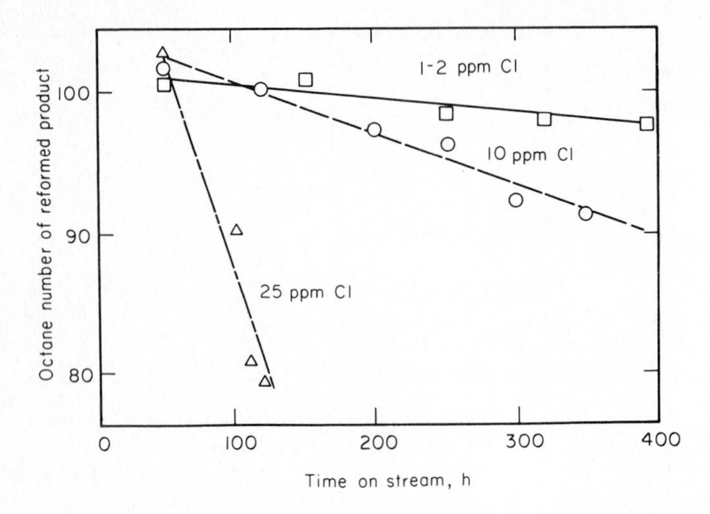

Figure 3-67*A* Effect of feed chlorine content on activity and stability of a Pt/Al_2O_3 catalyst during reforming (water content in recirculated hydrogen was 50-100 ppm by volume). Conditions of pilot plant study: temperature = 496°C, LHSV = 3.0 hr^{-1}, H_2/HC = 6, pressure = 150 atm [222]. (Permission to reprint given by the American Institute of Chemical Engineers.)

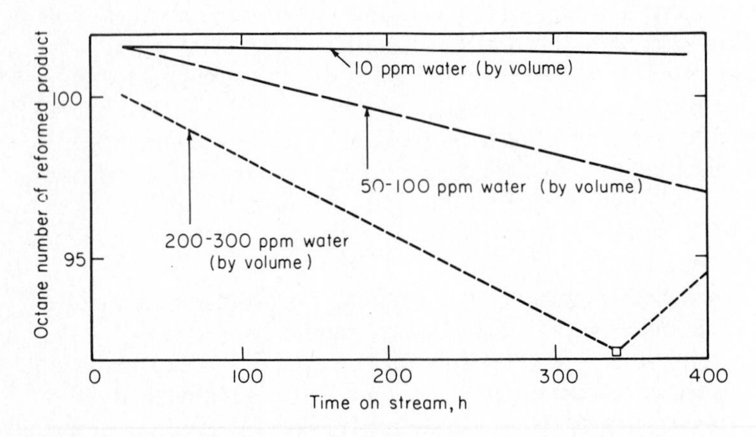

Figure 3-67*B* Effect of water content in recycle hydrogen stream on the activity and stability of a reforming catalyst containing 0.6 wt % Pt and 1.0 wt % Cl. Conditions of pilot plant study: temperature = 496°C, LHSV = 3.0 hr^{-1}, H_2/HC = 6, pressure = 150 atm [222]. (Permission to reprint given by the American Institute of Chemical Engineers.)

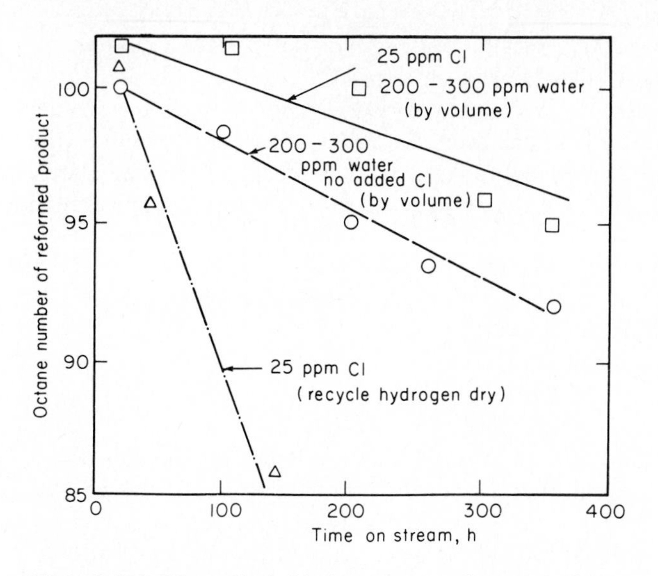

Figure 3-67*C* Effect of feed chlorine content and recycle hydrogen stream water content on reforming catalyst activity and stability. Conditions of pilot plant study: temperature = 496°C, LHSV = 3.0 hr^{-1}, $H_2/HC = 6$, pressure = 150 atm [222]. (Permission to reprint given by the American Institute of Chemical Engineers.)

poisoning of Pt with sulfur to reduce its initial hydrogenolysis activity. Without such pretreatment, a large fraction of the hydrocarbons undergo hydrogenolysis on the Pt to give light gases, with the liberation of much heat. The catalyst can be overheated, and the metal crystallites can be sintered to a useless state; temperatures have become high enough to destroy the reactor.

Since cyclization reactions are catalyzed more selectively by highly dispersed Pt, any procedure (such as high-temperature regeneration) which results in a loss

Table 3-14 Effect of Pt crystallite size change on reforming reactions of *n*-heptane[a]
Reaction conditions: 500°C, WHSV = 2.44 g/g · h, H_2/C_7 ratio = 25.3, 14 atm total pressure; catalyst was sintered by heating to 780°C for various lengths of time.

Sample no.	Average crystallite size, Å	nC_7 remaining, %	Dehydro-cyclization, %	Isomeriza-tion, %	Hydro-cracking, %
1	10	3.0	37.4	9.0	50.6
2	12	3.5	32.8	10.6	53.1
3	33	4.6	26.8	14.2	54.4
4	73	7.0	21.6	21.7	49.7
5	158	9.8	17.7	24.3	48.2
6	452	23.4	12.4	28.9	35.3
7	Pt-free support	74	0.4	0.1	25.5

[a] *n*-Heptane reforming on $Pt/Al_2O_3(Cl)$.

of metal surface area may be detrimental to both activity and selectivity in reforming. Table 3-14 shows how the balance between reforming reactions changes as the average Pt crystallite size of a Pt/Al_2O_3 catalyst is increased by sintering at high temperature. The dehydrocyclization rate decreases markedly, whereas the isomerization rate increases, as expected from studies of crystallite size effects [174]. These results emphasize the desirability of maintaining high degrees of metal dispersion.

REACTION ENGINEERING

Introduction

In the introduction to this chapter, it was shown qualitatively how the thermodynamics and relative rates of reforming reactions influence reactor design and operation. The reactor operation is now considered more specifically in terms of the catalytic chemistry. Hydrogen partial pressure is the key processing variable. When a given operating hydrogen partial pressure is chosen, the equilibrium conversion of naphthenes to aromatics as a function of temperature is defined (Fig. 3-1). Since high conversion to aromatics is desirable, the pressure sets lower limits on the temperature permissible in the catalyst bed. Once the hydrogen partial pressure and temperature are determined, the rate of catalyst deactivation and the required regeneration frequency are determined. Because deactivation at higher temperatures is severe, and because reaction rates and equilibrium conversions at lower temperatures are low, the optimal temperature range in the reactor is quite narrow, typically 40°C.

Higher hydrogen partial pressures decrease the catalyst deactivation rate (and therefore the regeneration frequency), and they decrease the yield of aromatics but increase the hydrocracking rate. A trade-off therefore exists between deactivation rate (regeneration frequency), aromatics production, and hydrocracking rate; this trade-off dictates the hydrogen partial pressure, which is usually chosen to be about 20 atm. Hydrocracking rates are also amenable to partial control by controlling the acidity of the alumina.

Older reformers were typically high-pressure units operated with low regeneration frequencies. The development of new, more stable bimetallic and multimetallic catalysts and requirements for increased aromatics yields have dictated the use of markedly lower hydrogen partial pressures and much higher frequencies of regeneration. Regeneration frequencies as high as once a week have been used [221].

Reactor Modeling

Modeling of catalytic reforming is complicated by the large number of reactions involved. Industrial models have been developed which predict reforming performance very well, but only simple versions of these models have been published [6,

224, 225]. We approach the reaction engineering here by starting with a detailed set of kinetics equations then introducing simplifications which lead ultimately to the quantitative modeling of a four-reactor reforming system.

We begin by considering the diffusional influence that can exist in catalytic reforming, returning to the aforementioned studies with mechanical mixtures of the two separate catalyst components (Figs. 3-59 and 3-60). The reaction network is approximated as

$$\mathrm{P} \underset{k_2}{\overset{k_1}{\rightleftharpoons}} \mathrm{O} \xrightarrow{k_3} \mathrm{IO} \tag{52}$$

$$n\text{-Paraffin} \underset{\text{metal function}}{\rightleftharpoons} \text{olefin} \underset{\text{acid function}}{\rightleftharpoons} \text{isoolefin}$$

Assuming that excess hydrogen is present and writing all reactions as pseudo-first-order reactions, we write the steady-state mass balance for a spherical metal-containing catalyst particle as

$$\frac{D_\mathrm{M}}{R^2}\frac{\partial}{\partial R}\left(R^2\frac{\partial C_\mathrm{P}}{\partial R}\right) - k_1 C_\mathrm{P} + k_2 C_\mathrm{O} = 0 \qquad \text{for P} \tag{53}$$

and

$$\frac{D_\mathrm{M}}{R^2}\frac{\partial}{\partial R}\left(R^2\frac{\partial C_\mathrm{O}}{\partial R}\right) + k_1 C_\mathrm{P} - k_2 C_\mathrm{O} = 0 \qquad \text{for O} \tag{54}$$

The boundary conditions are

At $R = \mathbf{R}_\mathrm{M}$: $\qquad C_\mathrm{P} = C_\mathrm{P}^b \qquad C_\mathrm{O} = C_\mathrm{O}^b \qquad (55)$

At $R = 0$: $\qquad C_\mathrm{P}$ and C_O are finite $\qquad (56)$

In the acidic pellet the balance on O is

$$\frac{D_\mathrm{A}}{R^2}\frac{\partial}{\partial R}\left(R^2\frac{\partial C_\mathrm{O}}{\partial R}\right) - k_3 C_\mathrm{O} = \mathrm{O} \tag{57}$$

with boundary conditions similar to (55) and (56). The two systems are connected by two conditions: (1) the steady-state rate of formation of O in the metal-containing particles equals the rate of consumption of O in the acidic particles,

$$\frac{1}{V_C}\frac{d\bar{N}}{dt} = -D_\mathrm{M}\frac{A_\mathrm{M}}{V_C}\left(\frac{\partial C_\mathrm{O}}{\partial R}\right)_{\mathbf{R}_\mathrm{M}} = -D_\mathrm{A}\frac{A_\mathrm{A}}{V_C}\left(\frac{\partial C_\mathrm{O}}{\partial R}\right)_{\mathbf{R}_\mathrm{A}} \tag{58}$$

where A_M = total exterior surface area of metal-containing particles
A_A = total exterior surface area of acidic particles
V_C = total volume of catalyst

and (2) the concentration of O at both particle surfaces is $C_{\mathrm{O},\mathbf{R}_\mathrm{M}}^b = C_{\mathrm{O},\mathbf{R}_\mathrm{A}}^b$.

Solution of these equations leads to an effectiveness-factor expression of the usual form [229]. For the metal-containing particles,

$$\phi_M = \mathbf{R}_M \sqrt{\frac{k_1}{KD_M}} \tag{59}$$

where $$K = \frac{k_1}{k_2} = \left(\frac{C_O}{C_P}\right)_{eq} \quad \text{and} \quad \eta_M = \frac{3}{\phi_M}\left(\frac{1}{\tanh \phi_M} - \frac{1}{\phi_M}\right) \tag{60}$$

Similar forms result for the acidic particles. Diffusional resistance is unimportant when $\phi < 1$ and thus when

$$\phi = \left(\frac{-1}{V_{C,M}}\frac{d\bar{N}}{dt}\right)\left(\frac{1}{C_{O,eq} - C_O}\right)\frac{\mathbf{R}_M^2}{D_M} < 1 \tag{61}$$

for the metal-containing particle, and

$$\phi = \left(\frac{-1}{V_{C,A}}\frac{d\bar{N}}{dt}\right)\frac{1}{C_O}\frac{\mathbf{R}_A^2}{D_A} < 1 \tag{62}$$

for the acidic particle, where

$$\frac{-1}{V_C}\frac{d\bar{N}}{dt}$$

is the steady-state rate of conversion of n-paraffin per unit (total) volume of catalyst. In the case where the first reaction is very rapid ($C_O \to C_{O,eq}$), Eq. (62) is the one of design interest, and it becomes

$$\phi = \left(\frac{-1}{V_{C,A}}\frac{d\bar{N}}{dt}\right)\frac{1}{C_{O,eq}}\frac{\mathbf{R}_A^2}{D_A} < 1 \tag{63}$$

Similarly, when the first reaction is rate-limiting, $C_O \ll C_{O,eq}$, and Eq. (61) becomes

$$\phi = \left(\frac{-1}{V_{C,M}}\frac{d\bar{N}}{dt}\right)\frac{1}{C_{O,eq}}\frac{\mathbf{R}_M^2}{D_M} < 1 \tag{64}$$

For isomerization of n-heptane in an integral reactor, it can be shown for pseudo-first-order kinetics that

$$\left(\frac{-1}{V_{C,A}}\frac{d\bar{N}_P}{dt}\right)\frac{1}{C_P} = \frac{1}{\tau}\ln\frac{1}{1 - X_e} \tag{65}$$

where τ is the contact time and X is the observed fractional conversion, which cannot exceed the equilibrium value. Equation (62) can be written

$$\left(\frac{-1}{V_{C,A}}\frac{d\bar{N}}{dt}\right)\frac{1}{C_P}\frac{C_P}{C_{O,eq}}\frac{\mathbf{R}_A^2}{D_A} = \frac{C_P}{C_{O,eq}}\frac{\mathbf{R}_A^2}{D_A}\frac{1}{\tau}\ln\frac{1}{1 - X} < 1 \tag{66}$$

from which it can be seen that

$$\phi \approx 7 \text{ to } 30 \times 10^4 \mathbf{R}_A^2 \tag{67}$$

for $D_M = 2 \times 10^{-3}\ \text{cm}^2/\text{s} \qquad X_e = 0.40 \qquad \tau = 17\ \text{s}$

and $$\frac{C_O}{C_{O,\,eq}} = (1 \text{ to } 4) \times 10^{-3} \text{ at } 40\% \text{ conversion}$$

Thus for ϕ to be less than 1, that is, for diffusional limitations to be unimportant, $\mathbf{R}_A$ must be less than 20 to 40 μm; this conclusion is in good agreement with the results of Fig. 3-59.

Diffusion can also affect selectivity in mixed-particle systems when more complex reaction networks are operative, as in methylcyclopentane conversion (Fig. 3-58 and Table 3-13). The data of Table 3-15 show how reduction of diffusional resistance by reduction of the particle size improves the selectivity. Reduction in particle size allows the acidic function to intercept the dehydrogenation products and isomerize them to a C_6 ring before they can undergo hydrogenolysis on the Pt function. Selectivity to C_6-ring products was also improved by poisoning the Pt with sulfur, which shows that the C_1 to C_6 paraffins were formed on the metal function.

The kinetics of each of the reactions of methylcyclopentane and of each of the reactions of its products in the presence of Pt/SiO_2 and of SiO_2–Al_2O_3 have been determined in a differentially operated, isothermal flow reactor under conditions of no mass-transfer influence (Fig. 3-68 and Table 3-16). The first-order rate constants were incorporated into a set of flow-reactor mass-balance equations describing the formation and disappearance of all species, and the conversion to benzene was predicted as a function of the fraction of the catalyst mixture which was Pt/SiO_2 (for constant space velocity). Figure 3-69 shows that the predicted behavior agrees well with the experimentally observed behavior.

Table 3-17 gives first-order rate constants for methylcyclopentane reactions catalyzed by a Pt/SiO_2–Al_2O_3. The reaction network appears different from that

Table 3-15 Selectivity in methylclopentane conversion in the presence of mixed-catalyst particles [2]

Reaction conditions: 380°C; 10.3 atm hydrogen partial pressure; 1.1 atm methylcyclopentane partial pressure; 7.5 s contact time.

Mixed catalyst	Catalyst particle size, μm	Products, wt %	
		Benzene + cyclohexane	C_1–C_6 paraffins
Pt/Al_2O_3 + SiO_2–Al_2O_3	500	3.3	12.5
	5	20.0	5.6

Source: Reprinted with permission from *Advances in Catalysis*. Copyright by Academic Press.

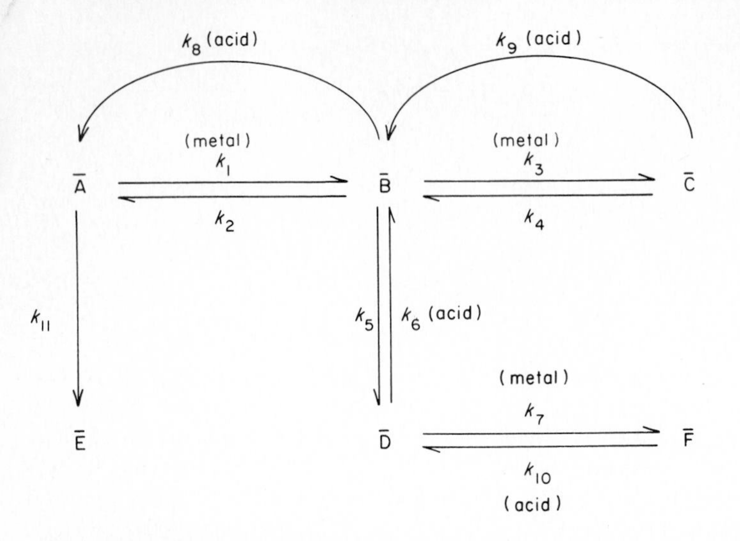

Figure 3-68 Reaction network for methylcyclopentane conversion catalyzed by a mixture of large, monofunctional catalyst particles; $\bar{A}$ = methylcyclopentane, $\bar{B}$ = methylcyclopentene, $\bar{C}$ = methylcyclopentadiene, $\bar{D}$ = cyclohex-1-ene, $\bar{E}$ = hydrogenolysis products, $\bar{F}$ = benzene [226, 227].

found with the mixed catalysts, as indicated by the absence of $\bar{C}$ and the apparent occurrence of the reaction $\bar{B} \rightarrow \bar{F}$; the kinetics parameters also differed significantly from those for the individual catalyst functions. Some of the differences may be real; others probably reflect the difficulty of clearly defining kinetics for single reactions on a bifunctional catalyst. The solution of the component equations for all components (including mass transfer) and the particle energy balance show

Table 3-16 First-order rate constants for methylcyclopentane reactions on separate catalyst functions[a] [226, 227]

Reaction	Catalyst function	Rate constant: Designation	Rate constant: Value at 500°C, s^{-1}	E_{act}, kcal/mol	A', s^{-1}
$\bar{A} \longrightarrow \bar{B}$	Metal	k_1	0.293	13.0	4.56×10^2
$\bar{B} \longrightarrow \bar{A}$	Metal	k_2	0.370	10.0	2.39×10^2
$\bar{B} \longrightarrow \bar{C}$	Metal	k_3	1.03	10.7	1.13×10^3
$\bar{C} \longrightarrow \bar{B}$	Metal	k_4	0.760	8.8	4.56×10^2
$\bar{B} \longrightarrow \bar{D}$	Acid	k_5	8.53	9.7	2.34×10^2
$\bar{D} \longrightarrow \bar{B}$	Acid	k_6	4.43	12.3	1.39×10^4
$\bar{D} \longrightarrow \bar{F}$	Metal	k_7	16.6	9.7	9.31×10^3
$\bar{B} \longrightarrow \bar{A}$	Acid	k_8	5.65	12.3	1.75×10^3
$\bar{C} \longrightarrow \bar{B}$	Acid	k_9	6.12	13.6	4.49×10^4
$\bar{D} \longrightarrow \bar{F}$	Acid	k_{10}	0.043	17.9	3.47×10^3

[a] The Pt/SiO_2 and SiO_2–Al_2O_3 catalysts were investigated separately.

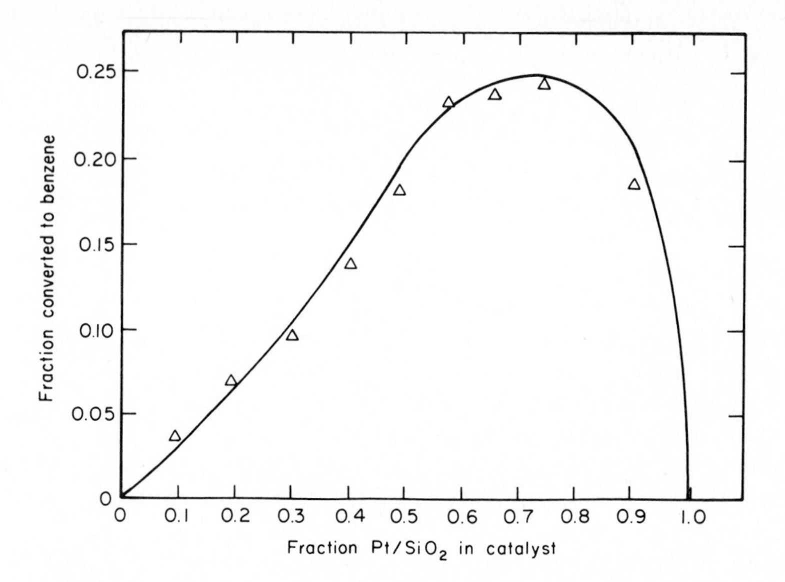

Figure 3-69 Comparison of predicted conversion of methylcyclopentane to benzene with that observed for mixed catalysts with the indicated fractions of Pt/SiO_2 [226, 227].

that the effectiveness factor is about 0.8 at 450°C and about 0.75 at 500°C for 0.4-cm-diameter particles. Although significant concentration gradients existed in the catalyst particles, there were no significant temperature gradients within them. The major resistance to heat transfer existed between the catalyst surface and the surrounding fluid, which indicates that the energy-balance equation for the catalyst particles can be replaced by an algebraic energy balance over the fluid film surrounding the apparently isothermal particle, which greatly reduces the calcula-

Table 3-17 First-order rate constants for methylcyclopentane reactions on Pt/SiO_2–Al_2O_3 [228]

Reaction	Catalyst function	Rate constant: Designation	Rate constant: Value at 500°C, s^{-1}	E_{act}, kcal/mol	A', s^{-1}
$\bar{A} \longrightarrow \bar{B}$	Metal	k_1	0.326	40.0	5.85×10^{10}
$\bar{B} \longrightarrow \bar{A}$	Metal	k_2	0.182	8.99	63.4
$\bar{B} \longrightarrow \bar{D}$	Acid	k_5	0.020	8.66	5.73
$\bar{D} \longrightarrow \bar{B}$	Acid	k_6	0.256	9.81	152
$\bar{D} \longrightarrow \bar{F}$	Metal	k_7	2.74	34.3	1.39×10^{10}
$\bar{B} \longrightarrow \bar{F}$	Metal + acid	k_{12}	0.72	47.7	2.25×10^{13}
$\bar{A} \longrightarrow \bar{E}$	Metal	k_{11}	0.025	49.0	1.85×10^{12}

Source: Reprinted with permission from *Advances in Chemistry Series.* Copyright by the American Chemical Society.

tional effort. Similarly, concentration differences between the catalyst surface and the bulk fluid can in most cases be ignored.

Combining the energy balance about each catalyst particle with that for the entire flow reactor and combining the kinetics equations including mass-transfer influences with the reactor mass balances gave the following results:

1. Because of heat- and mass-transfer limitations, the composite catalyst Pt/SiO_2–Al_2O_3 gives better performance than a mixture of catalyst particles each containing one of the two catalyst functions.
2. The highest yields occur in an isothermal reactor, and the lowest yields occur in an adiabatic reactor. For a typical set of conditions, the predicted benzene yield was about 70 percent of the methylcyclopentane fed for the isothermal case, but it was less than 5 percent for the adiabatic case.
3. Reaction selectivity and overall rate are affected by varying the amount of Pt present, but the effect is not large over a fairly broad range about the optimum.

The complexity of naphtha reforming makes the analysis of the reaction process by the techniques discussed above very complex, and only simplified analyses can readily be carried out. The analysis requires that the naphtha be broken down into its separate components or that these components be lumped into classes of components with similar properties and kinetic behavior. Krane et al. [6] have reported first-order rate constants for 53 individual reactions characteristic of naphtha reforming (Table 3-18). The rate constant for each reaction is defined by

$$\frac{\partial \bar{N}_i}{\partial(1/\mathrm{SV})} = -k_i \bar{N}_i \tag{68}$$

where N_i denotes moles of reacting component i and SV denotes weight liquid hourly space velocity. The temperature dependence of k_i was not determined.

These data can be combined into a set of 20 coupled, linear, first-order ordinary differential equations which account for the rate of formation and of disappearance of each paraffin with 1 to 10 carbon atoms and of each naphthene with 6 to 10 carbon atoms; the isomer distribution within each group is not accounted for. Thus all isomeric species have been lumped into one pseudocompound of a given carbon number and type. The appropriate equation for the C_7 paraffins, for example, is

$$\frac{dC_{\mathrm{P}_7}}{d(1/\mathrm{SV})} = 0.0109C_{\mathrm{P}_{10}} + 0.0039C_{\mathrm{P}_9} + 0.0019C_{\mathrm{P}_8} + 0.0020C_{\mathrm{N}_7} + 0.0016C_{\mathrm{A}_7} - 0.0122C_{\mathrm{P}_7} \tag{69}$$

where P, N, and A have the same meanings as in Table 3-18. This equation describes the formation of heptanes from hydrocracking of higher-carbon-number paraffins (the first three terms) and from hydroisomerization of naphthenes and aromatics. It also includes the disappearance of heptanes by hydrocracking to

Table 3-18 First-order rate constants for naphtha-reforming reactions[a] [6]
Typical reaction conditions: 496°C, hydrogen-to-naphtha molar ratio = 5 : 1, 6.8 to 30 atm total pressure, and 0.30 wt % Pt on alumina catalyst

Reaction	$k \times 10^2$, h^{-1}
Dehydrocyclization k_1:	
$P_{10} \longrightarrow N_{10}$	2.54
$P_9 \longrightarrow N_9$	1.81
$P_8 \longrightarrow N_8$	1.33
$P_7 \longrightarrow N_7$	0.58
$P_6 \longrightarrow N_6$	0
Dehydrogenation k_2:	
$N_{10} \longrightarrow A_{10}$	24.50
$N_9 \longrightarrow A_9$	24.50
$N_8 \longrightarrow A_8$	21.50
$N_7 \longrightarrow A_7$	9.03
$N_6 \longrightarrow A_6$	4.02
Hydrocracking of paraffins k_3:	
$P_{10} \longrightarrow P_9 + P_1$	0.49
$P_{10} \longrightarrow P_8 + P_2$	0.63
$P_{10} \longrightarrow P_7 + P_3$	1.09
$P_{10} \longrightarrow P_6 + P_4$	0.89
$P_{10} \longrightarrow 2P_5$	1.24
$P_9 \longrightarrow P_8 + P_1$	0.30
$P_9 \longrightarrow P_7 + P_2$	0.39
$P_9 \longrightarrow P_6 + P_3$	0.68
$P_9 \longrightarrow P_5 + P_4$	0.55
$P_8 \longrightarrow P_7 + P_1$	0.19
$P_8 \longrightarrow P_6 + P_2$	0.25
$P_8 \longrightarrow P_5 + P_3$	0.43
$P_8 \longrightarrow 2P_4$	0.35
$P_7 \longrightarrow P_6 + P_1$	0.14
$P_7 \longrightarrow P_5 + P_2$	0.18
$P_7 \longrightarrow P_4 + P_3$	0.32
$P_6 \longrightarrow P_5 + P_1$	0.14
$P_6 \longrightarrow P_4 + P_2$	0.18
$P_6 \longrightarrow 2P_3$	0.27
Hydroisomerization of aromatics k_4:	
$A_{10} \longrightarrow P_{10}$	0.16
$A_9 \longrightarrow P_9$	0.16
$A_8 \longrightarrow P_8$	0.16
$A_7 \longrightarrow P_7$	0.16
Hydrocracking of naphthenes k_5:	
$N_{10} \longrightarrow N_9$	1.34
$N_{10} \longrightarrow N_8$	1.34
$N_{10} \longrightarrow N_7$	0.80
$N_9 \longrightarrow N_8$	1.27
$N_9 \longrightarrow N_7$	1.27
$N_8 \longrightarrow N_7$	0.09
Hydrogenation k_6:	
$A_6 \longrightarrow N_6$	0.45
Hydrocracking of aromatics k_7:	
$A_{10} \longrightarrow A_9$	0.06
$A_{10} \longrightarrow A_8$	0.06
$A_{10} \longrightarrow A_7$	0.00
$A_9 \longrightarrow A_8$	0.05
$A_9 \longrightarrow A_7$	0.05
$A_8 \longrightarrow A_7$	0.01
Hydroisomerization of naphthenes k_8:	
$N_{10} \longrightarrow P_{10}$	0.54
$N_9 \longrightarrow P_9$	0.54
$N_8 \longrightarrow P_8$	0.47
$N_7 \longrightarrow P_7$	0.20
$N_6 \longrightarrow P_6$	1.48

[a] P, N, and A denote paraffin, naphthene, and aromatic, respectively, and the subscript is the carbon number.

lighter paraffins and by dehydrocyclization to naphthenes; the last rate constant is the sum of four individual rate constants from Table 3-18.

The good comparison between calculated and experimental product compositions as a function of 1/SV for a typical whole naphtha is shown in Fig. 3-70. Light and intermediate paraffin yields increase with increasing contact time (due to hydrocracking); heptane yield passes through a weak maximum; and heavier paraffin yield decreases gradually. (This last result indicates that reactivity increases with carbon number.) Aromatic yields increase sharply then level off, and the higher-molecular-weight products are preferentially formed.

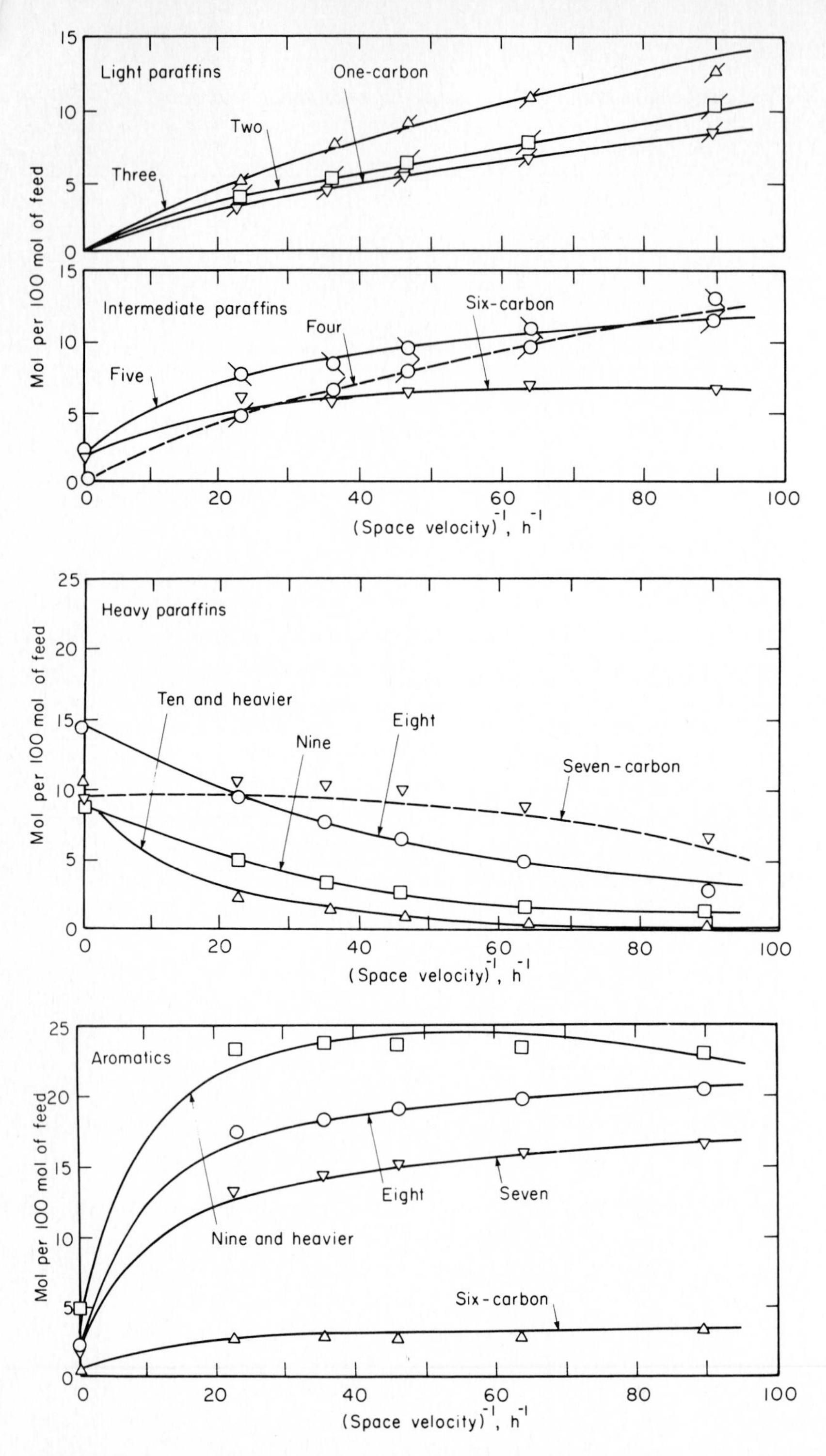

Figure 3-70 Comparison between predicted (lines) and observed (points) conversions for reforming of a whole naphtha [6]. Reaction conditions are those of Table 3-18.

To further simplify the calculation, more lumping can be carried out, reducing the number of equations requiring solution and also affecting the accuracy of predictions. For example, the above reactions (Table 3-18) can be lumped into dehydrocyclization of paraffins, dehydrogenation of naphthenes to aromatics, hydrocracking of paraffins, hydrocracking of naphthenes, hydrocracking of aromatics, hydroisomerization of aromatics, hydroisomerization of naphthenes, and hydrogenation of aromatics. There are now only 8 lumped components instead of 53, and the number of first-order coupled linear ordinary differential equations to be solved is reduced to three:

$$\frac{dC_P}{d(1/SV)} = -k_1 C_P - k_3 C_P + k_4 C_A + k_8 C_N \tag{70}$$

$$\frac{dC_N}{d(1/SV)} = k_1 C_P - k_2 C_N - k_5 C_N + k_6 C_A - k_8 C_N \tag{71}$$

$$\frac{dC_A}{d(1/SV)} = k_2 C_N - k_4 C_A - k_6 C_A - k_7 C_A \tag{72}$$

To demonstrate the effect of the lumping and of the values of the rate constants used, the above equations were solved using the average value, the minimum value, and the maximum value of the rate constant for each reaction class for isothermal operation (Table 3-19). Figure 3-71 shows the effect of the k values on the conversion of paraffins, naphthenes, and aromatics. Figure 3-72 compares the predictions of the lumped-parameter model with those of Krane et al. [6]. The paraffin pool in this case consists of only C_6 and higher hydrocarbons. The composition of the whole naphtha was not the same for the two cases, but when this correction is made, the lumped-parameter model is in good agreement with the more complex model [6] except at high conversions. If nonisothermal behavior were included (and temperature dependencies would be required) (Table 3-17), lumping procedures would become more difficult.

As a final example, we consider the modeling of the four-reactor reforming system shown in Fig. 3-73 and include temperature effects. Table 3-20 gives the feedstock composition and assigned operating parameters. The feed is lumped into paraffins, naphthenes, and aromatics, and the reactions to be considered include conversion of naphthenes into aromatics, conversion of naphthenes into paraffins, hydrocracking of paraffins, and hydrocracking of naphthenes. At the high pressure of operation (40 atm), hydrocracking can be considered more important than dehydrocyclization.

The rate equations are similar to those above, but an energy balance must now be added. Heats of reaction are nearly independent of the component molecular weights over the range normally encountered in reforming when they are based on the number of moles of hydrogen entering into the reaction. The energy balance can be written around a differential mass of catalyst in the reactor dW_C; for adiabatic operation it is

$$Q\rho_m c_P \, dT = r_{N\to P} \, \Delta H_{r,N\to P} \, dW_C + r_{N\to A} \, \Delta H_{r,N\to A} \, dW_C + r_{PC} \, \Delta H_{r,PC} \, dW_C + r_{NC} \, \Delta H_{r,NC} \, dW_C \tag{73}$$

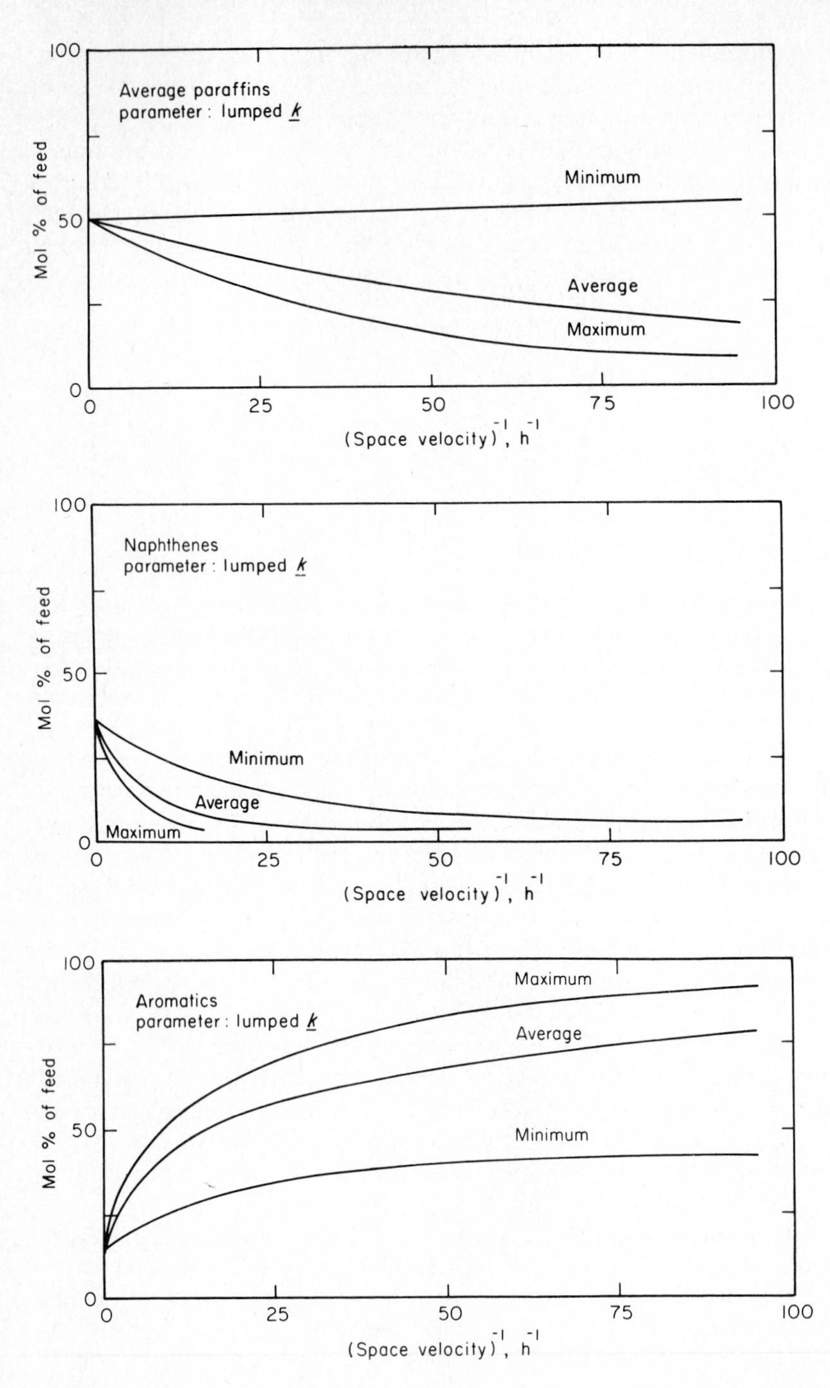

Figure 3-71 Calculated conversion of an average whole naphtha using average, maximum, and minimum values for the first-order rate constants in each reaction class. Reaction conditions are those of Table 3-18.

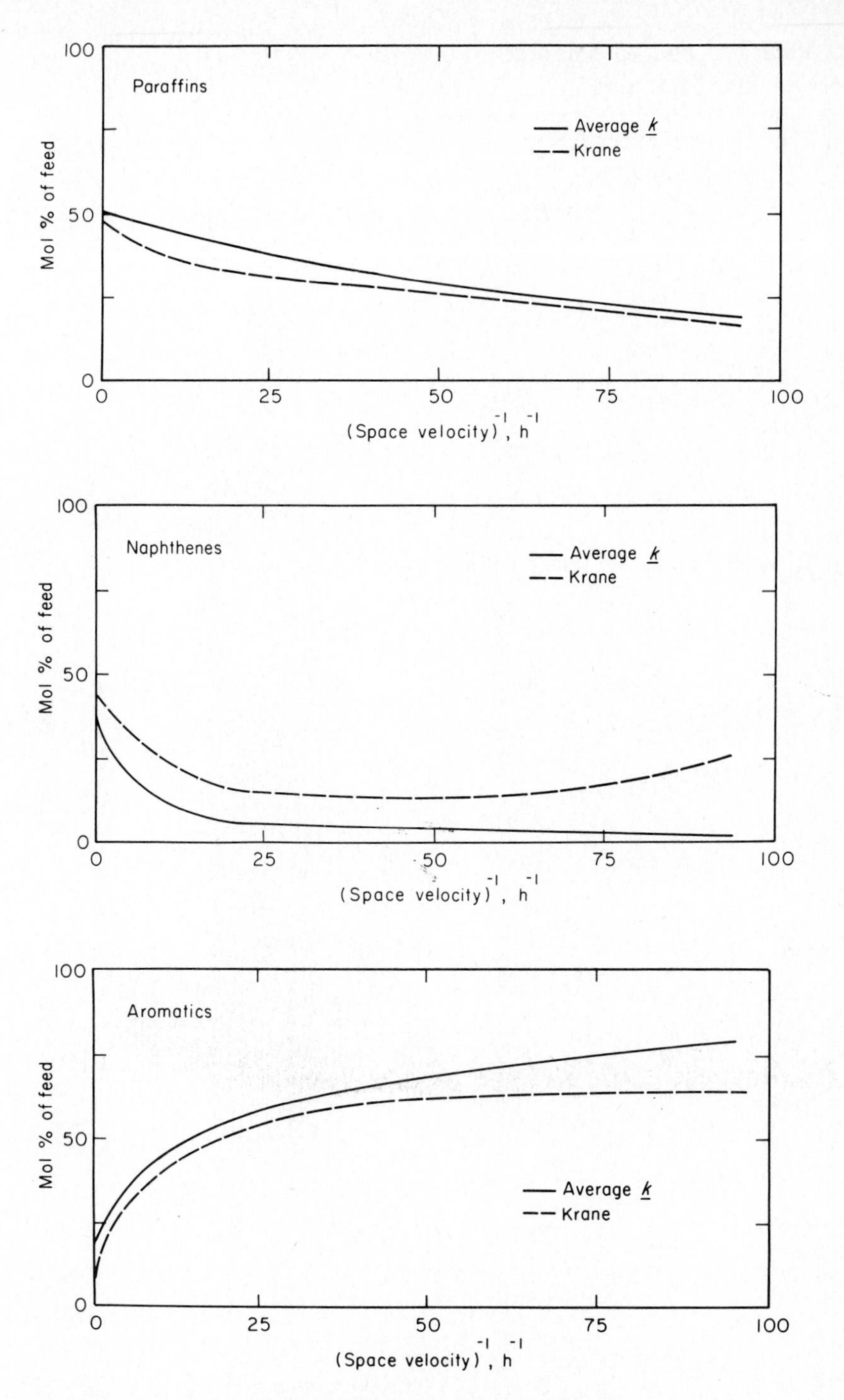

Figure 3-72 Comparison of conversion predicted by lumped-parameter model using average k value for each reaction group with that predicted by Krane et al. [6].

Table 3-19 Values of lumped first-order rate constants for naphtha reforming

Rate constant[a]	$10^2 \cdot$ First-order rate constant, h^{-1}		
	Maximum	Minimum	Average
k_1	2.54	0.00	1.27
k_2	24.50	4.02	14.26
k_3	1.09	0.14	0.62
k_4	0.16	0.16	0.16
k_5	1.34	0.09	0.72
k_6	0.45	0.45	0.45
k_7	0.06	0.00	0.03
k_8	1.48	0.20	0.84

[a] k_i defined in Table 3-18.

where $Q\rho_m/W_C$ = molar hourly space velocity (MSV) and where in general

$$r_i = k_{f,i} C_{R,i} P_{H_2}^{n_1} - k_r C_{P,i} P_{H_2}^{n_2} \tag{74}$$

Equation (73) becomes

$$\frac{dT}{d(1/\text{MSV})} = \frac{1}{c_P}(r_{N \to P}\, \Delta H_{r,N \to P} + r_{N \to A}\, \Delta H_{r,N \to A} + r_{PC}\, \Delta H_{r,PC} + r_{NC}\, \Delta H_{r,NC}) \tag{75}$$

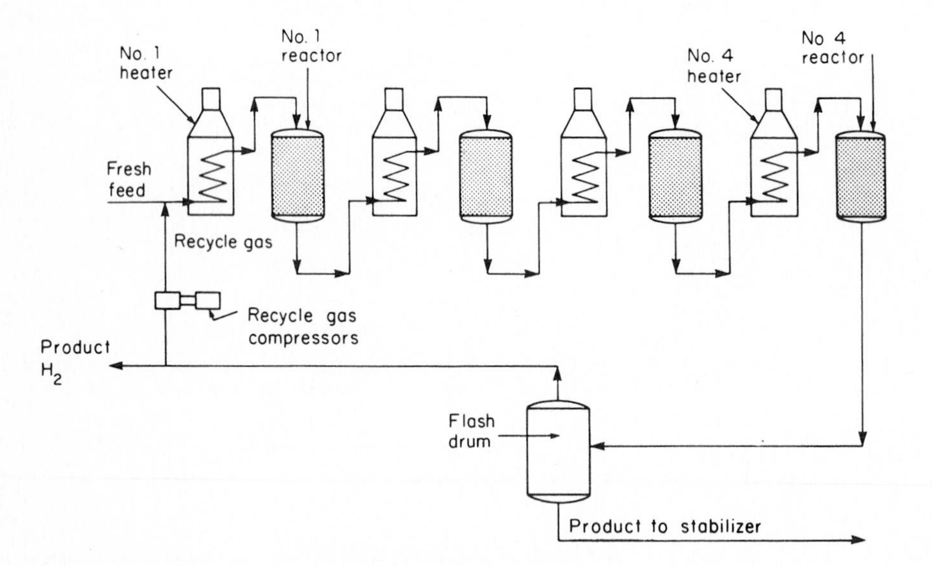

Figure 3-73 Schematic of four-reactor reforming system.

Table 3-20 Feedstock analysis and system operating parameters [224]

Feedstock analysis

Specific gravity = 54.3°API
ASTM distillation, °C

Initial boiling point	83
10%	114
50%	129
90%	155
Endpoint	180
Average molecular weight	114.8

Feedstock components (basis = 100 bbl)

Component	mol %	Average molecular weight	lb	mol	mol/bbl	bbl
Paraffins	31.0	$C_{8.2}H_{18.4} = 116.8$	8,410	72.0	2.19	32.9
Naphthenes	58.3	$C_{8.2}H_{16.4} = 114.8$	15,544	135.4	2.33	58.4
Aromatics	10.7	$C_{8.2}H_{10.4} = 108.8$	2,709	24.9	2.86	8.7
Total	100.0	114.8[a]	26,663	232.3	2.32[a]	100.0

Operating parameters

	Reactor			
	1	2	3	4
Catalyst distribution	1	1	2.75	3.5
WHSV (overall WHSV = 2.75)	22.69	22.69	8.25	6.48
Molar hourly space velocity	0.198	0.198	0.072	0.056
Reactor inlet temperature, °C	506	506	506	506
Reactor pressure, atm	39.9	39.5	39.1	38.1
Recycle ratio, mol/mol fresh feed	6.9			
H_2 in recycle, mol %		85.3		
Recycle molecular weight		6.0		
Flash-drum pressure, atm		35.0		
Flash-drum temperature, °C		38		

[a] Average values

The required equations and data are as follows [224, 230] (rate constants are for fresh catalyst):

1. For naphthene + $H_2 \rightleftharpoons$ paraffin

$$\text{Equilibrium constant } K_p = \frac{P_{\text{P}}}{P_{\text{N}} P_{\text{H}_2}} \tag{76}$$

$$K_p = 8.09 \times 10^{-4} \exp \frac{-4900}{RT} \quad \text{atm}^{-1} \tag{77}$$

Rate equation:

$$\rho_m r_{N\to P} = \frac{dC_N}{d(1/\mathrm{MSV})} = \frac{k_{N\to P} P_T}{K_p} P_{H_2}\left(C_N K_p - \frac{C_P}{C_N P_{H_2}} \right) \tag{78}$$

$$k_{N\to P} = 4.23 \times 10^{15} \exp \frac{-36{,}500}{RT} \qquad \text{mol/g catalyst} \cdot \text{atm}^2 \cdot \text{h} \tag{79}$$

Energy balance:
$\Delta H_r = -10{,}500$ cal/mol

$$r_{N\to A}\, \Delta H_{r,\, N\to A} = \frac{dC_N}{d(1/\mathrm{MSV})} \frac{-10{,}500}{\rho_m} \qquad \text{cal/mol} \tag{80}$$

2. For naphthene $\rightleftharpoons$ aromatic + $3H_2$

$$\text{Equilibrium constant } K_p = \frac{P_A P_{H_2}^3}{P_N} \tag{81}$$

$$K_p = 1.10 \times 10^{20} \exp \frac{-27{,}600}{RT} \qquad \text{atm}^3 \tag{82}$$

Rate equation:

$$\rho_m r_{N\to A} = \frac{dC_N}{d(1/\mathrm{MSV})} = \frac{k_{N\to A} P_T}{K_p} C_N \left(K_p - \frac{C_A}{C_N} P_{H_2}^3 \right) \tag{83}$$

$$k_{N\to A} = 1.20 \times 10^{10} \exp \frac{-21{,}300}{RT} \qquad \text{mol/g catalyst} \cdot \text{atm} \cdot \text{h} \tag{84}$$

Energy balance:
$\Delta H_r = 16{,}900$ cal/mol of H_2 liberated

$$r_{N\to A}\, \Delta H_{r,\, N\to A} = \frac{dC_N}{d(1/\mathrm{MSV})} \frac{50{,}800}{\rho_m} \qquad \text{cal/mol} \tag{85}$$

3. For hydrocracking of paraffins, the rate equation is

$$\frac{dC_P}{d(1/\mathrm{MSV})} = -k_{PC} C_P \tag{86}$$

$$k_{PC} = 4.59 \times 10^{18} \exp \frac{-38{,}200}{RT} \qquad \text{mol/g catalyst} \cdot \text{h} \tag{87}$$

Energy balance:
$\Delta H_r = -13{,}500$ cal/mol H_2 consumed

$$r_{PC}\, \Delta H_{r,\, PC} = \frac{dC_P}{d(1/\mathrm{MSV})} \frac{-13{,}500(n-3)/n}{\rho_m} \tag{88}$$

4. For naphthene cracking the rate equation is

$$\frac{dC_N}{d(1/\text{MSV})} = -k_{NC}C_N \tag{89}$$

$$k_{NC} = 4.5 \times 10^{18} \exp\frac{-38{,}200}{RT} \qquad \text{mol/g catalyst} \cdot \text{h} \tag{90}$$

Energy balance:

$$\Delta H_r = -12{,}400 \text{ cal/mol } H_2 \text{ consumed}$$

$$r_{NC}\,\Delta H_{r,NC} = \frac{dC_N}{d(1/\text{MSV})}\frac{-12{,}400n/3}{\rho_m} \tag{91}$$

The component mass balances and the energy balance are solved by progressively stepping down the reactor; the results of this solution are shown in Figure 3-74. The model, simple as it is, predicts reactor-system performance quite well (Table 3-21). It does not require information about diffusional resistances because the kinetics constants already contain this information implicitly, having been obtained under reaction conditions and with catalyst particle sizes similar to those of commercial operation. The model does not predict product composition in sufficient detail to allow adequate prediction of the octane number of the product.

The reactor inlet temperature used in the calculation was set at a low value because the kinetics data used in the calculation were for a fresh catalyst and the commercial system involved a partially deactivated catalyst. The large temperature drop in the first reactor is indicative of dehydrogenation of the naphthenes

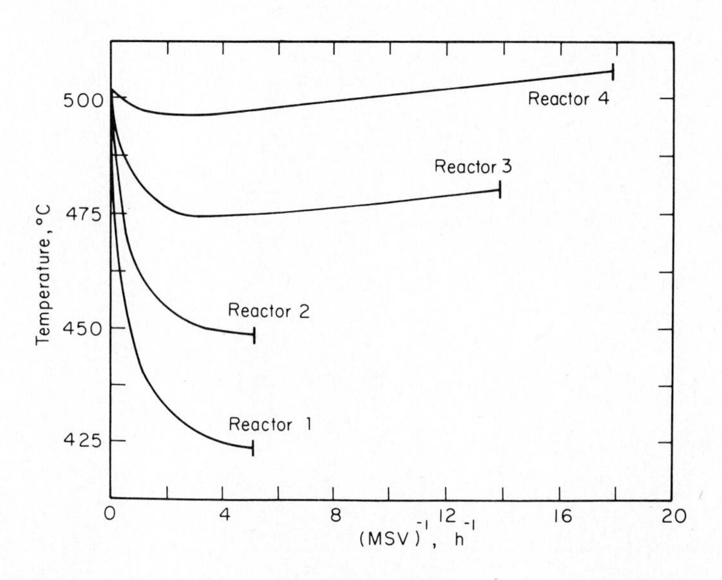

Figure 3-74 Calculated temperature profiles in the four reactors [224].

Table 3-21 Comparison of calculated and commercial results of a four-reactor system [224]

	Calculated results				Results of commercial run			
	Reactor				Reactor			
	1	2	3	4	1	2	3	4
Individual reactor results								
Catalyst distribution	1	1	2.75	3.5	1	2	2.75	3.5
WHSV (Overall WHSV = 2.75)	22.7	22.7	8.3	6.5	22.7	22.7	8.3	6.5
Inlet temp., °C	506	506	506	506	511	511	514	511
Outlet temp., °C	436	459	487	509	439	472	489	504
Temp. drop, °C	70	47	19	−3	72	39	25	7
C_{6+} paraffins leaving reactor, mol	76.0	79.2	75.0	42.5				
C_{6+} naphthenes leaving reactor, mol	80.0	39.6	12.7	4.9				
Aromatics leaving reactor, mol	74.7	109.5	129.4	140.1				
Total C_{6+} leaving reactor, mol	230.7	228.3	217.1	187.5				

Overall results

	Calculated results		Results of commercial run
Total reactor temp. drop, °C	133		143
	Yield		
	mol[c]	bbl[c]	mol[c]
H_2	252.5		279.6
C_1	24.5		
C_2	24.5		
C_3	24.5		
iC_4	12.2[a]		2.5
nC_4	12.2[a]		
iC_5	16.3[b]	5.4	2.5
nC_5	8.2[b]	2.7	1.6
C_{6+}:			
Paraffins	42.5	19.4	21.3
Naphthenes	4.9	2.1	2.0
Aromatics	140.1	49.9	51.3
Total	562.4	79.5	81.2[d]
Aromatics in C_5 to endpoint reformate, vol %	62.8	...	63.2
Reformate research octane number, (clear)	95.4	...	95.5

[a] Moles of iC_4 = (moles total C_4's)/2.
[b] Moles of iC_5 = 2(moles total C_5's)/3.
[c] Basis of calculation = 100 bbl fresh feed.
[d] Total barrels of C_5 to endpoint, vol % yield.

present in the feed. This large temperature decrease almost quenches all reaction, and it is therefore undesirable to have additional catalyst in the first reactor because reheating of the gas stream is required. In the third and fourth reactors, the temperature increase following conversion of naphthenes to give equilibrium with aromatics, which causes an initial temperature drop, is a result of exothermic hydrocracking. Hydrocracking occurs predominantly in the fourth reactor.

NOTATION

A	area, l^2
A'	preexponential factor, t^{-1}
A_S	activity in arbitrary units for total conversion of cyclopropane
A_H	activity in arbitrary units for hydrogenolysis of cyclopropane
A	aromatic
$\bar{\mathrm{A}}$	methylcyclopentane
$\bar{\mathrm{B}}$	methylcyclopentene
b	integer
C	concentration, mol/100 mol of feed
c_P	heat capacity, E/mol T
$\bar{\mathrm{C}}$	methylcyclopentadiene
c	integer
D	effective diffusion coefficient, l^2/t
$\bar{\mathrm{D}}$	cyclohex-1-ene
E_{act}	activation energy, E/mol
E	energy, magnitude of bond energy
$\bar{\mathrm{E}}$	hydrogenolysis products
$\bar{\mathrm{F}}$	benzene
H_{ij}	bond energy of bond between atoms i and j
IO	isoolefin
K	equilibrium constant, variable dimensions
k	first-order rate constant, variable dimensions; Boltzmann's constant, E/mT
L	any ligand; for metal complexes, PR_3
LHSV	liquid hourly space velocity, t^{-1}
l	fraction of nearest neighbors in the plane of the surface atom
M	metal
MSV	molar hourly space velocity; moles fresh feed per gram of catalyst per hour
m	fraction of nearest neighbors below plane of surface atom
n	integer; reaction order
$N(E)$	density of states
$\bar{N}_i$	number of moles of i
N	naphthene
n	average number of carbon atoms in feed per molecule; fraction between 0 and 0.6 for water molecules in Al_2O_3; integer
O	olefin
P	pressure, F/l^2
P	paraffin, product
Q	volumetric flow rate, l^3/t
q	atomic charge
R	radial distance, l
R	particle radius, l
R	alkyl group, reactant
r_i	rate of reaction i, moles per mass of catalyst per unit of time

S	selectivity, fraction of total conversion to a given reaction product
SV	weight liquid hourly space velocity, t^{-1}
T	temperature, °C or K
T_R	temperature required for 50 percent conversion
V	volume, l^3; atomic potential
W	weight
X	fractional conversion; atomic fraction
X	halogen ligand
*	metal adsorption site

Dimensional

E = energy
F = force
l = length
m = mass
t = time
T = temperature

Greek

α	number of hydrogen atoms adsorbed at 293 K per xenon atom adsorbed at 78 K
α'	constant between 0 and 1
β	irreversible hydrogen adsorption per square centimeter of metal area at 293 K
γ	specific work required to form new surface, surface free energy, E/l^2
ΔH_{ad}	heat of adsorption, E/mol
ΔH_r	heat of reaction, E/mol
ΔH_{sub}	heat of sublimation, E/mol
ΔS°	standard entropy change of reaction, $E/\text{mol} \cdot \text{K}$
η	effectiveness factor
ρ_m	mass density, mol/l^3
τ	contact time, t
Φ	modified Thiele modulus
ϕ	Thiele modulus
Ω	regular solution parameter, heat of formation of the alloy AB from A and B, E

Subscripts

A	acid, aromatic
AA	bonding between species A
AB	bonding between species A and species B
BB	bonding between species B
C	catalyst
cr	cracking
cy	cyclization
eq	equilibrium
f	formation; Fermi; forward
i	species i
is	isomerization
M	metal
m	molar, integer
NC	naphthene cracking
n	integer
O	olefin

P paraffin
PC paraffin cracking
p pressure
r reaction; reverse
T total

Superscripts

° standard, degrees
b bulk
s surface

REFERENCES

1. Ciapetta, F. G., R. M. Dobres, and R. W. Baker, in P. H. Emmett (ed.), "Catalysis," vol. 6, Reinhold, New York, 1958.
2. Weisz, P. B., *Adv. Catal.*, **13,** 137 (1962).
3. Sinfelt, J. H., *Adv. Chem. Eng.*, **5,** 37 (1964).
4. Ciapetta, F. G., and D. N. Wallace, *Catal. Rev.*, **5,** 67 (1971).
5. Sterba, M. J., and V. Haensel, *Ind. Eng. Chem. Prod. Res. Dev.*, **15,** 2 (1976).
6. Krane, H. G., A. B. Groh, B. L. Schulman, and J. H. Sinfelt, *Proc. 5th World Pet. Cong., New York, 1959*, sec. III, p. 39.
7. Pauling, L., "The Nature of the Chemical Bond," Cornell University Press, Ithaca, N.Y., 1960.
8. Many, A., Y. Goldstein, and N. B. Grover, "Semi-conductor Surfaces," North-Holland, Amsterdam, 1965.
9. Grimley, T. B., in E. Drauglis, R. D. Gretz, and R. J. Jaffee (eds.), "Molecular Processes on Solid Surfaces," p. 299, McGraw-Hill, New York, 1968.
10. Ertl, G., and J. Koch, in F. Ricca (ed.), "Adsorption Desorption Phenomena," p. 345, Academic, New York, 1972.
11. Bond, G. C., *C. R. Sem. Etud. Catal.* (*Extr. Mem. Soc. R. Sci. Liège*), 6ᵉ ser., tome I, fasc. 4, p. 49, 1971.
12. Trapnell, B. M. W., and D. O. Hayward, "Chemisorption," Butterworth, London, 1964.
13. Sachtler, W. M. H., G. J. H. Dorgelo, and R. Jongepier, "Basic Problems in Thin Film Physics," p. 218, van der Hoek and Ruprecht, Göttingen, 1966.
14. Tanaka, K., and K. Tamaru, *J. Catal.*, **2,** 366 (1963).
15. Ugo, R., *Proc. 5th Int. Cong. Catal.*, p. B-19, North-Holland, Amsterdam, 1973.
16. Ugo, R., *Catal. Rev. Sci. Eng.*, **11,** 225 (1975).
17. Eastman, D. E., and J. K. Cashion, *Phys. Rev. Lett.*, **27,** 1520 (1971).
18. Hagstrom, H. O., and G. E. Becker, *J. Chem. Phys.*, **54,** 1015 (1971).
19. Fahrenfort, J., L. L. van Reijen, and W. M. H. Sachtler, *Z. Elektrochem.*, **64,** 216 (1960).
20. Holscher, A. A., and W. M. H. Sachtler, in E. Drauglis, R. D. Gretz, and R. J. Jaffee (eds.), "Molecular Processes on Solid Surfaces," p. 317, McGraw-Hill, New York, 1969.
21. Selwood, P. W., "Adsorption and Collective Paramagnetism," Academic, New York, 1962.
22. Benard, J., *Catal. Rev.*, **3,** 93 (1970).
23. Sabatier, P., *Ber. Dtsch. Chem. Ges.*, **44,** 2001 (1911).
23*a*. Rootsaert, W. J. M., and W. H. M. Sachtler, *Z. Phys. Chem.*, **26,** 16 (1960).
24. Boudart, M., "Kinetics of Chemical Processes," Prentice-Hall, Englewood Cliffs, N.J., 1968.
25. Balandin, A. A., *Adv. Catal.*, **10,** 120 (1958).
26. Balandin, A. A., "Surface Chemical Compounds and Their Adsorption Phenomena" (in Russian), p. 277, Moscow University Press, 1957.
27. Peri, B., *Discuss. Faraday Soc.*, **41,** 121 (1966).
28. Beeck, O., *Rev. Mod. Phys.* **17,** 61 (1945).
29. Beeck, O., *Discuss. Faraday Soc.*, **8,** 118 (1950).

30. Kouskova, A., J. Adamek, and V. Ponec, *Collect. Czech. Chem. Commun.*, **35,** 2538 (1970).
31. McCarty, J., and R. J. Madix, *J. Catal.*, **38,** 402 (1975).
32. Inglis, H. S., and D. Taylor, *Inorg. Phys. Theor. J. Chem. Soc.* **1969,** A2985.
33. Duell, J. J., and A. J. B. Robertson, *Trans. Faraday Soc.*, **57,** 1416 (1961).
34. Sinfelt, J. H., *Adv. Catal.*, **23,** 91 (1973).
35. Plank, P. van der, and W. M. H. Sachtler, *J. Catal.*, **12,** 35 (1968).
35*a*. Vecher, A. A., and J. I. Gerasimov, *Russ. J. Phys. Chem.*, **37,** 254 (1963).
35*b*. Rapp, R. A., and R. Maak, *Acta. Metall.*, **10,** 62 (1962).
36. Ponec, V., *Catal. Rev. Sci. Eng.*, **11,** 41 (1975).
37. Bouwman, R., and W. M. H. Sachtler, *J. Catal.*, **19,** 127 (1970).
38. Ollis, D. F., *J. Catal.*, **23,** 131 (1971).
39. Sinfelt, J. H., *J. Catal.*, **29,** 308 (1973).
40. Williams, F. L., and D. Nason, *Surf. Sci.* **45,** 377 (1974).
41. Sinfelt, J. H., J. L. Carter, and D. J. C. Yates, *J. Catal.*, **24,** 283 (1974).
42. Burton, J. J., E. Hyman, and D. G. Fedak, *J. Catal.*, **36,** 114 (1975).
43. Lang, N. D., and H. Ehrenreich, *Phys. Rev.*, **168,** 605 (1968).
44. Kouvel, J. S., and J. B. Comly, *Phys. Rev. Lett.*, **24,** 598 (1970).
45. Seib, D. H., and W. C. Spicer, *Phys. Rev. Lett.*, **20,** 1441 (1968).
46. Hüfner, S., G. C. Wertheim, and J. H. Wernick, *Phys. Rev.*, **B8,** 4511 (1973).
47. Sinfelt, J. H., *Adv. Catal.*, **23,** 91 (1973).
48. Ponec, V., and W. M. H. Sachtler, *J. Catal.*, **24,** 250 (1972).
49. Beelen, J. M., V. Ponec, and W. M. H. Sachtler, *J. Catal.*, **28,** 376 (1973).
50. Merta, R., and V. Ponec, *Proc. 4th Int. Cong. Catal.*, vol. 2, p. 53, Akadémiai Kiadó, Budapest, 1971.
51. Sinfelt, J. H., D. J. C. Yates, and W. F. Taylor, *J. Phys. Chem.*, **69,** 1827 (1965).
52. Visser, C., J. G. P. Zuidwijk, and V. Ponec, *J. Catal.*, **35,** 457 (1974).
53. Bouwman, R., and W. M. H. Sachtler, *J. Catal.*, **19,** 127 (1971).
54. Schaik, J. R. H. van, R. P. Dessing, and V. Ponec, *J. Catal.*, **38,** 273 (1975).
54*a*. Johansson, G. H., and I. O. Linde, *Ann. Phys.*, **1930,** 5, 762.
55. Kubaschewski, O. and J. A. Catterall, "Thermochemical Data of Alloys," Pergamon, London, 1956.
56. Stephen, J. J., V. Ponec, and W. M. H. Sachtler, *Surf. Sci.*, **47,** 403 (1975).
57. McKervey, M. A., J. J. Rooney, and N. G. Samman, *J. Catal.*, **30,** 330 (1973).
58. Gray, T. J., N. G. Masse, and H. G. Oswin, *Actes* 2me *Cong. Int. Catal.*, p. 1697, Edition Technip, Paris, 1961.
59. Barneveld, W. A. A. van, and V. Ponec, *Rec. Trav. Chim.*, **93,** 243 (1974).
60. Kittel, C., "Introduction to Solid State Physics," Wiley, New York, 1976.
61. Anderson, P. W., *Phys. Rev.*, **124,** 41 (1961).
62. Slater, J. C., and K. H. Johnson, *Phys. Today*, **27,** 34 (1974).
63. Slater, J. C., "Quantum Theory of Molecules and Solids," vol. IV, McGraw-Hill, New York, 1974.
64. Schrieffer, J. R., and P. Soven, *Phys. Today*, **28,** 24 (1975).
65. Gadzuk, J. W., in J. W. Blakeley (ed.), "Surface Physics of Crystalline Materials," vol. II, p. 339, Academic, New York, 1975.
66. Messmer, R. P., in C. A. Segal (ed.), "Modern Theoretical Chemistry," vol. 4, Plenum, New York, 1976.
67. Grimley, T. B., in F. Ricca (ed.), "Adsorption Desorption Phenomena," p. 215, Academic, New York, 1972.
68. Grimley, T. B., *Proc. Int. Sch. Phys. Enrico Fermi*, course LVIII, p. 298, Ed. Comp., Bologna, 1974.
69. Fassaert, D. J. M., Ph.D. thesis, University of Nymegen, The Netherlands, 1976.
70. Weinberg, W. H., *J. Vac. Sci. Technol.*, **10,** 89 (1973).
71. Weinberg, W. H., H. A. Deans, and R. P. Merrill, *Surf. Sci.*, **41,** 312 (1974).
72. Baetzold, R. C., *Surf. Sci.*, **36,** 123 (1973).

73. Baetzold, R. C., *J. Catal.*, **29,** 129 (1973).
74. Baetzold, R. C., *J. Chem. Phys.*, **55,** 4355, 4363 (1971).
75. Fassaert, D. J. M., H. Verbeek, and A. van der Avoird, *Surf. Sci.*, **29,** 501 (1972).
75*a*. Davenport, J. W., T. L. Einstein, and J. R. Schreiffer, *Proc. 2d Int. Conf. Solid Surf., Kyoto, 1974.*
76. Fassaert, D. J. M., and A. van der Avoird, *Surf. Sci.*, **55,** 291 (1976).
77. Deuss, H., and A. van der Avoird, *Phys. Rev.*, **B8,** 2441 (1973).
78. Tamm, J., *J. Phys.*, **76,** 843 (1932).
79. Shockley, W., *Phys. Rev.*, **5b,** 317 (1939).
80. Grimley, T. B., *Adv. Catal.*, **12,** 1 (1960).
81. Paulson, R. H., and J. R. Schrieffer, *Surf. Sci.*, **48,** 329 (1975).
82. Grimley, T. B., *Proc. Phys. Soc.*, **90,** 751 (1967).
82*a*. Newns, D. M., *Phys. Rev.*, **178,** 1123 (1969).
82*b*. Gadzuk, J. W., *Surf. Sci.*, **43,** 44 (1974).
83. Ciapetta, F. G., and C. J. Plank, in P. H. Emmett (ed.), "Catalysis," vol. I, p. 315, Reinhold, New York, 1954.
84. Gil'debrand, E. I., *Int. Chem. Eng.*, **6,** 449 (1966).
85. Marisic, M. M. (ed.), *Ind. Eng. Chem.*, **49,** 240 (1957).
86. Morikawa, K., T. Shirasaki, and M. Okada, *Adv. Catal.*, **20,** 98 (1967).
87. Harriott, P., *J. Catal.*, **14,** 43 (1969).
88. Tauster, S. J., *J. Catal.*, **18,** 358 (1970).
89. Furuoya, I., T. Yanagihara, and T. Shiraski, *Int. Chem. Eng.*, **10,** 333 (1970).
90. Benesi, H. A., R. M. Curtis, and H. P. Studer, *J. Catal.*, **10,** 328 (1968).
91. Wilson, G. R., and W. K. Hall, *J. Catal.*, **24,** 306 (1972).
92. Dorling, T. A., B. W. J. Lynch, and R. L. Moss, *J. Catal.*, **20,** 190 (1971).
93. Dalla Betta, R. A., and M. Boudart, *Proc. 5th Int. Cong. Catal.*, p. 1329, North-Holland, Amsterdam, 1973.
94. Freel, J., *J. Catal.*, **25,** 149 (1972).
95. Boudart, M., A. W. Aldag, L. D. Ptak, and J. E. Benson, *J. Catal.*, **11,** 35 (1968).
96. Spenadel, L., and M. Boudart, *J. Phys. Chem.*, **64,** 204 (1960).
97. Adams, C. R., H. A. Benesi, R. M. Curtis, and R. G. Meisenheimer, *J. Catal.*, **1,** 336 (1962).
98. Moss, R. L., *Platinum Met. Rev.*, **11,** 1 (1967).
99. Wilson, G. R., and W. K. Hall, *J. Catal.*, **17,** 190 (1970).
100. Dorling, T. A., and R. L. Moss, *J. Catal.*, **7,** 378 (1967).
101. Klug, H. P., and L. E. Alexander, "X-ray Diffraction Procedures," Wiley, New York, 1954.
102. Whyte, T. E., Jr., P. W. Kirklin, R. W. Gould, and H. Heinemann, paper presented at the *Spring Symp. Philadelphia Catal. Club*, May 8, 1972.
103. Bramberger, H., "Small Angle X-ray Scattering," p. 450, Gordon and Breach, New York, 1967.
104. Gregg, S. J., and K. S. W. Sing, "Adsorption, Surface Area and Porosity," Academic, New York, 1967.
105. Dorling, T. A., C. J. Burlace, and R. L. Moss, *J. Catal.*, **12,** 207 (1968).
106. Ostermaier, J. J., J. R. Katzer, and W. H. Manogue, *J. Catal.*, **33,** 457 (1974).
107. Dorling, T. A., and R. L. Moss, *J. Catal.*, **7,** 378 (1967).
108. Lyon, H. B., and G. A. Somorjai, *J. Chem. Phys.*, **46,** 2538 (1967).
109. McLean, M., and H. Kykura, *Surf. Sci.*, **5,** 466 (1966).
110. Cormack, D., and R. L. Moss, *J. Catal.*, **13,** 1 (1969).
111. Brennan, D., and F. H. Hayes, *Phil. Trans. R. Soc. (Lond.)*, **258,** 347 (1965).
112. Taylor, H. S., *Proc. R. Soc.*, **A108,** 105 (1925).
113. Boudart, M., *Adv. Catal.*, **20,** 153 (1969).
114. Hoare, M. R., and R. Pal, *J. Cryst. Growth*, **17,** 77 (1972).
115. Allpress, J. G., and J. V. Sanders, *Aust. J. Phys.*, **23,** 23 (1970).
116. Burton, J. J., *Catal. Rev. Sci. Eng.*, **9,** 209 (1974).
117. Pashley, D. W., *Adv. Phys.*, **5,** 173 (1956); **14** 327 (1965).
118. Boswell, F. W. C., *Proc. Phys. Soc.*, **A64,** 465 (1951).
119. Schroeer, D., and R. C. Nininger, *Phys. Rev. Lett.*, **19,** 632 (1967).

120. May, C. W., J. S. Vermaak, and D. Kuhlmann-Wilsdorf, *Surf. Sci.*, **12,** 134 (1968).
121. Hardeveld, R. van, and F. Hartog, *Surf. Sci.* **15,** 189 (1969).
122. Poltorak, O. M., and V. S. Boronin, *Russ. J. Phys. Chem.*, **40,** 1436 (1966).
123. Hardeveld, R. van, and A. van Montfoort, *Surf. Sci.*, **4,** 396 (1966).
124. Hardeveld, R. van, and F. Hartog, *Proc. 4th Int. Cong. Catal.*, vol. 2, p. 295, Akadémiai Kiadó, Budapest, 1971.
125. Somorjai, G., *Catal. Rev.*, **7,** 87 (1973).
126. Boudart, M., A. Aldag, J. E. Benson, N. A. Dougharty, and C. Harkins-Given, *J. Catal.*, **6,** 92 (1966).
127. Poltorak, O. M., V. S. Boronin, and A. N. Mitrofanova, *Proc. 4th Int. Cong. Catal.*, vol. 2, p. 276, Akadémiai Kiadó, Budapest, 1971.
128. Aben, D. C., J. C. Platteeuw, and B. Stouthamer, *Rec. Trav. Chim.*, **89,** 449 (1970).
129. Dixon, G. M., and K. Singh, *Trans. Faraday Soc.*, **64,** 1128 (1968).
130. Nikolajenko, V., V. Bosacek, and V. Danes, *J. Catal.*, **2,** 127 (1963).
131. Taylor, W. F., and H. K. Staffin, *Trans. Faraday Soc.*, **63,** 2309 (1967).
132. Cusumano, J. A., G. W. Dembinski, and J. H. Sinfelt, *J. Catal.*, **5,** 471 (1966).
133. Kraft, M., and H. Spindler, *Proc. 4th Int. Cong. Catal.*, vol. 2, p. 286, Akadémiai Kiadó, Budapest, 1971.
134. Hughes, T. R., R. J. Houston, and R. P. Sieg, *Ind. Eng. Chem. Process Des. Dev.*, **1,** 96 (1962).
135. Dautzenberg, F. M., and J. C. Platteeuw, *J. Catal.*, **24,** 364 (1972).
136. Dautzenberg, F. M., and J. C. Platteeuw, *J. Catal.*, **19,** 41 (1970).
137. Benson, J. E., and M. Boudart, *J. Catal.*, **4,** 704 (1965).
137*a*. Coenen, J. W. E., R. Z. C. van Meerten, and H. T. Rijnten, *Proc. 5th Int. Cong. Catal.*, p. 671, North-Holland, Amsterdam, 1973.
138. Taylor, W. F., J. H. Sinfelt, and D. J. C. Yates, *J. Phys. Chem.*, **69,** 3857 (1965).
139. Carter, J. L., J. A. Cusumano, and J. H. Sinfelt, *J. Phys. Chem.*, **70,** 2257 (1966).
140. Corolleur, G., F. G. Gault, D. Juttard, G. Maire, and J. M. Muller, *J. Catal.*, **27,** 466 (1972).
141. Krug, S., M.Ch.E. thesis, University of Delaware, 1975.
141*a*. Joyner, R. W., B. Lang, and G. A. Somorjai, *J. Catal.*, **27,** 405 (1972).
142. Maxted, E. B., and A. G. Walker, *J. Chem. Soc.*, **1093,** (1948).
143. Dilke, M. H., D. D. Eley, and E. B. Maxted, *Nature*, **161,** 804 (1948).
144. Reijen, L. L. van, thesis, Technical University of Eindhoven, The Netherlands, 1964.
145. Lippens, B. C., Ph.D. thesis, Technical University of Delft, The Netherlands (1961).
146. Peri, J., *J. Phys. Chem.*, **69,** 211, 220, 231 (1965).
147. Peri, J., *J. Catal.*, **41,** 227 (1976).
148. Bonhoeffer, K. F., and A. Farkas, *Z. Phys. Chem.*, **B12,** 231 (1931).
149. Sinfelt, J. H., H. Hurwitz, and R. A. Shulman, *J. Phys. Chem.*, **64,** 1559 (1960).
150. Madden, W. F., and C. Kemball, *J. Chem. Soc.*, **1961,** 302.
151. Anderson, J. R., and C. Kemball, *Proc. R. Soc.*, **A226,** 472 (1954).
152. Bond, G. C., *C. R. Sem. Etud. Catal.* (*Extr. Mem. Soc. R. Sci. Liege*), 6ᵉ ser., tome I, fasc. 4, p. 61, 1971.
153. Demuth, J. E., and D. Eastman, *Phys. Rev. Lett.*, **32,** 1123 (1974); *Jap. J. Appl. Phys. Suppl. 2*, pt. 2, **1974,** 827.
154. Turkevich, J., F. Bonner, D. O. Schissler, and A. P. Irsa, *Discuss. Faraday Soc.*, **8,** 352 (1950).
155. Hartog, F., and P. Zwietering, *J. Catal.*, **2,** 79 (1963).
156. Eischens, R. P., and W. A. Pliskin, *Adv. Catal.*, **10,** 1 (1958).
157. Kemball, C., *Adv. Catal.*, **9,** 223 (1959).
158. Kemball, C., *Trans. Faraday Soc.*, **50,** 1344 (1954).
159. Kemball, C., *Proc. R. Soc.*, **A207,** 539 (1951).
160. Kemball, C., *Proc. R. Soc.*, **A217,** 376 (1953).
161. Beeck, O., *Adv. Catal.*, **2,** 151 (1950).
162. Ragaini, V., *J. Catal.*, **34,** 1 (1974).
163. Germain, J. E., "Catalytic Conversion of Hydrocarbons," Academic, New York, 1969.
164. Galwey, A. K., and C. Kemball, *Trans. Faraday Soc.*, **55,** 1959 (1959).

165. Keulemans, A. I. M., and H. H. Voge, *J. Phys. Chem.*, **63,** 476 (1959).
166. Maurel, R., and J. E. Germain, *J. Am. Chem. Soc.*, **77,** 2819 (1955).
167. Pines, H., E. F. Jenkins, and V. N. Ipatieff, *J. Am. Chem. Soc.*, **75,** 6226 (1953).
168. Pines, H., and J. Marcechal, *J. Am. Chem. Soc.*, **77,** 2819 (1955).
169. Zelinskii, N. D., *Ber. Dtsch. Chem. Ges.*, **44,** 3121 (1911); **45,** 3678 (1912).
170. Muller, J. M., and F. G. Gault, *J. Catal.*, **24,** 361 (1972).
171. Anderson, J. R., R. J. MacDonald, and Y. Shimoyama, *J. Catal.*, **20,** 147 (1971).
172. Anderson, J. R., *Adv. Catal.*, **23,** 1 (1973).
173. Tomanova, D., C. Corolleur, and F. G. Gault, *C. R. Acad. Sci.*, **C269,** 1605 (1969).
174. Corolleur, C., S. Corolleur, D. Tomanova, and F. G. Gault, *J. Catal.*, **24,** 385, 401 (1972).
175. Dartigues, J. M., A. Chambellan, and F. G. Gault, *J. Am. Chem. Soc.*, **98,** 856 (1976).
176. O'Cinneide, A., and F. G. Gault, *J. Catal.*, **37,** 311 (1975).
177. Barron, Y., G. Mairr, J. M. Mullher, and F. G. Gault, *J. Catal.*, **5,** 428 (1966).
178. Anderson, J. R., and N. R. Avery, *J. Catal.*, **5,** 446 (1966).
179. Anderson, J. R., and B. G. Baker, *Proc. R. Soc.*, **A271,** 402 (1963).
180. Anderson, J. R., and N. R. Avery, *J. Catal.*, **2,** 542 (1963).
181. Anderson, J. R. (ed.), "Chemisorption and Reactions on Metallic Films," vol. 2, Academic, New York, 1971.
182. Maire, G., G. Plouidy, J. C. Prudhomme, and F. G. Gault, *J. Catal.*, **4,** 556 (1965).
183. Davis, B. H., and P. B. Venuto, *J. Catal.*, **15,** 363 (1969).
184. Lam, Y., and J. Sinfelt, *J. Catal.*, **42,** 319 (1976).
185. Roberti, A., V. Ponec, and W. M. H. Sachtler, *J. Catal.*, **28,** 381 (1973).
186. Ponec, V., and W. M. H. Sachtler, *Proc. 5th Int. Cong. Catal.*, p. 645, North-Holland, Amsterdam, 1973.
187. Goble, A. G., and P. A. Lawrence, *Proc. 3d Int. Cong. Catal.*, p. 320, North-Holland, Amsterdam, 1965.
188. Kouwenhoven, H. W., *Adv. Chem. Ser.*, **121,** 529 (1973).
189. Pines, H., and R. C. Wackher, *J. Am. Chem. Soc.*, **68,** 595 (1946).
190. Callender, W. L., S. G. Brandenberger, and W. K. Meerbott, *Proc. 5th Int. Cong. Catal.*, p. 1265, North-Holland, Amsterdam, 1973.
191. Brandenberger, S. G., W. L. Callender, and W. K. Meerbott, *J. Catal.*, **42,** 282 (1976).
192. Badische Anilin und Sodafabrik, German patent, 1,167,824 (1964) [*Chem. Abstr.*, **61,** 1776 (1964)].
193. Weisz, P. B., and E. W. Swegler, *Science*, **126,** 31 (1957).
194. Hettinger, W. P., C. D. Kieth, J. L. Gring, and J. W. Teter, *Ind. Eng. Chem.*, **47,** 719 (1955).
195. Mills, G. A., H. Heinemann, T. H. Milliken, and A. G. Oblad, *Ind. Eng. Chem.*, **45,** 134 (1953).
196. Sinfelt, J. H., H. Hurwitz, and J. C. Rohrer, *J. Phys. Chem.*, **64,** 892 (1960).
197. Sinfelt, J. H., H. Hurwitz, and J. C. Rohrer, *J. Catal.*, **1,** 481 (1962).
198. Starnes, W. C., and R. C. Zabor, *Symp. Div. Petrol. Chem. Am. Chem. Soc., Boston, Mass., April 5–10, 1959*: see also Ref. 3.
199. Maslyanskii, G. N., N. R. Bursian, S. A. Barkan, *Zhr. Prikl. Khim.*, **39**(3), 650 (1966).
200. Silvestri, A. J., P. A. Naro, and R. L. Smith, *J. Catal.*, **14,** 386 (1969).
201. Pollitzer, E. L., J. C. Hayes, and V. Haensel, paper presented at *Am. Chem. Soc. Refining Petrol. Chem. Symp., New York, Sept. 7–12, 1969.*
202. Weisz, P. B., *Actes 2me Cong. Int. Catal.*, p. 937, Edition Technip, Paris, 1961.
203. Rohrer, J. C., H. Hurwitz, and J. H. Sinfelt, *J. Phys. Chem.*, **65,** 1458 (1961).
204. Sinfelt, J. H., and J. C. Rohrer, *J. Phys. Chem.*, **65,** 978 (1961).
205. Myers, C. G., W. H. Lang, and P. B. Weisz, *Ind. Eng. Chem.*, **53,** 299 (1961).
206. Kouwenhoven, H. W., and W. C. van Zijll Langhout, *Chem. Eng. Prog.*, **67**(4), 65 (1971).
207. Chick, D. J., J. R. Katzer, and B. C. Gates, p. 515 in Molecular Seives: II, *Am. Chem. Soc. Symp. Ser* 40, 1977.
208. Schlatter, J. C., and M. Boudart, *J. Catal.*, **25,** 93 (1972).
209. Benson, J. E., H. W. Kohn, and M. Boudart, *J. Catal.*, **5,** 307 (1966).
210. Boudart, M., A. W. Aldag, and M. A. Vannice, *J. Catal.*, **18,** 46 (1970).

211. Neikam, W. C., and M. A. Vannice, *Proc. 5th Int. Cong. Catal.*, p. 609, North-Holland, Amsterdam, 1973.
212. Vannice, M. A., and W. C. Neikam, *J. Catal.*, **20,** 260 (1971).
213. Weisz, P. B., and C. D. Prater, *Adv. Catal.*, **9,** 583 (1957).
214. Kluksdahl, H. E., U.S. Patent 3,415,737 (1968).
215. Jacobson, R. L., H. E. Kluksdahl, C. S. McCoy, and R. W. Davis, *Proc. Am. Petrol. Inst.*, **1969,** 504.
216. Anon., *Chem. Eng. News*, July 3, 1972, p. 18.
217. Cecil, R. R., W. S. Kmak, J. H. Sinfelt, and L. W. Chambers, *Oil Gas J.*, **70**(32), 50 (1972).
218. Johnson, M. F. L., *J. Catal.*, **39,** 487 (1975).
219. Webb, N. W., *J. Catal.*, **39,** 485 (1975).
220. Anon., *Oil Gas J.*, **64**(33), 87 (1966).
221. Socony Mobil Oil Co., DAS Patent 1,144,862 (1957).
222. Svajgl, O., *Int. Chem. Eng.*, **12,** 55 (1972).
223. Daniels, L. J., P. Sperling, and A. G. Rouquier, *Oil Gas J.*, May 8, 1972, p. 78.
224. Smith, R. B., *Chem. Eng. Prog.* **55**(6), 76 (1959).
225. Zhorov, Yu. M., G. M. Panchenkov, and S. P. Zel'tser, *Kinet. Katal.*, **6,** 986 (1965).
226. Thomas, W. J., *Trans. Inst. Chem. Eng.*, **49,** 204 (1971).
227. Jenkins, B. C., and W. J. Thomas, *Can. J. Chem. Eng.*, **48,** 179 (1970).
228. Al-Samadi, R. A., P. R. Luckett, and W. J. Thomas, *Adv. Chem. Ser.*, **133,** 316 (1974).
229. Satterfield, C. N., "Mass Transfer in Heterogeneous Catalysis," MIT Press, Cambridge, Mass., 1970.
230. Rossini, F. D., K. S. Pitzer, R. L. Arnett, R. M. Braun, and G. C. Pimintel, "Selected Values of Physical and Thermodynamic Properties of Hydrocarbons and Related Compounds," American Petroleum Institute Project 44, Carnegie Press, Pittsburgh, 1953.

PROBLEMS

3-1 It has been suggested that the catalytic activities of a series of metals are well represented by a correlation with the stability of a surface compound (or the most stable surface intermediate) whereas correlations with percentage *d* character are lacking fundamental validity. Test this suggestion by showing (*a*) whether the heat of sublimation and/or the heat of formation of the most stable oxide correlates well with the percentage *d* character of the transition metals and (*b*) whether the rate of formic acid decomposition correlates well with the percentage *d* character.

3-2 (*a*) If a reforming catalyst consists of 12-Å crystallites on an alumina support having a surface area of 200 m^2/g, what fraction of the total surface is metal?

(*b*) If the metal crystallites are uniformly spaced on the alumina surface, what is the average distance between them?

3-3 Reforming takes place in a reactor with an inlet temperature of 482°C and a pressure of 34 atm. The feed is mostly hydrogen. The outlet temperature is 454°C. With a new catalyst, the reactor is proposed to operate with an inlet temperature of 510°C at a total pressure of 13.6 atm (with the hydrogen partial pressure equal to 10.2 atm). Consider the reactant hydrocarbon to be composed of four classes of compounds (aromatics, cycloparaffins, isoparaffins, and normal paraffins) and predict the effect of the change of conditions on the equilibrium, the reaction kinetics, and the temperature profile in the reactor bed.

3-4 When a reforming catalyst prepared by ion exchange is calcined, the metal complex is decomposed and the metal ion probably remains atomically dispersed, experiencing strong coulombic interactions with the support surface. It does not migrate to form clusters. After reduction, the zero-valent metal atom experiences only weak interactions of a van der Waals nature (about 3 kcal/mol) with the support surface. Since there is a large energy change associated with the formation of metal dimers and trimers (about 40 kcal/mol or more), a large driving force exists for cluster formation, which occurs if surface diffusion is sufficiently rapid. Consider the surface diffusion of a zero-valent metal atom having an activation energy of about 3 kcal/mol. Estimate the surface-diffusion coefficient at 700 K and the root-mean-square migration distance in 10 min. What do the estimates suggest about agglomeration or crystallite formation?

3-5 Weisz [3] studied the conversion of *n*-hexane into isohexanes using the mixed-catalyst technique with Pt/SiO_2 and silica-alumina, with the following results:

n-Hexane isomerization in the presence of coarse catalyst mixtures

Reaction conditions: 373°C, 1 atm total pressure, 17.2 g of *n*-hexane/h, H_2 to *n*-hexane molar ratio 5.0

Catalyst charged to the reactor	Conversion to isohexanes, %
10 cm^3 of Pt/SiO_2	0.9
10 cm^3 of SiO_2–Al_2O_3	0.3
Mixture of 10 cm^3 Pt/SiO_2 and 10 cm^3 of SiO_2–Al_2O_3	6.8

The maximum (equilibrium) conversion of *n*-hexane into hexene is 0.04 percent if only straight-chain olefin is formed and 0.6 percent if dehydrogenation occurs and all olefin isomers are formed. If the average particle diameter is 0.6 mm, and if the diffusivity is determined to be 2×10^{-3} cm^2/s in the catalyst particles, determine to what extent diffusion may be limiting reaction.

3-6 Sinfelt, Carter, and Yates [41] measured the specific activity of Cu–Ni alloy catalysts for the hydrogenolysis of ethane to give methane (see Fig. 3-27). Sinfelt [*J. Catal.*, **27,** 468 (1972)] has shown that the rate-determining step in hydrogenolysis of ethane involves scission of the carbon-carbon bond in an adsorbed intermediate. Hydrogenolysis probably involves only Ni; Cu is essentially inactive.

(*a*) Assuming that the metal forms a homogeneous alloy with a difference in the heat of sublimation of $10kT$, calculate and plot the surface composition as a function of the bulk composition.

(*b*) Develop a model representing the rate of ethane hydrogenolysis on a Cu–Ni alloy surface as a function of alloy composition and determine the number of Ni atoms which form an active site for this reaction.

3-7 Ethylene streams from naphtha crackers contain low concentrations of acetylene, which must be removed before the ethylene can be used. This removal is typically carried out by hydrogenation of the acetylene with a metal catalyst in the presence of the ethylene. Selectivity in acetylene hydrogenation is defined by

$$S = \frac{r_{C_2H_2}}{r_{C_2H_2} + r_{C_2H_4}}$$

where r_i is the rate of hydrogenation of species *i*, given in Table 3-10.

(*a*) If it is assumed that the rates of hydrogenation of adsorbed acetylene and of adsorbed ethylene are the same and that there is a fourfold excess of hydrogen to hydrocarbon and an ethylene-to-acetylene ratio of 20, what must the difference in ΔH_{ad} between ethylene and acetylene be to account for the observed selectivity differences between Pd, Pt, and Ir? State your assumptions.

(*b*) The selectivity for hydrogenating acetylene in an acetylene-ethylene mixture can be significantly improved by the addition of small amounts of sulfur to the feed. Explain.

3-8 Ethylene deuteration (conversion with D_2) catalyzed by Pt and by Ru gives the following product distributions:

Observed distributions of deuterated products from reaction of C_2H_4 with D_2, mole fraction [152]

Product	Conditions	
	Pt at 102°C	Ru at 32°C
$C_2H_3D_3$	7.5	2.1
$C_2H_4D_2$	26.8	41.6
C_2H_5D	24.3	10.4
C_2H_6	9.5	2.2
$C_2H_2D_2$	4.5	0.9
C_2H_3D	19.6	42.8

What conclusions can you draw about olefin interactions with the surface of the metal? What would you propose as a plausible reaction mechanism to explain the data?

3-9 Refer to the model of reforming with a mechanical mixture of the two kinds of catalyst particles (Figs. 3-59 and 3-60) and to the diffusion-reaction analysis for the isomerization of a paraffin [reaction (52)].

(*a*) Derive the expression for the Thiele modulus for the acidic particles.

(*b*) Derive Eq. (65).

(*c*) From the results for the smallest particles in Fig. 3-59, predict the conversion expected for catalyst particles having diameters of 70, 500, 1000, and 5000 μm. Compare the results with those of Fig. 3-59.

3-10 Ethylene hydrogenation on a metal surface has been postulated to proceed via the following sequence of steps:

$$2* + C_2X_4(g) \underset{k_2}{\overset{k_1}{\rightleftharpoons}} \underset{*\quad *}{X_2C{-}CX_2} \underset{k_4(-X)}{\overset{k_3(+X)}{\rightleftharpoons}} \underset{*}{X_2C{-}CX_3} \xrightarrow{k_5(+X)} C_2X_6(g)$$

where X represents H or D. Devise an analysis to obtain ratios of the rate constants from isotopic distributions in the products of ethylene deuteration. The analysis requires assigning probabilities that X is D in steps 3, 4, and 5. Using the data given in Prob. 3-8, determine the ratios k_3/k_2 and k_5/k_4.

3-11 Figure P3-11 shows the fractional surface coverage of Ni foil by S as a function of the ratio of partial pressures of H_2S/H_2 and the temperature [M. Perderau and J. Oudor, *Surf. Sci.*, **20**, 80 (1970)].

(*a*) Estimate the standard free energy of formation of surface NiS at 700 and 900°C.

(*b*) Are the data thermodynamically consistent?

(*c*) Rosenquist [*J. Iron Steel Inst.*, **176**, 37 (1954)] reported for formation of bulk Ni_3S_2

$$\tfrac{3}{2}Ni + H_2S \rightleftharpoons \tfrac{1}{2}Ni_3S_2 + H_2 \qquad \Delta G_f^\circ = -18{,}040 + 7.70T$$

where the standard free energy of formation, ΔG_f°, is in calories and T is in Kelvins. How does the bulk sulfide stability compare with that of the surface sulfide? Can you explain the differences?

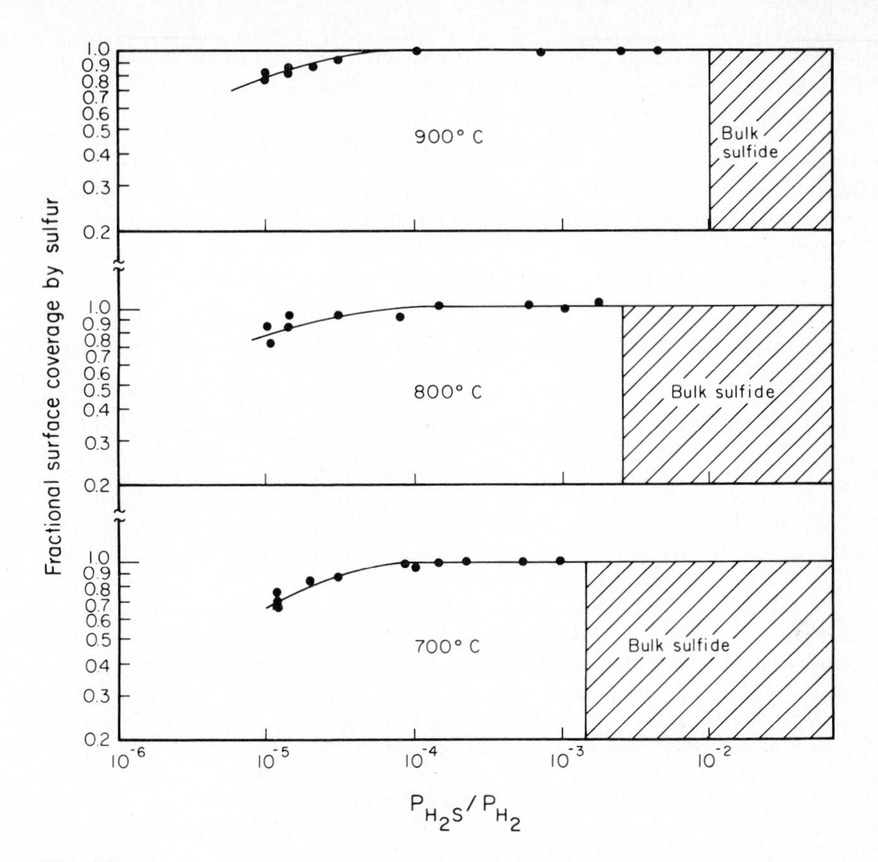

Figure P3-11

3-12 There are no data for Pt similar to those of Prob. 3-11 for Ni. The following rough estimate, however, may be helpful, giving the standard Gibbs free energy of formation of bulk PtS:

$$2\text{Pt} + \text{S}_2(g) \rightleftharpoons 2\text{PtS} \qquad \Delta G_f^\circ = -68{,}271 + 44.53T$$

for

$$2\text{H}_2 + \text{S}_2(g) \rightleftharpoons 2\text{H}_2\text{S}$$

Rosenquist [*J. Iron and Steel Institute*, **176,** 37 (1954)] reported that

$$\Delta G_f^\circ = -43{,}160 + 23.16T$$

The dimensions are as given in Prob. 3-11. The value of ΔS_f° can be assumed to be -25 cal/mol K for a reaction involving solid Pt, H_2S, and H_2.

(*a*) Assuming that the free energy of formation of surface PtS is 10,000 cal/mol less than that of the bulk, plot the surface coverage of Pt by S at 450°C in the presence of a 10 : 1 hydrogen-to-hydrocarbon ratio as a function of feed S concentration, assuming that all S is converted into H_2S and that S adsorption is not affected by the hydrocarbons present. At what feed S concentration do you expect severe S poisoning (monolayer coverage)? At what concentration do you expect minimal poisoning (less than 10 percent of a monolayer)?

(*b*) Assuming a liquid hourly space velocity of 2 h^{-1} and a 0.6 weight percent Pt/Al_2O_3 catalyst

(70 percent dispersion) operating with a 10 : 1 hydrogen-to-hydrocarbon ratio at 450°C, estimate the time required to poison all the catalyst to the equilibrium value with each of the H_2S concentrations of part (*a*).

(*c*) Interpret the results in view of the older industrial specification of 1 to 2 ppm S in the feed to a catalytic reformer and the newer specification of < 0.4 ppm S.

3-13 Equations (49) and (50) apply when the metal surface area is sufficient to assure an equilibrium concentration of olefin. Develop a more general analysis to apply when (*a*) the rate of dehydrogenation on the metal surface is rate-determining, when (*b*) the reactions on both the metal and the acid functions are kinetically important, and (*c*) when the reaction on the acid function is rate-determining.

3-14 Weisz [2] has classified bifunctional catalysis as "trivial" or "nontrivial." In the former case, the first reaction in a sequence, for example, $A \xrightarrow{k_1} B$, can proceed to a high conversion, as can the second, for example, $B \xrightarrow{k_3} C$. In the latter case, equilibrium severely limits the conversion of the first reaction, for example, $A \underset{k_2}{\overset{k_1}{\rightleftharpoons}} B$, where $K = k_1/k_2 \ll 1$. Demonstrate the assertion that if the two catalyst functions are intimately mixed, high conversions of A to C can be achieved for a nontrivial reaction whereas the intimate mixing provides no advantage in the other case. Derive an expression giving the concentration of B and the overall rate of reaction as a function of K, k_1, k_2, and C_A, assuming the reactions are first order. Plot rate as a function of k_2 (for k_2 between 0.01 and 1) with $K = 0.0001$, 0.001, 0.01, 0.1, and 1.

3-15 For the four-reactor reforming system (Fig. 3-73 and Table 3-20) calculate the concentration profiles of aromatics, naphthenes, paraffins, and light paraffins as a function of fractional length through the first reactor. Use the temperature profile of Fig. 3-74. State your assumptions.

3-16 The main advantage of the new multimetallic reforming catalysts appears to be their improved dehydrocyclization selectivities (Fig. 3-66). To demonstrate this assertion, assume that catalyst improvements lead to a 50 percent increase in the dehydrocyclization rate constant, that is, k_1 (Table 3-19) is increased to 1.89 h^{-1}, and to a 50 percent reduction in the hydrocracking rate constant, that is, $k_3 = 0.31\ h^{-1}$ and $k_5 = 0.36\ h^{-1}$. Calculate the concentrations of paraffins, naphthenes, and aromatics as a function of inverse space velocity using the rate constants of Table 3-19 with the improved values of k_1, k_3, and k_5 and compare the results with those of Fig. 3-72.

3-17 Gaseous hydrogen molecules are not measurably chemisorbed on gold at temperatures less than 200°C, but when formic acid decomposes on gold, the gold surface becomes partly covered by hydrogen atoms. Consequently, when a mixture of $HCOOH + D_2$ is contacted with a gold catalyst, only H is chemisorbed whereas D is not chemisorbed. The following experimental results were obtained [W. M. H. Sachtler and J. H. de Boer, *J. Phys. Chem.*, **64**, 1579 (1960)] using gold in powder form and dispersed on a support.

Hydrogen exchange at 150°C:

$$H_2 + D_2 \longrightarrow \text{no formation of HD} \tag{1}$$

$$HCOOH + D_2 \longrightarrow \text{only } H_2 \text{ and } CO_2 \text{ formed; HD not formed} \tag{2}$$

$$HCOOH + DCOOD \longrightarrow H_2,\ D_2, \text{ and HD formed in an equilibrium distribution} \tag{3}$$

$$HCOOD \longrightarrow H_2,\ D_2, \text{ and HD formed in an equilibrium distribution} \tag{4}$$

The rate of HCOOH (or DCOOH) decomposition in (3) and (4) was 6.4×10^{-3} molecules per site per second. The equilibrium constant $K = [HD]^2/[H_2][D_2]$ had a value of 3.5 ± 0.1 for all experiments when mixtures of HCOOH + HCOOD + DCOOD were decomposed. The theoretical equilibrium constant, as given by Farkas [A. Farkas, "Orthohydrogen, Parahydrogen, and Heavy Hydrogen," Cambridge University Press, London, 1935.] is $K = 3.52$.

(*a*) What can you infer about the mechanism of the reaction?

The results of a second set of experiments for hydrogen oxidation at 120°C are

$$2H_2 + O_2 \longrightarrow \text{very slow formation of } H_2O \tag{5}$$

$$2HCOOH + O_2 \longrightarrow \text{fast formation of } H_2O \tag{6}$$

$$2HCOOH + O_2 + \text{excess } D_2 \longrightarrow \text{fast formation of } H_2\text{, containing only traces of DOH or } D_2O\text{; no } H_2 \text{ present in the gas phase} \tag{7}$$

(*b*) What do these data imply about the mechanism of the reaction in the presence of molecular oxygen? What can you conclude about the interaction of the reactants dioxygen, dihydrogen, and formic acid with gold under these conditions?

3-18 Considering the results of Prob. 3-17, derive a rate expression for the reaction of formic acid on gold by postulating a sequence of elementary steps consistent with the observations. Assume that the step

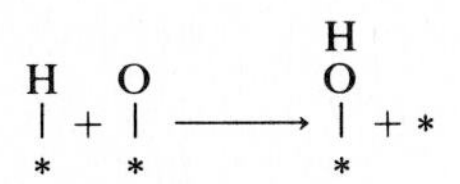

is rate-limiting, where * is a surface site.

3-19 Ruthenium and copper are almost completely immiscible in the bulk since they have different bulk crystal structures. These metals have been used in a study of ethane hydrogenolysis. The catalysts were bulk bimetallic powders which had a dispersion of about 1 percent (> 1000-Å particles) and supported bimetallic particles which had dispersions > 80 percent. The results for both sets of catalysts are shown schematically in Fig. P3-19.

(*a*) Explain the behavior observed for the bimetallic powders and the highly dispersed supported clusters in terms of the phases present (phase separation), phase compositions, surface compositions, and effects of dispersion (metal particle size) as indicated by both the hydrogen-chemisorption and hydrogenolysis-activity data.

(*b*) Develop a correlation between catalytic activity and the amount of hydrogen chemisorbed. What information about the number of metal atoms constituting an active site or the nature of an active site can be obtained from this correlation?

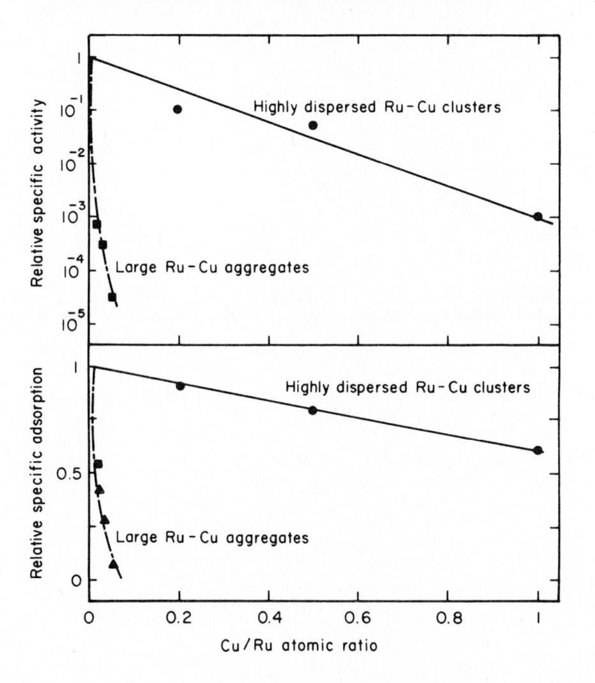

Figure P3-19 (Data from Sinfelt et al., *J. Catal.*, **42**, 227 (1976). Reprinted with permission from Academic Press.)

3-20 In the catalytic hydrogenolysis of ethane on metals, ethane is chemisorbed with breaking of carbon-hydrogen bonds, and the resulting hydrogen-deficient species undergo carbon-carbon bond rupture. Monocarbon species are then rapidly hydrogenated off the surface. Assume that the reaction proceeds via the following steps:

$$C_2H_6 \rightleftharpoons C_2H_{5ad} + H_{ad} \tag{1}$$

$$C_2H_{5ad} + H_{ad} \rightleftharpoons C_2H_{x,\,ad} + aH_2 \tag{2}$$

$$C_2H_{x,\,ad} \longrightarrow \text{adsorbed } C_1 \text{ fragments} \tag{3}$$

where a is equal to $(6 - x)/2$. If equilibrium is established in steps (1) and (2) and if $C_2H_{x,\,ad}$ is the only surface species present in significant concentration, derive a rate expression for ethane hydrogenolysis. The following table summarizes the effects of ethane and of H_2 partial pressure on the rate of hydrogenolysis:

		Reaction order†	
Catalyst	Temp., °C	n for ethane	m for H_2
Co/SiO_2	219	1.0	−0.8
Ni/SiO_2	177	1.0	−2.4
Rh/SiO_2	214	0.8	−2.2
Ir/SiO_2	210	0.7	−1.6
Pt/SiO_2	357	0.9	−2.5

† Exponents are for the power-law rate expression, $r = kp^n_{C_2H_6}p^m_{H_2}$ [J. H. Sinfelt, *J. Catal.*, **27**, 468 (1972)].

Interpret these results in terms of the chemical nature of the predominant adsorbed species $C_2H_{x,\,ad}$.

CHAPTER
FOUR

PARTIAL OXIDATION OF HYDROCARBONS: THE AMMOXIDATION OF PROPYLENE

INTRODUCTION

Partial-oxidation reactions catalyzed by surfaces of transition-metal oxides are some of the most widely applied processes for converting hydrocarbons into valuable chemical intermediates, such as those incorporating the groups

$$-\text{CHO} \qquad -\text{COOH} \qquad -\overset{\displaystyle O}{\text{C}\!-\!\text{C}}- \qquad \text{and} \qquad -\text{CN}$$

Similarly, partial-oxidation reactions catalyzed by metal-containing enzymes such as cytochromes are involved in the sequences of reactions responsible for metabolism and efficient energy conversion in living organisms.

A catalyst for a partial-oxidation process is designed to provide a limited amount of oxygen to a reactant, allowing formation of the desired product but restricting further oxidation that would give CO and CO_2. The industrial catalysts which are successful in this way are usually complex solid oxides, the surfaces of which donate oxygen to adsorbed hydrocarbon reactants. Catalyst selectivity, measured by the rate of formation of desired products relative to CO, CO_2, and other undesired side products, is often sensitive to reaction conditions, especially temperature; reactor design is difficult since the reactions are strongly exothermic and reactor temperature profiles are difficult to predict.

Reactor operation can be illustrated easily in a qualitative way for a fixed bed. For simplicity, only two reactions are considered, partial oxidation to give the

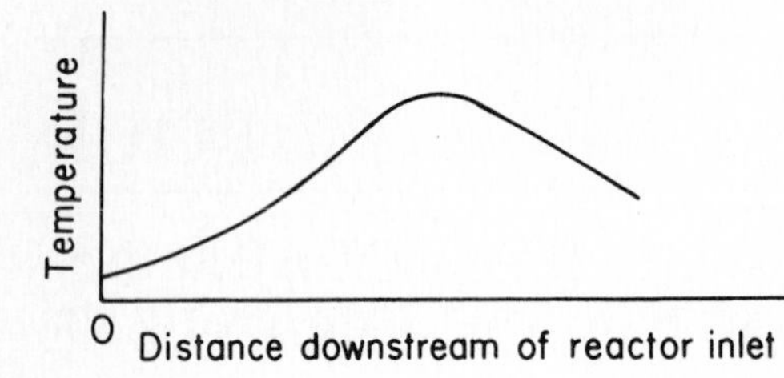

Figure 4-1 Axial temperature profile in a fixed-bed oxidation reactor.

desired product and further oxidation of the desired product to give CO_2. Both reactions are usually highly exothermic, total oxidation being the more so. As a consequence of the heat of reaction, the reactor temperature tends to increase with increasing distance downstream of the inlet. But as the reactants proceed through the reactor, they are depleted, so that the rate of reaction (and heat evolution) may decrease toward the exit. Consequently, the steady-state axial temperature profile may be expected to resemble that shown in Fig. 4-1. The important characteristic of this profile is the temperature maximum, the location of which is referred to as a *hot spot*. The shape of the profile in a fixed-bed reactor with heat transfer to the wall and the temperature of the hot spot may depend on the following variables (but the phenomena are complicated, and the pattern is not general):

1. *The reaction rate at the inlet.* If the rate is low, little energy is evolved and the maximum temperature may be low and near the bed exit. If the rate is high, the peak may be narrow and the maximum located nearer the bed entrance.
2. *The heat of reaction.* If an excess of oxygen is present so that complete oxidation is possible, the heat evolved by the partial oxidation can raise the reactor temperature sufficiently for the complete combustion reaction to be ignited at some position in the reactor.
3. *The rate of heat transfer out of the reactor.* If this rate is high, as in a narrow tubular reactor, the temperature profile may be nearly flat. Since total combustion in the presence of a selective catalyst becomes rapid only at high temperatures, the flat profile corresponds to maximum rates of partial oxidation relative to total oxidation. (Since one method of reducing temperature gradients in a reactor is to mix the contents rapidly, fluidized beds are commonly applied as partial-oxidation reactors.)

To understand the operation of the fluidized-bed reactor, we consider the effects of a gradual increase in average temperature imposed upon the reactor, to which a feed of hydrocarbon and oxygen is maintained unchanged. At a low reactor temperature, the rate of heat evolution and the difference between the average reactor temperature and the temperature of the surroundings may both be small, as illustrated in Fig. 4-2.

Increasing the reactor temperature so that the temperature difference increases from ΔT_1 to ΔT_2 causes an increase in the rate of heat generation; this must be balanced in the steady state by an increased rate of heat removal, which is

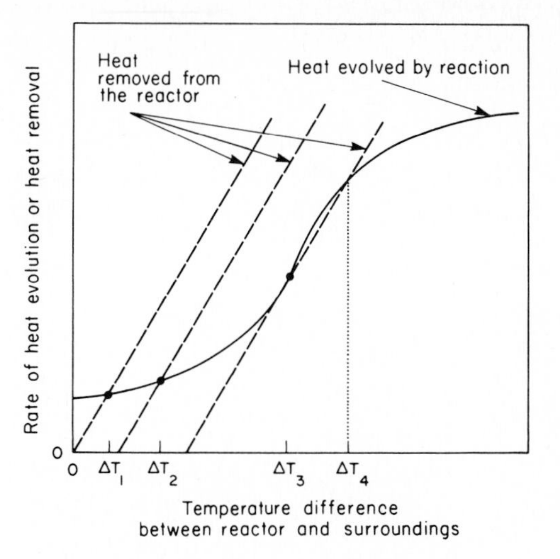

Figure 4-2 Energy balance of a fluidized-bed reactor. At steady state, the rate of heat evolution is equal to the rate of heat removal, as indicated by the intersections of the curved and straight lines.

approximately proportional to the temperature difference, as shown in Fig. 4-2. At a still greater value of the temperature difference ΔT_3, the reaction rate, which is exponentially dependent on temperature according to the Arrhenius relation, may become so great that the rate of heat removal cannot keep up with the rate of heat generation. Then the reactor temperature increases still more, and a steady state is reached only when the oxidation has become virtually complete and the temperature difference equals ΔT_4. The result is then that at a temperature difference of ΔT_3, the reaction rate and temperature appear to jump suddenly from a low to a high level; we say that the reaction "runs away." Reducing the reactor temperature after the temperature jump does not immediately lead to a similar decrease; the decrease is observed only at a considerably lower temperature, and what is observed is a kind of hysteresis.

Now if a selective catalyst is used so that the temperature ranges in which partial and complete combustion occur are separated by a wide interval, the reaction for temperature differences less than ΔT_3 is confined to partial oxidation, whereas for greater temperature differences, more or less complete oxidation takes place, as illustrated in Fig. 4-3.

These considerations lead us to the identification of two subjects central to the chemistry and engineering of catalytic partial-oxidation processes:

1. The understanding and design of selective catalysts having a sufficient separation of temperature ranges for partial and complete oxidation
2. The analysis of reactor performance and the design of reactors in which these highly exothermic reactions occur

These considerations pertain to many industrial partial-oxidation processes, some of the most important of which are listed in Table 4-1. Some generalizations can be made about the reactions and the catalysts applied in these processes:

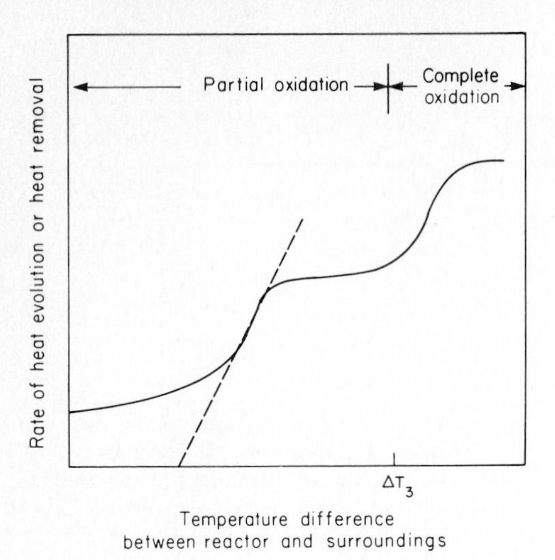

Figure 4-3 Energy balance of a fluidized-bed reactor with a highly selective oxidation catalyst.

1. In the various examples mentioned, the applicability of the catalysts is restricted to a particular reaction or group of reactions. If a catalyst selective for one group is applied to another group, the result is either that complete combustion predominates or that there is almost no conversion. If we denote the selective reactions of aromatics as group I, the oxidation of olefins as group II, and the oxidation of methanol as group III, we can summarize by saying that catalysts for group I reactions lead to excessive combustion in group II and III reactions and the catalysts for group II and III reactions are only slightly active for group I reactions. This generalization appears to indicate a trend in the tendencies of the various catalysts to donate oxygen, which is a central point of discussion in this chapter.
2. Many of the catalysts are not simple oxides but binary mixtures or compounds, and the binary systems have catalytic properties strikingly different from those of the components.
3. All the catalysts are characterized by complex and poorly defined surface structures, and the reaction networks and mechanisms are hardly understood. Our lack of understanding limits this chapter to the single partial-oxidation process for which a relatively coherent picture of catalyst structure and reaction mechanism has emerged, namely, the ammoxidation of propylene to give acrylonitrile:

$$H_2C{=}CHCH_3 + NH_3 + \tfrac{3}{2}O_2 \longrightarrow H_2C{=}CHCN + 3H_2O \qquad (1)$$

This process is applied in the production of more than 1.4 billion kilograms of acrylonitrile per year, most of which is polymerized for the manufacture of acrylic fibers [1].

This chapter proceeds with a general introduction to the chemistry of oxidation followed by a summary of the chemistry of ammoxidation. The concluding

Table 4-1 Industrial partial-oxidation processes

Reaction	Catalyst	Reactor
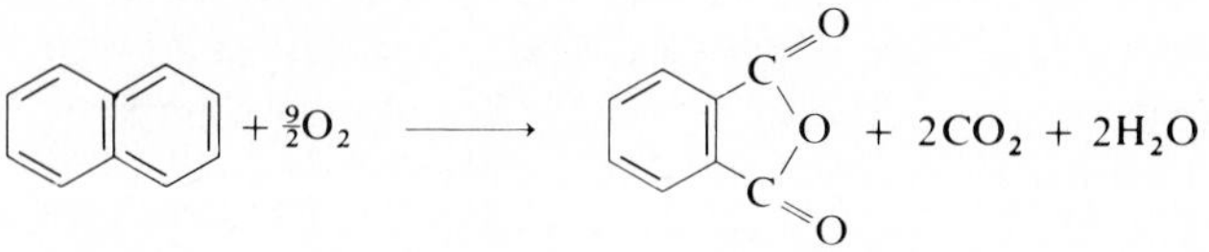 (naphthalene) $+ \frac{9}{2}O_2 \longrightarrow$ (phthalic anhydride) $+ 2CO_2 + 2H_2O$	Supported V_2O_5	Fluidized bed
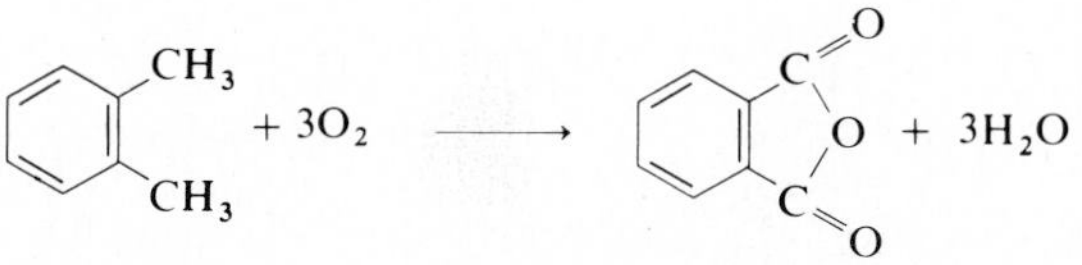 (o-xylene, CH_3, CH_3) $+ 3O_2 \longrightarrow$ (phthalic anhydride) $+ 3H_2O$	Supported V_2O_5	Fluidized or fixed bed
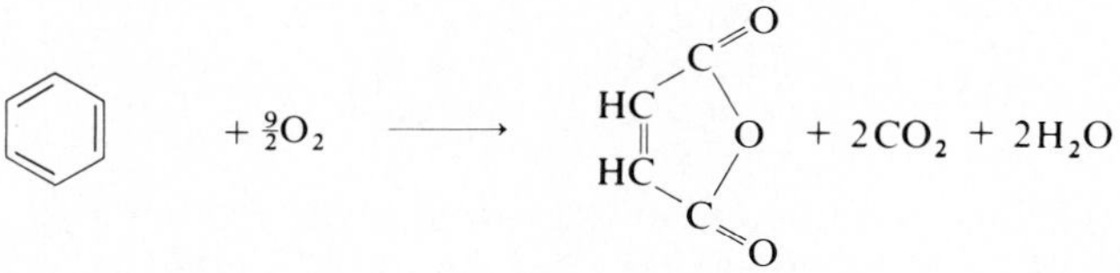 (benzene) $+ \frac{9}{2}O_2 \longrightarrow$ (maleic anhydride, HC=HC, C=O, O, C=O) $+ 2CO_2 + 2H_2O$	Supported $V_2O_5 + MoO_3$	Fixed bed
$CH_3OH + \frac{1}{2}O_2 \longrightarrow HCHO + H_2O$	$Fe_2(MoO_4)_3 + MoO_3$ or unsupported Ag	Fixed bed
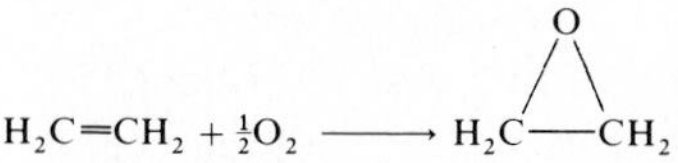 $H_2C{=}CH_2 + \frac{1}{2}O_2 \longrightarrow H_2C{-}CH_2$ (with bridging O)	Supported Ag	Fixed bed
$H_2C{=}CHCHO + \frac{1}{2}O_2 \longrightarrow H_2C{=}CHCO_2H$	Mixed molybdates	Fixed bed
$H_2C{=}CHCH_3 + \frac{3}{2}O_2 \longrightarrow H_2C{=}CHCO_2H$	Mixed molybdates (two-step conversion)	Fixed bed
$H_2C{=}CHCH_3 + NH_3 + \frac{3}{2}O_2 \longrightarrow H_2C{=}CHCN + 3H_2O$	See text	Fluidized or fixed bed

section describing ammoxidation processes and summarizing the process engineering is much less quantitative than the comparable sections of preceding chapters since almost all the pertinent information is proprietary and too little is available to allow meaningful reactor-design estimates.

THE CHEMISTRY OF OXIDATION

NONCATALYTIC OXIDATION

Oxidation is part of a general reaction class involving the incorporation of elements such as N, P, O, S, F, Cl, and Br into reactants such as hydrocarbons. Oxidation typically proceeds via the breaking of bonds in the reactant molecule along with the incorporation of oxygen into it. Clearly, bonds like O=O (or N≡N or F—F) must also be broken, so that reactions in this class require the breaking of bonds in both the reactants. Since the bond strength measured by the dissociation energy decreases in the series N_2 (226 kcal/mol) > O_2 (119 kcal/mol) > F_2 (38 kcal/mol), it is not surprising that rates of reaction usually increase in the series proceeding from N_2 to F_2. This pattern of bond strengths is readily explained in qualitative terms from the following pattern of electron occupancy:

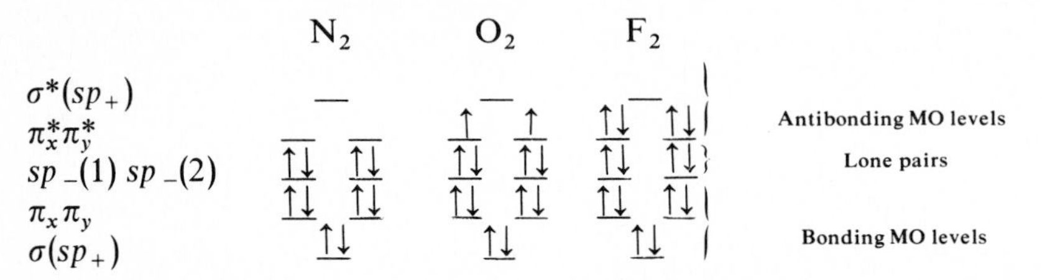

The N_2 molecule has only its bonding and nonbonding orbitals filled, whereas O_2 has two antibonding electrons and F_2 four. Therefore, the F—F bond is weaker than the O=O bond, which in turn is weaker than the relatively unreactive N≡N bond.

In reactions involving F_2, the first step is usually the breaking of the weak F—F bond to give two free atoms. These react many times in a chain sequence like the following before they are removed by reaction with another free radical or a solid surface:

$$F_2 \longrightarrow 2F\cdot \qquad \text{chain initiation} \tag{2}$$

$$F\cdot + H_2 \longrightarrow HF + H\cdot \tag{3}$$

$$H\cdot + F_2 \longrightarrow HF + F\cdot \qquad \text{chain propagation} \tag{4}$$

$$H_2 + F_2 \longrightarrow 2HF \qquad \text{net reaction} \tag{5}$$

The O=O bond, in contrast to the F—F bond, is stronger than most other bonds in the reactant to be oxidized, and oxidation reactions are commonly initiated with the breaking of a bond in the reactant molecule. For example, in the oxidation of H_2, the reaction may be initiated with the dissociation of H_2, which has a bond energy of 104 kcal/mol:

$$H_2 \longrightarrow 2H\cdot \tag{6}$$

The H atom then attacks an O_2 molecule at one of the lone pairs. To form an H—O bond, an electron from the lone pair must be promoted to a π antibonding level, whereby the O=O bond may become sufficiently weakened for bond breaking to occur:

$$H\cdot + O_2 \longrightarrow (H{-}O{-}O) \longrightarrow \cdot OH + O\cdot \tag{7}$$

In this reaction step, one free radical produces two free radicals, and the result is a branched chain reaction, which can proceed at an increasing rate, possibly until it becomes explosive.

Saturated hydrocarbons are oxidized by similar mechanisms, but the first bond to be broken is usually a C—C bond like the following, for which the dissociation energy is 88 kcal/mol:

$$R_2\overset{H}{\overset{|}{C}}-\overset{H}{\overset{|}{C}}R_2 \longrightarrow R_2\overset{H}{\overset{|}{C}}\cdot + \cdot\overset{H}{\overset{|}{C}}R_2 \tag{8}$$

Again, the radicals attack O_2 via the lone pairs:

$$R_2\overset{H}{\overset{|}{C}}\cdot + O_2 \longrightarrow R_2\overset{H}{\overset{|}{C}}OO\cdot \tag{9}$$

At higher temperatures the peroxy radical undergoes bond rupture just as HO_2 does, but at lower temperatures it may be sufficiently stable to undergo a second attack such as

$$R_2\overset{H}{\overset{|}{C}}OO\cdot + R_2\underset{H}{\underset{|}{C}}-\underset{H}{\underset{|}{C}}R_2 \longrightarrow R_2\overset{H}{\overset{|}{C}}OOH + R_2\dot{C}-\overset{H}{\overset{|}{C}}R_2 \tag{10}$$

Consequently, the O—O bond is further weakened, being now about as strong as a F—F bond, and it eventually ruptures:

$$R_2\underset{H}{\underset{|}{C}}OOH \longrightarrow R_2\underset{H}{\underset{|}{C}}O\cdot + \cdot OH \tag{11}$$

The result is called a *degenerate branched chain reaction.* At this stage, more extensive oxygen additions can occur by reactions like

$$RO^{\bullet} + O_2 \longrightarrow \overset{O}{\overset{/\!/}{R}}-O-O^{\bullet} \tag{12}$$

A thorough account of the many gas-phase oxidation reactions proceeding via free-radical mechanisms has been given by Semenov [2], who was awarded the Nobel prize for his work on chain reactions.

OXIDATION CATALYZED BY COORDINATION COMPLEXES OF TRANSITION METALS

Reactions such as

$$C_6H_5-CH_3 + \tfrac{3}{2}O_2 \xrightarrow{\text{Co coordination complex}} C_6H_5-COOH + H_2O \tag{13}$$

and

$$H_3C-C_6H_4-CH_3 + 3O_2 \xrightarrow{\text{Co or Mn complexes}} HOOC-C_6H_4-COOH + 2H_2O \tag{14}$$

are examples of oxidation reactions catalyzed by transition-metal complexes.

In this class of oxidation reactions, a different mechanism prevails in the breaking of the O=O bond. The transition-metal ion transfers electrons to this bond by formation of *dp* π bonds between the cation and the O_2 molecule. For example,

$$Co^{2+} + O_2 + Co^{2+} \longrightarrow Co^{3+}-O_2^{2-}-Co^{3+} \tag{15}$$

Such a reaction can occur by σ bonding via the lone pairs of O_2 as well as by *dp* π bonding involving antibonding oxygen π^* orbitals and t_{2g} orbitals of the cation, as illustrated in Fig. 4-4*A*. An alternative possibility involves the bonding and antibonding oxygen π orbital in combination with the *d* orbitals of the cation, as shown in Fig. 4-4*B*.

To continue transferring electrons to the oxygen molecule, the transition metal must be engaged in a reaction enabling it to accept electrons. The required electron acceptance may often result from a reaction with the molecule to be oxidized, as in

$$Co^{3+} + C_6H_5-CH_3 \longrightarrow Co^{2+} + H^+ + C_6H_5-\dot{C}H_2 \tag{16}$$

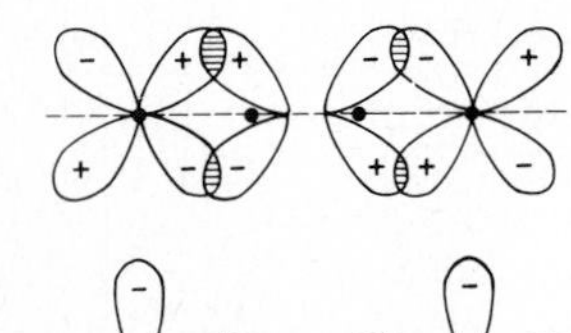

Figure 4-4 Bonding in complexes of Co and O_2.

This reaction is facilitated by resonance stabilization of the product

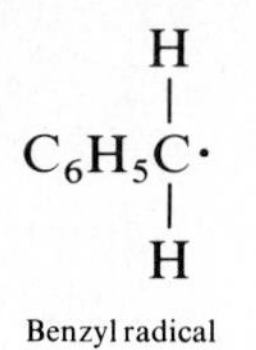

Benzyl radical

Consistent with the foregoing explanation, we can suggest a plausible sequence of steps for toluene oxidation:

$$Co^{3+} + C_6H_5CH_3 \longrightarrow Co^{2+} + H^+ + C_6H_5\dot{C}H_2 \quad (17)$$

$$C_6H_5\dot{C}H_2 + O_2 \longrightarrow C_6H_5CH_2OO\cdot \quad (18)$$

$$Co^{2+} + C_6H_5CH_2OO\cdot \longrightarrow Co^{3+} + OH^- + C_6H_5CH{=}O \quad (19)$$

$$H^+ + OH^- \longrightarrow H_2O \quad (20)$$

$$Co^{3+} + C_6H_5CH{=}O \longrightarrow Co^{2+} + C_6H_5\dot{C}{=}O + H^+ \quad (21)$$

$$2C_6H_5\dot{C}{=}O + O_2 \longrightarrow C_6H_5C({=}O){-}O{-}O{-}C({=}O)C_6H_5 \quad (22)$$

$$2Co^{2+} + C_6H_5C({=}O){-}O{-}O{-}C({=}O)C_6H_5 \longrightarrow 2Co^{3+} + 2C_6H_5C({=}O){-}O^- \quad (23)$$

$$C_6H_5C({=}O){-}O^- + H^+ \longrightarrow C_6H_5C({=}O){-}OH \quad (24)$$

This mechanism is only one example of a number of proposed mechanisms having several common characteristics. They all involve nonbranched chain reactions in which the intermediate species are free radicals and a transition-metal cation exhibiting two oxidation states differing by 1 unit. Transfer of electrons occurs in single steps, either via the change in oxidation state of the cation or via free radicals. Parenthetically, we note that a direct transfer of one electron from the cation to O_2 is an expected reaction in this context. Therefore, O_2^- is a possible intermediate. It is also clear that the catalytic chain-reaction sequence producing free radicals may initiate a thermal branched-chain process like the ones just discussed, e.g., via noncatalytic rupture of the O—O bond in peroxides.

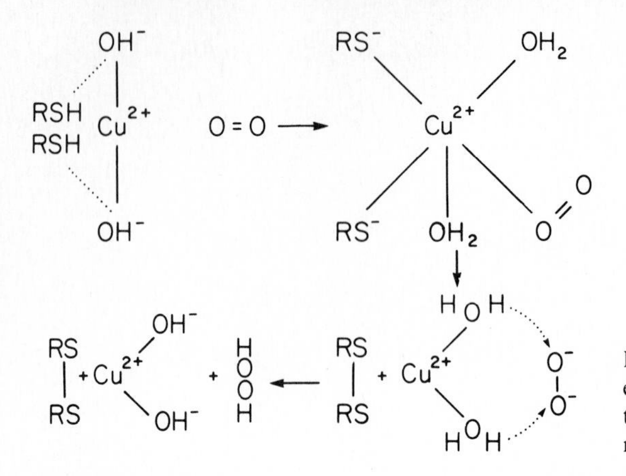

Figure 4-5 Oxidation of mercaptans by electron transfer and proton transfer in the coordination sphere of a transition-metal ion.

There is reason to believe that free radicals are not always intermediates in oxidation reactions catalyzed by metal complexes. Alternatively, the electron and proton transfers might proceed entirely in the coordination sphere of the metal ion. For one particular reaction, the oxidation of mercaptans catalyzed by Cu^{2+} ions in aqueous solutions at high pH, the occurrence of a non-free-radical mechanism appears to be well established [3, 4].

Since the mercaptan oxidation is similar to that of ascorbic acid, we might suggest a mechanism similar to that given by Taqui Khan and Martell [5] for the latter reaction, as shown in Fig. 4-5. This mechanism accounts for the formation of RSSR + H_2O_2 from RSH and O_2 in a concerted, two-electron + two-proton transfer. Alternatively, the occurrence of binuclear Cu complexes, with each of the two Cu^{2+} ions bonded to two S ligands and O_2 bridging the two Cu^{2+} ions, could even allow a fully concerted, four-electron + four-proton transfer for incorporating O_2 into two H_2O molecules. Since the kinetics of the reaction, however, does not favor this suggestion [4], consistent with Fig. 4-5, H_2O_2 is inferred to be an intermediate. Hydrogen peroxide is rapidly decomposed in the presence of transition-metal cations, either by a one-electron + one-proton transfer process (a free-radical mechanism) or by a concerted two-electron + two-proton transfer. Since no radicals have been observed, the latter possibility appears more probable. One H_2O_2 molecule then must be the donor of electrons and protons and the other the acceptor. A possible mechanism is suggested in Fig. 4-6.

A transition-metal-complex-catalyzed oxidation reaction, the Wacker reaction, already considered in Chap. 2, proceeds via the sequence

$$Pd^{2+} + C_2H_4 + H_2O \longrightarrow Pd^0 + CH_3C\begin{matrix} {}^{\nearrow\!O} \\ {}_{\searrow H} \end{matrix} + 2H^+ \tag{25}$$

$$Pd^0 + 2Cu^{2+} \longrightarrow Pd^{2+} + 2Cu^+ \tag{26}$$

$$4Cu^+ + O_2 + 4H^+ \longrightarrow 4Cu^{2+} + 2H_2O \tag{27}$$

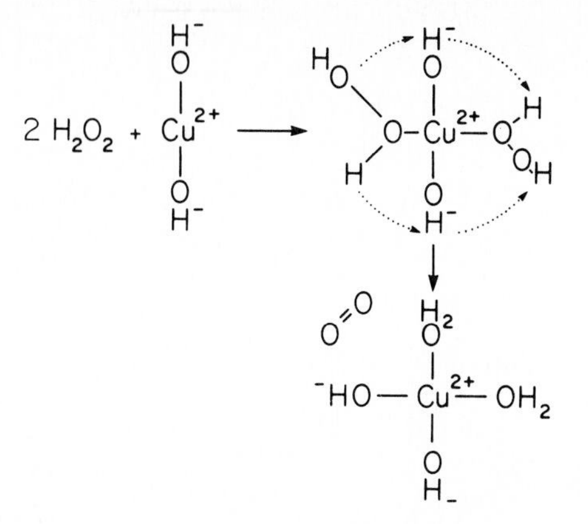

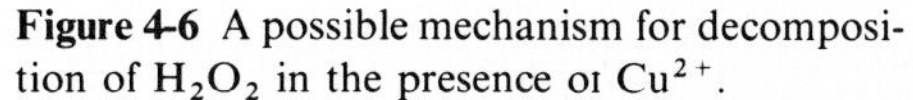

Figure 4-6 A possible mechanism for decomposition of H_2O_2 in the presence of Cu^{2+}.

It is important that in reaction (25) the oxygen is actually supplied to the hydrocarbon by H_2O, which is an intermediate oxidation product formed from O_2 in a separate step. Reaction (25) is believed to proceed via a cis-insertion mechanism, but the preceding reduction of O_2 is sometimes referred to as a noninsertion reaction. The transport of electrons to O_2 occurs via Cu^{2+}. Protons are furnished by the H_3O^+ ions in the solution. As has just been discussed, two steps are probably required in the transfer of the four electrons. The first step can be formulated as shown in Fig. 4-7.

By way of summary of the foregoing discussion of oxidation chemistry, the following conclusions are reemphasized because of their importance to the subsequent discussion of surface-catalyzed oxidation:

1. Many catalytic oxidation reactions involve the transfer of single electrons and protons via intermediates such as free radicals, O_2^-, H_2O_2, and transition-metal ions. Such reactions are nonbranched chain reactions, but occasionally the free radicals may start thermal (noncatalytic) chain reactions, including branched chain reactions. On the other hand, the reactions sometimes proceed via pairwise transfer of electrons and protons in the coordination sphere of a transition-metal ion; these reactions do not show the branching characteristics of the noncatalytic oxidation reactions.
2. It is sometimes advantageous to apply two different metal ions, one to attack the molecule to be oxidized and the other to interact with oxygen and facilitate electron transfer between the cations.

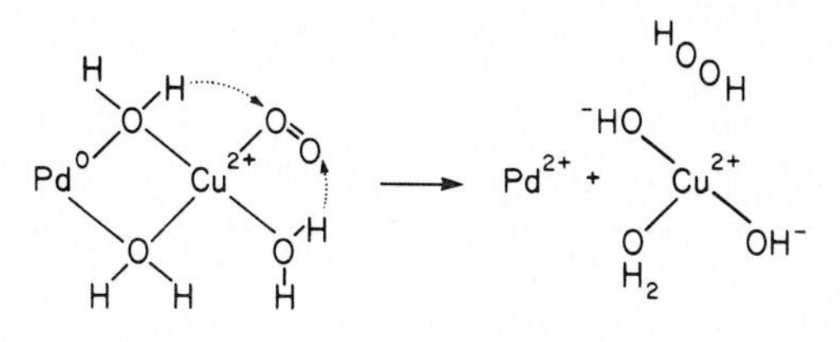

Figure 4-7 Reoxidation of Pd^0 in the presence of Cu^{2+} and O_2.

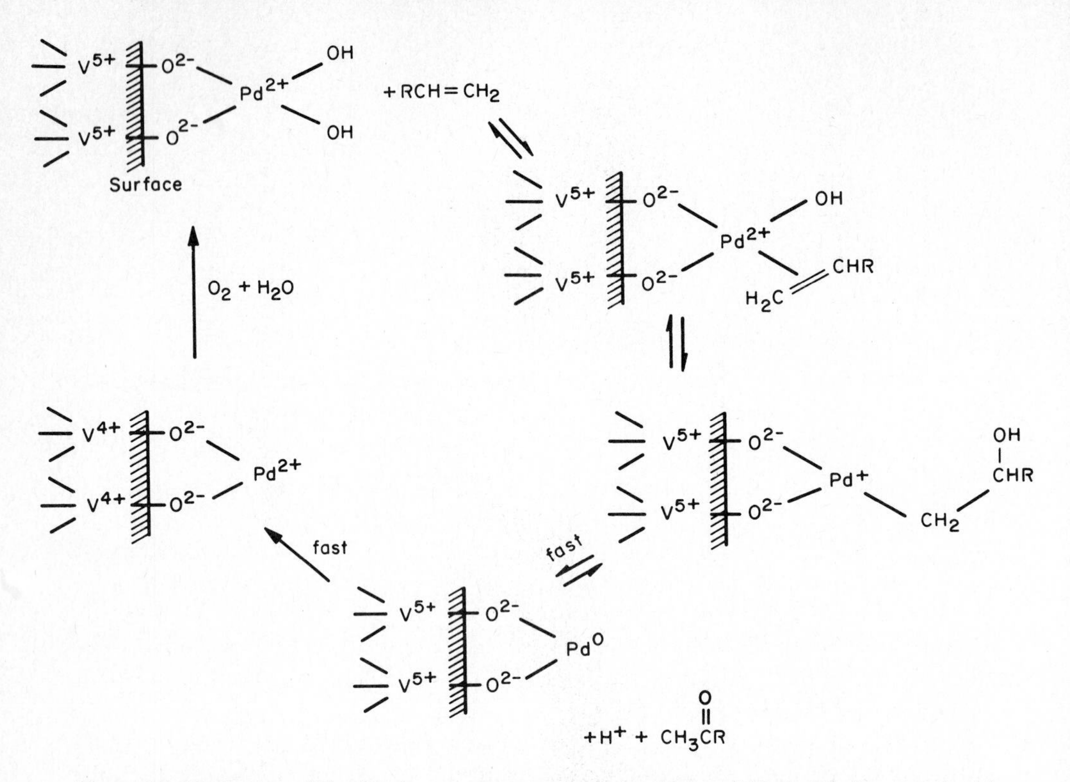

Figure 4-8 Mechanism of the Wacker reaction catalyzed by Pd-doped V_2O_5 [6]. (Reprinted with permission from *Journal of Catalysis*. Copyright by Academic Press.)

3. The oxygen introduced into the molecule which is oxidized need not necessarily be derived directly from O_2 but may instead come from an intermediate such as H_2O.

These characteristics help explain the patterns of oxidation reactions catalyzed by solid surfaces; in particular, many solid oxidation catalysts serve as the sources of oxygen transferred to the reactant molecules, and further, some of the best selective oxidation catalysts contain two types of cations.

To illustrate some common features of catalytic oxidation in solutions and on surfaces, it is useful to reconsider the solid Wacker catalyst of Evnin et al. [6], which was described in Chap. 2. ESR spectra showed that the Wacker reaction leads to the formation of V^{4+} on the catalyst surface, and the mechanism shown in Fig. 4-8 was suggested to account for this observation. The catalytic site shown here is similar to the bimetallic complex of Fig. 4-7; again, the cis insertion is envisioned as taking place in the coordination sphere of Pd, and the electron transfers are similar; however, the suggested catalytic site does not provide enough electrons to convert the O_2 molecule completely. More electrons must be supplied by the semiconducting bulk of the catalyst.

In the following pages these patterns of catalytic oxidation reactions are discussed as they specifically apply to the reactions of olefins on surfaces of solid oxides. The discussion focuses on the ammoxidation of propylene.

OXIDATION CATALYZED BY SURFACES OF OXIDES

Hydrocarbon Surface Intermediates

In consideration of the interaction of an olefin with the surface of an oxide, it is appropriate to ask what intermediate structures can be formed. Or, in terms of the preceding paragraphs, how is the reactant olefin molecule attacked first? The example of Fig. 4-8 shows a surface transition-metal complex, but much more important for the following discussion of ammoxidation is a surface radical formed from a dissociative adsorption of the olefin. It is called an *allylic intermediate* and is formed by the dissociation of a H atom from a C atom next to a double bond:

$$H_2C{=}CHCH_3 \longrightarrow H_2C\overset{..}{=}CH\overset{..}{=}CH_2 + H\cdot \tag{28}$$

As a radical, this allylic intermediate is stabilized by resonance, and consequently a C—H bond next to a C—C double bond is more easily dissociated than one on a carbon atom in a paraffin.

The identification of the allylic intermediate is based on results of experiments with isotopically marked olefins. For example, Sachtler and de Boer [7] performed experiments on the oxidation of propylene catalyzed by bismuth molybdate, the structure of which is discussed later. The reaction was the formation of acrolein:

$$O_2 + H_2C{=}CHCH_3 \longrightarrow H_2C{=}CHCHO + H_2O \tag{29}$$

The following tagged olefin molecules were used as reactants:

$$\underset{\text{I}}{H_2{}^{14}C{=}CHCH_3} \qquad \underset{\text{II}}{H_2C{=}CH{-}{}^{14}CH_3} \qquad \underset{\text{III}}{H_2C{=}{}^{14}CHCH_3}$$

The ^{14}C isotope is radioactive, and so the position of ^{14}C in the product acrolein could be found after the acrolein was photochemically decomposed:

$$H_2C{=}CHCHO \xrightarrow{h\nu} H_2C{=}CH_2 + CO \tag{30}$$

The experiment required determination of the radioactivity of C_2H_4 and CO. The results showed that species I and II gave products in which the ratio of radioactive C_2H_4 to radioactive CO was 1, whereas species III produced exclusively radioactive C_2H_4 and no radioactive CO. The similarity in behavior of reactants I and II demonstrated that there was an intermediate, formed before O was attached to the molecule, in which the terminal C atoms were equivalent. A path leading to this equivalence involves the removal of an H atom from the CH_3 group, presumably giving the allyl.† The results for reactant III confirm that there was no randomization of the hydrogens, since if this had occurred, radioactive CO would also have been formed.

Measurements of isotope effects in kinetics have also been used in studying reactions involving C—H bond breaking. The C—H bond is slightly weaker than the C—D bond, the difference resulting from the lower zero-point energy of bonds containing the heavier D atom. The isotope effect is therefore useful in the present context, provided that a molecule is oxidized in a sequence of steps, the step involving the breaking of a C—H bond being rate-determining. In this case, reaction of the C—H compound is faster and characterized by a lower activation energy than reaction of the C—D compound.

Adams and Jennings [8] reported the following rates of bismuth molybdate–catalyzed oxidation of propylene containing D at various positions:

Reactant	Relative rate of oxidation
$H_2C{=}CHCH_3$	1.00
$H_2C{=}CHCH_2D$	0.85
$DHC{=}CHCH_3$	0.98
$D_2C{=}CDCD_3$	0.55

The observed isotope effect leads to the conclusion that the rate-determining step does involve the breaking of a C—H bond. Furthermore, since there is an isotope effect only if the D is located in the CH_3 group rather than the CH_2 group, the CH_3 group must be the one from which the H atom is abstracted and this step clearly leads to the formation of an allylic intermediate. There is now evidence, considered in detail later, for the formation of allylic intermediates on bismuth molybdates [7–10], Cu_2O [11, 12], and USb_3O_{10} [13].

Adams and Jennings suggested that knowledge of the isotope effect derived from relative rates might allow a prediction of the position of the isotope in the final product (acrolein or acrylonitrile). If the first step alone is rate-determining, the relative rates of the C—H and C—D splitting determine how much D is present in the various positions. Consider, for example, the oxidation of $H_2C{=}CHCH_2D$ in the presence of NH_3 by the following scheme [8], considered in more detail later:

† Alternatively, the result could be explained by addition of hydrogen to the olefin, e.g., a protonation to give a secondary carbonium ion, which is another species having equivalent terminal carbon atoms. The isotope effects described on the following page, however, are not consistent with a carbonium-ion intermediate.

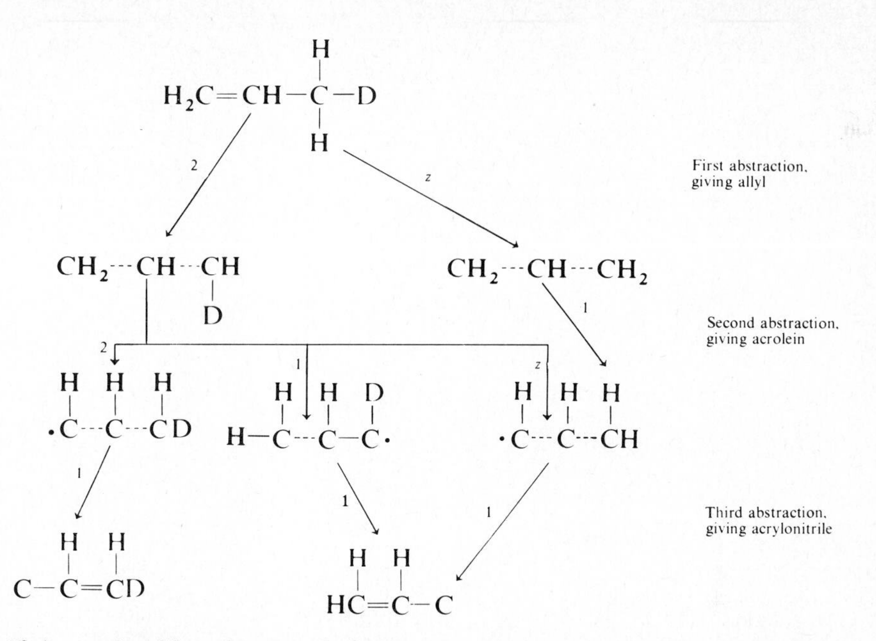

If the probability of removal of a D atom relative to an H atom is denoted as z (and this probability is already known from data for the first abstraction), then the probability that D is bonded to a certain C atom in acrolein or acrylonitrile can be calculated straightforwardly for the scheme given above. The agreement between the prediction and experiment is satisfactory.

The concept of the resonance-stabilized intermediate can also be used to explain the oxidation of compounds which do not form allylic intermediates. For example, the rate-determining step in toluene oxidation would be expected to be

$$C_6H_5{-}CH_3 \longrightarrow C_6H_5{-}\dot{C}H_2 + H^{\bullet} \tag{31}$$

The next steps would be expected to be similar to those of propylene oxidation, the abstraction of a second H atom and the addition of an O atom:

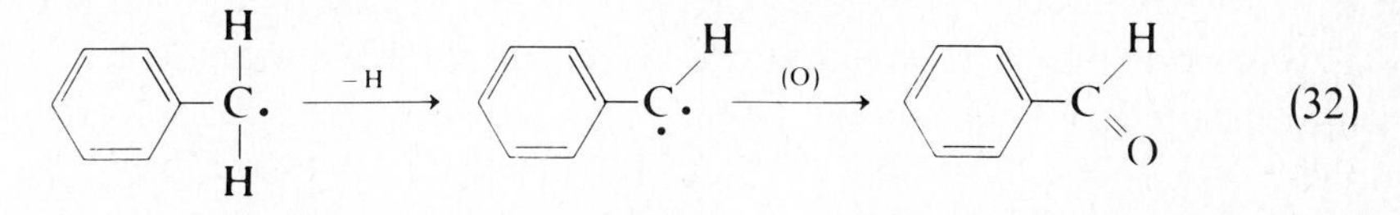

The further oxidation of the reaction product, benzaldehyde, has been studied by Sachtler and coworkers [14, 15], who determined infrared spectra of the surface intermediates on a SnO_2–V_2O_5 catalyst. After benzaldehyde contacted the catalyst, it was present in the form of a complex resembling a benzoic acid salt. Its formation can be visualized as shown in Fig. 4-9. The process is similar to that

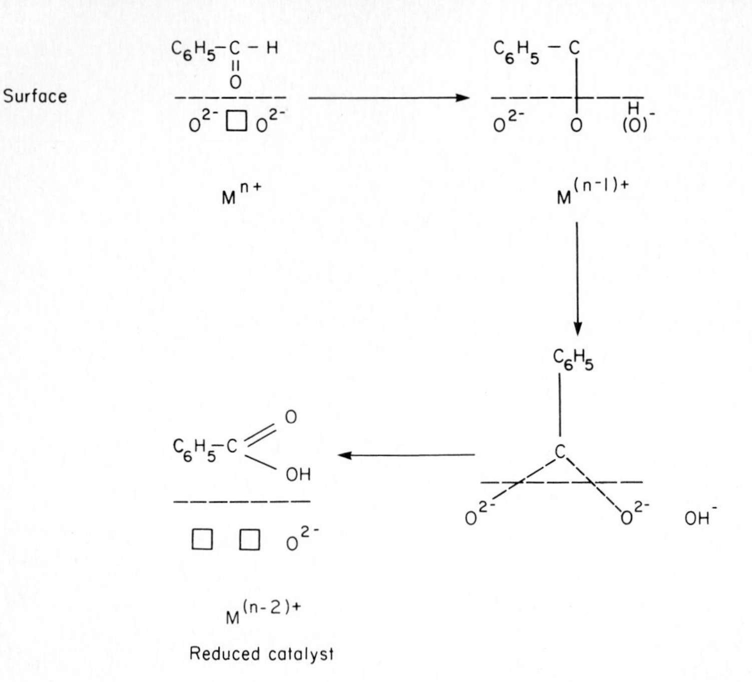

Figure 4-9 Oxidation of benzaldehyde on SnO_2–V_2O_5 (adapted from refs. [14] and [15]). The symbol □ denotes a surface anion vacancy.

taking place during the formation of an allylic intermediate: a proton is donated to an O^{2-} ion at the surface, and the residue is then bonded to a metal ion at an anion vacancy.

A different surface species has been postulated to account for the oxidation of propylene to acetone catalyzed by mixed oxides such as SnO_2–MoO_3 [16–19]. This reaction occurs only at relatively low temperatures and in the presence of steam. Since H_2O reacts with oxide surfaces to form hydroxyls,

$$\square + O^{2-} + H_2O \longrightarrow 2{-}OH^- \tag{33}$$

it might be supposed that the $-OH^-$ groups are involved in the oxidation reaction. It has been shown in Chap. 1 that, provided the surface $-OH^-$ groups are sufficiently acidic, an $-OH^-$ group can react with an olefin leading to the formation of a carbonium ion:

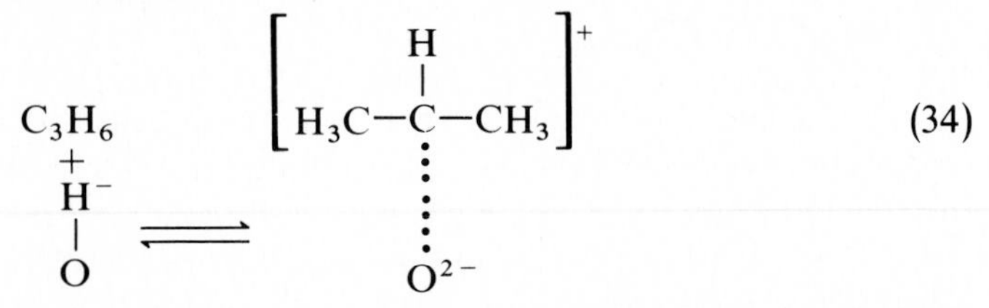

(34)

The carbonium ion might then decompose with abstraction of a proton and simultaneous dislocation of an oxygen atom from the surface and donation of an electron pair to the solid:

$$\left[\begin{array}{c} \mathrm{H} \\ | \\ \mathrm{H_3C{-}C{-}CH_3} \end{array}\right]^{+} \quad \begin{array}{c} \vdots \\ \mathrm{O^{2-}} \quad \mathrm{O^{2-}} \end{array} \longrightarrow \begin{array}{c} \mathrm{CH_3{-}C{-}CH_3} \\ \| \\ \mathrm{O} \\ + \quad \mathrm{H^-} \\ | \\ \square + 2e^- + \mathrm{O} \end{array} \tag{35}$$

Such a reaction can occur only if the oxide can be reduced. A confirmation of this suggested mechanism of propylene oxidation was given by Moro-Oka et al. [19], who showed that addition of $H_2{}^{18}O$ to the reaction mixture led to the formation of $CH_3C^{18}OCH_3$ product.

Bonding of Surface Intermediates

The foregoing discussion of organic surface intermediates has glossed over the question of how the intermediates are bonded to the surface. The only specific suggestion has been shown in Fig. 4-9, depicting the adsorption of benzaldehyde. Here, a proton is shown being donated to a surface oxygen ion, with the benzoyl radical (or carbanion) forming a bond with a metal cation. It can be supposed that a similar reaction occurs in the binding of an allylic intermediate, as follows. (To facilitate the discussion, the intermediates are referred to as ionic, but this usage does not imply that the reactions proceed exclusively via ionic intermediates.)

$$C_3H_6 \longrightarrow \underset{\text{Allyl}}{\underline{C_3H_5^-}} + \underline{H^+} \tag{36}$$

$$\underline{H^+} + \underline{O^{2-}} \longrightarrow \underline{OH^-} \tag{37}$$

$$\underline{C_3H_5^-} + \underset{\text{Catalyst cation}}{\underline{M^{n+}}} \longrightarrow \underline{C_3H_5^+} + \underset{\text{Reduced catalyst cation}}{\underline{M^{(n-2)+}}} \tag{38}$$

$$\underline{C_3H_5^+} + \underline{O^{2-}} \longrightarrow \underline{C_3H_5O^-} \tag{39}$$

The structure of the surface-bonded $C_3H_5O^-$, if it is considered to be a σ-bonded system, should be either

$$\underset{\text{I}}{\overset{1}{\mathrm{H_2C}}{=}\overset{2}{\mathrm{CH}}{-}\underset{\mathrm{O}}{\overset{3}{\mathrm{CH_2}}}} \qquad \text{or} \qquad \underset{\text{II}}{\underset{\mathrm{O}}{\overset{1}{\mathrm{H_2C}}}{-}\overset{2}{\mathrm{CH}}{=}\overset{3}{\mathrm{CH_2}}}$$

Since the H abstraction occurs at carbon atom 3, addition of O might be assumed to occur preferentially at atom 3. But it has already been mentioned that there is no such preference. Moreover, the electron pair on the carbanion should

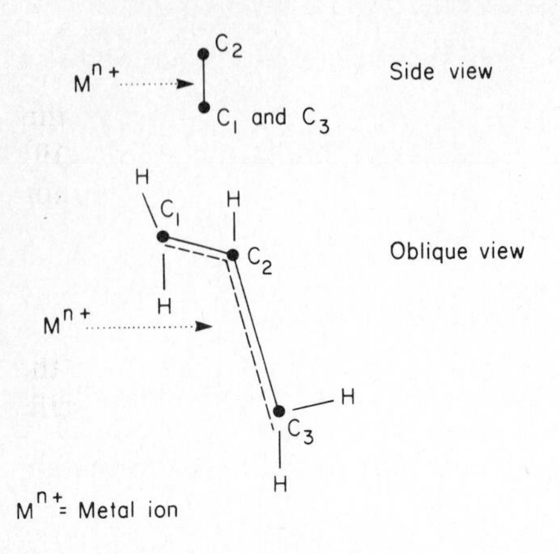

Figure 4-10 Bonding in a π-allyl complex.

somehow be transferred to the cation, but transfer via O^{2-} does not appear feasible. Therefore, it is inferred that there is an intermediate type of bonding of the carbanion which makes a transfer of electrons to the cation possible, carbon atoms 1 and 3 being equivalent. The requirements are satisfied by a hypothetical π-allyl bond with the cation. It has already been shown in Chap. 2 that olefins can be bonded to transition-metal ions, provided that the C=C axis is perpendicular to the bond between the cation and the center of the π bond. Similarly, it is expected that allyl groups can be bonded to metal ions or metal atoms in structures having the plane of the carbon atoms roughly perpendicular to the axis connecting the metal and the allyl. We refer then to a π-allyl bond, which is represented in Fig. 4-10.

The delocalized π-bond system of an allylic species as derived from the simple Hückel MO theory has three orbitals (ϕ_1, ϕ_2, and ϕ_3 are p orbitals on atoms C-1, C-2 and C-3):

$$\psi_B = \tfrac{1}{2}(\phi_1 + \sqrt{2}\,\phi_2 + \phi_3) \qquad \text{bonding}$$

$$\psi_{NB} = \frac{1}{\sqrt{2}}(\phi_1 - \phi_3) \qquad \text{nonbonding}$$

$$\psi_{AB} = \tfrac{1}{2}(\phi_1 - \sqrt{2}\,\phi_2 + \phi_3) \qquad \text{antibonding}$$

If the allylic species is situated in a ligand position of an octahedrally surrounded transition-metal ion as indicated in Fig. 4-11, it can be dp-π-bonded via the MO formation with the cation d orbitals:

$$d_{z^2} + \psi_B \qquad (d_{yz} \text{ or } d_{xz}) + \psi_{NB} \qquad \text{where } z = \text{axis from cation to center of allyl}$$

Electrons placed in the $d_{xz} + \psi_{NB}$ orbital are equally distributed over atoms C-1 and C-3, so that these two positions are equivalent.

It is clear that if the cation has no d electrons, for example, Mo^{6+}, the bonding electrons are all derived from the allylic species (carbanion). Consequently, the electrons are partly or even completely transferred to the cation in the complex. The complete transfer occurs preferentially with the electrons in the original

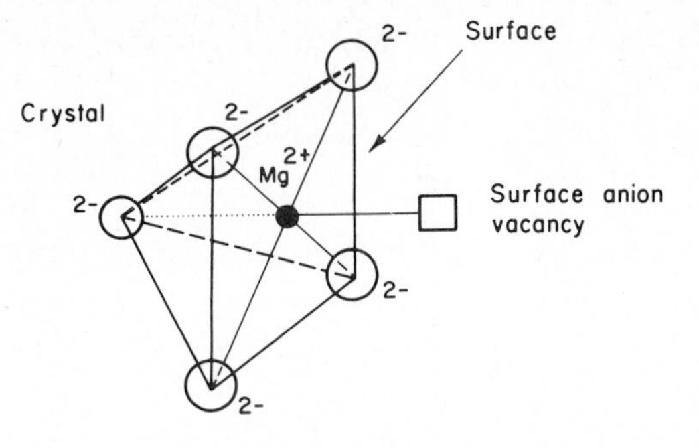

Figure 4-11 Anion vacancy at the surface of a MgO crystal.

nonbonding allyl orbital which are concentrated on the terminal C atoms. The resulting positive charge is therefore concentrated on these carbon atoms in a symmetrical distribution. Any of these atoms may interact with a neighboring O^{2-} to form a σ bond. The π-allyl structure therefore meets our requirements as a starting point for the discussion of organic surface intermediates.

Recent molecular orbital calculations by Haber et al. [20] support the qualitative model given above. Cations studied were Fe^{2+}, Fe^{3+}, Co^{2+}, Ni^{2+}, Mo^{6+}, Mo^{5+}, Mo^{4+}, and Mg^{2+}, all octahedrally surrounded, the sixth ligand position being occupied by an allyl radical. For all transition-metal cations, a considerable shift of electrons from allyl to cation was predicted; however, hardly any shift occurred for Mg^{2+}.

An important restriction is that the allylic species should find a ligand position (an anion vacancy) which allows it to interact with the cation. The purpose of the following section is to justify the assertion that such positions are present on the surfaces of oxides.

Structures of Oxide Surfaces

Consider, for example, a MgO crystal cleaved along a crystal face. If this face were such that it contained just as many Mg^{2+} ions as O^{2-} ions [it would have to be the (100) or (110) face], the electric charge on the two smaller crystals formed by the cleavage would remain zero. If the crystal were cleaved along a face containing only one of the two ions [the (111) face], half these ions should go to one and the remaining half to the other piece of crystal. On the resulting surfaces one should therefore expect to find cations lacking one of their coordinating ligands. For instance, instead of being octahedrally coordinated, a cation would now have a tetragonal-pyramidal coordination, as shown in Fig. 4-11.

Surface oxygen vacancies are necessary to satisfy the requirement of electrical neutrality on each crystal. This conclusion is important, for it leads to identification of the site where the cation can interact with a molecule from the gas phase. It is at such a site that the allylic species can form its π bond. In Chap. 2 a similar argument was made to identify the site allowing an olefin molecule to become π-bonded to Ti^{3+}, where it starts the Ziegler-Natta polymerization.

In general, the number of anion vacancies (or coordinatively unsaturated sites) and the number of fully coordinated sites are expected to be in the ratio of about 1 : 1 because cleavage of a crystal would leave about half the O^{2-} ions on each side of the plane of cleavage. This estimate is rough, however, since the actual ratio depends on the type of surface plane.

Many ligands can become bonded to the cations at the oxygen vacancies. One of the most common of these is water, a product of almost all oxidation reactions. Surface hydration can lead to a subsequent reaction, the formation of surface hydroxyl groups, which are familiar from Chap. 1:

$$\begin{array}{ccc} & \overset{H\quad H}{\diagdown\!\diagup} & \\ & O & \\ & | & \\ O & Me^{n+} & O \end{array} \longrightarrow \begin{array}{ccc} & H & \\ & | & \\ H & O & \\ | & | & \\ O & Me^{n+} & O \end{array} \qquad (40)$$

Coverage of surface anion vacancies by water would reduce the interaction of olefins with the surface cations, and therefore water can inhibit catalytic oxidation reactions of the kind described here. On the other hand, water is essential for the previously mentioned type of oxidation proceeding via carbonium ions.

The Introduction of Oxygen: The Mars–van Krevelen Mechanism

From results of kinetics experiments on the oxidation of naphthalene catalyzed by V_2O_5, Mars and van Krevelen [21] concluded that the reaction takes place in two steps: (1) a reaction between the oxide and the hydrocarbon in which the hydrocarbon is oxidized and the oxide reduced and (2) a reaction of the reduced oxide with O_2 to give back its initial state. The species directly responsible for the oxidation is generally assumed to be the O^{2-} ion on the surface of the oxide catalyst, and the reaction is therefore an example of the noninsertion type.

The validity of this pattern has now been established for numerous oxidation reactions and catalysts. In particular, it has been shown to be applicable to the oxidation of olefins on Bi_2O_3–MoO_3 catalysts [7], to the ammoxidation of propylene on Bi_2O_3–MoO_3 [22] and USb_3O_{10} [13], and to the oxidation of methanol on Fe_2O_3–MoO_3 [23]. We conclude that the Mars–van Krevelen mechanism prevails with some generality in hydrocarbon oxidation reactions catalyzed by oxide surfaces.

The generality of the Mars–van Krevelen mechanism led Sachtler and de Boer [7] to postulate that the tendency of an oxide (or oxide combination) to donate its oxygen should be of major importance in determining whether it is a selective oxidation catalyst. If reduction of the oxide is easy (in other words, if the free energy of O_2 dissociation and presumably also the enthalpy of dissociation are small), then O can easily be donated to a molecule from the gas phase and the catalyst is expected to be active and nonselective. On the other hand, if it is difficult to dissociate O_2 because the metal-oxygen bond is strong, the oxide is expected to have low catalytic activity. In the intermediate range, oxides might be moderately active and still be selective.

The foregoing hypothesis has been tested in studies of the interaction between compounds such as 1-butene and a series of oxides having known enthalpies of reduction. The results of such studies are well illustrated by the data of Simons et al. [24], who investigated the reactions of 1-butene and butadiene in a pulse microreactor with a series of catalysts including CuO, MnO_2, V_2O_5, Co_3O_4, Fe_2O_3, NiO, SnO_2, ZnO, and Cr_2O_3. By measuring the conversion of butene into butadiene and $CO_2 + CO$ as well as the conversion of butadiene into $CO_2 + CO$ at various temperatures, Simons observed that the first reaction of 1-butene was usually double-bond isomerization, which was followed by conversion of butene into butadiene and finally into $CO_2 + CO$.

The data show that for a particular catalyst the rate of production of 2-butenes passed through a maximum with increasing temperature; at a higher temperature there was a maximum for butadiene, and at a still higher temperature there was an increase in the rate of $CO_2 + CO$ formation. As a characteristic measure of the activity of each catalyst, we consider the temperature at which 50 percent of the reactant was converted into $CO_2 + CO$ at a particular set of flow conditions. This measure allows a convenient comparison of the various catalysts. The data (Fig. 4-12) show that the temperature for 50 percent conversion to $CO_2 + CO$ increases in roughly linear fashion with the heat of reaction Q_0, defined as

$$MO_n \longrightarrow MO_{n-1} + \tfrac{1}{2}O_2 - Q_0 \tag{41}$$

The maximum selectivity to butadiene was found to increase through a maximum with increasing values of Q_0.

These data therefore conform to the pattern expected from the hypothesis of Sachtler and de Boer and provide a confirmation of it. The data also show that the oxides which are generally combined in the best selective oxidation catalysts (such as Bi_2O_3 and MoO_3 for bismuth molybdates) all have Q_0 values of about 50 to 60 kcal/mol.

The Mars–van Krevelen mechanism is rather generally applicable in partial-oxidation reactions catalyzed by oxide surfaces, but it is important to consider the limitations of this mechanism and the nature of oxygen held on catalyst surfaces. In the Mars–van Krevelen mechanism, the oxidizing agent is presumably O^{2-} from the lattice, which is abstracted by a reducing agent as electrons are donated to a lattice cation. Reoxidation by O_2 is usually assumed to lead to formation of O^{2-} in the lattice again:

$$O_2 + 2\square + 4e^- \longrightarrow \underline{2O^{2-}} \tag{42}$$

This reaction is presumed to take place with the simultaneous transfer of four electrons.

It is surprising that this pattern of oxidation is followed so often, since on the basis of the earlier discussion one might expect a stepwise donation of electrons

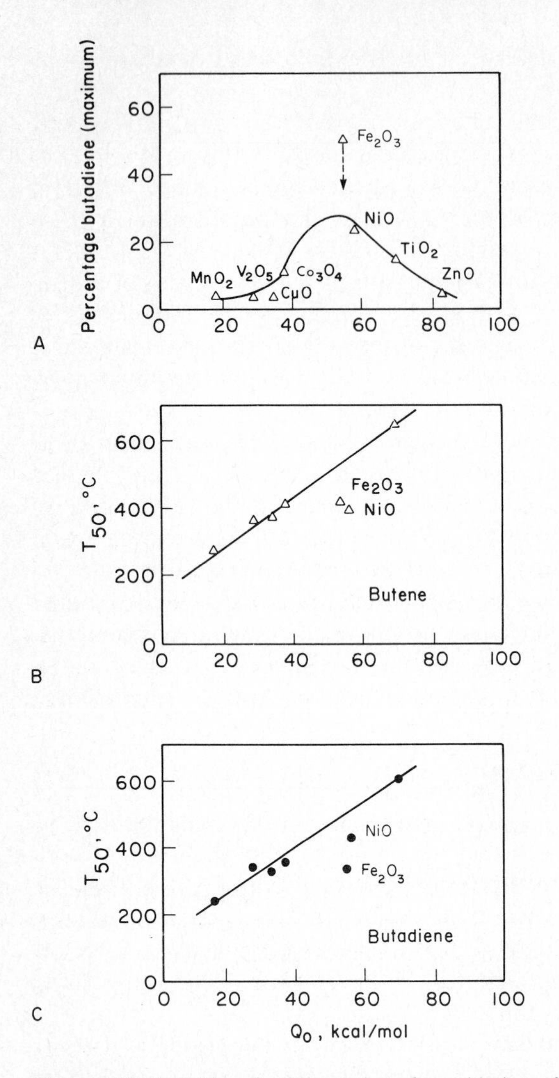

Figure 4-12 Activities of oxide catalysts for oxidation of butene and butadiene in a pulse reactor. Maximal butadiene production from butene occurred at intermediate values of Q_0 [see Eq. (41)] [24]. (Reprinted with permission from *Advances in Chemistry Series*. Copyright by the American Chemical Society.)

and reactions like those previously cited which proceed through free-radical intermediates, e.g.,

$$O_2 + \square + e^- \longrightarrow O_2^- \tag{43}$$

$$O_2^- + e^- \longrightarrow O_2^{2-} \tag{44}$$

$$O_2^{2-} + \square \longrightarrow 2O^- \tag{45}$$

$$O^- + e^- \longrightarrow O^{2-} \tag{46}$$

Indeed, recent work has indicated that even for typical ammoxidation catalysts, the Mars–van Krevelen mechanism does not apply at lower temperatures. Boreskov and his coworkers [25] found that as lower temperatures were approached, the rates of propylene oxidation catalyzed by bismuth molybdate in the absence of O_2 began to become less than rates of oxidation in the presence of O_2. Boreskov postulated the occurrence of an associative low-temperature reaction characterized by a low frequency factor and a low activation energy; at higher temperatures this reaction was negligibly slow compared with the common stepwise reaction characterized by a higher frequency factor and activation energy. Only the stepwise reaction is selective. Sancier et al. [26] studied propylene oxidation catalyzed by commercial bismuth molybdate in the presence of $^{18}O_2$. Their results, in agreement with those of Boreskov and in contrast to those obtained at higher temperatures by Keulks and coworkers [27, 27*a*], showed that the initial acrolein product contained ^{18}O and not exclusively ^{16}O from the oxide lattice. Therefore, Sancier et al. postulated that the low-temperature reaction involved some form of adsorbed oxygen, which at higher temperatures was converted into O_2^-. Boreskov's low frequency factor indicates, however, that the reaction observed is a concerted process between O_2 and olefin at the surface and might be related to the two-electron processes postulated earlier in this chapter.

Under favorable conditions the intermediate surface species postulated in Eqs. (43) to (45) might be detected by infrared absorption measurements, and O_2^- and O^- might be observed and even determined quantitatively by ESR spectroscopy. The evidence of the existence of O^- and O_2^- as well as of more complicated species such as O_3^- and perhaps O_4^- is now firmly established [28]. It is significant that signals indicating these species often disappear when the sample is brought into contact with an olefin or an aldehyde. These oxygen intermediates have not been observed on the selective oxidation catalysts of interest in this chapter, but they have been observed on some oxides, such as SnO_2 and V_2O_5, which tend to be unselective and to catalyze complete oxidation. It has been suggested [29] that the presence of these intermediates leads to chain reactions beginning at the surface and proceeding into the gas phase.

The variation in number and character of the oxygen species on oxide surfaces is reflected in the intricacies of ^{18}O–^{16}O exchange reactions as reported by several authors [30–35]. According to Stone [34], there are at least three different processes (denoted as R, R′, and R″), and more often than not, they occur simultaneously:

R′ is the exchange of oxygen atoms between gaseous $^{18}O_2$ and $^{16}O_s$ from the solid:

$$^{18}O_2 + {}^{16}O_s \longrightarrow {}^{18}O^{16}O + {}^{18}O_s$$

R″ is the multiple exchange between a molecule $^{18}O_2$ from the gas phase and a pair of ^{16}O ions from the solid:

$$^{18}O_2 + 2\,{}^{16}O_s \longrightarrow 2\,{}^{18}O_s + {}^{16}O_2$$

R is an exchange between $^{18}O_2$ and $^{16}O_2$, both in the gas phase in contact with an oxide surface.

The R process has been observed in the presence of ZnO [34], but it is uncommon and occurs only in association with well-outgassed surfaces [33]. This process might be related to the concerted oxidation reactions. The prevalence of the R′ and R″ processes for various oxides has been summarized as follows:

$$R': MgO, ZnO, TiO_2, Cr_2O_3, NiO$$

$$R'': V_2O_5, MoO_3, Bi_2O_3, WO_3, PbO, CuO$$

The most commonly applied catalysts and catalyst components for selective oxidation are found in the R″ class; they usually show an almost complete exchange of the oxygen in the bulk. Bismuth molybdate catalysts, however, do not give evidence of any exchange with O_2 unless the temperature exceeds 500°C [27, 36]. These catalysts do undergo exchange with $H_2{}^{18}O$ [19], which is accounted for as

$$H_2{}^{18}O + {}^{16}O\square{}^{16}O{}^{16}O \longrightarrow {}^{16}O{}^{18}OH{}^{16}OH{}^{16}O \qquad (47)$$

followed by dissociation of $H_2{}^{16}O$ after migration of H. These results suggest that bismuth molybdates do not have the required pairs of sites from which $^{16}O_2$ can be readily desorbed and $^{18}O_2$ adsorbed.

In summary, the oxygen-exchange and ESR evidence points to the existence of species like O_2^-, O_3^-, and O_4^- on oxide surfaces, but the evidence is that they are almost entirely absent from the selective commercial olefin oxidation catalysts like Bi–Mo or U–Sb oxides. This conclusion, however, does not imply that catalytic oxidation never occurs via surface species such as O_2^- or O^-. Indeed work by Kazansky and his collaborators [37], Tarama and his collaborators [38], and Akimoto and Echigoya [39], and others gives strong evidence that these species occur on, for instance, V_2O_5 and that they react with benzene, CO, and other oxidizable molecules.

There is even a commercial oxidation process for which the presence of O_2^- is essential for selective oxidation, namely, the oxidation of ethylene to ethylene oxide catalyzed by Ag. Ethylene is clearly a reactant from which an allylic intermediate cannot be formed. Infrared studies [40] and adsorption-rate measurements [41] showed that C_2H_4O formation occurs by the interaction of C_2H_4 with a bimolecular oxygen species on the Ag surface. The data indicate the presence of two types of adsorbed oxygen, which have been identified as those produced in the reactions

$$O_2 + 4Ag_{adj} \longrightarrow 4Ag^+_{adj} + 2O^{2-}_{ads} \qquad (48)$$

where Ag_{adj} refers to a member of a surface cluster of Ag atoms, and

$$O_2 + Ag \longrightarrow Ag^+ + O^-_{2,\,ads} \qquad (49)$$

Summary

The attack by a catalyst surface on an oxidizable olefin proceeds via several mechanisms, the most important and versatile being the fission of a hydrogen atom or ion to give an allylic surface species. Other mechanisms include the cis insertion and proton acceptance leading to the formation of oxidizable carbonium ions. In surface-catalyzed oxidation reactions, the introduction of oxygen into the reactant hydrocarbon proceeds primarily by noninsertion reactions, the species donating oxygen being an oxygen anion derived from the lattice of the oxide catalyst. For reactions at low temperatures or in the presence of high O_2/olefin ratios, reactions involving O_2 directly or surface species like O_2^- may predominate. Except for the oxidation of ethylene to ethylene oxide, data permitting development of a firmly based mechanistic model are lacking for reactions in this class.

AMMOXIDATION OF PROPYLENE

CATALYSTS

The ammoxidation reaction, as mentioned previously, is a conversion of mixtures of propylene and ammonia, and the main product is acrylonitrile

$$H_2C{=}CHCH_3 + NH_3 + \tfrac{3}{2}O_2 \longrightarrow H_2C{=}CHCN + 3H_2O \qquad (1)$$

Catalysts which are selective for this reaction are all oxide combinations or compounds containing at least two different metals. One of the metals always belongs to the later row 5*a* elements, including P, As, Sb, and Bi; the last two elements in this group give the most active catalysts. The second metal is almost always a transition metal. The best-known examples of ammoxidation catalysts are

$$\underset{\text{Bismuth molybdates}}{Bi_2O_3 + MoO_3} \qquad Fe_2O_3 + Sb_2O_4\ (FeSbO_4)$$

$$UO_3 + Sb_2O_4\ (USb_3O_{10}) \qquad SnO_2 + Sb_2O_4$$

In practice, these catalysts are supported on silica to give them attrition resistance in fluidized beds. The catalysts also contain small amounts of other elements such as K and P or (in the case of $FeSbO_4$ catalysts) W and Te.

Even more complicated catalysts have been described in the recent patent literature; they are mixtures of a number of molybdates, always containing bismuth molybdate as a component. An example is

$$Co_6^{2+}Ni_2^{2+}Fe_3^{3+}Bi^{3+}(MoO_4)_{12}O_2,$$

which is also supported on SiO_2 and contains small amounts of K and P.

The members of the group of ammoxidation catalysts have a number of characteristics in common, the most important being that they are highly selective for partial oxidation, having little activity for total combustion even at temperatures as high as 500°C. The oxidizing agent is always O^{2-} ions in the oxide lattice, and in the absence of O_2 the catalysts can be reduced.

Although the ammoxidation catalysts are active for the partial oxidation of olefins and are readily reduced by olefins at temperatures appreciably less than 400°C, they have low activity for the oxidation of H_2 or CO and they are reduced neither by H_2 nor by CO unless the temperature is considerably higher than 400°C. On the other hand, NH_3 reacts readily in the presence of any of these catalysts to give N_2, and NH_3 is a potent reducing agent, provided no olefins are present.

These patterns point to the conclusion that reduction of the catalyst and oxidation on its surface are similar reactions, taking place via mechanisms involving the fission of C—H and N—H bonds. The formation of allylic intermediates and NH_2 or NH surface species therefore appears to be characteristic of the ammoxidation catalysts. The details of catalyst surface structure and reaction mechanism are discussed later.

REACTIONS

Reactions possibly accompanying the ammoxidation of propylene include that giving acrolein

$$H_2C{=}CHCH_3 + O_2 \longrightarrow H_2C{=}CHCHO + H_2O \tag{50}$$

Side products in ammoxidation also include HCN and acetonitrile; the latter is a reaction product characteristic of molybdate catalysts.

In the absence of propylene, NH_3 is also oxidized to N_2:

$$2NH_3 + \tfrac{3}{2}O_2 \longrightarrow N_2 + 3H_2O \tag{51}$$

In the presence of some bismuth molybdates, olefins such as 1-butene undergo double-bond isomerization and oxidative dehydrogenation, giving, for example, butadiene.

The reaction network in propylene ammoxidation is far from being completely known. To a rough approximation, it has been written [42]

$$\begin{array}{ccc} H_2C{=}CHCH_3 & \xrightarrow{O_2,\ NH_3} & H_2C{=}CHCN + 3H_2O \\ & \searrow^{O_2} \quad {}^{NH_3}\nearrow & \\ & H_2C{=}CHCHO & \end{array}$$

Here, for simplicity, the previously cited side reactions are neglected. It is not known which of the above species can be converted directly into CO and CO_2, and in principle it must be assumed that all of them can be. A number of more complex reaction networks have also been suggested and have been compiled by Hucknall [43].

Good catalysts for ammoxidation are generally good catalysts for acrolein formation in the absence of NH_3. In the presence of NH_3, acrolein is observed only in small amounts in reaction products; evidently in the presence of the best ammoxidation catalysts, surface species which might give acrolein are for the most part converted into acrylonitrile before they can desorb as acrolein [42]. Therefore, acrolein itself is not a true intermediate in acrylonitrile formation on commercial catalysts.

KINETICS

The bismuth molybdate-catalyzed oxidation of propylene to give acrolein [42] and oxidative dehydrogenation of butene to give butadiene [44] have both been found to be first order in olefin and zero order in O_2 partial pressure at temperatures less than 400°C. Each reaction is inhibited by its product. The butene reaction is also inhibited by NH_3 [45]. These results point to the strong adsorption of butadiene, acrolein, and NH_3 on bismuth molybdate in competition with the reactants.

At temperatures exceeding 400°C, a first-order dependence on olefin partial pressure has been observed not only for the aforementioned oxidation reactions but for ammoxidation as well. Activation-energy data are collected in Table 4-2.

Recent experiments by Lankhuyzen et al. [47], however, led to the surprising result that the bismuth molybdate-catalyzed acrylonitrile-formation reaction may be zero order in the partial pressures of NH_3, O_2, and even propylene. To account for these kinetics, the authors suggested that oxygen was transported to the reaction site by diffusion through the solid catalyst. Oxygen adsorption would

Table 4-2 Apparent activation energies of bismuth molybdate-catalyzed oxidation reactions, each first order in olefin partial pressure[a]

Reaction	Catalyst	Apparent activation energy, kcal/mol	Ref.
$H_2C{=}CHCH_2CH_3 \longrightarrow H_2C{=}CH{-}CH{=}CH_2$	Bi_2MoO_6	12	44
$H_2C{=}CHCH_3 + \frac{1}{2}O_2 \longrightarrow H_2C{=}CHCHO$	Various bismuth molybdates	19–21	42
$H_2C{=}CHCH_3 + NH_3 + \frac{3}{2}O_2 \longrightarrow H_2C{=}CHCN + 3H_2O$	Various bismuth molybdates	19–21	46
	Commercial catalyst	17–19	42

[a] Further kinetics data have been compiled by Hucknall [43].

then be expected to take place at sites different from the catalytic sites and perhaps even on different surface planes [48].

Double-bond and cis-trans isomerization occur readily on molybdate catalysts at low temperatures (150 to 300°C), but they are kinetically insignificant at the higher temperatures used in ammoxidation. Since rates of the isomerization reactions increase with increasing partial pressures of water, it is probable that the absence of these reactions at the higher temperatures is a result of low concentrations of adsorbed water.

ADSORPTION

The adsorption of ammoxidation reactants and products on surfaces of oxidic catalysts has been studied by Matsuura [49–52]. His results show that in the pressure range 0.01 to 1 torr, the adsorption isotherms are well represented by the Langmuir form:

$$\theta = \frac{P_i^n}{P_i^n + P_{i,0}^n} \tag{52}$$

Here $P_{i,0}$ is a constant at constant temperature, and n is either 1 or $\frac{1}{2}$. Since $P_{i,0}$ is an equilibrium constant,

$$P_{i,0} = P_{i,0}^0 \exp - \frac{Q}{RT} \tag{53}$$

where $P_{i,0}^0$ is a constant defined by the entropy of adsorption and Q is the heat of adsorption. Adsorption equilibrium data conforming to Eq. (52) can be extrapolated easily to give the maximum (monolayer) surface coverages V_m.

On surfaces of oxides such as Fe_2O_3 or UO_3, the adsorption of 1-butene or butadiene is virtually irreversible, and desorption results only after the sample has been heated to temperatures so high that the adsorbed species decompose with evolution of CO and CO_2. The catalysts known to give this behavior are all highly active and nonselective.

Ammoxidation catalysts show a different pattern. Fully oxidized catalysts do not adsorb either O_2 or H_2O, but (partially) reduced catalysts adsorb O_2, H_2O, and NH_3. The adsorption of hydrocarbons and NH_3 on fully oxidized catalysts shows two types of behavior, called A type and B type.

A-type adsorption is characterized by rather high activation energies. It is observed only for butadiene, acrolein, and NH_3, the compounds which are inhibitors of ammoxidation and related reactions. Heats of adsorption are usually 20 to 30 kcal/mol, but for some catalysts the heat of adsorption may be so large that desorption occurs only at temperatures at which oxidation begins to take place. The V_m values indicating maximum surface coverages are in the ratio $V_m(NH_3)/V_m(\text{acrolein})/V_m(\text{butadiene}) = 2 : 1 : 1$.

The density of the sites responsible for A-type adsorption is low, corresponding to only approximately 10^{17} to 10^{18} molecules adsorbed per square meter; this value is independent of the surface area of a typical catalyst but varies from one catalyst to another. Prereduction of the catalyst causes the A-type adsorption to decrease, and there is a linear relation between the decrease in maximum adsorption and the degree of reduction.

The sum of butadiene A-site adsorption and H_2O adsorption has been found to remain constant with increasing reduction, and the same result has been found for the sum of A-site adsorption and half of the O_2 adsorption. Therefore, an A site is believed to incorporate one oxygen ion (O_A). To account for the adsorption of NH_3 and acrolein (for both of which $n = \frac{1}{2}$) and butadiene (for which $n = 1$), it is further assumed that O_A has two oxygen vacancies as neighbors. The vacancies are considered to be located on a metal ion (V_M). The description of an A site then becomes $V_M O_A V_M$, and the strongly adsorbed NH_3, acrolein, and butadiene are considered to have the following structures:

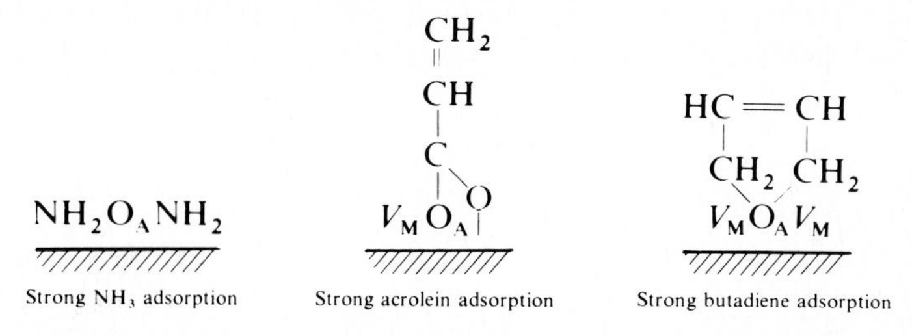

B-type adsorption, in contrast to the A type, is characterized by low activation energies and rather small heats of adsorption. It is common for olefins, butadiene, NH_3, and acrolein. The B-site density is invariably higher than the A-site density,

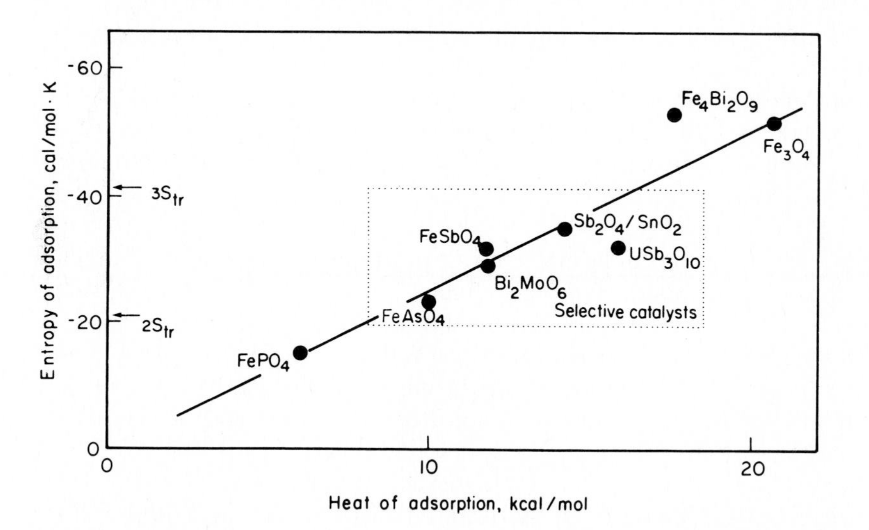

Figure 4-13 Adsorption of 1-butene on oxides: dependence of entropy of adsorption on heat of adsorption [53]. Reprinted with permission of the Chemical Society (London); copyright by the Chemical Society (London).

the ratio usually being 2. The maximum coverage is the same for almost all compounds adsorbed. The values of the exponent n are $\frac{1}{2}$ for most compounds except NH_3, which leads to the suggestion that an olefin may be dissociatively adsorbed as H plus a σ- or π-bonded allyl. B sites are not removed by previous reduction if the reduction occurs at temperatures less than 400°C.

Matsuura suggested that a B site is formed from an oxygen vacancy on a metal ion different from the one forming the A site. He has argued that B sites are associated with the transition-metal cations and A sites with the $5a$ row elements, but the identities of the sites are not resolved. Matsuura also observed [53] that the heats and entropies of adsorption of olefins on B sites are related linearly, as shown in Fig. 4-13. The catalysts for which the entropy of adsorption of 1-butene corresponds to a complete immobilization were found to be highly active but nonselective. Catalysts with low heats of adsorption, and therefore with low entropies of adsorption and high surface mobilities of butene, were inactive. Only catalysts that allowed an intermediate degree of mobility of adsorbed butene were found to be selective.

THE SELECTIVITY PROBLEM

In the preceding pages the selectivity of an oxidation catalyst has been identified as its most important characteristic, but the selectivity of ammoxidation catalysts has still not been explained. In seeking an explanation for selectivity, we must recognize that when an olefin reacts with an oxygen anion at a catalyst surface, it forms products which are usually more easily oxidized than the original reactant.

For example, acrolein in the presence of most catalysts is oxidized faster than propylene. In the presence of an ammoxidation catalyst, acrolein is formed at the surface, where it is surrounded by oxygen anions other than the one which formed it from propylene. Why then does it not react further? The first answer to come to mind is that, contrary to the original assumption, there are no available oxygen ions in its immediate environment. It could happen, for instance, that reoxidation of the catalyst would be relatively slow, so that the surface in operation would be extensively reduced.

Callahan and Grasselli [54] examined this possibility and came to the conclusion that it is entirely plausible for catalysts such as copper oxide, for which reoxidation is slow in comparison with reduction. This suggestion is compatible with the kinetics of copper oxide-catalyzed oxidation—first order in oxygen and zero order in propylene [55]. It is also in agreement with the observation that this catalyst in the fully oxidized state gives complete combustion, whereas a partially reduced catalyst is selective.

For most ammoxidation catalysts, however, the kinetics is different—zero order in oxygen and first order in hydrocarbon. Further, catalysts in the fully oxidized state are selective in ammoxidation. Callahan and Grasselli therefore accounted for selectivity by suggesting a mechanism of site isolation. The implication is that the oxygen anions on the surface fall into two distinct groups, the smaller group consisting of weakly bound oxygens and the larger group consisting of strongly bound oxygens, which are not readily available for hydrocarbon oxidation.

One might reason a step further by analogy with the Wacker reaction and suggest that the active sites, once reduced, could be replenished by diffusion of oxygen from the sites lacking catalytic activity. Reoxidation could then take place at positions derived from the inactive oxygen sites.

There are experimental observations which support the idea of islands of active oxygens in a sea of inactive oxygens. Evidence is given by Matsuura's observation that A sites occur only in low concentrations and his result that the A sites are adsorption sites for compounds such as butadiene and acrolein, which are reaction inhibitors.

Sleight and coworkers [56, 57] showed that defect scheelite structures containing Bi^{3+} and vacant cation sites are active and selective ammoxidation catalysts, although the particular scheelite matrix is nearly inactive. Defect structures may give rise to differences in properties of the oxygen anions, and for that reason the defect scheelite structures are worthy of a more extensive discussion even if they do not appear to be applied commercially.

The mineral scheelite, $CaWO_4$, has a structure typical of a number of compounds (such as $PbMoO_4$) having the composition AMO_4, where A is a divalent cation in an eightfold surrounding of oxygen anions and M is a tetrahedrally

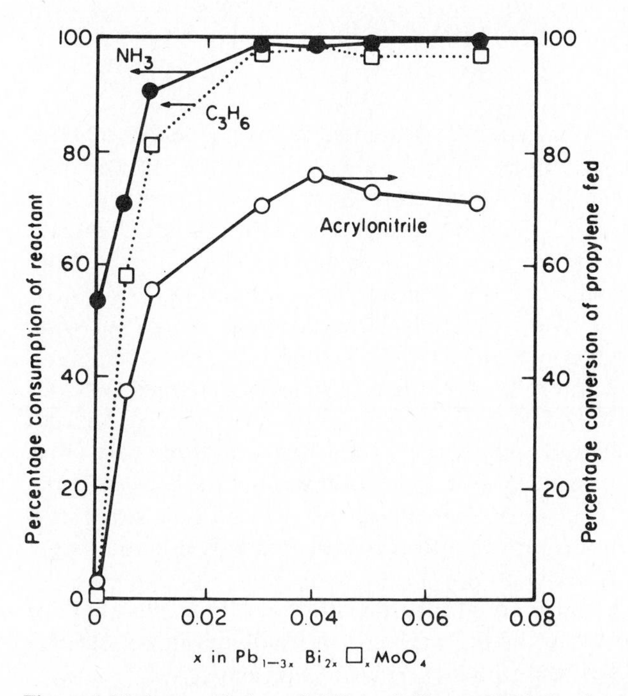

Figure 4-14 Ammoxidation of $NH_3 + C_3H_6$ on $PbMoO_4$ doped with Bi^{3+} [57]. (Reprinted with permission from *Journal of Catalysis*. Copyright by Academic Press.)

coordinated cation of high valency (such as Mo^{6+} or W^{6+}). Each A cation has eight MO_4 groups surrounding it, with each oxygen directly connected to A belonging to a different MO_4 group. Divalent A cations can be replaced by trivalent cations to give $A^{3+}_{2/3}\square_{1/3}(MO_4)$, where $\square$ is a cation vacancy; alternatively, two A^{2+} ions can be replaced by one trivalent and one monovalent cation to give $A^{3+}_{1/2}A^{+}_{1/2}(MO_4)$, but this replacement occurs only to a limited extent. The scheelite structure consequently contains defects that are not ordered but randomly distributed. Examples are given by $A^{3+} = Bi^{3+}$ and $A^{3+} + A^{+} = Bi^{3+} + Na^{+}$.

The catalytic behavior of scheelites for ammoxidation as a function of composition is given for $Pb_{1-3x}Bi_{2x}\square_x(MoO_4)$ in Fig. 4-14. The activity and selectivity sharply increased with x. The catalyst $Na_{1/2}Bi_{1/2}(MoO_4)$ was hardly active, and the conclusion is that activity for ammoxidation is associated with the simultaneous presence of Bi^{3+} ions and a cation vacancy. Catalysts with these types of defects are easily reduced and reoxidized. Evidently oxygen anions associated with cation vacancies are less tightly bonded than others and are easily removed by reduction. The presence of Bi^{3+} seems to be necessary for the oxidation step on the surface. If it is presumed that surface layers have defects similar to those in the bulk, it is clear from the structure that there are a few active oxygens on the surface in a sea of less active ones.

REACTION MECHANISM: A SIMPLIFIED INTERPRETATION

The foregoing background information, complemented by recent results of work on the incorporation of oxygen into the solid catalyst, allows us to formulate a picture of the ammoxidation reaction mechanism. As mentioned earlier, experiments reported by Keulks et al. [27] and by Wragg et al. [36] did not show any exchange of $^{18}O_2$ with ^{16}O from the bulk of the catalyst in the absence of reaction. When $^{18}O_2$ was allowed to react with propylene in the presence of $Bi_2Mo^{16}O_6$ or $Bi_2Mo_3{}^{16}O_{12}$, however, ^{18}O was gradually incorporated in the product (Fig. 4-15). Once the catalyst becomes reduced, ^{18}O can enter into its structure. As the data of the figure show, however, there is no immediate changeover from ^{16}O to ^{18}O in the product, as would be expected if the entire reaction process were confined to surface layers. What is apparently needed before ^{18}O appears in an appreciable amount in the product is a substantial exchange of the oxygen in the bulk. Keulks et al. [27] even concluded that the rate of the catalytic reaction is determined by the rate of diffusion of ^{18}O into the bulk, which is in good agreement with the results of Lankhuyzen et al. [47].

It is noteworthy that a similar gradual incorporation of ^{18}O is observed not only in the acrolein product but in the CO_2 formed in small amounts; selective oxidation and complete combustion evidently occur at the same site.

Otsubo et al. [58] prepared Bi_2MoO_6 with the ^{18}O concentration in the molybdenum layers or in the Bi_2O_2 layers. Their results indicate that propylene is

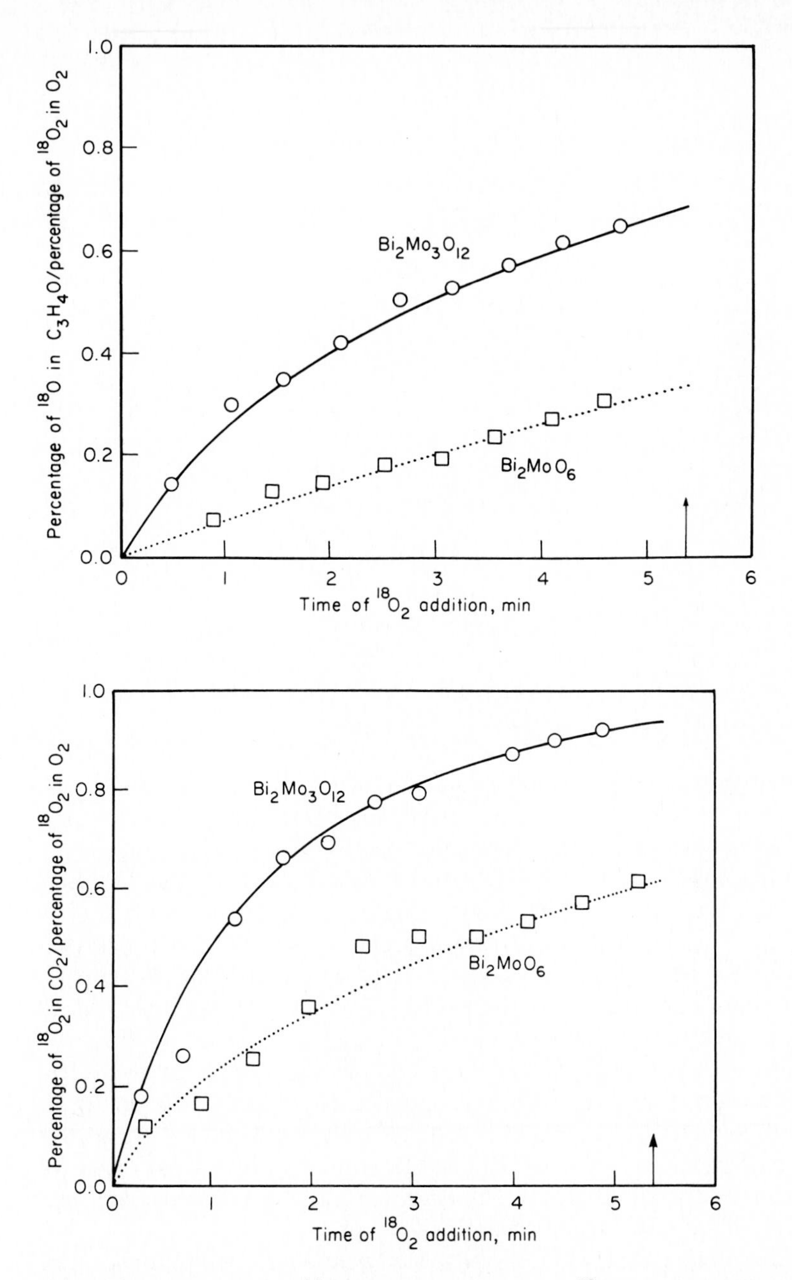

Figure 4-15 Incorporation of ^{18}O in products of reaction of ^{18}O and propylene on bismuth molybdate catalyst at 430°C. At the time indicated by the arrow, the ^{18}O feed was stopped [27*a*]. [Reprinted with permission of the Chemical Society (London): copyright by the Chemical Society (London).]

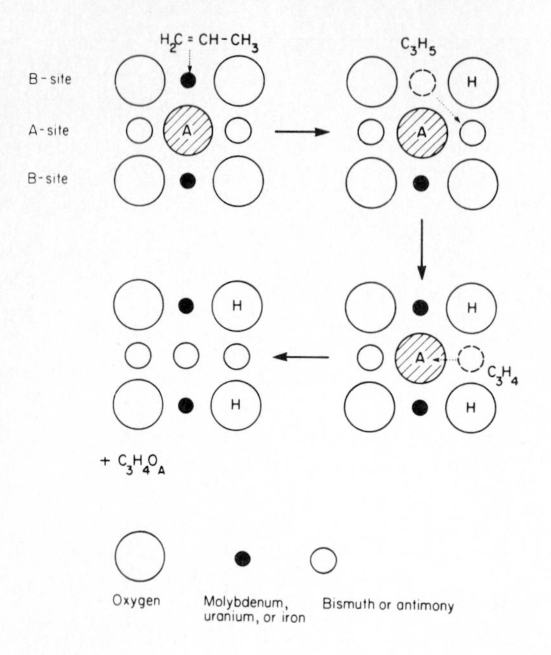

Figure 4-16A Mechanism of propylene oxidation to acrolein on an ammoxidation catalyst according to Matsuura.

oxidized by the Bi_2O_2 layers while bulk migration occurs from the MoO_2 layers. The migration of oxygen ions produces anion vacancies in the MoO_2 layers which can then serve as the adsorption sites for the oxidation.

The nature of the actual site at which the oxidation starts is still not clear. Matsuura's oxidation mechanism [50] is based on his adsorption studies and is given in Fig. 4-16. The reaction steps include a migration of the hydrocarbon intermediate from a B site (Mo) to an A site (Bi) and again to a B site. There is a stepwise dissociation of the hydrogens on the B sites and incorporation of O (or of NH when NH_3 is present) on the A site.

A model presented by Haber and Grzybowska [59] offers an alternative interpretation. Formation of the allylic species is assumed to occur directly on the 5*a* cation with the dissociation of a second H atom requiring a B site (a vacancy on Mo in the case of bismuth molybdate). This model explains why formation of the allylic species occurs also on Bi_2O_3 (a dimerization catalyst), whereas acrolein formation occurs only on Bi–Mo combinations.

Sleight [56] and Linn and Sleight [56*a*] proposed a model related to those given above but incorporating some features that merit special attention. It is summarized in Fig. 4-17 for the oxidation of propylene to acrolein. Propylene is assumed to be adsorbed on a MoO_4^{2-} group associated with a Bi-cation vacancy, leading to formation of an allylic intermediate and donation of a proton to a neighboring and equivalent MoO_4^{2-} group. The allyl then donates a second proton to this group, desorbs from the surface as acrolein, and converts the first MoO_4 group into

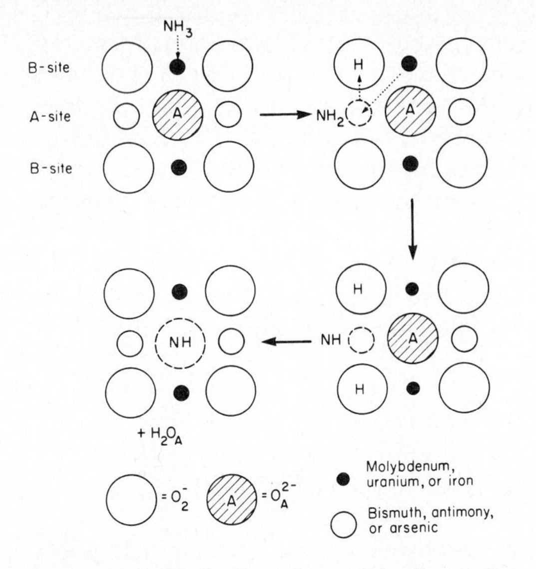

Figure 4-16B Mechanism of ammonia dissociation on an ammoxidation catalyst according to Matsuura. The NH group can react further with the allyl of Fig. 4-16A to give acrylonitrile.

MoO_3; the second group splits off H_2O and is also left with one oxygen missing:

$$C_3H_6 + 2MoO_4^{2-} \longrightarrow (C_3H_5 \cdot MoO_4)^{2-} + H(MoO_4)^{2-} \tag{54}$$

$$(C_3H_5 \cdot MoO_4)^{2-} + H(MoO_4) \longrightarrow C_3H_4O + H_2O + 2MoO_3 + 4e^- \tag{55}$$

The electrons are donated to an overlapping system of an empty Bi 6p conduction band and Mo 4d states. They are subsequently donated to an incoming oxygen molecule. The removal of two lattice oxygens associated with Mo serves to lower

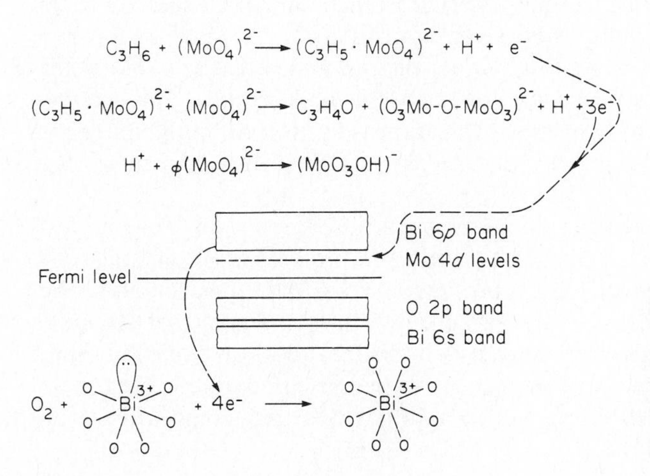

Figure 4-17 Redox mechanism for selective propylene oxidation on defect scheelites (adapted from [56a] and printed with permission of the New York Academy of Sciences).

the oxygen surrounding of a Bi^{3+} ion from eight to six; the O_2 molecule taking up four electrons of the conduction band to give two O^{2-} ions fills the two anion vacancies and restores both the MoO_4 configurations and the Bi^{3+} surrounding.

The model resembles Matsuura's in that the first attack occurs on a molybdenum site. However, migration of oxygen occurs from Bi to Mo, and in this respect it is closer to Keulks' model. It is noteworthy in this regard that the kinetics of the oxidation on the doped $PbMoO_4$ differs considerably from those on the bismuth molybdates since there is a marked dependence of the rate on the oxygen partial pressure and apparent activation energies are of the order of 30 to 35 kcal/mol. Oxygen adsorption and migration are apparently more difficult than on bismuth molybdates.

STRUCTURES OF AMMOXIDATION CATALYSTS

Since a detailed interpretation of the ammoxidation reaction mechanism requires incorporation of catalyst surface structures, the structures of uranium-antimony and bismuth molybdate catalysts are discussed in the following section. The uranium-antimony catalysts were applied industrially, and the interpretation of catalyst structure and reaction mechanism is due to Grasselli and his coworkers of Sohio, who discovered the catalysts. The older bismuth molybdate ammoxidation catalysts, the first to find industrial application, have been investigated thoroughly in several laboratories. Although long obsolete, these catalysts are sufficiently similar to modern catalysts to provide much of the foundation for interpreting their behavior.

Industrial ammoxidation catalysts are intricate combinations consisting of (1) a binary oxide system forming the actual catalyst, (2) a silica support, and (3) compounds such as K_2O, P_2O_5, or TeO_2 added in minor amounts. The structural studies reported in the literature have been focused on the simple unsupported binary oxides, which are unquestionably similar in their catalytic character to the more complex supported oxides. These structural studies have led to the unraveling of phase diagrams, isolation of individual phases, and, when possible, determination of their structures from experiments with single crystals. This strategy has provided considerable insight into such catalysts as bismuth molybdates and USb_3O_{10}, but the catalysts currently applied are much more complicated than these binary compounds.

Catalysts are usually prepared by precipitation from solution. They are obtained as powders giving x-ray diagrams similar to those of single crystals. The x-ray lines are usually broader than those of single crystals, however, either because of the small particle size or the presence of defects. Since defects may be important to the catalytic performance, it is important to know their structures and the extent to which they may be present. In particular, surfaces of catalysts may act as defects, and therefore surface symmetries and compositions are important.

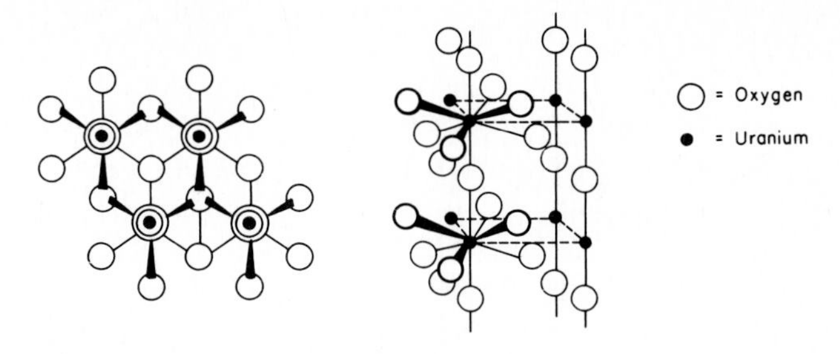

Figure 4-18 Structure of UO_3 [61].

The UO_3–Sb_2O_4 Catalyst

The composition of the catalyst which has been used in industrial ammoxidation processes is given as $USb_{4.6}O_{13.2}$ on SiO_2 in a weight ratio of 60 : 40 [42]. Grasselli et al. [60] reported the presence of two binary oxides of U and Sb in this catalyst. Later work by Grasselli and Suresh [13] and Aykan and Sleight [60*a*] established the compositions as $USbO_5$ and USb_3O_{10}.

The crystal structures of these compounds are related to that of UO_3. The key building element is the U^{6+} ion in an eightfold surrounding of O^{2-} ions, as shown in Fig. 4-18. The eight O^{2-} ions form two groups, a set of two designated as O_I forming a linear arrangement, $O_I—U—O_I$, and a group of six designated as O_{II} forming a distorted octahedron with the $O_I—U—O_I$ axis parallel to one of the c_3 axes. (The c_3 axis is perpendicular to the plane of the paper.)

The modifications of the UO_3 structure originate in slightly different ways which lead to oxygen sharing between UO_8 polygons. There is a one-phase region in the composition range between U_2O_5 and UO_3, in which the various structures continuously transform from one to another [61]. As the oxygen content is lowered until the composition becomes UO_2, another eightfold surrounding, similar to that in CaF_2, is formed.

The structures of the antimony oxides Sb_2O_3 and Sb_2O_4 are known, but that of Sb_2O_5 is not; Sb_2O_4 has a rutile structure, that is, $Sb^{3+}Sb^{5+}O_4$, with both antimony cations in octahedral surroundings. There are at least two forms of Sb_2O_3. In the high-temperature form, which is cubic (senarmontite), there are Sb_4O_6 entities similar to those in the gas phase; they have four cations at the corners of a tetrahedron, each edge of the tetrahedron being bridged by an oxygen ion. The low-temperature form (valentinite) is a double-chain structure in which chains of $(SbO)_n$ are cross-linked by O between two Sb cations from two chains.

The structures of UO_3, $USbO_5$, and USb_3O_{10} can be illustrated approximately if UO_8 and SbO_6 are represented as polygons in a direction perpendicular to the $U_I—O—U_I$ axis, as follows:

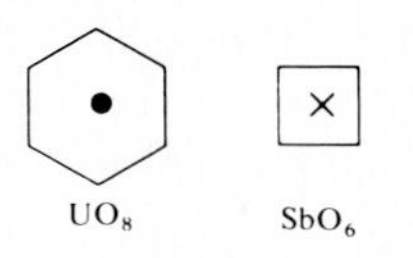

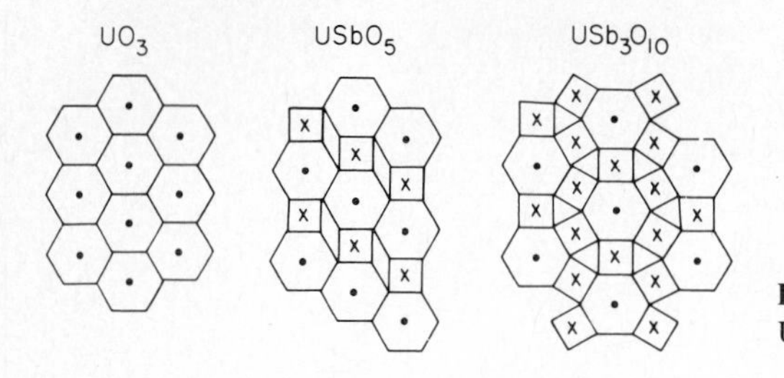

Figure 4-19 Structures of UO_3, $USbO_5$, and USb_3O_{10}.

The structures UO_3, $USbO_5$ and USb_3O_{10} are then the ones shown in Fig. 4-19.

The cation positions in USb_3O_{10} in a projection perpendicular to the aforementioned axis are shown in Fig. 4-20. The complete unit cell for USb_3O_{10} is given in Fig. 4-21. The unit cell has eight formula weights of USb_3O_{10}, each containing one type of uranium, two types of antimony, and four types of oxygen. The structure is composed of layers; alternating with layers containing heavy atoms and oxygens are layers containing only oxygen.

Of these compounds only USb_3O_{10} is an active and selective oxidation catalyst; Sb_2O_4 is inactive, and $USbO_5$ and UO_3 catalyze complete combustion, as also found by Simons et al. [62] for butene oxidation. Grasselli and Suresh [13] prepared $USbO_5$ with a small amount of extra Sb_2O_3 by impregnation in the presence of nitric acid, forming a catalyst with the composition $USb_{1.036}O_{5.09}$, about sufficient to give three monolayers of USb_3O_{10} on $USbO_5$. Short calcination periods provided a relatively selective catalyst, but after longer times the catalyst became considerably less selective, presumably because of diffusion of the surface components into the interior. These results confirm the identification of USb_3O_{10} as the actual catalyst.

Grasselli and Suresh also prereduced a catalyst with propylene, NH_3, or H_2, which removed oxygen. Removal of up to 6.5 percent of the lattice oxygen (corresponding to a reduction of USb_3O_{10} to $USb_3O_{9.5}$) led to no change in catalytic activity, and the catalyst was easily reoxidized. Higher degrees of reduction, however, led to a decrease in activity, which was accompanied by a decrease in selectivity for partial oxidation to acrolein. Grasselli and Suresh inferred from

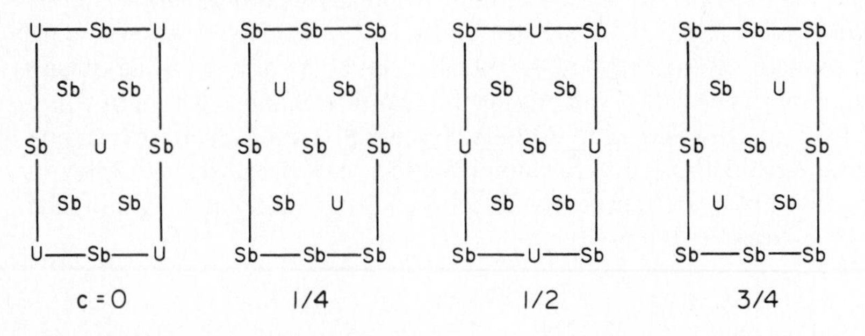

Figure 4-20 The cation positions in USb_3O_{10} [13]. (Reprinted with permission from *Journal of Catalysis*. Copyright by Academic Press.)

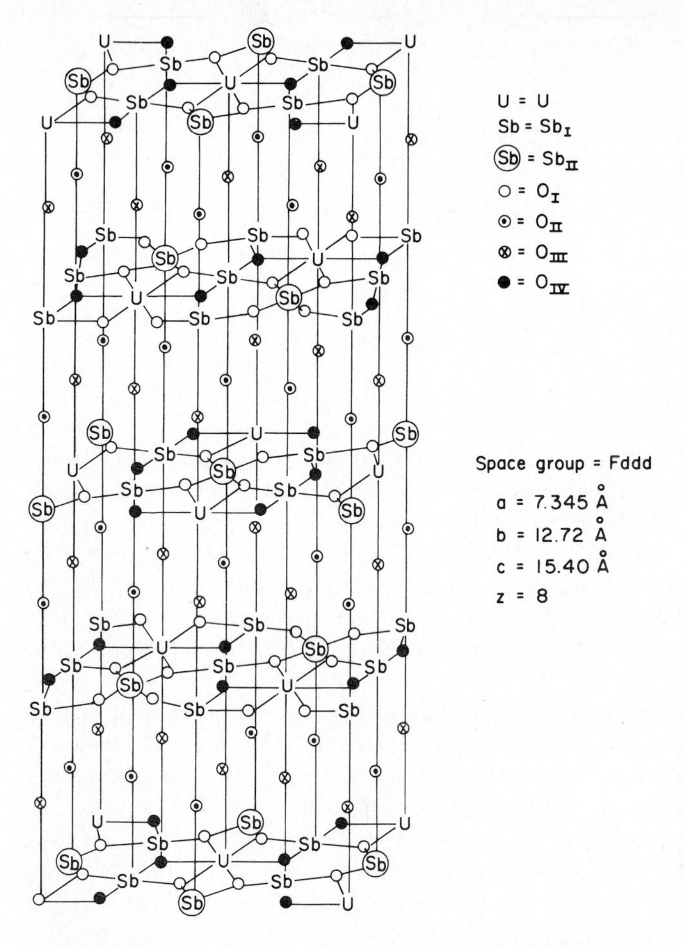

Figure 4-21 Unit cell of the USb_3O_{10} phase in the U-Sb ammoxidation catalyst [13]. (Reprinted with permission from *Journal of Catalysis*. Copyright by Academic Press.)

these results that there were two types of lattice oxygen, one giving fast and selective oxidation of propylene and the other corresponding to a thousandfold lower activity and giving CO_2 and CO. From crystallographic plausibility, Grasselli and Suresh suggested that the active oxygen is O_{IV} and the less active oxygen is O_I, as shown in Fig. 4-21.

Further information about the surface structures can be inferred from the adsorption equilibrium data of Matsuura, already mentioned. For USb_3O_{10}, Matsuura [52] observed A sites (giving strong, activated adsorption of butadiene and NH_3 in the volume ratio of 1 : 2) and B sites (giving weak, fast adsorption of butene, butadiene, etc.). The ratio of B sites to A sites was 2, as mentioned previously. The density of A sites on USb_3O_{10}, as calculated from butadiene adsorption data, is even less than on Bi_2MoO_6, corresponding to 1 site per 1800 Å^2.

Table 4-3 Adsorption of 1-butene on oxides of U and Sb

Adsorbent	Type of adsorption	Surface area per adsorbed molecule, Å^2	Surface area per oxygen removed, Å^2
UO_3	Strong, irreversible	230	19
U_3O_8	Strong, irreversible	220	19
$USbO_5$	Strong, irreversible	400	41
USb_3O_{10}	Weak, reversible, dissociative (B sites)	880	None removed

Source: Adapted from [52].

UO_3, $USbO_5$, and Sb_2O_4 are quite different in their adsorption characteristics. Hardly any adsorption was observed in experiments with Sb_2O_4, but with UO_3 and $USbO_5$, every molecule adsorbed was found to be almost irreversibly bound. For example, 1-butene, which was weakly and reversibly bound to the USb_3O_{10} surface, was strongly and irreversibly adsorbed on UO_3 and on $USbO_5$ and could be removed from the surface only at high temperatures after it was converted to $CO_2 + H_2O$. The foregoing adsorption results are summarized in Table 4-3.

These results, combined with the preceding discussion, strongly suggest that butene adsorption occurs on an anion vacancy associated with a U cation and that it is weak and reversible if these vacancies are isolated, as expected on USb_3O_{10}, but becomes strong and irreversible when pairs of the vacancies are formed, as would be expected on the surfaces of UO_3 and $USbO_5$. Since B sites are different from A sites on USb_3O_{10}, and since A sites are supposed to be responsible for the selective oxidation, the latter can be identified with surface oxygens bonded to Sb rather than U.

Reaction Mechanisms on USb_3O_{10}

Mechanisms of the conversion of propylene to acrolein and acrylonitrile catalyzed by USb_3O_{10} have been proposed by Grasselli and Suresh [13], as shown in Fig. 4-22. Dehydration of the structure shown in the upper left-hand corner is supposed to form a surface structure with an oxygen anion linking a U^{5+} ion to a Sb^{5+} ion. This structure can rearrange to leave an anion vacancy over the Sb^{5+} ion and an oxygen anion on U^{5+}. Allyl formation is then envisioned to occur on Sb^{5+}, the first hydrogen being donated to the oxygen on U. After an electron transfer, hydrogen is donated to the anion common to U and Sb. Subsequently, H_2O is dissociated, leaving an anion vacancy situated at the U—O—Sb position. This vacancy can now be filled again by the transfer of O (which leads to acrolein formation) or by NH from NH_3 and the transfer of N (which leads to acrylonitrile formation). The C_3H_4 residue can react directly with the transferred oxygen to

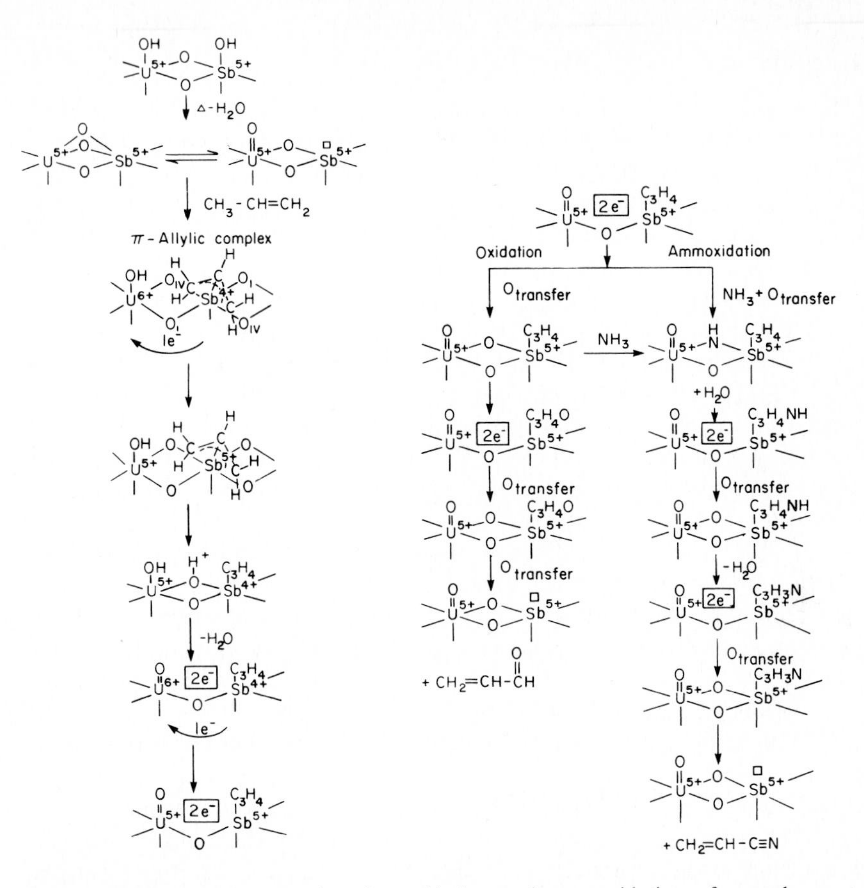

Figure 4-22 Proposed mechanism for oxidation and ammoxidation of propylene catalyzed by USb_3O_{10} [13]. (Reprinted with permission from *Journal of Catalysis*. Copyright by Academic Press.)

form acrolein, and the formation of acrylonitrile requires donation of NH to C_3H_4 followed by transfer of O to the anion vacancy, dissociation of H_2O to form C_3H_3N, another transfer of O, and finally dissociation of product acrylonitrile. Grasselli and Suresh gave no suggestion about how O_2 is incorporated into the lattice of the catalyst, from which it is then transferred to the catalytic site.

The most important details of the proposed mechanisms are the hypothesis of the π-allyl attachment to the Sb ion and its transformation into a highly hydrogen-deficient C_3H_4 moiety, which may be a vinyl carbene, $CH_2{=}CH{-}\dot{C}H$. The carbene is also supposed to be bonded to the Sb ion, where an O atom is added (by O transfer) or an NH group is added.

The preceding interpretation of reaction mechanism has strong parallels to that of Matsuura, but there are differences, and the Grasselli-Suresh mechanism more closely resembles that of Haber and Grzybowska.

$FeSbO_4$ Catalysts

The $FeSbO_4$ catalyst has been applied commercially for ammoxidation in Japan [63–66]. Its composition is $Fe_{10}Sb_{25}Si_{50}$, and the active components are $FeSbO_4$, Sb_2O_4, and SiO_2. Some promoter components are added, such as W (which seems to facilitate the reoxidation), Te (which provides better activity), and P. The properties of the commercial catalyst appear to be similar to those of the U–Sb catalyst.

The catalyst offers the possibility of Fe and Sb Mössbauer spectroscopy for investigation of the relations between bulk structure and catalytic activity. Skalkina et al. [67] investigated the Mössbauer spectra of a series of catalysts containing octahedral Fe^{3+}. The important parameter was found to be the quadrupole shift (QS). A small QS was indicative of active but nonselective catalysts. An increase in QS corresponded to a decrease in activity for complete oxidation and an increase in activity for formation of acrylonitrile. Further increases in QS corresponded to loss of activity.

Matsuura [53] found that the heats of adsorption of butene on B sites decreased in the same direction with increasing QS; the catalysts moved from the upper right corner to the lower left corner of Fig. 4-13. The QS parameter is expected to increase with increasing distortion of the oxygen surrounding of the cation in the bulk. The correlation of the bulk property with surface properties indicated by adsorption and catalysis experiments provides, at least for this one example, a confirmation of the idea that surface structures bear some relation to bulk structures.

Kriegsmann et al. [67*a*] investigated both ^{57}Fe and ^{121}Sb Mössbauer spectra of Fe_2O_3–Sb_2O_5 catalysts together with their activities for the acrylonitrile synthesis. The maximum acrylonitrile yield occurred for a quadrupole splitting of 0.83 mm/s, indicating the necessity of a moderately distorted lattice or change in M—O bond strength. The maximum activity was found at a Sb/Fe ratio of 5. In good catalysts, both Sb^{3-} and Fe^{2+} must be present.

Bismuth Molybdate Catalysts

Structures of the active catalysts The first catalyst applied to the commercial synthesis of acrylonitrile from propylene and NH_3 had the composition 50 weight percent $Bi_9PMo_{12}O_{52}$ + 50 weight percent SiO_2 [68, 69]. This catalyst evidently derives its activity and selectivity from some compound of Bi_2O_3 and MoO_3, since Bi_2O_3, MoO_3, and their mixtures all have catalytic properties significantly different from those of the catalyst which was used commercially by Sohio. Since the oxidative dehydrogenation of 1-butene to give butadiene is a convenient and simple model reaction, much was done with this reaction to identify the catalytically active component in the catalyst. It was found that there was a range of Bi/Mo atomic ratios, 2 : 1 to 2 : 3, exhibiting high activity combined with good selectivity [48]. In this range three compounds were found, one of these occurring in two modifications:

Compound	Ref.
$Bi_2Mo_3O_{12}$(α, 2 : 3)	48, 71, 72
$Bi_2Mo_2O_9$(β, 1 : 1, Erman phase)	70, 71
Bi_2MoO_6[two modifications, 2 : 1, γ (koechlinite) and γ']	70–73

The α phase is stable with a melting point around 700°C. The β phase is incongruent and stable only between approximately 550 and 670°C. At temperatures less than 550°C, it decomposes slowly into α and γ and at temperatures exceeding 670°C, it decomposes rapidly into α and γ'. The Bi_2MoO_6 occurs in two modifications. One has an x-ray diagram similar to that of the mineral koechlinite; it is metastable, and heating to temperatures in excess of 660°C produces an irreversible transition to the γ' phase.

Oxidation catalysts are prepared by precipitation from mixed solutions of bismuth nitrate and ammonium molybdate. Depending on conditions such as the Bi/Mo ratio, pH, temperature, and the time of interaction, precipitates are formed, which after calcination at 500°C, may be α, β, γ, or their mixtures [74–77]. The γ' phase is not usually formed.

The crystal structures of the various bismuth molybdates are best understood after an introduction to the symmetry of the oxygen surroundings of Mo^{6+} and Bi^{3+} ions. The hexavalent Mo cation occurs in oxygen coordinations varying from 4 to 6. These coordinations are related to the tetrahedral, trigonal bipyramidal, or octahedral surroundings, often in a strongly distorted form. In the sequence ReO_3, WO_3, MoO_3, CrO_3, the predominant coordination changes from pure octahedral to pure tetrahedral; MoO_3, occupying an intermediate position in this series, adopts both coordinations. In MoO_3 the distortion is so strong that it is difficult to decide whether the surrounding is better described as tetrahedral or octahedral. The crystal structure is about as well represented by infinite chains of corner-shared tetrahedra as by a system of octahedra (Fig. 4-23). In the latter case these octahedra form sheets in which there is edge sharing with two adjacent octahedra and corner sharing with two others [78].

Edge sharing of octahedra is common in the oxides of Mo. The always present but often variable distortion of the octahedra does not mean that there is only one type of surrounding that can be interpreted in different ways. For polymorphic compounds such as $CoMoO_4$ (to which we return in Chap. 5) the various modifications differ because the MoO polyhedra are either more nearly octahedral (with edge sharing) or tetrahedral.

A partial reduction of MoO_3 leads to a loss of oxygen and the formation of nonstoichiometric compounds with the general formula MoO_{3-x}. Some of these share blocks of corner-sharing octahedra, which share edges with similar blocks (*shear structures*). They have the general formula Mo_mO_{3m-1}; a picture of

MoO_3

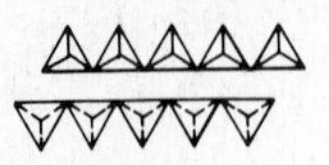

(a) represented as infinite chains of corner-shared tetrahedra

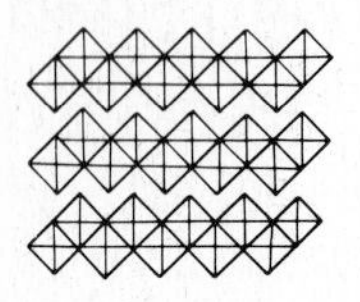

(b) represented as sheets formed from octahedra sharing edges with two adjacent octahedra (in the plane) and corners with two others (above and below plane).

Figure 4-23 Crystal structures of MoO_3 [78]. (Reprinted with permission of Academic Press; copyright by Academic Press.)

Mo_8O_{23} is given in Fig. 4-24. Others contain six- and seven-coordinated metal cations, as in Mo_5O_{14}, for which MoO_6 octahedra are linked to MoO_7 pentagonal bipyramids. In Mo_4O_{11}, slabs of octahedra are joined by tetrahedra. In MoO_2, the structure is similar to the rutile structure, but the octahedral surrounding is disturbed because of the presence of short Mo–Mo distances indicating cation-cation bonding.

In summary, it is known that Mo^{6+} adopts various oxygen coordinations, none of which is particularly well represented as tetrahedral or octahedral. The oxygen polyhedra can be linked in different ways, e.g., by corner or by edge sharing. Molybdenum oxide-containing lattices have a remarkable flexibility toward reduction: loss of oxygen from the lattice does not result in collapse but in a lattice restructuring to make it more spacious.

The crystal chemistry of Bi^{3+} is also complex and less than completely understood. There are at least two different oxygen surroundings. One is derived from a

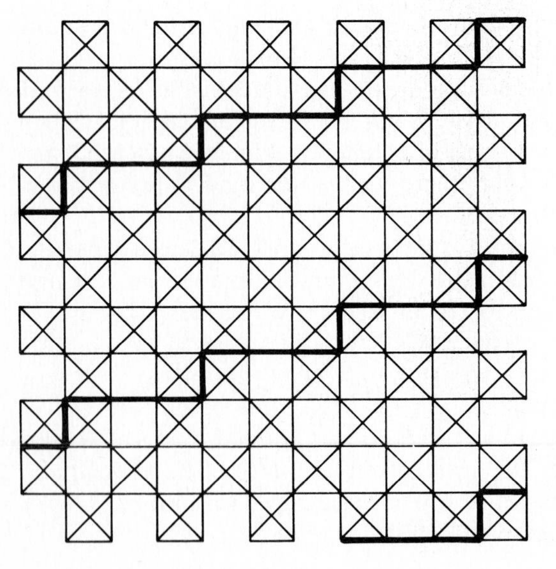

Figure 4-24 Structure of Mo_8O_{23} (shear structure). The heavy lines are shear planes [78]. (Reprinted with permission of Academic Press; copyright by Academic Press.)

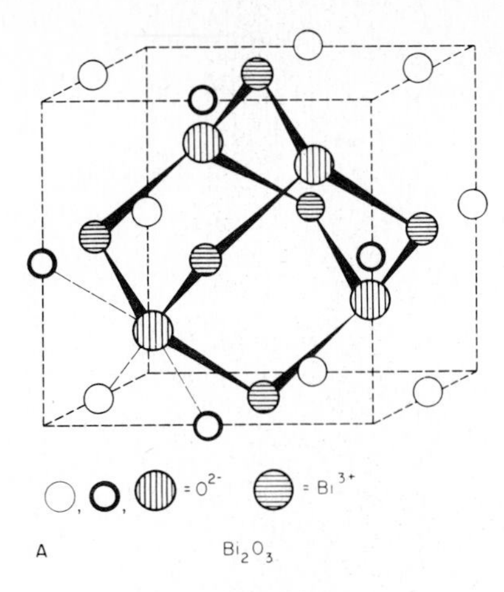

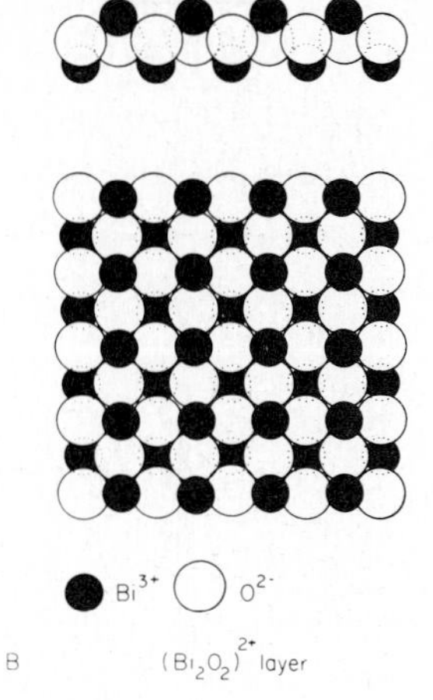

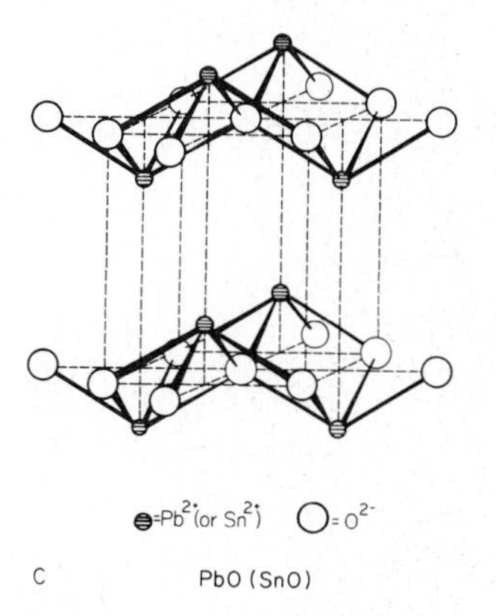

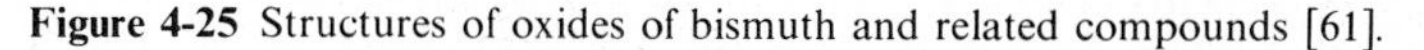

Figure 4-25 Structures of oxides of bismuth and related compounds [61].

cubic eightfold surrounding by the removal of two oxygens at opposite corners of the cube and is therefore a special type of six surrounding (Fig. 4-25*A*). Another is found in compounds such as BiOCl, in which the Bi^{3+} ions are connected to oxygen layers arranged in a square pattern with the Bi^{3+} ions placed alternatively above and below the centers of the oxygen squares (Fig. 4-25*B*), thereby forming a $(BiO)_n^+$ sheet. Similar arrangements are found in SnO or red PbO, for which the sheets are stacked on top of each other (Fig. 4-25*C*). Each cation is then present in a cubic oxygen arrangement but displaced from the center toward one of the faces.

Zemann [79] proposed a structure for the mineral koechlinite (γ-Bi_2MoO_6) as represented in Fig. 4-26. It was later worked out in more detail by van den Elzen and Rieck [80]. Zemann's model is that of a layer structure with alternating sheets of Bi_2O_2 layers as found in BiOCl and layers consisting of Mo^{6+} ions in octahedral surrounding, the octahedra sharing corners in the sheets and their apices directed toward the Bi_2O_2 layers. In this model the oxygens in the sheets have a square-planar arrangement, but Zemann reported a strong distortion of the squares. Van

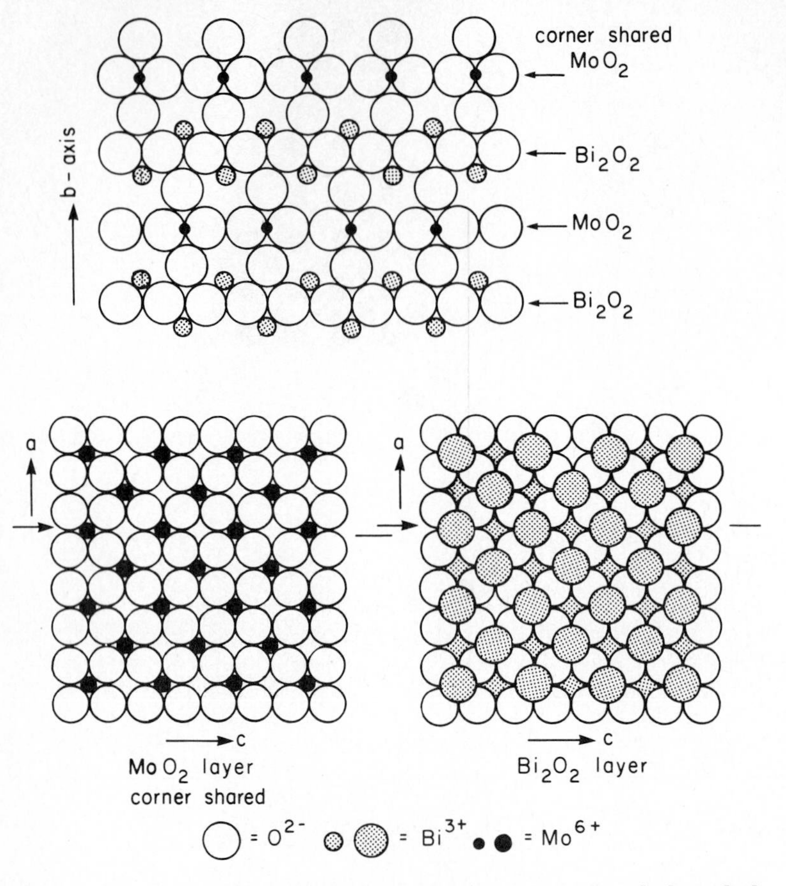

Figure 4-26 Structure of γ'-Bi_2MoO_6 (koechlinite) as inferred from [79] and [80].

den Elzen found the sheets somewhat warped and gave the following bond lengths. In the sheets there are two types of Mo—O bonds, two at 1.75 Å and two at 2.25 Å, and the Mo—O bonds to the apical octahedral oxygens are 1.8 Å. The Bi—O bonds in the sheets also occur in two pairs, two shorter than 2.3 Å and two longer; the distances to the apical oxygens of the Mo octahedra are 2.3 and 2.7 Å. The structure is therefore intermediate between one having Bi_2O_2 and a two-dimensional ReO_3-type of corner-sharing MoO_6 octahedra, on the one hand, and one having a $(Bi_2O_2)(MoO_4)$ structure with slightly distorted MoO_4 tetrahedra ordered in an infinite two-dimensional cluster, on the other hand.

Blasse [73] suggested that in γ-Bi_2MoO_6 (the high-temperature modification), the MoO_4 tetrahedra are separate.

The structure of $Bi_2Mo_3O_{12}$ was clarified by Cesare et al. [81] and van den Elzen and Rieck [82]. It can be derived from the structure of scheelite (see [56]). To arrive at the bismuth molybdate structure, three A cations have to be replaced by two Bi^{3+} ions, leaving one cation vacancy. The cation vacancies are ordered as

$Bi_2Mo_3O_{12}$

A

Projection along b-axis of Bi-sites. Large solid circles: occupied Bi-sites; small open circles, empty; broken lines; unit cell of $CaWO_4$; solid lines: unit cell of $Bi_2(MoO_4)_3$

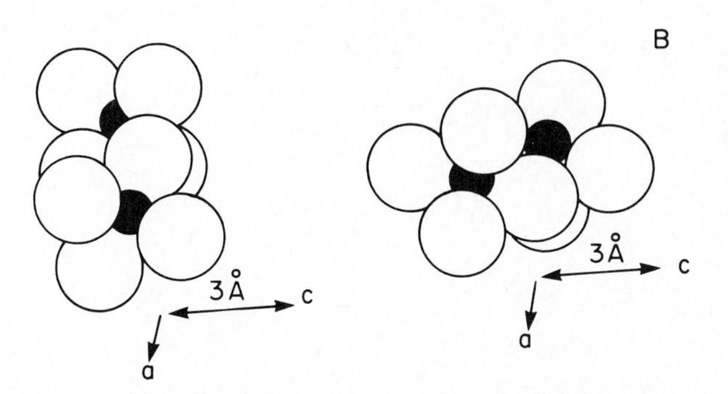

Two Mo_2O_8 arrangements. Black circles: Mo; Large circles: oxygen

Figure 4-27 Structure of $Bi_2Mo_3O_{12}$ [82]. (Reprinted with permission from *Acta Crystallographica*. Copyright by the International Union of Crystallography.)

shown in Fig. 4-27*A*. Moreover, all $(MoO_4)^{2-}$ ions now occur as pairs in two different arrangements (Fig. 4-27*B*). The MoO_4 tetrahedra have two bonds at 1.72 and two others at 1.87 Å; the pair formation occurs through a fifth oxygen from the neighboring tetrahedron at 2.2 Å. Each Bi^{3+} ion still occurs in an eightfold surrounding, all oxygens being shared with Mo cations. The Bi–O distances are far from equivalent, however, four being between 2.12 and 2.35 Å and the other four between 2.60 and 2.93 Å.

The structure of $Bi_2Mo_2O_9$, as given by van den Elzen and Rieck [83], is not certain, since it is based on x-ray powder diagrams; the positions of the heavy elements seem to be fairly well defined, however. The structure can best be visualized from the *xz* planes in which oxygens and cations alternate. Figure 4-28*A* gives the cation positions in successive *xz* planes with their oxygens above and below, and Fig. 4-28*B* shows a projection along the *y* axis. The remarkable feature of this structure is the rows of oxygens that are connected only to Bi^{3+}. There are four Bi sites per two oxygens, but only three are filled.

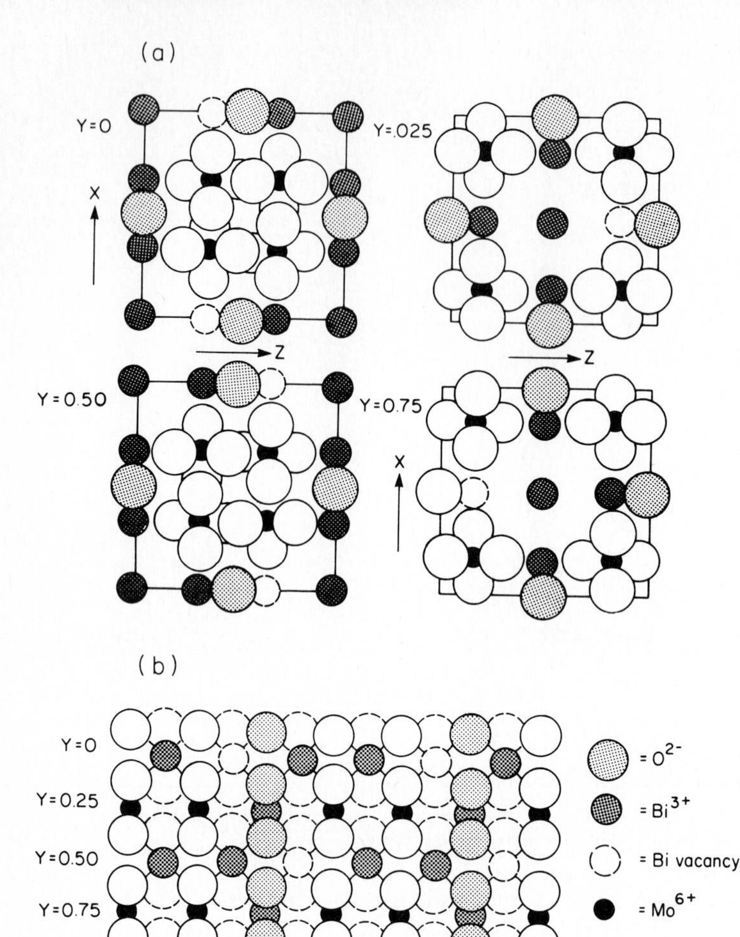

Figure 4-28 Structure of $Bi_2Mo_2O_9$ [83].

A description of the compound would then be

$$Bi(Bi_3\,\square O_2)(Mo_4O_{16})$$

where □ is a cation vacancy, the first Bi is associated only with the (Mo_4O_{16}) units, and the Bi cations in the parentheses are bonded to the oxygen associated with Bi (also in parentheses) and to oxygens in the (Mo_4O_{16}) units.

The changes occurring as the structure is transformed from Bi_2MoO_6 through $Bi_2Mo_2O_9$ into $Bi_2Mo_3O_{12}$ are indicated first by the change in Bi coordination; in the Bi-rich compound it is essentially given by the Bi_2O_2-layer struc-

Table 4-4 Structures of bismuth molybdates

Formula	Number of oxygens bound only to Bi (per formula)	Number of Bi^{3+} vacancies	Degree of clustering of MoO_4
$Bi_2Mo_3O_{12}$	...	1	2
$Bi_2Mo_2O_9$	1	$\frac{1}{2}$	4
Bi_2MoO_6	2	...	Infinite

ture, whereas in $Bi_2Mo_3O_{16}$ there is a cubic oxygen surrounding with all oxygens shared by Bi and Mo. A second indication of the change is in the degree of clustering of the Mo—O polyhedra, which pass from an infinite two-dimensional ReO_3-like structure via Mo_4O_{16} to Mo_2O_8. A third indication is the gradual formation of empty cation sites potentially available for Bi, as shown in Table 4-4.

An important question concerns the products of reduction of the bismuth molybdates upon interaction with olefins and/or NH_3. There is hardly any doubt that the oxidizing agent for the reaction partners in ammoxidation is an oxygen from the solid, and it is therefore necessary to ascertain what happens to the solid during reduction in the absence of O_2.

Reduction by olefins occurs even at relatively low temperatures ($< 250°C$), although the rate is low. Reduction by NH_3 and especially by H_2 or CO is a more difficult reaction, requiring temperatures in excess of 400°C; NH_3 is then converted into N_2. Aykan [22] and Batist et al. [84–86] studied the reduction, and the latter showed that olefin reduction at temperatures less than 400°C leads to partial reductions such as

$$Bi_2Mo_3O_{12} \longrightarrow Bi_2Mo_3O_{11.5}$$

$$Bi_2Mo_2O_5 \longrightarrow Bi_2MoO_8$$

and

$$Bi_2MoO_6 \longrightarrow Bi_2MoO_{5.5}$$

At temperatures greater than 400°C, reduction becomes far more extensive and generally leads to formation of Bi and MoO_2.

Reoxidation of partly reduced Bi_2MoO_6 and $Bi_2Mo_2O_5$ is fast even at temperatures less than 200°C [49, 50, 84–86]. The more extensively reduced systems (those reduced at temperatures greater than 400°C) are much more difficult to reoxidize and require temperatures exceeding 400°C; this statement also applies to $Bi_2Mo_3O_{12}$ reduced at low temperatures. The kinetics of reoxidation is also different for the two cases: for the partly reduced systems, the reaction has a zero-order dependence on oxygen partial pressure and a square-root dependence on time, suggesting a diffusion limitation; for more extensively reduced systems, the reaction has a first-order dependence on oxygen partial pressure and on the degree of reduction.

The composition of the product of partial reduction of $Bi_2Mo_2O_9$ can be accounted for by the removal of the rows of oxygens which are bonded only to Bi:

$$Bi(Bi_3 \square_c O_2)(Mo_4O_{16}) \longrightarrow Bi(Bi_3 \square_c \square_{a_2})(Mo_4O_{16})[= 2Bi_2Mo_2O_8] \quad (56)$$

where $\square_c$ and $\square_a$ are cation and anion vacancies, respectively.

The compositions of partly reduced Bi_2MoO_6 and $Bi_2Mo_3O_{12}$ are less easily explained. Reduction of the former is probably related to defects in the structure, since sintering at 600°C makes it virtually irreducible [50]. Such defects may arise because of the following rearrangements:

$$[(Bi_2O_2)O(MoO_2)O(Bi_2O_2)O(MoO_2)O] \longrightarrow (Bi_2O_2)O(Bi_2Mo_2O_9) \quad (57)$$

i.e., two Mo layers and one Bi_2O_2, with the intermediate oxygen layers, form a $Bi_2Mo_2O_9$ pattern. This is virtually an epitaxic arrangement of Bi_2O_3 and $Bi_2Mo_2O_9$. Removing one oxygen from every $Bi_2Mo_2O_9$ would produce the observed stoichiometry:

$$2Bi_2MoO_6 \rightleftharpoons Bi_2O_3 \cdot Bi_2Mo_2O_9 \rightleftharpoons Bi_2O_3 \cdot Bi_2MoO_8 + O \quad (58)$$

Such a rearrangement requires considerable mobility of the cations in the lattice.

Since reduction of $Bi_2Mo_3O_{12}$ and reoxidation of the reduced structure are comparatively slow, both reactions are probably different from those described above. They could be expected to be different because of the different types of bonding of the oxygens in the $Bi_2Mo_3O_{12}$ lattice. The observed degree of reduction can be accounted for provided that the reduction of $Bi_2Mo_3O_{12}$ is considered to be

$$Bi_4 \square_2(Mo_2O_8)_3 \longrightarrow Bi_4 \square_2(Mo_2O_8)_2(Mo_2O_7 \square_a) + O \quad (59)$$

and only one of the Mo_2O_8 groups, presumably one associated with two cation vacancies, can be reduced, thereby losing one oxygen.

The bismuth molybdates therefore appear to allow two different types of oxygen donation, one specific to $Bi_2Mo_3O_{12}$ and the other to $Bi_2Mo_2O_9$ and probably also to Bi_2MoO_6.

The catalytically active bismuth molybdate Wragg et al. [87] studied the ammoxidation and oxidation of propylene at temperatures around 400°C and pressures around 1 torr. One catalyst (UBM4) had a Bi/Mo ratio of 3 : 4 and was therefore similar to the commercial catalyst except for the absence of P and SiO_2. A second catalyst having the composition Bi_2MoO_6 was synthesised by slurrying basic bismuth nitrate and hydrated MoO_3 in water [76, 77], and it was shown to be active and selective for butene oxidation. Rates per unit surface area and the rate equations for the two catalysts were similar at 400°C, but the selectivity of UBM4

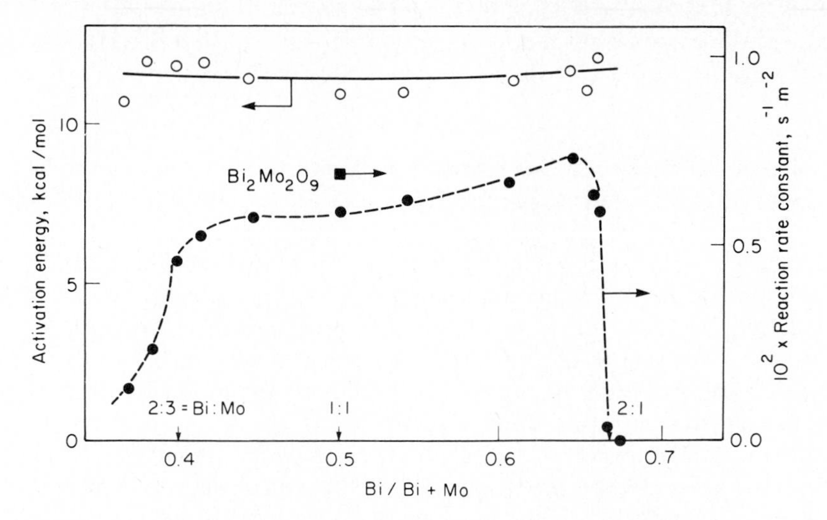

Figure 4-29 Butene oxidation activities of bismuth molybdate catalysts [88].

was much lower than that of the Bi_2MoO_6 catalyst. Ammoxidation on the latter was twice as fast as oxidation, suggesting that the adsorbing propylene molecule reacted immediately with a surface species containing nitrogen, rather than reacting first with the oxide lattice to form allyl and hydroxyl radicals, as it evidently did on UBM4. An intermediate formation of nitrogen oxides was shown to be unlikely, and the main course for ammoxidation through acrolein on UBM4 could also be ruled out.

Matsuura et al. [88] prepared a series of catalysts with varying Bi/Mo ratios in the range 2 : 3 to 2 : 1 and measured their activities for the oxidation of butene to butadiene. Some of the samples were also investigated by van Oeffelen and Sawatzky [89], who used ESCA (electron spectroscopy for chemical analysis or x-ray photoelectron spectroscopy), and by Brongersma [90], who used low-energy ion scattering. All the catalysts were found to be selective, but their activities varied considerably (Fig. 4-29).

The sharp transition in activity shown at a Bi/Mo ratio of 2 : 1 was investigated in detail. A catalyst having a Bi/Mo ratio of 2.04 was found to be inactive, and it could not even be reduced. The ESCA data for this catalyst showed a Bi/Mo ratio of 2.2, and the ion-scattering results showed a value of 4.4. In contrast, a catalyst with Bi/Mo ratio of 1.96 was highly active and easily reduced. ESCA data indicated a surface Bi/Mo ratio of 1.6, and ion-scattering data a ratio of 1.2 to 1.6. Since ion scattering is confined to the surface layer and ESCA probes somewhat more deeply into the solid, the difference between the two results is explained by a gradient in concentration from the surface toward the interior. The catalytic activity is defined primarily by the outer-layer composition, provided that the Bi/Mo ratio is in the range 2 : 3 to 2 : 1. Activities did not exceed that of pure $Bi_2Mo_2O_9$ (Fig. 4-29). Grasselli [91] proposed that a surface consists of small two-dimensional domains, each with structural properties comparable to those of the three-dimensional structures actually observed. This hypothesis conforms to the foregoing results, provided that the active domains are similar to $Bi_2Mo_2O_9$.

A comprehensive model of the reaction mechanism A model of a bismuth molybdate surface ammoxidation site is suggested in Fig. 4-30. It is derived from the *xz* bulk plane of Fig. 4-28 when half of the top-layer oxygens are left out. At the center of the figure is an oxygen belonging to the linear row of oxygens bonded only to Bi^{3+}; it has two Bi^{3+} ions next to it and a Bi^{3+} ion and a cation vacancy in a deeper layer. This oxygen forms the O_A site. It is surrounded by Mo_4 clusters lacking some of the oxygens of the bulk Mo_4O_{16} composition; these are the B sites. There is one O_A site per unit mesh, which accounts for the concentration measured by butadiene adsorption, about 7×10^{17} m^{-2}.

In terms of the Matsuura reaction mechanism given earlier, the formation of the allylic intermediate occurs on the B sites, i.e., on the oxygen vacancies on Mo. After NH_3 becomes adsorbed on a Bi^{3+} ion next to O_A, it donates two protons to O_A; this is followed by desorption of H_2O and a shift of the NH residue to the A vacancy. A shift of the allyl to a Bi^{3+} ion neighboring the A vacancy brings the two radicals next to each other, permitting the formation of a C—N bond. The combination must dissociate three hydrogens, which might occur by another shift to a Mo_4 group. All hydrogens dissociated migrate to other O_A sites and leave the surface as H_2O.

After the reaction sequence, the site has lost its O_A, and the surface has also lost two O_A's of neighboring sites. We now postulate that oxygens migrate from the Mo_4 clusters to O_A. During the reduction electrons are donated to Bi^{3+}, and

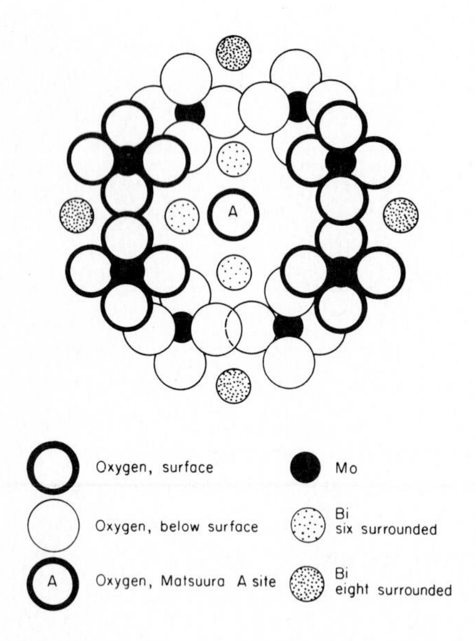

Figure 4-30 Postulated structure of the catalytic site for ammoxidation on bismuth molybdate.

these electrons are transferred to Mo^{6+}. The reduced and oxygen-deficient Mo_4 clusters can accept O_2 and give the original surface structure again.

Some details of the site structure and reaction mechanism remain uncertain, but the essential part of the interpretation is well established: the reaction steps take place on a complicated, multifunctional surface structure which is related to a bulk structure. It would have been virtually impossible to design such an intricate catalyst a priori, and the discovery of bismuth molybdate as a selective ammoxidation catalyst is a major event in the history of catalysis.

The foregoing interpretation leaves two important questions open. The first concerns the rate of formation of the allylic intermediate; in the original theory this step was considered rate-determining, whereas in the present interpretation it is considered to be fast. The high rate of formation of the allylic intermediate compared with the rate of overall oxidation was recently confirmed by H–D exchange in the olefin when D_2O was added to the reaction mixture [92, 93]. The explanation might be that the reaction actually proceeds in steps,

$$\text{Surface} + C_3H_6 \rightleftharpoons \underset{\text{B site}}{C_3H_5 + H} \longrightarrow \underset{\text{A site}}{C_3H_5} \tag{60}$$

the second step being rate-determining.

The second question arises [87, 94] since in the absence of catalyst there is a fast nonselective gas-phase reaction with the characteristics of a branched chain reaction. The catalyst apparently suppresses this reaction by elimination of the chain-propagating radicals. An explanation for this chain termination is still lacking.

Multicomponent Molybdate Catalysts

An ammoxidation catalyst with superior properties—presumably high selectivity—was announced by Sohio in 1972. The catalyst composition has not been disclosed, but it seems likely that it is related to the molybdates mentioned in patents [95–98]. These patents do not always refer to the ammoxidation reaction, but we shall nevertheless consider them in this regard. The catalyst description is usually in such terms as

$$M_a^{2+}, Fe_b^{3+}, Bi_c^{3+}, Mo_d^{4+}, Q_e, R_f, T_g, O_x,$$

where M^{2+} = metal ion (Ni^{2+} or Co^{2+})
Q = alkali metal
R = alkaline-earth metal
T = phosphorus, arsenic, or antimony

There are apparently applications in which the catalyst is supported on silica. One of the most recent patents [97] describes a special method of preparation in which a matrix, which does not contain Fe, Bi, or Te, is impregnated with solutions of the latter and subsequently dried.

The patent information implies that there is always a molybdate system present consisting of $CoMoO_4$, sometimes combined with $NiMoO_4$, $Fe_2(MoO_4)_3$, and some bismuth molybdate. To simplify matters the following discussion is restricted to compositions such as

$$M_8^{2+}Fe_3^{3+}Bi^{3+}(MoO_4)_{12}O_2^{2-}$$

where M^{2+} is Ni^{2+}, Co^{2+}, or Mg^{2+}, since this is the only composition reported outside the patent literature [99–102]. The presence of the alkali metals in some patented catalysts has not been discussed in the literature.

As usual, this section begins with discussion of the bulk structures of $Me^{2+}(MoO_4)$ and $Me_2^{3+}(MoO_4)_3$ and their combinations.

The $Me^{2+}(MoO_4)$ compounds occur primarily in two forms; one is α-$CoMoO_4$; it was analyzed by Smith and Ibers [103], who found both Co^{2+} and Mo^{6+} in octahedral surroundings, with edge sharing between the Co and Mo octahedra. These form rows which share corners. The other form is β-$CoMoO_4$ [104, 105], having a structure essentially like that of α-$CoMoO_4$ except that Mo is now tetrahedrally surrounded. $MnMoO_4$, $MgMoO_4$, and $FeMoO_4$ are isomorphous with β-$CoMoO_4$, and $NiMoO_4$ is isomorphous with α-$CoMoO_4$. Many of these molybdates can be prepared with an excess of MoO_3, and according to Oganowski et al. [106], the presence of the excess MoO_3 is essential for the catalytic activity of $MgMoO_4$ in the oxidative dehydrogenation of ethylbenzene to give styrene. The excess MoO_3 is octahedrally surrounded with some "active" centers present as Mo^{5+} in a distorted pyramidal structure.

$Fe_2(MoO_4)_3$, a catalyst for the oxidative dehydrogenation of methanol to give formaldehyde [107, 108], has a structure in which Fe^{3+} is octahedrally surrounded and Mo^{6+} tetrahedrally surrounded. The pure compound lacks catalytic activity unless it contains some excess of MoO_3, which is believed to be incorporated by replacement of Fe^{3+} by Mo^{6+} in some octahedral surroundings [108]. To compensate for the increase in positive charge, extra oxygen anions are built into interstices in the structure.

It was recognized early that Fe^{3+} could become incorporated in $Bi_2(MoO_4)_3$ [109]. Sleight and Jeitschko [110] proposed a structure for $Bi_3(FeO_4)(MoO_4)_2$ in which Fe^{3+} occupies a tetrahedral site in a distorted scheelite structure. A simplified way to understand the latter structure is to start from the following structure and replace the bivalent cation by a trivalent one,

$$Pb(MoO_4) \longrightarrow Bi_{2/3}\square_{1/3}(MoO_4) = Bi_2\square(MoO_4)_3 \tag{61}$$

and to follow by replacement of one Mo^{6+} by Fe^{3+} and simultaneous filling of the cation vacancy by Bi^{3+},

$$Bi_2^{3+}\square(MoO_4)_3 \longrightarrow Bi_3^{3+}(FeO_4)(MoO_4)_2 \tag{62}$$

This compound does not seem to be the only Bi–Fe–Mo compound [111], but little is known about the structures of the others.

Wolfs and Batist [99] found that the selectivity and activity of the multicomponent ammoxidation catalysts used in the oxidative dehydrogenation of butene were primarily dependent upon the presence of bismuth molybdates. Wolfs found similar results for the oxidation of propylene to acrolein [100]. The activity of the multicomponent systems per unit surface area was higher than that for the pure bismuth molybdates: x-ray studies of the catalyst before reaction showed the presence of $Bi_2Mo_3O_{12}$, Bi_2MoO_6, $Fe_2(MoO_4)_3$, and β-$CoMoO_4$. Most of the $Fe_2(MoO_4)_3$, however, was found to have been reduced to $Fe(MoO_4)$ after the catalytic reaction; this $Fe(MoO_4)$ could be reoxidized only with difficulty.

To account for these facts Wolfs and Batist suggested that a catalyst particle contained an inner core consisting of β-$CoMoO_4$ and an outer shell, about 100 Å thick, consisting of a mixture of bismuth molybdates, the $Fe_2(MoO_4)_3$ being present between the core and outer shell in the fresh catalyst but fused as $FeMoO_4$ with the $CoMoO_4$ in the used catalyst. ESCA measurements by Matsuura [101] and Wolfs [100] appeared to confirm that there was an outer shell consisting only of bismuth molybdate. Later experiments tended to cast some doubt on these results, but recent ESCA experiments by van Oeffelen and Sawatzky [89] and ion-scattering experiments by Brongersma [90] confirmed that the outer layers indeed consisted mainly, although probably not exclusively, of bismuth molybdates.

Studies of the fully oxidized catalyst $Mg_{11-x}Fe_xBiMo_{12}O_n$ showed a strong, irreversible adsorption of 1-butene, which has never been found for the simpler bismuth molybdates [100]. Otherwise, no differences from the previously described A- and B-type adsorptions were observed. Careful reduction of the catalyst surface by adsorption of butene at room temperature followed by pumping off the weakly adsorbed butene and a subsequent increase in the temperature to 250°C and evacuation of the oxidation products removed the capacity of the surface for strong adsorption of both butene and, to the same extent, butadiene.

The ratios of the various adsorption types were as follows:

Weakly adsorbed butene/strongly adsorbed butadiene after reduction/strongly adsorbed butene = 2 : 2 : 1

The surface density of sites for weak butene adsorption (B sites) was about 10^{18} m^{-2}, a typical value for the bismuth molybdates. The density of A sites was relatively high, about twice that found for the simpler bismuth molybdates; this observation explains the high catalytic activity. Some new component is inferred to have been present in the outer layers of the catalyst, but so far it has not been identified; it could initially be Fe^{3+} which could later be reduced to Fe^{2+}. [$Fe_2(MoO_4)_3$ shows no strong adsorption.]

The available information is not sufficient for formulation of a consistent model, but there is no question that the multicomponent catalysts are important and worthy of a more extensive investigation.

SUMMARY OF REACTION CHEMISTRY

The selective oxidation of olefins on catalysts of the type Bi_2MoO_6 occurs via oxygen anions from the bulk of the catalyst. The selectivity reflects the fact that bulk oxygens of a special type are easily abstracted and active for oxidation whereas the majority are relatively inactive.

The reaction mechanism is reasonably well understood: it consists of a stepwise abstraction of hydrogen from the hydrocarbon via allylic intermediates, followed by the acceptance of oxygen from the bulk structure. If NH_3 is present, the oxygen is first replaced by a nitrogen moiety, which then is incorporated in the hydrocarbon. There is less certainty concerning the rate-determining step. The supposition that this is the abstraction of the first hydrogen is less probable than originally assumed. At least for the case of the Bi–Mo systems, it is possible that the rate of reoxidation or the rate of diffusion of oxygen through the lattice is rate-determining. In the latter case, the point of introduction of O_2 must be at some distance from the surface catalytic site. For these catalysts there is good reason to assume that the oxygen donated comes from a Bi—O—Bi bond. The actual attack on the hydrocarbon occurs via the Mo—O—Mo bonds. The combination of Mo and Bi then might take care of the diffusion, since Mo–O layers might donate O to the Bi–O layers via nonstoichiometric structures such as shear structures.

For catalysts such as $Fe_2Sb_2O_7$ and USb_3O_{10}, the diffusion mechanism has not been envisaged. The main strength of the arguments is directed at the explanation of the activation of Fe—O—Sb or U—O—Sb oxygens by the structural properties of the compounds.

So far, no ready explanation has been offered for the observation that selective catalysts necessitate a special pairing of two cations. Bi_2UO_6 is very active but nonselective; USb_3O_{10}, $Fe_2Sb_2O_7$, and Bi–Mo combinations are active and selective, but $Fe_2(MoO_4)_3$ is not active. Nor is the promotional effect of Fe in bismuth molybdates well understood.

Although replenishment of anion vacancies by oxygen usually occurs by diffusion from the bulk, some oxygen may be introduced from the gas phase; it then gives rise to a nonselective reaction with a low activation energy but also a low frequency factor.

No explanation has been given for the observation that the homogeneous gas-phase oxidation, a nonselective branched-chain reaction, is inhibited. It has been suggested that chain termination occurs at the catalyst surface, but the actual nature of this reaction remains obscure.

PROCESS DESIGN

Processes and Operating Conditions

The most commonly applied ammoxidation process is the one licensed by Sohio, shown schematically in Fig. 4-31. The reactor is a fluidized bed which is fed with a mixture of air, ammonia, propylene, and possibly water. The reactor operates at pressures of 1 to 3 atm and temperatures between 400 and 510°C [1]. The catalyst is Sohio's so-called catalyst 41, which might be inferred to be one of the complex molybdates such as $Co_6^{2+}Ni_2^{2+}Fe_3^{3+}Bi^{3+}(MoO_4)_2$, discussed in the preceding section.

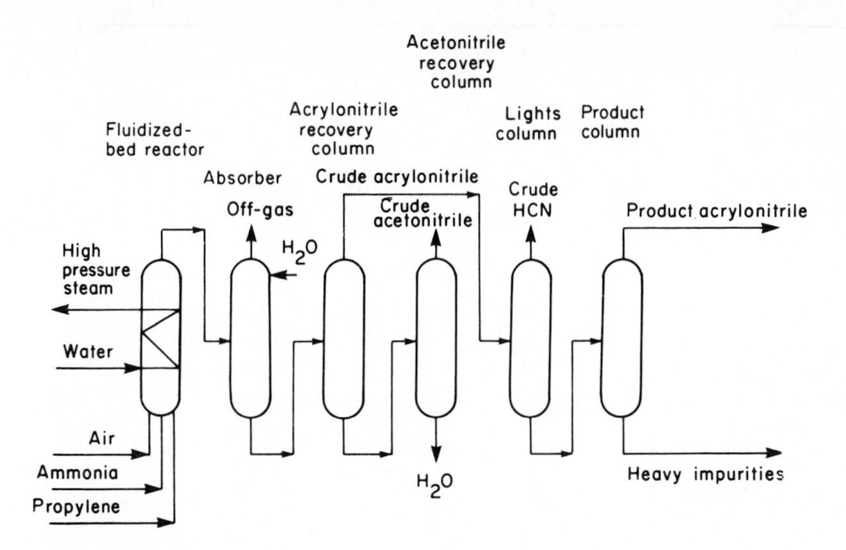

Figure 4-31 Schematic diagram of the Sohio ammoxidation process [1].

The apparent contact time (defined as the bulk catalyst volume divided by the volumetric flow rate of feed gas at reactor conditions) is about 1 to 15 s [42]. The preferred feed compositions, cited in patents by Callahan and Gertisser [112] for the U–Sb catalysts and by Callahan et al. [42] for the more recently developed catalysts, are

Oxygen to propylene, 1 : 1 to 3 : 1
Ammonia to propylene, 0.5 : 1 to 5 : 1
Water to propylene, 1 : 1 to 4 : 1

The inclusion of water in the feed is not necessary, but, according to the patents, water increases the selectivity for formation of acrylonitrile; its effect is probably partly to modify the catalyst and partly to provide heat capacity to moderate the temperature rise in the reactor.

The primary product in ammoxidation is acrylonitrile, and the patents cite values of conversion of propylene to acrylonitrile ranging from 40 to more than 80 percent. Some product-distribution data are collected in Table 4-5. The table refers to the U–Sb catalyst, which has been discussed in the preceding section, but which is no longer applied; comparable data are lacking for the recently developed catalysts, which are more likely to be representative of current practice. The patent data [95–99] show that newer catalysts, in comparison with the U–Sb catalyst, give lower conversions to the undesired products, which are HCN, acetonitrile, CO_2, CO, and traces of acrolein. A comparison of the literature describing the newer and older catalysts [42, 112, 113] also indicates that the newer catalysts are more active, which suggests that lower temperatures may be applied in the newer processes. The patents suggest that ammoxidation catalysts lose activity

Table 4-5 Yields from a bench-scale ammoxidation reactor [112]

Reaction conditions[a]	
Reactor	Bench-scale fluidized bed
Catalyst	U–Sb supported on SiO_2[b]
Weight, g	650
Particle size range, μm	15 wt % > 88
	30 wt % < 44
Reaction temperature, °C	482
Reaction pressure, atm	1.1
Apparent contact time, s	8.0
Reactant feed rates, std m^3/s:	
Propylene	2.20×10^{-6}
Ammonia	2.65×10^{-6}
Air	24.3×10^{-6}
Percentage conversion of propylene to:	
Acrylonitrile	82.1
HCN	5.7
Acetonitrile	Trace
Acrolein	Trace
CO	6.7
CO_2	3.4
Unconverted propylene, % of feed	0.3
Unconverted ammonia, % of feed	21.1

[a] Data based on a run of 1800 s, preceded by an equally long operating period.

[b] Details of preparation are given in Ref. 112.

and/or selectivity during operation and therefore require regeneration, but no information is available from which to infer the nature or rate of the aging process.

Besides the Sohio fluidized-bed process, there is a small commercial operation with a fixed-bed reactor, as depicted in Fig. 4-32. Except for the reactor, the process is almost the same as the Sohio process; the reactor operates at 440 to 470°C and about 2 atm. The yield of acrylonitrile is reportedly high and the catalyst life long, but quantitative information is lacking.

Since acrylonitrile is highly toxic, the process must be carried out in a tightly closed system, with ample provisions for good ventilation and safe procedures for handling materials.

Separation Processes

Purification of reaction products is generally carried out by distillation, as shown in Figs. 4-31 and 4-32. The patent literature points to several difficulties in the purification: (1) removal of unreacted ammonia may require scrubbing the product gas with acid [115], although this step is not shown in the preceding process flow diagrams; (2) separation of the products does not proceed straightforwardly

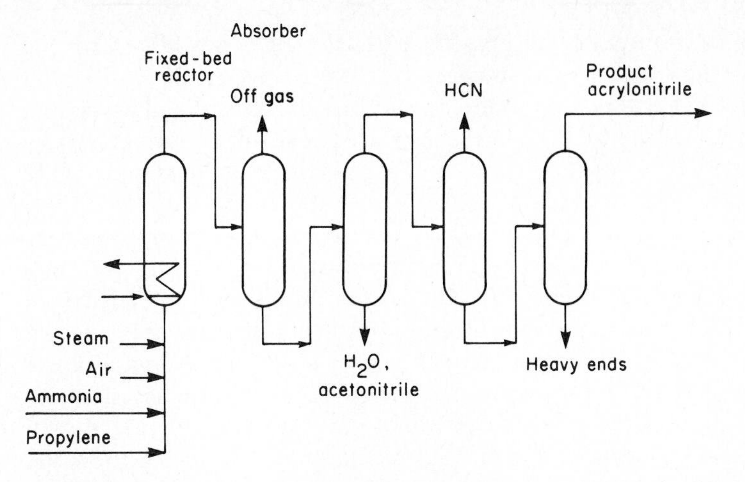

Figure 4-32 Schematic diagram of an ammoxidation process with a fixed-bed reactor [114].

by distillation; e.g., acrylonitrile must be recovered from acetonitrile as an azeotrope with water in a column requiring 50 theoretical plates [115].

The acrylonitrile is produced for use as an intermediate in a free-radical polymerization, and it is therefore not surprising that polymerization in the processing equipment must be considered as a side reaction. Compounds such as hydroquinone may be added as free-radical scavengers to prevent the polymerization, especially in the purification train, where the acrylonitrile concentration is highest.

The products of reaction may also combine to give unstable products. The formation of cyanohydrin in the purification train can be inhibited, for example, by added oxalic acid [116].

Reactor Design

Because of the lack of a well-characterized reaction network and of quantitative kinetics, the discussion of reactor design cannot be carried very far. An initial question is whether ammoxidation reaction rates are significantly influenced by mass transfer. Data given by Callahan et al. [42] for ammoxidation catalyzed by bismuth molybdate show that the rate constant was unchanged as the catalyst in the form of pellets of unspecified size was replaced by a mixture of 20- to 150-mm-diameter particles. These results imply that intraparticle mass transfer did not influence the rate; we infer that it is probable that catalyst particle sizes and pore geometries are generally chosen to ensure the absence of significant intraparticle mass-transfer resistance. Expecting that gas-phase mass-transfer resistance is also negligible, we therefore conclude tentatively that the rate of ammoxidation is determined only by the chemical reaction steps.

Most of the reactors in use are fluidized beds, which implies that the disadvantage of inefficient reactor utilization caused by backmixing of reactants is more than offset by the advantage of the high heat-transfer rates resulting from the rapid mixing. Since the reaction is so strongly exothermic (about 120 kcal/mol of propylene converted), it is clear that steep temperature gradients would exist in a fixed bed. It follows that the fixed-bed design would consist of a series of narrow tubular reactors probably cooled by a surrounding heat-transfer medium such as a molten salt solution.

On the basis of the introductory discussion of reactor operation, it is expected that ammoxidation reactors could operate in an unstable fashion and constitute an explosion hazard, but data to test this hypothesis are lacking. The development of reactor stability theory has outpaced its application, and the reader is directed to the book by Denn [117] summarizing this subject. A simple quantitative example of how catalyst selectivity in naphthalene oxidation determines the stability of reactor operation has been given by Kramers and Westerterp [118]. Recent work by Froment [119] illustrates the difficult techniques of data collection and interpretation required for determining the appropriate kinetics parameters to permit reactor design calculations accurate enough to predict instability in operation. In view of the difficulty of these procedures, it is not surprising that the literature is almost devoid of the quantitative data required to illustrate the instabilities in a process like ammoxidation.

Another reactor design criterion is the explosive limits of the gas-phase reaction mixtures, mentioned in Chap. 2 for the Wacker and vinyl acetate processes. The patent literature makes reference to this explosion hazard, and suggestions have been made, for example, to restrict the O_2 content in the feed to less than 10 percent, or, alternatively, to keep the propylene content outside the 2 to 10 percent range. It is possible, however, that fluidized-bed reactors are operated with reactant mixtures within the explosion limits, since the high heat capacity and high effective thermal conductivity of a fluidized bed minimize the chances of an explosive gas-phase reaction.

NOTATION

M	metal
n	an exponent
O_A	oxygen ion in an A site
P_i	partial pressure of compound i, F/l^2
$P_{i,0}$	inverse adsorption equilibrium constant, F/l^2
Q	Enthalpy of reaction, E/mol
R	gas constant
T	temperature, °C or K
V_m	amount of gas adsorbed to complete a monolayer, l^3 at STP
V_M	anion vacancy located over a metal ion M
z	probability of removal of a D atom from a hydrocarbon (relative to an H atom)

Dimensional

E = energy
F = force
l = length
m = mass
t = time

Greek

ΔT temperature difference
θ fraction of surface sites occupied
ϕ molecular orbital
ψ wave function

Subscript

s solid

REFERENCES

1. Anon., *Hydrocarbon Process.*, **52**(11), 99 (1973).
2. Semenov, N. N., "Chemical Kinetics and Chain Reactions," Clarendon Press, Oxford, 1935.
3. Cullis, C. F., J. D. Hopton, and D. L. Trimm, *J. Appl. Chem.*, **18,** 330 (1968); C. F. Cullis, J. D. Hopton, C. S. Swan, and D. L. Trimm, ibid. **18,** 335 (1968); C. S. Swan, and D. L. Trimm, ibid., 340 (1968).
4. Kuijpers, F. P. J., Ph.D. thesis, Technical University of Eindhoven, The Netherlands, 1974.
5. Taqui Khan, M. M., and A. E. Martell, "Homogeneous Catalysis by Metal Complexes," vol. 1, p. 117. Academic, New York, 1974.
6. Evnin, A. B., J. A. Rabo, and P. H. Kasai, *J. Catal.*, **30,** 109 (1973).
7. Sachtler, W. M. H., and N. H. de Boer, *Proc. 3d Int. Cong. Catal.*, p. 252, North-Holland, Amsterdam, 1965.
8. Adams, C. R. and T. Jennings, *J. Catal.*, **2,** 63 (1963).
9. Sachtler, W. M. H., *Rec. Trav. Chim.*, **82,** 243 (1963).
10. McCain, C. C., G. Gough, and G. W. Godin, *Nature*, **198,** 58 (1963).
11. Voge, H. H., C. D. Wagner, and D. P. Stevenson, *J. Catal.*, **2,** 58 (1963).
12. Sixma, F. L. J., E. F. S. Duinstee, and J. L. J. P. Hennekens, *Rec. Trav. Chim.*, **82,** 901 (1963).
13. Grasselli, R. K., and D. D. Suresh, *J. Catal.*, **25,** 273 (1972).
14. Sachtler, W. M. H., G. J. H. Dorgelo, J. Fahrenfort, and R. J. H. Voorhoeve, *Proc. 4th Int. Cong. Catal.*, Akadémiai Kiadó, Budapest, 1971.
15. Sachtler, W. M. H., *Catal. Rev.*, **4**(1), 27 (1970).
16. Buiten, J., *J. Catal.*, **10,** 188 (1968); **13,** 373 (1967); **27,** 232 (1972).
17. Moro-Oka, Y., Y. Takita, and A. Ozaki, *J. Catal.*, **27,** 177 (1972).
18. Moro-Oka, Y., S. Tan, and A. Ozaki, *J. Catal.*, **12,** 291 (1968); **17,** 125, 132 (1970).
19. Moro-Oka, Y., and Y. Takita, *Proc. 5th Int. Cong. Catal.*, p. 1025, North-Holland, Amsterdam, 1973.
20. Haber, J., M. Sochacka, B. Grzybowska, and A. Golzbiewski, *J. Molec. Catal.*, **1,** 35 (1975).
21. Mars, P., and D. W. van Krevelen, *Chem. Eng. Sci. Suppl.*, **3,** 41 (1954).
22. Aykan, K., *J. Catal.*, **12,** 281 (1968).
23. Jiru, P., B. Wichterlova, and J. Tichy, *Proc. 3d Int. Cong. Catal.*, p. 199, North-Holland, Amsterdam, 1965.
24. Simons, Th. G. J., E. J. Verheyen, P. A. Batist, and G. C. A. Schuit, *Adv. Chem. Ser.*, **76**(II), 261 (1968).

25. Boreskov, G. K., paper presented at *2d Jap.-Sov. Catal. Sem., Tokyo, 1973.*
26. Sancier, K. M., P. R. Wentrede, and H. Wise, *J. Catal.*, **39,** 141 (1975).
27. Keulks, G. W., *J. Catal.*, **19,** 232 (1970).
27*a*. Keulks, G. W., and L. D. Krenzke, *6th Int. Cong. Catal., London, 1976*, prepr. B-20.
28. Lunsford, J. H., *Catal. Rev.*, **8,** 135 (1973).
29. Hooff, J. H. C. van, Ph.D. thesis, Technical University of Eindhoven, The Netherlands, 1968.
30. Winter, E. R. S., *J. Chem. Soc.* **A1968,** 2889.
31. Klier, K., J. Nováková, and P. Jiru, *J. Catal.*, **2,** 479 (1963).
32. Nováková, J., *Catal. Rev.*, **4,** 77 (1970).
33. Boreskov, G. K., and V. S. Muzykantov, *Ann. N.Y. Acad. Sci.*, **213,** 137 (1973).
34. Stone, F. S., paper presented at *Int. Summer Sch. Fundam. Princ. Heterogeneous Catal., Venice, September–October, 1972.*
35. Barry, T. J., and F. S. Stone, *Proc. R. Soc.*, **A255,** 124 (1960).
36. Wragg, R. D., P. G. Ashmore, and J. A. Hockey, *J. Catal.*, **22,** 19 (1971).
37. Kazansky, V. B., V. A. Shvets, M. Ya. Kon, V. V. Nikisha, and B. N. Shelimov, *Proc. 5th Int. Cong. Catal.*, p. 1423, North-Holland, Amsterdam, 1973.
38. Yoshida, S., T. Matsuzaki, S. Ishida, and K. Tarama, *Proc. 5th Int. Cong. Catal.*, p. 1049, North-Holland, Amsterdam, 1973.
39. Akimoto, M., and E. Echigoya, *J. Catal.*, **29,** 191 (1973); **31,** 278 (1973).
40. Gerei, S. V., E. V. Rozhkova, and Y. B. Ghorokhavatsky, *J. Catal.*, **28,** 341 (1973).
41. Kilty, P. A., N. C. Rol, and W. M. H. Sachtler, *Proc. 5th Int. Cong. Catal.*, p. 929, North-Holland, Amsterdam, 1973.
42. Callahan, J. L., R. K. Grasselli, E. C. Milberger, and H. A. Strecker, *Ind. Eng. Chem. Prod. Res. Dev.*, **6,** 134 (1970).
43. Hucknall, D. J., "Selective Oxidation of Hydrocarbons," pp. 55–69, Academic, London, 1974.
44. Batist, P. A., H. J. Prette, and G. C. A. Schuit, *J. Catal.*, **15,** 267 (1969).
45. Batist, P. A., P. C. M. van der Heyden, and G. C. A. Schuit, *J. Catal.*, **22,** 411 (1971).
46. Kolchin, I. K., S. S. Belekov, L. Ya. Margolis, *Neftichimiya*, **4,** 301 (1964).
47. Lankhuyzen, S. P., P. M. Florack, and H. S. van der Baan, *J. Catal.*, **42,** 20 (1976).
48. Batist, P. A., A. H. W. M. der Kinderen, Y. Leeuwenburgh, F. A. M. G. Metz, and G. C. A. Schuit, *J. Catal.*, **12,** 45 (1968).
49. Matsuura, I., and G. C. A. Schuit, *J. Catal.*, **20,** 19 (1971).
50. Matsuura, I., and G. C. A. Schuit, *J. Catal.*, **25,** 314 (1972).
51. Matsuura, I., *J. Catal.*, **33,** 420 (1974).
52. Matsuura, I., *J. Catal.*, **35,** 452 (1974).
53. Matsuura, I., *6th Int. Cong. Catal., London, 1976*, prepr. B-21.
54. Callahan, J. L., and R. K. Grasselli, *Am. Inst. Chem. Eng., J.* **6,** 755 (1963).
55. Voge, H. H., and C. R. Adams, *Adv. Catal.*, **17,** 151 (1967).
56. Sleight, A. W., in J. J. Burton and R. L. Garten (eds.), "Advanced Materials in Catalysis," Academic, New York, 1977.
56*a*. Linn, W. J., and A. W. Sleight, *Ann. N.Y. Acad. Sci.*, **272,** 22 (1976).
57. Aykan, K., O. Halvorson, A. W. Sleight, and D. B. Rodgers, *J. Catal.*, **35,** 401 (1975).
58. Otsubo, T., H. Miura, Y. Morikawa, and T. Shirasaki, *J. Catal.*, **36,** 240 (1975).
59. Haber, J., and B. Grzybowska, *J. Catal.*, **28,** 489 (1975).
60. Grasselli, R. K., D. D. Suresh, and K. Knox, *J. Catal.*, **18,** 56 (1970).
60*a*. Aykan, K., and A. W. Sleight. *J. Am. Chem. Soc.*, **53,** 427 (1970).
61. Wells, A. F., "Structural Inorganic Chemistry," 3d ed., Oxford University Press, London, 1962.
62. Simons, T. G. J., P. N. Houtman, and G. C. A. Schuit, *J. Catal.*, **23,** 1 (1971).
63. Sasaki, T., Y. Nakamura, K. Moritani, A. Morri, and S. Saito, *Catalyst* (*Jap.*), **14,** 191 (1972).
64. Boreskov, G. K., S. A. Ven'yaminov, V. A. Dzis'ko, D. V. Tarasova, V. M. Dindoin, N. W. Sazanova, J. P. Olen'kova, and L. M. Kefeli, *Kinet. Katal.*, **10,** 1109 (1969).
65. Yoshino, T., S. Saito, and B. Sobukawa, Japanese Patent 7,103,438 (1971).
66. Yoshino, T., S. Saito, J. Ishikara, T. Sasaki, and K. Sofugawa, Japanese Patent 7,102,802 (1971).
67. Skalkina, L. V., L. P. Suzdalev, J. K. Kolchin, and L. Ya. Margolis, *Kinet. Katal.*, **10,** 378 (1969).

67a. Kriegsmann, H., G. Öhlmann, J. Scheve, and F. J. Ulrich, *6th Int. Cong. Catal., London, 1976*, prepr. B-23.
68. Veatch, F., J. L. Callahan, and E. C. Milberger, Actes 2me *Cong. Int. Catal.*, p. 2647. Edition Technip, Paris, 1961.
69. Veatch, F., J. L. Callahan, J. D. Idol, and E. C. Milberger, *Chem. Eng. Prog.*, **56**(10), 65 (1960).
70. Erman, L. Ya., E. L. Gal'perin, I. K. Kolchin, G. F. Dobrzhanskii, and K. S. Chernyshev, *Russ. J. Inorg. Chem.*, **9**, 1174 (1964).
71. Gelbstein, A. J., S. S. Stroeva, N. V. Kulkova, Yu. M. Vashkin, V. L. Lapidus, and N. Sevast-'yanov, *Neftichimiya*, **4**, 909 (1964).
72. Bleyenberg, A. C. A. M., B. C. Lippens, and G. C. A. Schuit, *J. Catal.*, **4**, 581 (1965).
73. Blasse, G., *J. Inorg. Nucl. Chem.*, **28**, 1124 (1966).
74. Trifiro, F., H. Hoser, and R. D. Scarle, *J. Catal.*, **25**, 12 (1972).
75. Grzybowska, B., J. Haber, and J. Komorek, *J. Catal.*, **25**, 25 (1972).
76. Batist, P. A., J. F. H. Bouwens, and G. C. A. Schuit, *J. Catal.*, **25**, 1 (1972).
77. Batist, P. A., and S. P. Lankhuyzen, *J. Catal.*, **28**, 496 (1973).
78. Kepert, D. L., "The Early Transition Metals," Academic, New York, 1972.
79. Zemann, J., *Heidelberger Beitr. Mineral. Petrogr.*, **5**, 139 (1956).
80. Elzen, A. F. van den, and G. D. Rieck, *Acta Crystallogr.*, **B29**, 2436 (1973).
81. Cesare, M., G. Perego, A. Zazzetta, G. Manare, and B. Notari, *J. Inorg. Nucl. Chem.*, **33**, 3595 (1971).
82. Elzen, A. F. van den, and G. D. Rieck, *Acta Crystallogr.*, **B29**, 2433 (1973).
83. Elzen, A. F. van den, and G. D. Rieck, *Mat. Res. Bull.*, **10**, 1163 (1975).
84. Batist, P. A., C. J. Kapteÿns, B. C. Lippens, and G. C. A. Schuit, *J. Catal.*, **7**, 33 (1967).
85. Batist, P. A., H. J. Prette, and G. C. A. Schuit, *J. Catal.*, **15**, 267 (1969).
86. Batist, P. A., J. F. H. Bouwens, and I. Matsuura, *J. Catal.*, **32**, 362 (1974).
87. Wragg, R. D., P. G. Ashmore, and J. A. Hockey, *J. Catal.*, **31**, 293 (1973).
88. Matsuura, I., K. Ishikawa, and R. Schut, to be published.
89. Oeffelen, D. J. van, and S. Sawatzky, to be published.
90. Brongersma, H., to be published.
91. Grasselli, R. K., discussion at *Manchester Conf. Oxid., April 1975.*
92. Kondo, T., S. Saito, and K. Tamaru, *J. Am. Chem. Soc.*, **96**, 6857 (1974).
93. Batist, P. A., to be published.
94. Cathala, M., and J. Germain, *Bull. Soc. Chim. Fr.*, **1970**, 2167.
95. Grasselli, R. K., G. H. Heights, and J. L. Callahan, U.S. Patent 3,414,631 (1968).
96. Grasselli, R. K., G. H. Heights, and H. F. Hardman, U.S. Patent 3,642,930 (1972).
97. Standard Oil Company of Ohio, Dutch patent application 7,401,362 (1974).
98. Nippon Kayaku Kabushishiki, Dutch patent application 7,006,454 (1970).
99. Wolfs, M. W. J., and P. A. Batist, *J. Catal.*, **32**, 25 (1974).
100. Wolfs, M. W. J., Ph.D. thesis, Technical University of Eindhoven, The Netherlands, 1974.
101. Matsuura, I., and M. W. J. Wolfs, *J. Catal.*, **37**, 174 (1975).
102. Wolfs, M. W. J., and J. H. C. van Hooff, paper presented at the *Int. Symp. Prep. Heterogeneous Catal., Brussels, 1975.*
103. Smith, G. W., and J. A. Ibers, *Acta Crystallogr.*, **19**, 269 (1965).
104. Sleight, A. W., and B. L. Chamberland, *Inorg. Chem.*, **7**, 1671 (1968).
105. Abrahams, S. C., and J. M. Reddy, *J. Chem. Phys.* **43**, 2533 (1965).
106. Oganowski, W., J. Hanuza, B. Jezowska-Trerbiatowska, and J. Wryszca, *J. Catal.*, **39**, 161 (1975).
107. Plyasova, L. M., and L. M. Kefeli, *Inorg. Mat.*, **3**, 812 (1967).
108. Fagherazi, G., and N. Pernicone, *J. Catal.*, **16**, 321 (1970).
109. Batist, P. A., C. G. M. van de Moesdijk, I. Matsuura, and G. C. A. Schuit, *J. Catal.*, **20**, 40 (1971).
110. Sleight, A. W., and W. Jeitschko, *Mat. Res. Bull.*, **9**, 951 (1974).
111. Lo Jacono, M., T. Notermann, and G. W. Keulks, *J. Catal.*, **40**, 19 (1975).
112. Callahan, J. L., and B. Gertisser, U.S. Patent 3,198,750 (1965).
113. Grasselli, R. K., and J. L. Callahan, *J. Catal.*, **14**, 93 (1965).
114. Anon., *Hydrocarbon Process.*, **52**(11), 100 (1973).

115. Hadley, D. J., and D. G. Stewart, British Patent 835,962 (1958).
116. Standard Oil Company (Ohio), British Patent 965,351 (1964).
117. Denn, M. M. "Stability of Reaction and Transport Processes," Prentice-Hall, Englewood Cliffs, N.J., 1974.
118. Kramers, H., and K. R. Westerterp, "Elements of Chemical Reactor Design and Operation," pp. 133–136, Academic, New York, 1963.
119. Froment, G. F., paper presented at *Manchester Conf. Oxid., April 1975.*

PROBLEMS

4-1 The value of z referred to on page 339 is determined by the data on page 338. What is this value? What are the yields of the two final products referred to in the scheme on page 339?

4-2 Use the data of Ref. 49 to determine V_m, the entropy of adsorption, and the enthalpy of adsorption for 1-butene on Bi_2O_3–MoO_3. What is the lowest pressure of 1-butene which would give monolayer coverage at each of the temperatures referred to in the figure?

4-3 Comparing the requirements of the solid Wacker catalyst and the methods for designing solid polymeric catalysts (chap. 2), suggest the structure of a polymeric Wacker catalyst and a synthetic route for its preparation.

4-4 The oxidation of Tetralin is apparently catalyzed by polyethylene. Mody et al. [*J. Catal.*, **31,** 372 (1973)] have suggested that a free-radical reaction takes place in solution with initiation by transfer of radicals from the polyethylene surface (p) to Tetralin (RH):

$$R_p\cdot + RH \longrightarrow R_pH + R\cdot$$

Suggest the remainder of the reaction mechanism to give Tetralin hydroperoxide as the product and kinetics showing zero-order reaction in O_2, $\frac{1}{2}$ order in the amount of polyethylene, and $\frac{3}{2}$ order in RH.

4-5 Use the data of Table 4-5 to prepare a rough design of a fluidized-bed reactor to produce 5 million kilograms of acrylonitrile per year. Assume that the reactor contains five stages, has a diameter of 20 cm, and is surrounded by a molten-salt heat-transfer medium.

4-6 It is still disputed whether the Cu-catalyzed oxidation of mercaptans is a concerted two-electron process or a radical, one-electron process. Suggest a mechanism for the latter. Suggest experiments to determine which mechanism prevails.

4-7 The introduction of D_2O in the reaction of propylene with O_2 in the presence of an ammoxidation catalyst leads to incorporation of D in the propylene [92, 93]. This result may be attributed to the reversibility of the allyl formation,

$$C_3H_6 \rightleftharpoons H_2C\text{=\!=}CH\text{=\!=}CH_2 + H \tag{1}$$

and a dissociative adsorption of D_2O,

$$D_2O \rightleftharpoons DO + D \tag{2}$$

with interchange of H and D.

(*a*) How many D atoms can be incorporated maximally if the allyl is symmetrically bonded as shown in Fig. 4-10? How many if it is σ-bonded, as shown on page 341 (I or II)?

(*b*) Is a combination of processes (1) and (2) above in agreement with the assumption that abstraction of the first H is rate-determining?

4-8 A more realistic way than the one suggested to determine a parameter for the activity of bulk oxygens (Fig. 4-10) would be to consider the reaction

$$MO_n + C_4H_8 \longrightarrow MO_{n-1} + C_4H_6 + H_2O$$

Use handbook data for standard enthalpies of formation to calculate the enthalpy of reaction and show that the sequence presented in the chapter remains unaltered.

4-9 Consider the use of H_2O as an oxygen donor instead of O_2 by writing the reaction

$$C_4H_8 \rightleftharpoons C_4H_6 + H_2$$

as

$$C_4H_8 + MO_{g+1} \longrightarrow C_4H_6 + MO_g + H_2O \tag{a}$$

$$MO_g + H_2O \longrightarrow MO_{g+1} + H_2 \tag{b}$$

Where MO_{g+1} is an oxide catalyst that can donate O to form MO_g.

Balandin's principle would lead us to expect that maximum activity would arise for equal rates of reactions (a) and (b). Express this criterion mathematically by assuming that the standard free energy changes of the two reactions are equal. These expressions contain $\Delta H_f^\circ(MO_{g+1}) - \Delta H_f^\circ(MO_g)[= -Q_0$ of eq (41)]; calculate this value and check which oxide would meet the criterion. Check whether this result is in agreement with the catalyst actually used.

4-10 A reaction temperature at which butadiene may be produced is 627°C. Calculate from the data in problem 4-9 the ratio of $C_{H_2O}/C_{C_4H_8}$ to give a 50% conversion at this temperature and a pressure of 1 atm. What would happen to the Fe_2O_3 catalyst if this ratio were 0?

4-11 In gasification of coke, steam is used as the oxygen donor to produce H_2. Use benzene as a model of coke and represent the reaction as

$$CH + 2H_2O \longrightarrow CO + \tfrac{5}{2}H_2$$

Use the method as given in problem 4-9 to find a possible oxidic catalyst. What is the reason why H_2O is not always used for oxidation?

4-12 The reaction $C_4H_8 + \frac{1}{2}O_2 \rightarrow C_4H_6 + H_2O$ at temperatures exceeding 400°C and in the presence of bismuth-molybdate catalysts is found to be first order in the partial pressure of butene with an activation energy of about 12 kcal/mole. At lower temperatures the rate decreases much faster than predicted by the Arrhenius equation, which is usually ascribed to inhibition by the product butadiene. Matsuura found the heat of adsorption of butadiene to be about 23 kcal/mole. Set up the Langmiur-Hinshelwood rate equation for the reaction and infer the kinetics at temperatures around 200°C, for which surface coverage by butadiene is almost complete; deduce the activation energy for this temperature. Is there an alternative explanation for the increase in activation energy? Hint: see the results of Keulks [27, 27*a*].

4-13 Keulks and Krenzke deduced from the ^{18}O concentration in the product acrolein of the oxidation of propylene catalyzed by Bi_2MoO_6 that $^{18}O_2$ adsorbs at a site different from that giving rise to the oxidation of propylene [27*a*]. They inferred that O^{2-} diffuses through the lattice or along the surface from one site to another. In the oxidative dehydrogenation of C_4H_8 to give C_4H_6 or the oxidation of $C_3H_6 + NH_3$ to give acrylonitrile, there is no oxygen incorporated in the hydrocarbon product; it is incorporated only in H_2O. Suggest an alternative mechanism for the transfer of oxygen from the adsorption site to the catalytic site. What are the consequences of this mechanism?

4-14 Matsuura computed an entropy of adsorption for the adsorption of 1-butene leading to a loss of three translational degrees of freedom [52]. Consult the reference and perform a calculation to check Matsuura's value of the entropy of adsorption.

4-15 Accepting Matsuura's number of active catalytic sites to be $10^{18}/m^2$, estimate the production rate of butadiene molecules per m^2 at a temperature of 700 K. Assume that the activation energy = 12 kcal/mole and that there is a loss of three translational degrees of freedom from the reactant in formation of the transition state.

CHAPTER

FIVE

HYDRODESULFURIZATION

INTRODUCTION†

Developments in petroleum refining technology in the last two decades have brought the reactions referred to as *hydroprocessing* to a level of economic importance matching cracking and reforming. In hydroprocessing applications, petroleum components react catalytically with hydrogen. *Hydrocracking*, one of the most important examples, involves cracking and hydrogenation of hydrocarbons to give refined fuels with smaller molecules and higher H/C ratios. The most important hydroprocessing application is *hydrodesulfurization*, which involves reactions leading to removal of sulfur from petroleum compounds by their conversion into H_2S and hydrocarbon products.

This chapter is devoted to hydrodesulfurization, beginning with an introductory statement of the reactions, catalysts, and processes and following with detailed accounts of the catalytic chemistry and process engineering. Hydrodesulfurization is one of the less well-developed topics of this book but is perhaps the one of greatest current industrial interest. The catalysts are comparable in their complexity to partial-oxidation catalysts, being formed from mixtures of transition-metal compounds. A coherent picture of catalyst structure and surface reaction mechanism has begun to emerge only in the last few years. The kinetics and process-engineering data are almost entirely qualitative since so much of the

† This chapter has been adapted from a review article in *AIChE J.*, **19**, 417 (1973).

essential information remains proprietary. Nonetheless, the engineering problems are now clearly defined, and they have a complexity matching that of the catalyst structures: the catalysts age by both chemical and physical processes; the reactors contain gas, liquid, and solid phases undergoing intricate flow patterns; and the exothermic reactions can sometimes constitute an explosion hazard.

Hydrodesulfurization reactions are those of the following class:

$$\text{Organic sulfur compound} + H_2 \longrightarrow H_2S + \text{desulfurized organic compound}$$

Catalytic processing of light petroleum feedstocks with hydrogen to remove sulfur is carried out with several objectives, among them pretreatment of catalytic reformer feeds to prevent sulfur poisoning of the platinum-containing catalysts, as discussed in Chap. 3, and treatment of gasoline formed in catalytic cracking to provide sweetened and stabilized products. Desulfurization of heavy petroleum fractions gives products including diesel and jet fuels, heating oils, and residual fuel oils. The incentive is now strong for removal of sulfur from fuel oils, since combustion of sulfur-containing fuels is the primary cause of SO_2 pollution of the atmosphere; this pollution is most severe in the eastern United States, Japan, and western Europe. The need for clean fossil fuels extends to coal, and in the last few years effort has been devoted to development of processes for converting coal into low-sulfur liquids to be burned in power plants.

FEEDSTOCKS

In this chapter we refer to light petroleum feeds, for which desulfurization technology is well established and routinely applied, and to heavy feeds, including petroleum residua and coal-derived oils, for which the relatively new technology is undergoing rapid evolution. Table 5-1 summarizes the approximate boiling ranges of light and heavy petroleum feeds used in hydrodesulfurization processes.

Petroleum feedstocks include the sulfur-containing compounds listed in Table 5-2; only the thiophenic compounds are found in coal-derived liquids. The compound classes are listed approximately in order of decreasing reactivity in

Table 5-1 Petroleum fractions

Petroleum feedstock	Approximate boiling range, °C
Light:	
Light gasoline	< 80
Naphtha (reformer feed)	80–160
Kerosene (jet fuel)	150–230
Gas oil (diesel and heating oil)	170–370
Heavy:	
Residual oil (fuel oil)	> 350

Table 5-2 Some sulfur-containing compounds in petroleum

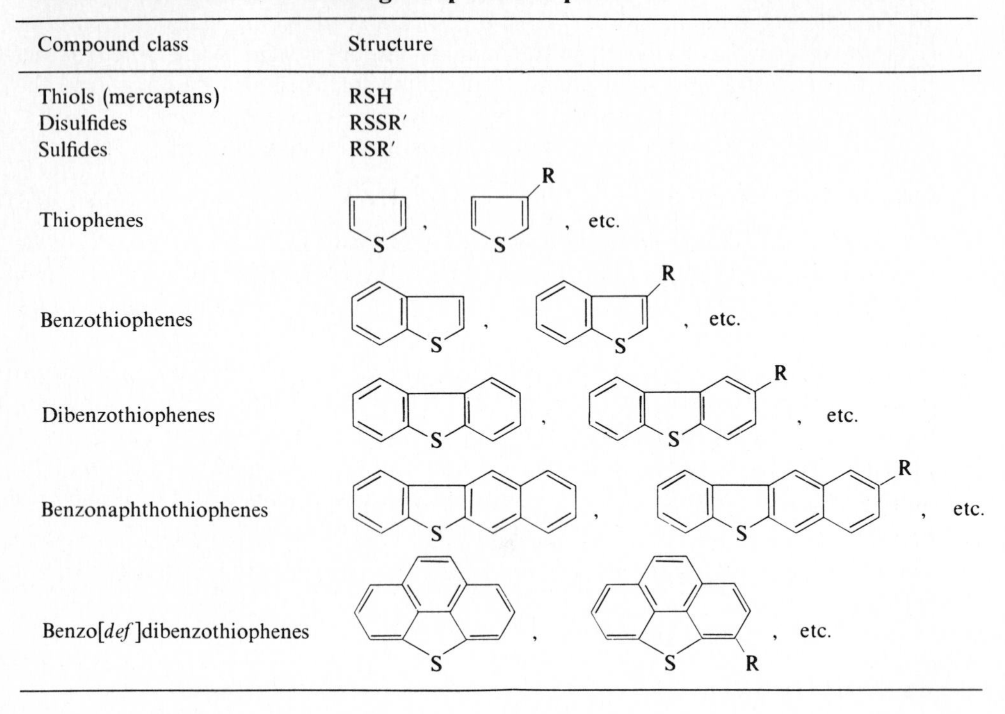

Compound class	Structure
Thiols (mercaptans)	RSH
Disulfides	RSSR′
Sulfides	RSR′
Thiophenes	S, S (R), etc.
Benzothiophenes	S, S (R), etc.
Dibenzothiophenes	S, S (R), etc.
Benzonaphthothiophenes	S, S (R), etc.
Benzo[*def*]dibenzothiophenes	S, S (R), etc.

hydrodesulfurization: thiols are very reactive, and compounds in the class of thiophenes are much less reactive. The kinetics and reaction networks in hydrodesulfurization of these compounds are considered in the following pages.

REACTIONS

A summary of the most important classes of reactions occurring in hydrodesulfurization processes is given here, to be followed in another section by an account of the hydrodesulfurization reactions themselves. The desired reactions are hydrogenolyses resulting in cleavage of a C—S bond; e.g.,

$$R{-}SH + H_2 \longrightarrow RH + H_2S \tag{1}$$

Under industrial reaction conditions, hydrogenolysis reactions resulting in breaking of C—C bonds also occur, e.g., the hydrocracking reaction

$$RCH_2CH_2R' + H_2 \longrightarrow RCH_3 + R'CH_3 \tag{2}$$

Another kind of hydrogenolysis reaction is hydrodenitrogenation; e.g.,

$$RNH_2 + H_2 \longrightarrow RH + NH_3 \tag{3}$$

Reactions of this type are effective in removing nitrogen from fuels, which is often necessary before refining by hydrocracking, since basic nitrogen-containing compounds are poisons for the acidic sites of bifunctional hydrocracking catalysts, which have both cracking and hydrogenation activity.

Hydrogenation of unsaturated compounds also occurs during hydrodesulfurization, and the reaction rates are significant compared with those of hydrodesulfurization; e.g.,

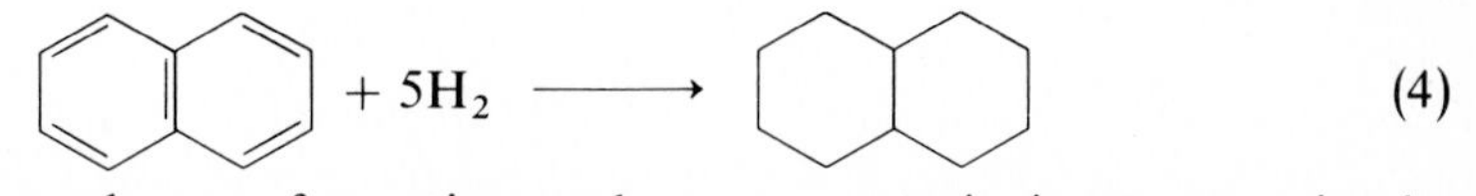

The last three classes of reactions take on economic importance in the processes by consuming expensive hydrogen without removing sulfur.

Thermal cracking reactions can also take place at relatively high temperatures to produce molecular-weight reduction.

Demetallization reactions are important in accompanying the hydrodesulfurization of residua. The heaviest fractions of some petroleum feedstocks contain significant concentrations of organometallic compounds, especially of V and Ni. These react to give solid metal sulfides, which can accumulate in a reactor and ultimately plug the pores of a catalyst or the interstices of a fixed bed of catalyst particles.

Coking reactions, common to virtually all hydrocarbon reaction processes as described in Chaps. 1 and 3, occur as well in hydrodesulfurization. Coke not only poisons catalyst surfaces but contributes to blocking of catalyst pores and fixed-bed interstices. In contrast to inorganic compounds of V and Ni, however, coke can be burned off catalysts, as in catalytic cracking processes, leading to successful regeneration.

The hydrodesulfurization reactions are virtually irreversible at temperatures and pressures ordinarily applied, roughly 300 to 450°C and up to 200 atm. The reactions are exothermic with heats of reaction of the order of 10 to 20 kcal/mol of hydrogen consumed, or roughly 50 to 100 Btu per standard cubic foot of hydrogen consumed.

CATALYSTS

The catalysts applied in hydrodesulfurization have evolved from those developed in prewar Germany for hydrogenation of coal and coal-derived liquids. They are formed from alumina-supported oxides of Co and Mo, and the surfaces are usually sulfided in operation. Catalysts of this type are often referred to as cobalt molybdate, which is a jargon term. Industrial catalysts may contain as much as 10 to 20 percent of Co and Mo. A number of related compositions have been applied, including, for example, Ni and W instead of Co and Mo. In contrast to the supported Pt and Pt-alloy catalysts used in reforming, hydrodesulfurization catalysts have hydrogenation activity in the presence of high concentrations of sulfur compounds; their activities, however, are characteristically less than that of Pt by several orders of magnitude.

Hydrodesulfurization catalysts are used as porous particles or extrudates, typically having dimensions of 1.5 to 3 mm. The particle size and pore geometry significantly influence catalyst performance, especially for the heaviest feeds, since intraparticle mass transport has a significant influence on reaction rates.

PROCESSES

In the processing of light petroleum fractions, gas or liquid-phase oil is contacted concurrently with hydrogen, flowing downward through solid catalyst particles in a fixed-bed reactor. The design of these reactors for new feedstocks is straightforward, since all the reactors are substantially the same, and a wealth of plant operating data is available. A simplified process flow diagram is shown in Fig. 5-1. Unconverted hydrogen from the product stream is scrubbed to remove H_2S and recycled. A rough guide to the severity of the processing conditions is given in Table 5-3, where light and heavy feeds are contrasted.

Temperature is increased gradually during operation to compensate for activity decline caused by coke accumulation. Coke is periodically burned off the catalyst, which may endure several operating cycles and have a lifetime of as much as 10 years. Catalyst costs in light-feed hydrodesulfurization processes account for less than about 10 percent of processing costs, and there is therefore little incentive for developing new catalysts; it seems unlikely that there will soon be significant changes in this well-established processing technology.

If the reacting petroleum in one of these processes is present as a liquid, such as a gas oil, then the reactor has gas, liquid, and solid phases and is referred to as a *trickle-bed reactor*. Hydrodesulfurization processes appear to be the most important applications of this reactor type.

Trickle-bed reactors are also used on an industrial scale in hydrodesulfurization of heavy petroleum feeds. Operating conditions are more severe with these less reactive feeds, as summarized in Table 5-3. The previously mentioned coking

Table 5-3 Approximate processing conditions in catalytic hydrodesulfurization[a]

	Feedstock		
Processing conditions	Light petroleum (distillates)	Heavy petroleum (residua)	Coal
Temp., °C	300–400	340–425	400–460
Pressure range, atm	35–70	55–170	135–270
LHSV, vol feed/vol catalyst · h	2–10	0.2–1	~ 1
H_2 recycle rate, std ft³/bbl	300–2000	2000–10,000	> 25,000
Catalyst life, years	~ 10	$\frac{1}{2}$–1	Undetermined

[a] Processing conditions have been inferred from many literature sources.

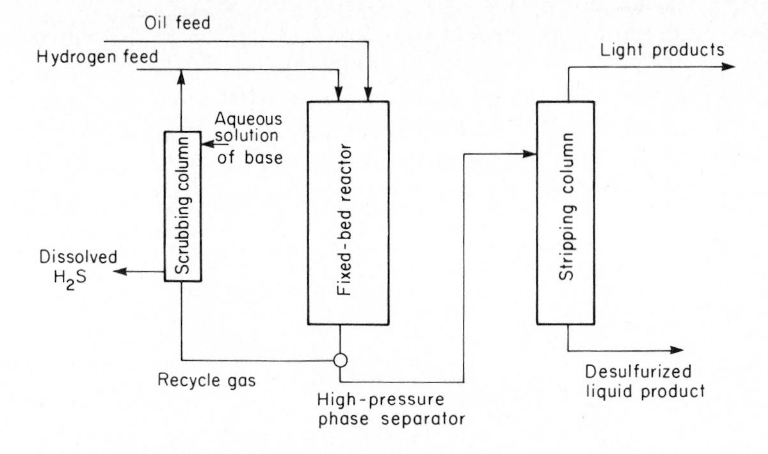

Figure 5-1 Simplified diagram of a hydrodesulfurization process.

and demetallization reactions are especially important for the petroleum residua. Deposits of coke and inorganic metal compounds are readily formed, and accumulation of the metal-containing deposits prevents regeneration of catalysts. The application of nonregenerable catalysts in residuum hydrodesulfurization processes is unique among large-scale catalytic processes for hydrocarbon conversion.

It is the problems of catalyst aging that set heavy-feed processes apart from the others. The engineering of these processes is emphasized in a later section of the chapter in a discussion of the problems of obtaining efficient fluid-solid contacting, uniform catalyst-bed temperatures, and efficient catalyst utilization in the presence of significant mass-transport resistances.

Another reactor used commercially in residuum hydrodesulfurization is a fluidized bed. This reactor is distinct from the gas-solid catalytic cracking reactor because it also contains liquid. It is alternatively referred to as an *ebullating-bed* or *slurry-bed reactor* since the catalyst particles are held in suspension by the upward velocity of the liquid reactant, through which hydrogen flows concurrently. There are several advantages to this reactor design: (1) the deactivating catalyst can be removed and replaced continuously (as in catalytic cracking); (2) small particles of catalyst can be used, and they are more effective than large particles when intraparticle mass-transport resistance is significant (particles with dimensions less than about 0.7 mm are avoided in fixed beds because pressure drops are unacceptably high); and (3) the fluidized-bed-reactor contents are expected to be very well mixed, which suggests that uniform temperatures can be maintained (recall that this advantage was especially important for partial-oxidation reactors).

There are compensating disadvantages to the slurry-bed reactor: (1) the high degree of mixing of reactants in the direction of flow necessitates higher temperatures or lower space velocities than in a fixed bed to achieve the same conversion; (2) corresponding to higher operating temperatures, undesired side reactions such as hydrocracking are more important, leading to higher hydrogen consumptions; and (3) the complex problems of reactor stability and control have not yet been solved for the three-phase fluidized system with exothermic reactions involving hydrogen at high temperatures and pressures.

CATALYTIC CHEMISTRY

REACTIONS

Introduction

Petroleum and coal-derived liquids contain many different compounds susceptible to hydrodesulfurization, and in many applications most of them must be converted. Sulfur may be bonded to one carbon, as in thiols and disulfides; to two carbons of alkyl groups, as in sulfides; or to two carbons in aromatic rings, as in thiophene and dibenzothiophene. There are several basic questions concerning hydrodesulfurization of these compounds: What stable intermediates are formed during hydrodesulfurization; i.e., what are the reaction networks? What are the reactivities of the various classes of sulfur-containing compounds, and what are the effects of substituent groups on reactivities within a given class? What are the structures of catalyst surfaces and adsorbed intermediates? These questions form an outline for the following section of the chapter.

Reaction Networks

Thiophenic compounds are the least reactive sulfur compounds in petroleum, and so the simplest and most easily obtained compound in this class, thiophene itself, has frequently been chosen for study, being regarded as a model reactant. Thiophene hydrodesulfurization was examined in a series of kinetics studies by Amberg and coworkers [1–6]. The catalysts were a commercial Co-Mo/Al_2O_3 (1.3 percent Co and 6.1 percent Mo), chromia, and several molybdenum disulfides. Kinetics data were obtained from a pulse microreactor and from a steady-state flow reactor operated at low conversions (< 0.5 percent). Reaction products were included in feeds to provide identification of reaction inhibitors and stable reaction intermediates. Some conversion and product-distribution data from a steady-state flow-reactor study [6] are collected in Tables 5-4 and 5-5.

Owens and Amberg [1] also used the microreactor containing chromia catalyst for determination of conversions of the individual compounds listed in Table 5-6. The data of Tables 5-5 and 5-6 lead to the reaction network suggested in Fig. 5-2 for thiophene hydrodesulfurization catalyzed by chromia and by Co-Mo/Al_2O_3.

Table 5-4 Catalyst activities in thiophene hydrodesulfurization [6]

Catalyst	BET area, m^2/g	Apparent reaction rate,[a, b] $mol/m^2 \cdot s$
MoS_2	3.3	0.9
MoS_2 + 1% Co	3.3	0.5
MoS_2 from MoS_3 heated to		
400°C	154	11.0
700°C	67	1.6
800°C	12	1.5
$Co\text{-}Mo/Al_2O_3$	241	1.6
Cr_2O_3	150	2.0

[a] Rate of reaction varied significantly between fractional conversions of 0 and 0.005. Tabulated values are average rates between these two conversions.

[b] Data were obtained with a steady-state flow reactor at 288°C. Partial pressures of hydrogen and thiophene were 1.00 and 0.03 atm, respectively. Flow rate of hydrogen was maintained at about 3×10^{-7} m^3/s. Amounts of catalyst were varied to give a surface area of 2 to 5 m^2 for each.

Source: Reproduced with permission from *Canadian Journal of Chemistry*. Copyright by the National Research Council of Canada.

The suggestion that the first reaction of thiophene in the primary reaction path is the C—S bond cleavage to form 1,3-butadiene, rather than hydrogenation of the C=C bond, is supported by further data of Desikan and Amberg [5]. Their results show that the hydrogenated compound with the C—S bonds intact

Table 5-5 Product distribution in thiophene hydrodesulfurization [6]

Products from a steady-state flow reactor at 0.5% conversion of thiophene; other conditions specified in Table 5-4

	Mol % in C_4 hydrocarbon product				
Catalyst	Butadiene	1-Butene	*cis*-2-Butene	*trans*-2-Butene	Butane
MoS_2	6.9	42.5	22.3	19.2	8.8
MoS_2 + 1% Co	8.4	55.6	14.0	17.4	4.7
MoS_2 from MoS_3, heated to:					
400°C	7.2	39.9	16.7	23.5	12.7
700°C	4.0	28.5	22.0	36.5	9.5
$Co\text{-}Mo/Al_2O_3$	2.2	47.5	19.8	24.3	6.2
Cr_2O_3	7.7	31.3	11.8	11.8	37.4

Source: Reproduced with permission from *Canadian Journal of Chemistry*. Copyright by the National Research Council of Canada.

Table 5-6 Product distribution in reactions catalyzed by chromia [1]

Conversions in a pulse microreactor at 415°C

	Mol % in C_4 hydrocarbon product				
Reactant	Isobutane	*n*-Butane	1-Butene + isobutene[a]	*trans*-2-Butene	*cis*-2-Butene + butadiene
Thiophene	0.0	8.6	30.1	27.4	33.9[b]
1,3-Butadiene	0.0	2.0	23.1	36.7	38.2[b]
1-Butene	0.0	8.9[c]	42.1	26.6	22.4
trans-2-Butene	0.0	1.9	22.3	52.4	23.5
cis-2-Butene	0.0	2.0	21.2	37.6	39.2
Isobutene	2.0	0.0	98.0	0.0	0.0
n-Butane	...	100.0	0.0	0.0	0.0
n-Butenes[d]	...	...	20	48	32

[a] Assumed to be absent when isobutane was absent.
[b] Butadiene present.
[c] Value possibly high.
[d] Gas-phase equilibrium attained.
Source: Reproduced with permission from *Advances in Chemistry Series*. Copyright by the American Chemical Society.

(tetrahydrothiophene) gives products in hydrodesulfurization different from those observed for thiophene.

Amberg and coworkers also observed that H_2S inhibited reaction of thiophene and hydrogenation of butene but had only little effect on cis-trans isomerization, double-bond shifts, or butadiene conversion to butenes. These results led

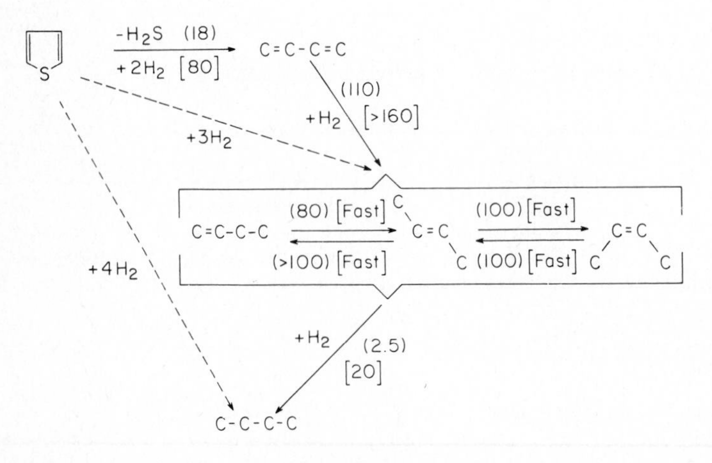

Figure 5-2 Thiophene hydrodesulfurization reaction network [1]. Numbers in parentheses are approximate rates [(mmol)/g · s] with chromia catalyst at 415°C; numbers in brackets are for Co-Mo/Al_2O_3 catalyst at 400°C. (Reprinted with permission from *Advances in Chemistry Series*. Copyright by the American Chemical Society.)

them to suggest that more than one kind of site is operative in hydrodesulfurization; we shall return to this suggestion.

A study of the reaction network involved in hydrodesulfurization of substituted benzothiophenes was reported by Givens and Venuto [7]. Conversion data were obtained with a steady-state flow reactor packed with particles of a commercial Co-Mo/Al_2O_3 catalyst containing 2.4 percent Co and 6.7 percent Mo. Product-distribution data for several feeds are collected in Table 5-7. Both desulfurized and undesulfurized products were found, along with some unidentified products.

Givens and Venuto observed that a hydrogenation-dehydrogenation equilibrium was established at a rate which was high compared with the rate of desulfurization:

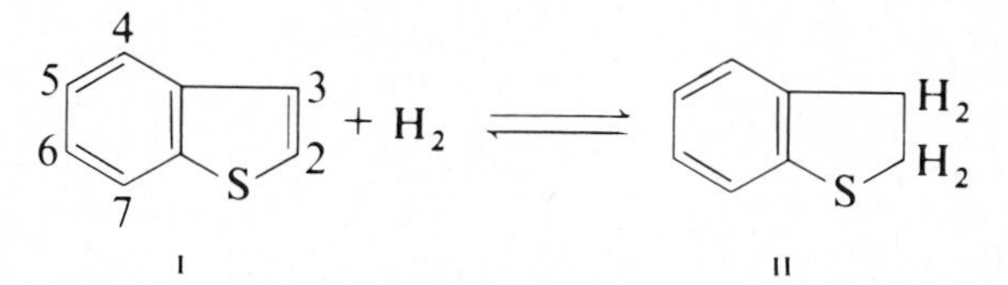

It was also observed that these two compounds were desulfurized at approximately equal rates, however, so that it was impossible to establish whether one was an intermediate in the desulfurization of the other.

This information and the result that the hydrogenated compound II (dihydrobenzo[*b*]thiophene) was found as a product of reaction of the benzothiophene (I) at low temperature (300°C) and high space velocity (1.1×10^{-3} vol/vol · s) is consistent with identification of the hydrogenated compound as an intermediate in hydrodesulfurization of compound I. A similar set of results was obtained when the reactant was 2,3,7-trimethylbenzo[*b*]thiophene; two products identified as *erythro*- and *threo*-2,3-dihydro-2,3,7-trimethylbenzo[*b*]thiophene were found:

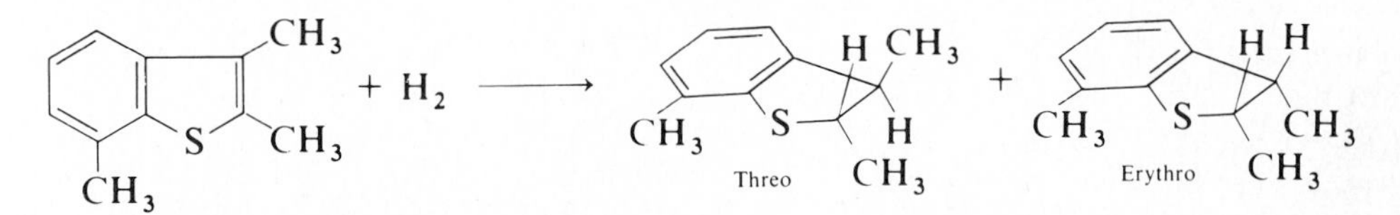

We therefore conclude that hydrogenated benzothiophenes (substituted and unsubstituted) are formed rapidly from benzothiophenes and hence that the former may be intermediates in the desulfurization although they cannot be unequivocally identified as such. This conclusion contrasts with Amberg's results, which show that hydrogenated compounds are not intermediates in hydrodesulfurization of thiophene.

Possible pathways for desulfurization of the hydrogenated compound 1-methyldihydrobenzo[*b*]thiophene are shown in Fig. 5-3*a*. Neither of the postulated mercaptan intermediates was detected, but neither pathway can be excluded, since at the reaction conditions mercaptans are converted about 10 to 100 times faster than the benzothiophene from which they might be formed.

Table 5-7 Product distribution in hydrodesulfurization of substituted benzothiophenes catalyzed by Co–Mo/Al_2O_3 at 400°C [7]

Data represent uncorrected peak areas by gas chromatography; catalyst was not presulfided but pretreated in hydrogen for 10,800 s at 600°C; feed ratio hydrogen to benzothiophene = 3 : 5; pressure assumed atmospheric

	Reactant							
	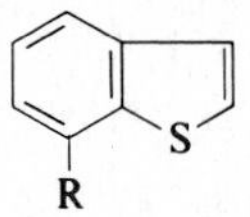 S, R		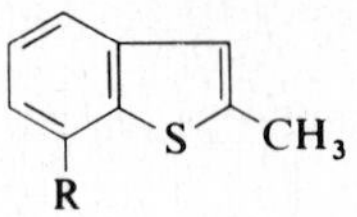 S, CH_3, R		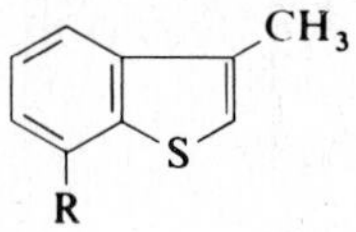 CH_3, S, R		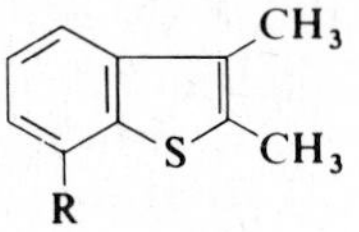 CH_3, S, CH_3, R	
Product, approximate mol %	R = H	R = CH_3	R = H	R = CH_3	R = H	R = CH_3	R = H	R = CH_3
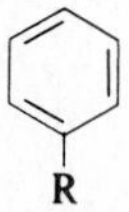 R	...	2	0.5	2	5	2	4	1
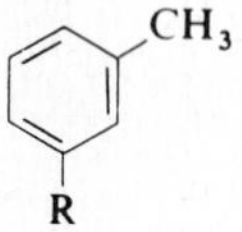 CH_3, R	...	4	0.5	0.5	0.5	...	...	...
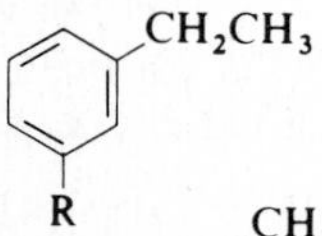 CH_2CH_3, R	96	86	6	7	10	10	3	1
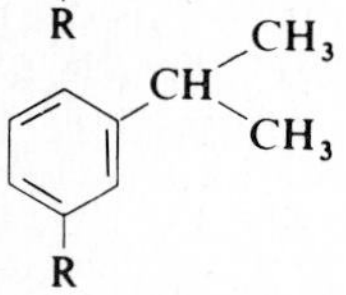 $CH(CH_3)_2$, R	...	...	5	3	42	22	2	2
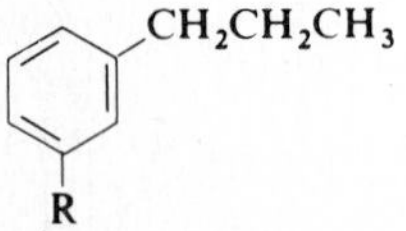 $CH_2CH_2CH_3$, R	...	...	75	54	18	10	8	4
Unidentified nonsulfur compounds	...	0.5	1	4	1	0.5	5	10
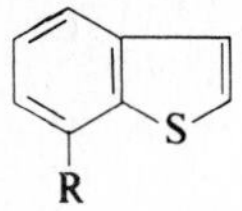 S, R	Charge		1	8	2	11	...	2
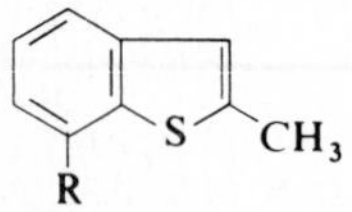 S, CH_3, R	...	...	Charge		18	26	15	14

Table 5-7 (*Continued*)

Product, approximate mol %	Reactant: S, R		S, CH_3, R		CH_3, S, R		CH_3, S, CH_3, R	
	R = H	R = CH_3	R = H	R = CH_3	R = H	R = CH_3	R = H	R = CH_3
CH_3, S, R	...	...	9	14	Charge		15	17
CH_3, S, CH_3, R	...	...	...	...	...	...	Charge	
Unidentified sulfur compounds	...	7	2	9	3	19	12	32

Source: Reproduced with permission from *Preprints, American Chemical Society Division of Petroleum Chemistry*. Copyright by the American Chemical Society.

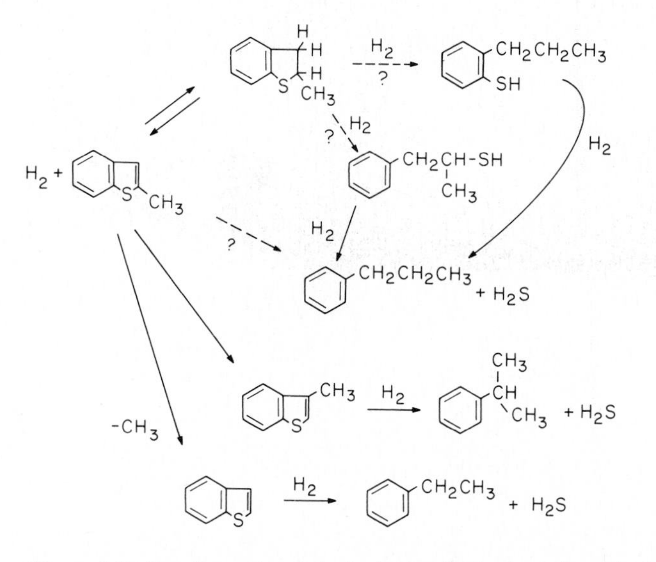

Figure 5-3*a* Reaction network in hydrodesulfurization of a substituted benzothiophene catalyzed by Co-Mo/Al_2O_3 [7]. (Adapted with permission from *Preprints, American Chemical Society Division of Petroleum Chemistry*. Copyright by the American Chemical Society.)

Givens and Venuto were able to rule out desulfurization reactions involving C—C bond breaking by showing that the corresponding products were not formed. Saturation of the aromatic ring was found not to be necessary for breaking of the bond between sulfur and the aromatic carbon to occur.

Using the appropriate intermediates shown in Fig. 5-3*a* as feeds, Givens and Venuto demonstrated that primary sulfur extrusion, alkyl migration on the thiophene ring, and dealkylation of the thiophene ring all occur during hydrodesulfurization of 2-methylbenzo[*b*]thiophene at 400°C and atmospheric pressure.

The data for related feeds (Table 5-7) are consistent with this network; they show further that neither dealkylation nor migration of methyl substituents on the benzene ring of benzo[*b*]thiophene occurs appreciably at the reaction conditions.

Hydrodesulfurization of dibenzothiophene, in contrast to benzothiophene, takes place with high selectivity at both low [8] and high [8*a*] pressures:

$$+ 2H_2 \longrightarrow H_2S + \qquad (5)$$

A detailed representation of the reaction network involving dibenzothiophene and hydrogen in the presence of sulfided $CoO\text{-}MoO_3/\gamma\text{-}Al_2O_3$ at about 300°C and 100 atm is shown in Fig. 5-3*b*.

Kinetics

Thorough quantitative studies of hydrodesulfurization kinetics, covering wide ranges of temperature and pressure and including effects of all reactants and products, are still generally lacking. A partial determination of kinetics of thio-

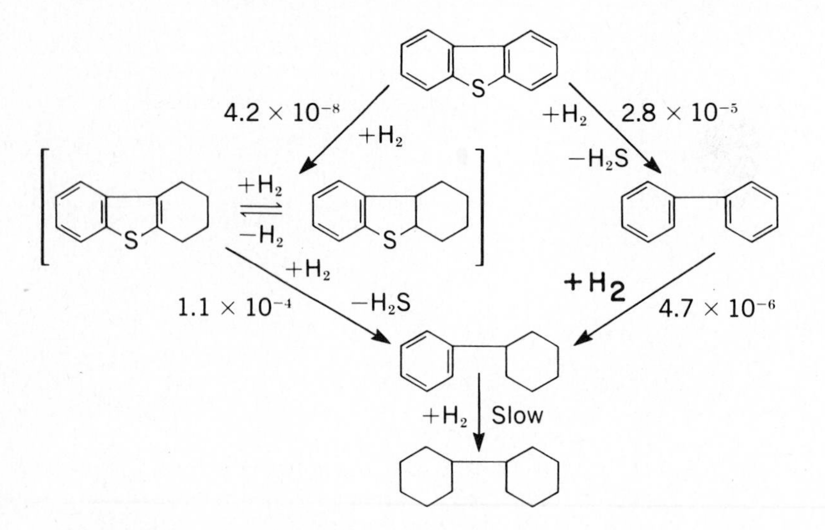

Figure 5-3*b* Reaction network in hydrodesulfurization of dibenzothiophene catalyzed by sulfided $Co\text{-}Mo/Al_2O_3$ at 300°C and 102 atm. The numbers over the arrows represent the pseudo first-order rate constants in units of m^3/kg of catalyst · s [8*a*].

phene hydrodesulfurization in the absence of mass-transfer influence was reported by Satterfield and Roberts [9], who used a Co–Mo/Al_2O_3 catalyst containing about 3 percent Co and 7 percent Mo. Reaction-rate data were determined from low conversions attained in a steady-state recirculation-flow reactor. Pressure was slightly in excess of atmospheric, temperature was 235 to 265°C, and feeds contained various concentrations of thiophene and H_2S; hydrogen partial pressure was varied only insignificantly. The products observed are consistent with Amberg's reaction network (Fig. 5-2). The data for rates of thiophene disappearance (hydrogenolysis) and rates of butane formation (butene hydrogenation) were represented by Langmuir-Hinshelwood rate equations as follows:

$$r_{\text{HDS}} = \frac{kP_T P_{\text{H}_2}}{(1 + K_T P_T + K_{\text{H}_2\text{S}} P_{\text{H}_2\text{S}})^2} \tag{6}$$

$$r_{\text{hyd}} = \frac{k' P_{\text{B}} P_{\text{H}_2}}{(1 + K'_B P_B + K'_{\text{H}_2\text{S}} P_{\text{H}_2\text{S}})} \tag{7}$$

Values of the rate-equation parameters are summarized in Table 5-8. Although the appropriateness of these equations is not firmly established by the data, several clearly important qualitative results are confirmed by the literature generally: (1) H_2S inhibits both the hydrogenolysis and hydrogenation reactions; (2) significant amounts of reactant thiophene and butene are adsorbed on the catalyst surface in competition with H_2S; (3) less clear, but more important, is the conclusion that the hydrogenolysis and hydrogenation reactions proceed on separate catalytic sites, consistent with the results of Amberg mentioned previously.

A kinetics study of hydrodesulfurization of light catalytic cycle oils (boiling range about 180 to 330°C) was reported by Frye and Mosby [10]. Their reactor was a trickle bed containing particles of commercial Co-Mo/Al_2O_3 catalyst of unspecified composition. The ranges of variables studied are given in Table 5-9. Feed and product analysis by gas chromatography provided data for evaluation of kinetics for individual sulfur compounds in the oil. Figure 5-4 includes chromatograms for feed and product. The results illustrate a common pattern: the low-molecular-weight compounds are more readily desulfurized than

Table 5-8 Thiophene hydrodesulfurization kinetics: parameter values for Eqs. (6) and (7) [9]

Temp., °C	$10^5 \times k$, mol/g · atm^2 · s	$10^{-1} \times K_T$, atm^{-1}	$10^{-1} \times K_{\text{H}_2\text{S}}$, atm^{-1}	$10^8 \times K'$, mol/g · atm · s	$10^{-2} \times K'_B$, atm^{-1}	$10^{-1} \times K'_{\text{H}_2\text{S}}$, atm^{-1}
235	1.5	4.3	3.1	3.0	9.8	9.1
251	1.6	2.3	1.3	12.2	1.2	1.9
265	1.8	2.5	0.56	...	~0	1.3

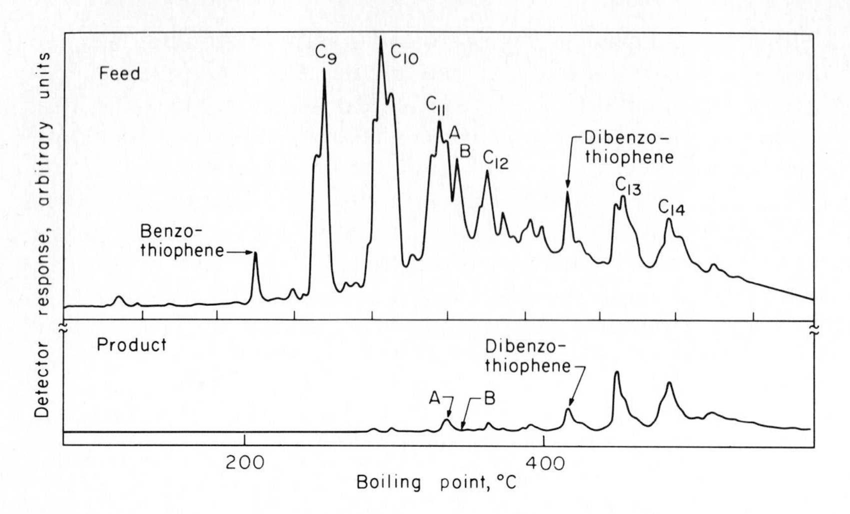

Figure 5-4 Gas chromatographic analysis of benzothiophene and dibenzothiophene in desulfurized and undesulfurized gas oil. Reaction was at 330°C and 20 atm [10].

high-molecular-weight compounds. Detailed data for three components, a trimethylbenzothiophene (A), another trimethylbenzothiophene (B), and dibenzothiophene, are shown in Fig. 5-5. For each sulfur-containing compound, the logarithm of the fraction remaining unconverted decreases in proportion to the inverse space velocity, demonstrating that the hydrodesulfurization reactions are first order in the concentration of the sulfur compound.

Frye and Mosby found that hydrodesulfurization was inhibited by H_2S and by aromatic hydrocarbons, including those undergoing hydrodesulfurization; these results are consistent with Eq. (6) and many other results in the literature. Frye and Mosby also found that hydrodesulfurization was first order in hydrogen partial pressure, confirming the assumption incorporated in Eq. (6).

Frye and Mosby's results are summarized in the following rate equation, for which a full set of parameter values was not given:

$$r_{\mathrm{HDS}} = \frac{kP_{\mathrm{S}}P_{\mathrm{H_2}}}{(1 + K_{\mathrm{ar}}P_{\mathrm{ar}} + K_{\mathrm{H_2S}}P_{\mathrm{H_2S}})^2} \tag{8}$$

In a more exact analysis, the term indicating competitive adsorption of aromatic hydrocarbons ($K_{\mathrm{ar}}P_{\mathrm{ar}}$) could be replaced by a sum of terms for specific

Table 5-9 Ranges of variables studied in hydrodesulfurization of cycle oil [10]

Temperature, °C	260–370
Partial pressure of hydrogen, atm	7–38
Hydrogen feed rate, std m^3/m^3 of oil	89–356
10^4 × space velocity, g oil/g catalyst · s	4.2–28
Sulfur content of oil, wt %	0.4–2.0

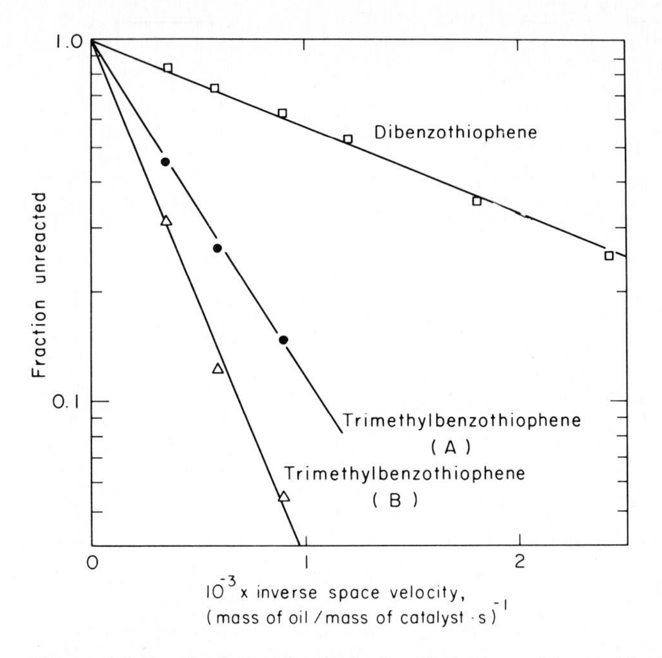

Figure 5-5 Pseudo first-order hydrodesulfurization of benzothiophene and dibenzothiophene in cycle oil at 290°C and 15 atm [10].

aromatic compounds, including those containing sulfur. This equation has the same form as Eq. (6), and its confirmation by independent investigators under different conditions provides strong support for its appropriateness.

Study of kinetics of hydrodesulfurization of compounds found in very light distillates has been reported by Phillipson [11]. The catalyst was Co-Mo/Al_2O_3 containing 2 percent Co and 8 percent Mo. Vapor feeds contained heptane mixed with dimethyl sulfide, phenyl mercaptan, diethyl sulfide, tetrahydrothiophene, or thiophene. Steady-state conversions were independent of linear velocity through the catalyst bed, demonstrating the absence of significant fluid-phase mass-transfer resistance. Intraparticle mass-transfer resistance might have been significant, however, but data to test the possibility were not obtained. The data summarized in Fig. 5-6 were represented by the following empirical rate equation, which shows a dependence of rate on total pressure and surprisingly indicates inhibition by the hydrocarbon heptane:

$$r_{\mathrm{HDS}} = \frac{k'' P_{\mathrm{S}} P_{\mathrm{H_2}}^{1/2} P_{\mathrm{Tot}}^{1/2}}{P_{HC}^{1/2}\left(1 + 0.21\,\dfrac{P_{\mathrm{H_2S}}}{P_{\mathrm{tot}}}\right)} \tag{9}$$

The kinetics is difficult to justify on theoretical grounds and should therefore be extrapolated with caution. The results of Fig. 5-6 are valuable, at least approximately, in demonstrating the relative lack of reactivity of thiophene compared with mercaptans, sulfides, and disulfides. Phillipson reported that the

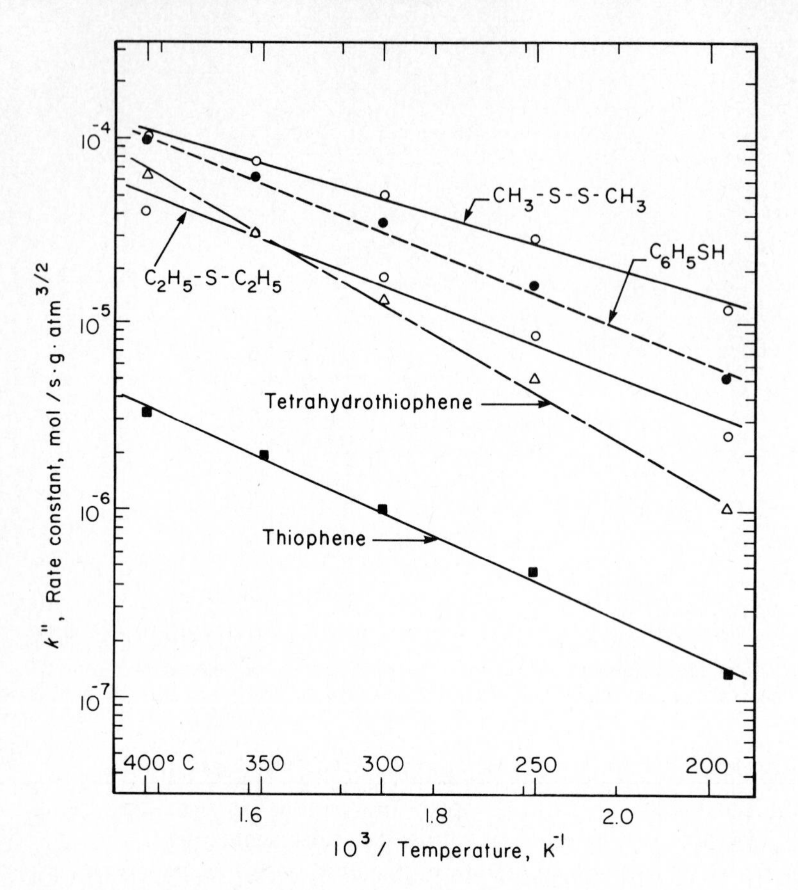

Figure 5-6 Arrhenius plot for hydrodesulfurization of compounds in light petroleum distillates. The rate constant k'' is defined by Eq. (9). The data are from [11] and the figure is from [12].

data are useful for design of light distillate desulfurization reactors, but they are of little use for heavier distillates such as kerosene and gas oils, which contain significant amounts of less reactive compounds.

Experiments are needed to determine kinetics of hydrodesulfurization of the less reactive compounds at industrial reaction conditions. Some results are available indicating relative reactivities: Frye and Mosby [10] reported up to fivefold variations in rate for compounds with the same molecular weight, as exemplified by substituted benzothiophenes. Similarly, Givens and Venuto [7] observed that the conversion of benzothiophenes depended on the position and number of methyl substituents, as shown by the data of Table 5-10. On the other hand, data obtained in a pulse microreactor operating at atmospheric pressure seem to contradict these results and show roughly equal conversions of thiophene, benzothiophene, methyl-substituted benzothiophenes, and dibenzothiophene [13]. The data obtained at low pressures are evidently not in accord with the reactivity patterns prevailing at practical processing conditions.

Table 5-10 Conversion and selectivity in desulfurization of substituted benzo[*b*]thiophenes [7][a]

Positions of reactant methyl group(s)	Fraction of reactant		Selectivity, fraction desulfurized/ fraction converted
	Converted	Desulfurized	
None	0.91–0.99	0.91–0.99	1.00
7	0.60	0.57	0.95
2	0.74	0.66	0.89
2, 7	0.54	0.43	0.80
3	0.43	0.32	0.75
3, 7	0.47	0.24	0.51
2, 3	0.39	0.15	0.38
2, 3, 7	0.47	0.16	0.34

[a] Reaction conditions assumed to be those specified in Table 5-7.

Source: Reproduced with permission from *Preprints, American Chemical Society Division of Petroleum Chemistry*. Copyright by the American Chemical Society.

The effect of methyl substituents on the rate of hydrodesulfurization of dibenzothiophene has recently been investigated [13, 14]. The results indicate a large decrease in reactivity as methyl groups are incorporated in the two positions nearest to S:

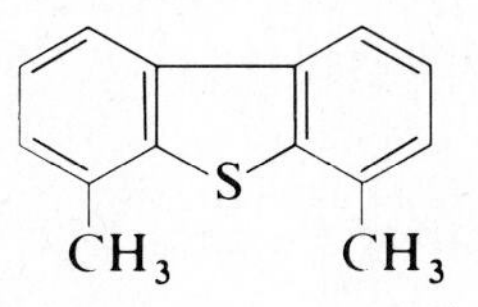

Table 5-11 Relative reactivities of dibenzothiophene (DBT) and methyl-substituted DBT [14]

Reactions catalyzed by fresh, sulfided $CoO–MoO_3/\gamma\text{-}Al_2O_3$ at 300°C and 103 atm; each reaction mixture contained DBT, a methyl-substituted DBT, *n*-hexadecane carrier oil, and dissolved hydrogen; each hydrodesulfurization reaction was first order in the sulfur-containing compound

Substituted DBT	$\dfrac{k_{\text{substituted DBT}}}{k_{\text{DBT}}}$
4-Methyl	0.16
4,6-Dimethyl	0.1
3,7-Dimethyl	1.5
2,8-Dimethyl	2.6

Source: Reproduced with permission from *Preprints, American Chemical Society Division of Petroleum Chemistry*. Copyright by the American Chemical Society.

Methyl groups in other positions have little effect on reactivity (Table 5-11). These results may be understandable in terms of steric hindrance of adsorption and they are discussed subsequently. They suggest that some of the more highly substituted dibenzothiophenes, especially those with bulky groups, may be much less reactive than the compounds for which data are available. These compounds with highly branched substituents may be more prevalent in petroleum residua than in coal-derived liquids.

The inhibiting effect of H_2S has been determined quantitatively by Metcalfe [15] for distillate hydrodesulfurization catalyzed by Co-Mo/Al_2O_3:

$$k_{eff} = k \frac{1}{1 + 21(P_{H_2S}/P_{tot})} \tag{10}$$

where k is a pseudo-first-order rate constant and k_{eff} is a corrected value incorporating the effect of H_2S inhibition. When the H_2S concentration becomes 0.3 mole percent in the reactant gas mixture, the reaction rate is reduced by about 5 percent; these values may be representative of complete reaction of nonthiophenic sulfur compounds in light distillates [11].

Inhibition of reaction of many feedstocks by H_2S has been represented by the same form of equation [16]. With an unidentified gas oil, the reaction rate was reduced about twofold when 10 mole percent of the treat gas was H_2S. Correspondingly, the rate would be reduced by about 5 percent when the gas contained 0.5 percent H_2S, which indicates that the results are quantitatively similar to Metcalfe's for light distillates.

Simplified Kinetics for Industrial Feedstocks

Many of the kinetics results cited for pure compounds are observed generally for hydrodesulfurization of distillates and residua. Reaction is inhibited by product H_2S and is first order in hydrogen partial pressure. The data of Cecil et al. [16] shown in Fig. 5-7, for example, demonstrate that with an unidentified catalyst, the rate of desulfurization of a Middle Eastern residuum is proportional to hydrogen partial pressure over a range of 0 to 140 atm; some data, however, indicate that the reaction becomes zero order in hydrogen partial pressure at a value of about 100 atm [17].

The rate of hydrodesulfurization of any feed is generally consistent with the first-order reaction of each of a series of sulfur-containing compounds, as is to be expected from the foregoing discussion. The results of Cecil et al. (Fig. 5-8) show a typical curve for total sulfur concentration as a function of inverse space velocity. The shape of the curve is close to that expected for a second-order dependence of rate on total feed sulfur content. This result has been observed often and is useful for process design, but the approximation becomes invalid at the highest fractional removals of sulfur, when the actual conversion is less than that predicted by extrapolation of the curve for second-order kinetics.

A more firmly based analysis involves considering the feed as a mixture of sulfur-containing compounds, each of which reacts at a rate proportional to its

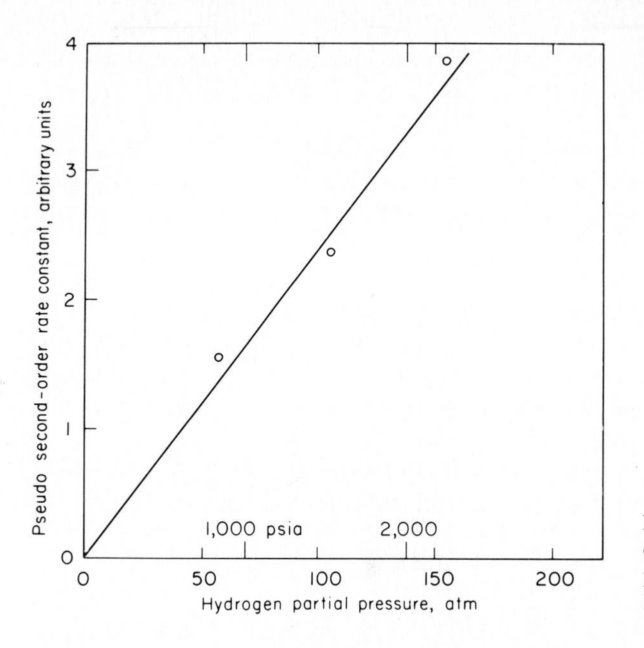

Figure 5-7 Dependence of hydrodesulfurization rate on hydrogen partial pressure for an unidentified Middle Eastern residuum [16].

concentration. Often the curve for total sulfur content vs. inverse space velocity is approximated as if there were only two reactive components, although a third may be required at the highest conversions. The two straight lines in Fig. 5-8 indicate first-order reaction of each of the two hypothetical components; these lines sum to a curve representing the data. We can write the following rate equation, which is useful for design:

$$r_{\mathrm{HDS}} = \alpha_1 k_1 c_{\mathrm{S}} + \alpha_2 k_2 c_{\mathrm{S}} \tag{11}$$

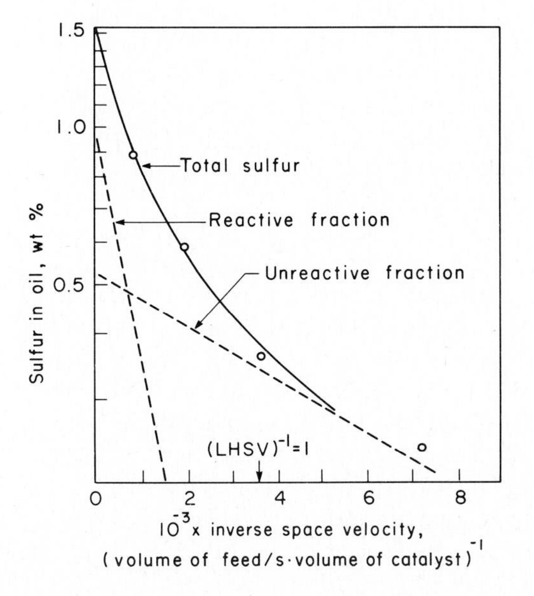

Figure 5-8 Hydrodesulfurization of Venezuelan vacuum gas oil: two-component, pseudo first-order kinetics [16].

This simplified representation of the kinetics is valid only when partial pressures of hydrogen and H_2S are held constant; these values are incorporated in k_1 and k_2.

For many practical reactor designs the more reactive fraction is almost completely removed from the product, and the design can be based on the other fraction (referred to as the unreactive fraction) alone. This simplification reduces the intrinsically complex kinetics to first order. The unreactive fraction is not to be confused with a single sulfur-containing compound. The adjustable parameters α_2 and k_2 are to be determined empirically for each feed and catalyst.

It is to be expected that α_2, the fraction of the sulfur-containing compounds which are relatively unreactive, will usually increase with increasing feed boiling range; correspondingly, k_2, the rate constant for the unreactive fraction, is expected to decrease with increasing feed boiling range.

For all feedstocks we can summarize the intrinsic kinetics in the following equation, which accounts approximately for all the cited effects:

$$r_{\mathrm{HDS}} = \frac{\alpha_2 k_2 P_{\mathrm{H_2}} c_{\mathrm{S}}}{1 + K_{\mathrm{H_2}} P_{\mathrm{H_2}} + K_{\mathrm{H_2S}} P_{\mathrm{H_2S}} + K_{\mathrm{ar}} P_{\mathrm{ar}}} \tag{12}$$

The term $K_{\mathrm{ar}} P_{\mathrm{ar}}$ represents the aromatic species, including sulfur-containing compounds. The equation might be better if the denominator were squared. Since it is written for the unreactive fraction of sulfur only, the equation is useful only for the high conversions usually encountered in industrial practice. Inhibition of reaction by H_2S and aromatic compounds is accounted for by the denominator term. The hydrodesulfurization is first order in hydrogen partial pressure at low values and zero order at high values.

The primary disadvantage of the foregoing representation of the kinetics is that it requires empirical determination of the parameters for each feed. In principle, we would expect a similar form of rate equation for each compound undergoing desulfurization, and each compound would be characterized by its own rate constant [10]. With knowledge of these rate constants, the rate of hydrodesulfurization of a new feed could be predicted if its composition were also known. Such detailed knowledge is surely obtainable for light feeds, but the difficulties of analysis make it impractical to obtain for heavy feeds.

An alternative analysis is based not on consideration of individual compounds but on characterization of feeds by components defined by their boiling ranges and solubility characteristics. This method was illustrated by Mosby et al. [18], who characterized a residuum by the sulfur-containing compounds in the following boiling ranges: 180 to 340, 340 to 540, and $> 540°C$. The residuum boiling at temperatures greater than 540°C was subdivided into classes separable by solvent extraction: oils, resins, and asphaltenes. (Oils are butane-soluble; resins are the butane-insoluble species which dissolve in pentane, and asphaltenes are the remaining benzene-soluble species.) The relative values of the rate constants for components in West Texas and Khafji residua are shown in Table 5-12. The least

Table 5-12 Relative values of rate constants for residuum components defined by boiling range and solubility [18][a]

Component	West Texas residuum	Khafji residuum
Fraction boiling between 340 and 540°C	1.56	1.00
Fraction boiling at temperatures > 540°C:		
Oils	1.00	0.98
Resins	1.00	0.82
Asphaltenes	0.25	0.16

[a] Neither the exact form of the kinetics nor the absolute value of any rate constant was given.

reactive component for each oil is the asphaltenes. The rate constants for each group are nearly the same for the two feeds, indicating that the approximations built into the analysis are good and that hydrodesulfurization rates for new feeds can be predicted from analysis of the amount of sulfur in each component.

Reaction rates predicted by equations like those suggested above are not sufficient for process-design calculations. Account must also be taken of mass-transfer effects and of catalyst aging, which influences both the mass transfer and the intrinsic kinetics. These effects are considered in the section on process engineering.

CATALYSTS

The hydrodesulfurization catalysts available from many commercial manufacturers have compositions like 4% CoO–12% MoO_3 on a γ-Al_2O_3 support. Some catalysts contain Ni instead of Co, and some contain W instead of Mo. The transition-metal oxides in hydrodesulfurization operations become reduced and partially or completely converted to sulfides, and some commercial catalysts are available in a presulfided form.

Molybdenum or tungsten appears to be a necessary constituent of a hydrodesulfurization catalyst, whereas neither cobalt nor nickel, if present alone, shows any significant activity. Combinations of Co (or Ni) and Mo (or W) are more active than Mo or W alone, and Co and Ni are therefore generally described as promoters.

To explain the action of hydrodesulfurization catalysts, we require the framework of a structural model that accounts not only for the hydrodesulfurization activity but also for the effects of the promoters and the necessity of γ-Al_2O_3 as the support.

The models which have been proposed fall into two categories:

1. Models starting from the structure of the oxidic catalyst and based on the assumption that the structure remains preserved after reduction and sulfiding, except for partial replacement of oxygen by sulfur. These models are usually derived from a model of the γ-Al_2O_3 structure, since it is uniquely applied as the support.
2. Models based on the assumption of complete conversion of the transition-metal oxide into a structure containing Co (or Ni) and Mo (or W) sulfides but not related to the γ-Al_2O_3 structure. These models must also explain why only γ-Al_2O_3 is used as the support.

Structures of Oxidic Catalysts

Physical methods for characterizing the solid state have been applied to hydrodesulfurization catalysts in a continuing series of investigations, beginning with that of Richardson in 1964 [19]. As a result, we are now able to present an essentially accurate model of the structure of the oxidic form of the catalyst.

To a good first approximation, the catalyst without a promoter, MoO_3/γ-Al_2O_3, is composed of a monolayer of MoO_3 on the surface of the γ-Al_2O_3 support [20]. The monolayer must be in registry with the underlying solid structure, as illustrated by the following simple model suggested by Dufaux et al. [21] for the monolayer formation reaction:

$$\gamma\text{-}Al_2O_3 \begin{array}{l} \text{AlOH} \\ \text{AlOH} \end{array} + \begin{array}{l} \text{HO} \\ \text{HO} \end{array}\!\!>\!\text{Mo}\!\!<\!\!\begin{array}{l} =\text{O} \\ =\text{O} \end{array} \longrightarrow \begin{array}{l} \text{Al}-\text{O} \\ \text{Al}-\text{O} \end{array}\!\!>\!\text{Mo}\!\!<\!\!\begin{array}{l} =\text{O} \\ =\text{O} \end{array} + 2H_2O \qquad (13)$$

Once the structure and surface plane of γ-Al_2O_3 are known, the monolayer structure can be inferred. The structure of γ-Al_2O_3 can be discussed conveniently from the starting point of the structure of spinels. The unit cell of the spinel $MgAl_2O_4$ is illustrated in Ref. 22 and Fig. 3-46; it consists of a cubic close packing of oxygen anions with one octahedral and two tetrahedral interstices per anion. Half the octahedral sites are occupied by Al^{3+} ions, and the Mg^{2+} ions are situated in tetrahedral positions (one Mg^{2+} per eight interstices).

Introducing a shorthand notation in which brackets represent octahedral positions and a lack of brackets represents tetrahedral positions, we write the structure of the spinel as $Mg[Al_2O_4]$. A description of γ-Al_2O_3 now follows if we replace one Mg by $\frac{2}{3}$ Al, to give $Al_{2/3}[Al_2O_4]$. (This structure of γ-Al_2O_3 has not been firmly established, however, and there may actually be more tetrahedrally and fewer octahedrally surrounded Al^{3+} ions). Different distributions of Al^{3+} ions in the two kinds of surroundings may account for differences in properties of various forms of alumina, such as η-Al_2O_3 and γ-Al_2O_3 [23, 24].

The preparation of γ-Al_2O_3 involves formation of well-crystallized boehmite and subsequent calcination, leading to lamellae, which are predominantly oriented along the (110) planes. The crystal can therefore be considered to be a stacking of two types of planes, C layers, containing all the tetrahedral sites next to octahedral sites, and D layers, containing only octahedral sites. This structure is represented in Fig. 3-44. The C layers are $Al_{4/3}[Al_2O_4]$, and the D layers are $[Al_2O_4]$. Surface planes might be either C or D layers, presumably covered by enough OH groups to allow for electrical neutrality.

A structure of the MoO_3 monolayer based on this model of γ-Al_2O_3 was postulated by Schuit and Gates [12]. They suggested that upon impregnation, two types of surface layers can be formed epitaxially to the γ-Al_2O_3 surface:

$$\begin{array}{ccc} O_2^{2-} & & O_4^{2-} \\ [Mo_2O_4] & \text{and} & Mo_{2/3}[Mo_2O_4] \\ Al_{4/3}[Al_2O_4] & & [Al_2O_4] \end{array}$$

The extra oxygen anions shown above the Mo layer are necessary to compensate for the higher cation charge of Mo.

The surface area per MoO_3 group based on this model can be computed to be about 18 $Å^2$, which is in good agreement with the experimental results of Sonnemans and Mars [20]. This agreement provides some confirmation of the monolayer model, but other data cast some doubt upon it. At high loadings of MoO_3, $Al_2(MoO_4)_3$ is always observed, which shows that the interaction between γ-Al_2O_3 and MoO_3 involves more than monolayer formation. The ultraviolet-visible reflectance spectra obtained by Ashley and Mitchell [25] gave evidence only of Mo in tetrahedral configuration, as occurs in $Al_2(MoO_4)_3$. Studies of the luminescence due to the presence of Cr^{3+} tracer ions in the Mo-containing alumina yielded spectra similar to those of Cr-doped $Al_2(MoO_4)_3$, even at MoO_3 concentrations as low as 3 percent. All these results suggest that there must be diffusion of some Al^{3+} ions into the monolayer, presumably compensated by diffusion of Mo^{6+} ions into the interior and accompanied by a partial rearrangement of the outer layers.

The structure of a catalyst containing not only Mo but also promoter ions such as Co is obviously more complicated. Understanding of the interactions between the support and promoter ions has been the object of many studies beginning with that of Richardson [19], who recognized that Co^{2+} upon impregnation becomes incorporated to some extent within the bulk of γ-Al_2O_3. A structure similar to that of the spinel $Co[Al_2O_4]$ is formed, with Co^{2+} ions located at tetrahedral interstices. Recent investigations with ultraviolet-visible reflectance spectroscopy, ESR spectroscopy, magnetic-susceptibility measurements, and infrared spectroscopy of adsorbed pyridine [25, 27–31] have led to the general conclusion that upon impregnation Co and Ni first become attached to the surface, forming structures akin to the aforementioned monolayer model. If only a small amount of promoter is added, all the ions are incorporated in the monolayer. When larger amounts are added, however, compounds such as Co_3O_4 are formed.

Calcination, especially at temperatures as high as 600°C, leads to diffusion of promoter ions into the interior, with Ni^{2+} diffusing more rapidly than Co^{2+}. Co^{2+} becomes positioned primarily at tetrahedral sites as in $Co[Al_2O_4]$, but more than 75 percent of the Ni^{2+} ions occupy octahedral sites.

On the basis of these results, we might suggest a monolayer model defined by structure A below, but it is evident that a more realistic model is that defined schematically by B, with Co^{2+} and Mo^{6+} ions becoming incorporated in the bulk of the γ-Al_2O_3:

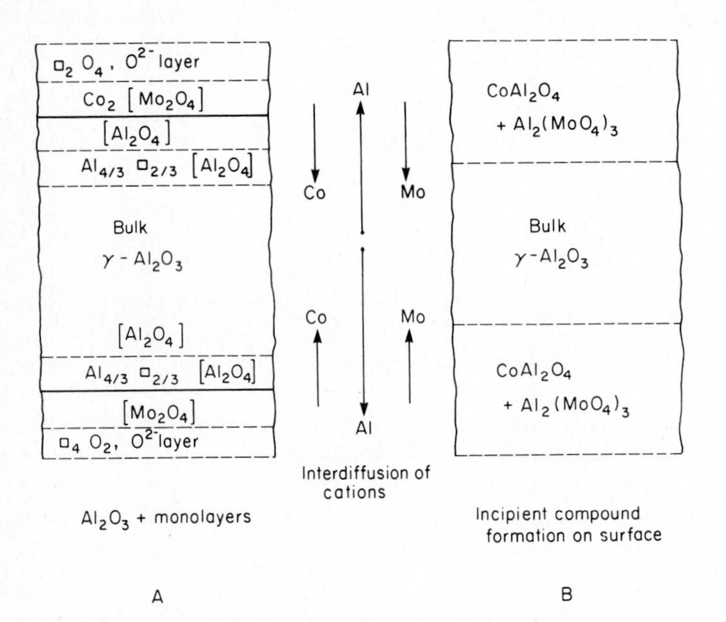

The process of counterdiffusion of the cations cannot change the number of oxygen anions necessary to maintain the balance of charge. Neither should it change the Co/Mo ratio of $\frac{1}{2}$ predicted in structure A, which is roughly equal to the value usually applied. The symmetry of the cations in the $Co[Mo_2O_4]$ layer is different from that of the cations in the compound $CoMoO_4$, which is cobalt molybdate. In the layer as given here, Mo is octahedrally surrounded and Co is tetrahedrally surrounded, whereas in $CoMoO_4$, Mo is octahedrally surrounded and in β-$CoMoO_4$ it is tetrahedrally surrounded, with Co remaining in octahedral sites. Heating $CoMoO_4$ with γ-Al_2O_3 leads to its decomposition, presumably because of a solid-state reaction like [27]

$$3CoMoO_4 + 4Al_2O_3 \longrightarrow 3CoAl_2O_4 + Al_2(MoO_4)_3 \qquad (14)$$

It is clear that there is actually no cobalt molybdate in the catalysts referred to by that name.

Structures of Sulfidic Catalysts

The central question of catalyst structure concerns the sulfided form, which is produced from the oxidic structure before or during the initial stages of operation. Pretreatment with $H_2 + H_2S$ under carefully controlled conditions or application of the catalyst in hydrodesulfurization causes partial reduction and replacement of O^{2-} ions by S^{2-} ions. If only the Mo (or W) ions were reduced, their oxidation state would ultimately become +3. If the previously discussed epitaxial arrangement remained intact, the O–S exchange would be restricted to the outer anion layer. Since S^{2-} ions (radius = 1.84 Å) are appreciably larger than O^{2-} ions (radius = 1.32 Å), the maximum occupancy by S would have to be two S atoms per unit mesh, that is, $0 < S/Mo < 1$. This restriction is exemplified in the structures proposed by Schuit and Gates (Fig. 5-9). These structures, presumably stabilized by promoter ions in the underlying support, were suggested as the catalytic sites for hydrodesulfurization.

In contrast to expectations based on this monolayer model, however, deBeer et al. [28] found the sulfided catalyst to have a sulfur content corresponding approximately to that of the compounds $Co_9S_8 + MoS_2$. This result therefore indicates the occurrence of an extensive destruction of the monolayer, and it also shows that cations which must have been embedded in the γ-Al_2O_3 support diffused back to the surface, where new structures were formed. It follows that the structure of the sulfidic catalyst may be entirely different from that of the oxidic catalyst from which it is formed.

Therefore, although the monolayer model is useful in accounting for the catalyst structure in the early stages of operation, it may not be appropriate for describing most hydrodesulfurization operations. We note, however, that hydrodenitrogenation reactions, in contrast to hydrodesulfurization reactions, take place on the oxidic as well as on the sulfidic forms of the catalyst, even though the sulfidic form is more often used for hydrodenitrogenation [20]. Therefore, a surface structure needed to form a catalytic site for heteroatom removal by hydrogenolysis does not require sulfiding, and the monolayer model may indeed account for some catalytic activity.

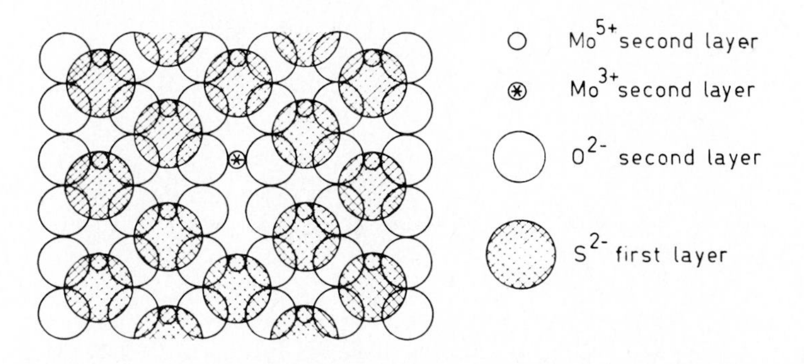

Figure 5-9a Sulfided monolayers on Co-promoted γ-Al_2O_3; the surface layer is a C-layer [12].

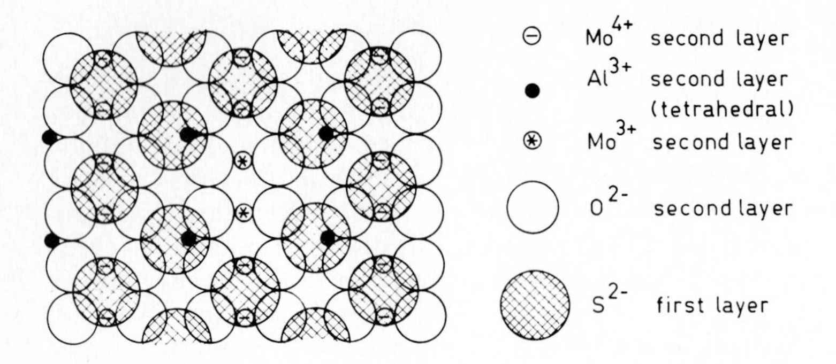

Figure 5-9b Sulfided monolayers on Co-promoted γ-Al_2O_3; the surface layer is a D layer [12]. The monolayers in these structures are depicted to be as dense as possible, except for single sites at which a sulfur anion vacancy exists. These sites have been suggested to be catalytically active sites.

An explicit structure of the sulfidic form of hydrodesulfurization catalysts was proposed by Voorhoeve in 1971 [32] and developed further by Farragher and Cossee [33]. The Voorhoeve-Farragher-Cossee model, in contrast to those just discussed, was derived from an understanding of the structures of a group of sulfide compounds known as *intercalated solids*.

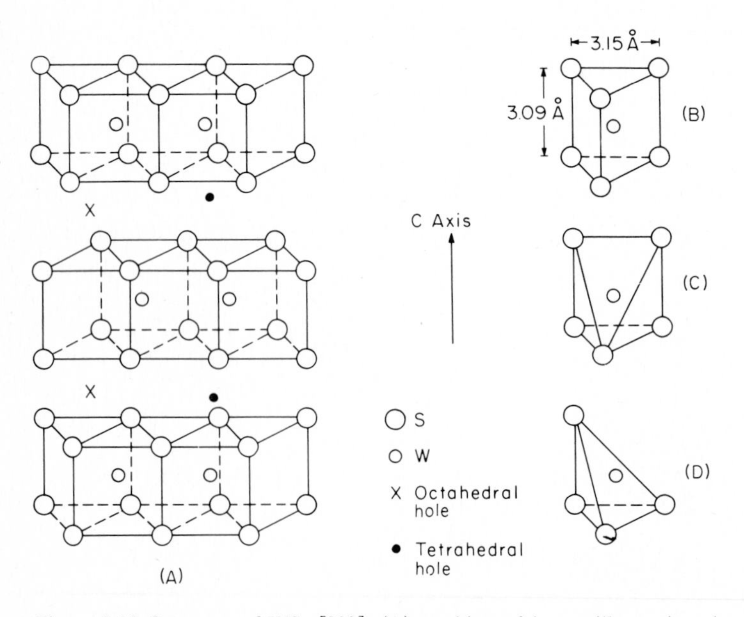

Figure 5-10 Structure of WS_2 [32b]: (A) stacking of layers illustrating the positions of octahedral holes which may be partly occupied by Ni; (B) site symmetry of W^{3+} ions in the bulk; (C) in the sideface; and (D) in an edge parallel to the *c* axis. (Reprinted with permission from *Journal of Catalysis*. Copyright by Academic Press.)

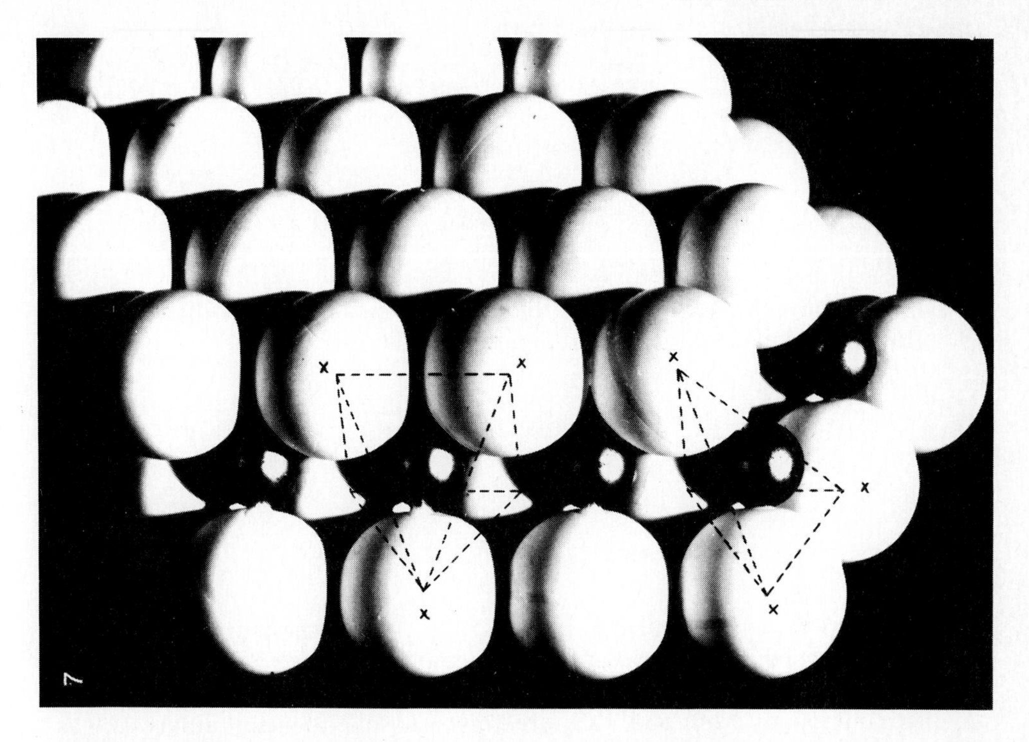

Figure 5-11 Model of a small WS_2 crystallite [32*b*]: the white spheres represent sulfur ions, the black, tungsten. Steric hindrance of adsorption is caused by sulfur ions marked by crosses. (Reprinted with permission from *Journal of Catalysis*. Copyright by Academic Press.)

Sulfides of tetravalent transition metals such as NbS_2, TaS_2, MoS_2, and WS_2 have layer structures as illustrated in Figs. 5-10 and 5-11. The cations are found in a sixfold trigonally prismatic surrounding. The sulfur anions in a layer are close-packed, and cation sites are present between two close-packed layers. Only half the interstices are filled, however, and the sequence of stacking is

————	S layer
× × × × ×	Trigonal prismatic interstices (filled)
————	S layer
	Octahedral interstices (empty)
————	S layer
× × × × ×	Trigonal prismatic interstices (filled)
————	S layer

The degeneracy of the *d* orbitals of transition-metal ions in the interstices is partly lifted but in different ways for trigonal prismatic and octahedral interstices [34, 35]:

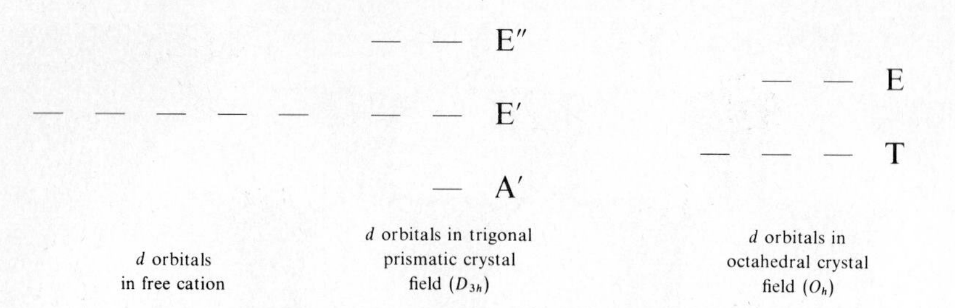

Because of interaction between orbitals of cations in neighboring interstices, the orbitals of similar symmetry form energy bands:

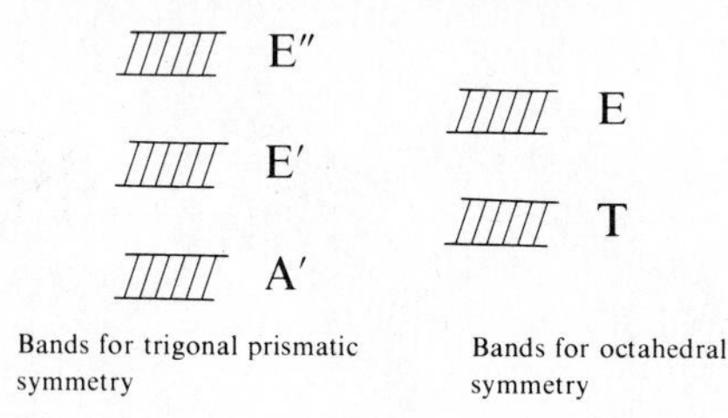

For cations such as Nb^{4+}, there is one electron per atom, and the A′ band is therefore half filled, the E bands remaining empty. For Mo^{4+} and W^{4+}, the A band is completely filled.

The compounds NbS_2 and TaS_2 are known to take up metal atoms such as Cu and Ag, which find positions in the empty octahedral interstices. The added atoms donate electrons to Nb^{4+} or Ta^{4+}, which then become reduced to the trivalent state, the electrons finding a place in the partly filled A band. This process, known as *intercalation*, can take place because there is a decrease in free energy as electrons of the intercalated atoms become positioned in the low-lying A band. Intercalation is evidently less likely to occur in MoS_2 and WS_2, since their A bands are already filled. The situation would be different, however, if all interstices were octahedral, since the triply degenerate T level would then offer sufficient possibilities for accepting electrons to allow formation of a (hypothetical) MoS_2 with Mo^{3+} in octahedral interstices. A compound, $CoMo_2S_4$, is even known, for which this situation occurs, but it is reported to be inactive for hydrodesulfurization [36].

Although intercalation between the sulfur layers does not occur in bulk MoS_2 and WS_2, it has been assumed by Voorhoeve, Farragher, and Cossee that it can occur at the layer edges, where the sulfur surrounding is incomplete and the site symmetry lower. Voorhoeve's arguments for this assumption are derived primarily from the existence of an ESR signal in WS_2 at 3250 Oe, which he assigned to W^{3+} ions. The signal had a very low intensity for stoichiometric WS_2, but it was clearly present in sulfur-deficient WS_2. The signal intensity increased with surface area, which shows that it was associated with some species at the surface, and it also increased with increasing amounts of metallic Ni incorporated in small doses, presumably because of the surface intercalation, leading to the reaction $Ni^{\circ} + 2W^{4+} \rightarrow Ni^{2+} + 2W^{3+}$. For example, a particular WS_2 sample exhibited

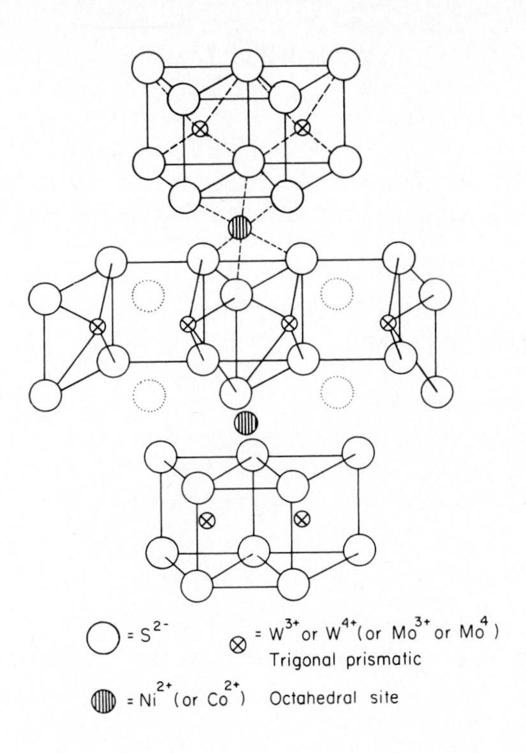

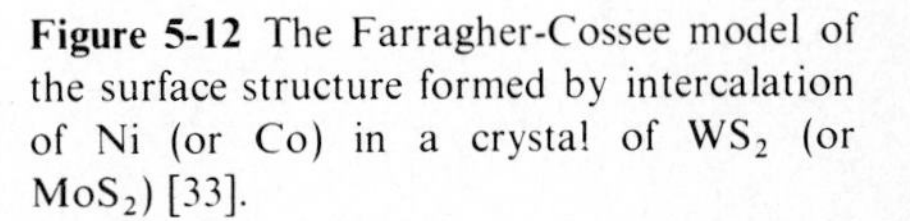
Figure 5-12 The Farragher-Cossee model of the surface structure formed by intercalation of Ni (or Co) in a crystal of WS_2 (or MoS_2) [33].

about 10^{14} to 10^{15} spins (W^{3+} ions) per square meter. Addition of only 0.16 atom of Ni per atom of W led to a hundredfold increase in spin intensity. Further, at high contents of Ni, the spin intensity was proportional to the surface area of the sample.

There is therefore a mass of evidence supporting Voorhoeve's hypothesis, but the spin assignment to W^{3+} remains uncertain, although it is entirely plausible. A conclusive demonstration that intercalation indeed occurs at layer edges was given by Farragher and Cossee [33], whose electron micrographs showed a definite change in the form of WS_2 crystal edges after incorporation of small amounts of Ni.

A model of the crystal-edge structure resulting from intercalation of Ni was suggested by Farragher and Cossee [33], as shown in Fig. 5-12.

Catalytic Sites

The structure shown in Fig. 5-12 has been suggested to be the catalytic site for hydrodesulfurization, but there is no experimental evidence to relate this surface species to the appropriate catalytic activity. Voorhoeve [32] and Farragher and Cossee [33] gave no data for hydrodesulfurization but referred instead to benzene hydrogenation, assuming it to be a suitable model reaction. Hydrogenation of cyclohexene as well as of benzene was studied by Voorhoeve and Stuiver [32], and the former reaction was about 10 times faster than the latter. The rate of each reaction increased upon addition of Ni to the catalyst, but the promoter action was much stronger for benzene hydrogenation. Voorhoeve observed a linear dependence of the benzene hydrogenation rate constant on the ESR spin intensity indicating W^{3+} ions in a series of supported and nonsupported Ni–WS_2 catalysts, as shown in Fig. 5-13. Since no such relation could be found for the cyclohexene hydrogenation, although this reaction was also promoted by Ni, Voorhoeve concluded that the two reactions occurred on different types of sites (recall that evidence was mentioned previously for two kinds of sites).

The site of both hydrogenation and hydrodesulfurization reactions is assumed to be associated with anion vacancies on the crystal edges, as suggested for the Ziegler-Natta polymerization sites discussed in Chap. 2. The cations become exposed to the gas phase when a sulfur atom is missing, and the formation of anion vacancies is easily accounted for by both the monolayer model (Fig. 5-9) and the intercalated system (Fig. 5-12). The anion vacancy can be filled by S from H_2S according to the reaction

$$2W^{3+} + H_2S + \Box_S \longrightarrow H_2 + 2W^{4+} + S^{2-} \tag{15}$$

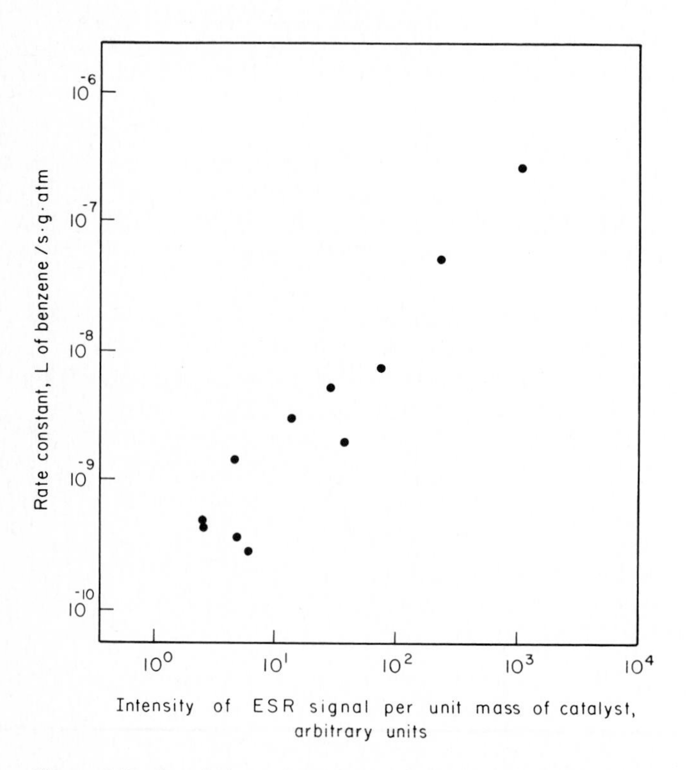

Figure 5-13 Correlation of catalytic activity for benzene hydrogenation with intensity of the tungsten ESR signal [32]. (Reprinted with permission from *Journal of Catalysis*. Copyright by Academic Press.)

where $\square_S$ represents a sulfur-ion vacancy. The inhibition by H_2S is now readily understood, and interaction with a sulfur-containing compound such as thiophene is also easily visualized as an analogous reaction.

All the evidence cited in support of the intercalation model is based on data for WS_2 catalysts. No data have been presented to demonstrate whether intercalation is also important for MoS_2, but recent results obtained by Furimsky and Amberg [37] suggest that it is. These authors measured conversions of thiophene catalyzed by unsupported MoS_2 crystals impregnated with $Co(NO_3)_2$ in liquid ammonia and then sulfided with $H_2 + H_2S$ at higher temperatures. The catalyst surface area was almost independent of Co content at low values, and there was a small deficiency of S compared with the composition $MoS_2 + Co_9S_8$. Further increases of Co content led to increased activity with simultaneous increases in surface area and almost constant activity per unit of surface area. When still more Co was added, activity and surface area both decreased strongly. The data obtained at low Co contents can be explained in terms of the intercalation model; increasing amounts of Co were intercalated and formed an increasing number of catalytic sites. The effect of Co at the high concentration is difficult to understand. Perhaps it indicates that the MoS_2 crystals disintegrate upon continued addition of Co.

The intercalation model rests on a firmer basis than any alternative yet suggested, but there are still some experimental results with which it is not easily reconciled. For example, Hagenbach and coworkers [38] observed that intimate mixtures of Co_9S_8 and MoS_2 were more active for thiophene hydrodesulfurization than combinations of the individual compounds not intimately mixed, and a maximum in activity was observed in the mixture at the ratio $Co/(Co + Mo) = 0.2$ to 0.3. It is also in this composition range that Furimsky and Amberg found the maximum activity and a relatively high surface area. No explanation has been offered for this increase in activity on mixing of Co_9S_8 and MoS_2, and in general the Co/Mo ratios for maximum promotion are far in excess of what might be expected from the edge-intercalation model.

In an attempt to reconcile these difficulties, we may speculate that a process resembling the spillover described in Chap. 3 could occur, involving the transfer of hydrogen atoms (or protons and electrons) from one of the solid compounds to the other, presumably because one compound (such as the promoter sulfide) would facilitate the activation of H_2 and the other (the actual catalyst) the acceptance of sulfur. If this hypothesis of spillover were valid, it would suggest that the ideal active site would be a juxtaposition of a H_2 activation site and a S acceptor site. A suggestion for the structure of such a site is given in Fig. 5-14. This hypothetical structure is derived from the Farragher-Cossee model (Fig. 5-12) by removal of two anions from the edge. The promoter and catalyst cations are in close proximity because of their orientation in the successive layer sites. One type of cation, say Co^{2+}, could accept electrons from the reductive adsorption of H_2, and the other cation, say Mo^{3+}, could transfer electrons to the reactant molecule, the reaction being completed by H^+ transfer to the reactant molecule and electron transfer from Co^0 to Mo^{4+}. If the proton mobility were sufficiently high and the

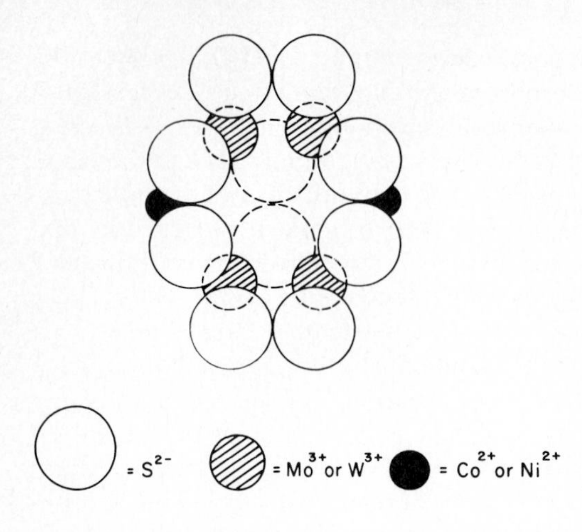

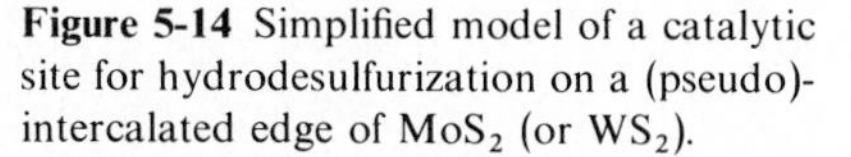

Figure 5-14 Simplified model of a catalytic site for hydrodesulfurization on a (pseudo)-intercalated edge of MoS_2 (or WS_2).

electron transfers rapid (as suggested by the semiconductor properties of the sulfides), separation of the two catalytic species would decrease the effect of the promoter but neither remove it entirely nor change the stoichiometric ratio $Co/Mo = \frac{1}{2}$.

An explanation is still needed for the exclusive use of γ-Al_2O_3 as a support if the intercalation model is to be regarded as a good representation of the actual catalyst. The important results were found by deBeer et al. [39], who used SiO_2 and other supports and observed the formation of $CoMoO_4$. This compound does not form on γ-Al_2O_3, as mentioned previously, but evidently it is formed on the other supports, and its sulfidation leads to formation of inactive sulfide systems such as $CoMo_2S_4$. Although it is not entirely understood why this compound is inactive, the demonstration of its formation on SiO_2 does explain the inactivity of the silica-supported catalysts. A demonstration of this conclusion was given by results of deBeer's experiments in modifying the preparation procedure. Successive impregnation of SiO_2 with $(NH_4)_2MoO_4$ and $Co(NO_3)_2$ (or vice versa) but with $H_2 + H_2S$ treatments between the impregnations prevented the formation of $CoMoO_4$ and led to the formation of catalysts as active as those supported on γ-Al_2O_3. These results suggest that such a decoupling of support and active catalyst systems might allow the application of synthesis methods to incorporate other catalytic functions in supports for hydroprocessing of coal-derived liquids, for which hydrocracking, hydrodesulfurization, and hydrodenitrogenation are all needed, although it is doubtful that such catalysts would retain their cracking activity for long in operation.

The Mechanisms of Reaction on Promoted Catalysts

The first reaction to take place during thiophene hydrodesulfurization, as shown in Fig. 5-2, is

$$C_4H_4S + 2H_2 \longrightarrow 2C_4H_6 + H_2S \qquad (16)$$

The available information about reaction kinetics, e.g., Eq. (8), and catalyst structure suggests that the adsorbed intermediates include hydrogen, H_2S, and sulfur-containing compounds, with the latter two occupying anion vacancies. We suggest that the reaction proceeds through at least three elementary surface-reaction steps:

$$2(H_2 \longrightarrow 2\underline{H}) \tag{17}$$

$$C_4H_4S + \square_S + 2\underline{H} + 2e^- \longrightarrow C_4H_6 + S^{2-} \tag{18}$$

$$S^{2-} + 2\underline{H} \longrightarrow H_2S + \square_S + 2e^- \tag{19}$$

where $\underline{H}$ = adsorbed H atom
$\square_S$ = anion vacancy
e^- = electron

We assume now, to be more specific, that the adsorbed H is formed by a reductive adsorption:

$$Co^{2+} + H_2 + 2S^{2-} \longrightarrow Co + 2SH^- \tag{20}$$

or $$2Co^{2+} + H_2 + 2S^{2-} \longrightarrow 2Co^+ + 2SH^- \tag{21}$$

i.e., the adsorbed H is present in SH^- groups. Electron donation can be ascribed to the reaction

$$2Mo^{3+} \longrightarrow 2Mo^{4+} + 2e^- \tag{22}$$

It is understood that Ni^{2+} can replace Co^{2+} and W^{3+} can replace Mo^{3+} in these equations.

In choosing this representation, we suggest that the hydrodesulfurization reaction consists of two separate redox steps, one supplying H atoms, the other donating electrons. The thiophene molecule is adsorbed at a sulfur-anion vacancy, as discussed previously, where it can accept electrons and protons from Mo^{3+} ions and SH^- groups, respectively. Electron transfer can then occur from Co° to Mo^{4+}. The reaction mechanism formulated in this way is shown schematically in Fig. 5-15, which incorporates a one-dimensional version of the site model suggested in Fig. 5-14. The manner of transferring electrons and protons separately is convenient but arbitrary; alternatively, the transfer might involve H atoms. Experiments reported by Smith et al. [40], who observed that thiophene reacting in the presence of D_2 was deuterated at the α-positions, and experiments reported by Mikovsky et al. [41], who observed that when thiophene reacted with D_2, the first product was H_2S, indicate that the reduction occurs in a multistep sequence, i.e., with the transfer of one H atom (or one proton and one electron) at a time, the organic intermediate remaining at the surface site, where it accepts at least two H atoms. A multistep mechanism would also offer a possible explanation for the occurrence of an ESR signal assigned to Co^+ ($g = 2.17$) in Co–Mo catalysts [38, 42].

Referring to Fig. 5-15, we recognize that a compound like dibenzothiophene might adsorb in a configuration similar to that shown for thiophene, and a process similar to the one shown would account for formation of H_2S and biphenyl.

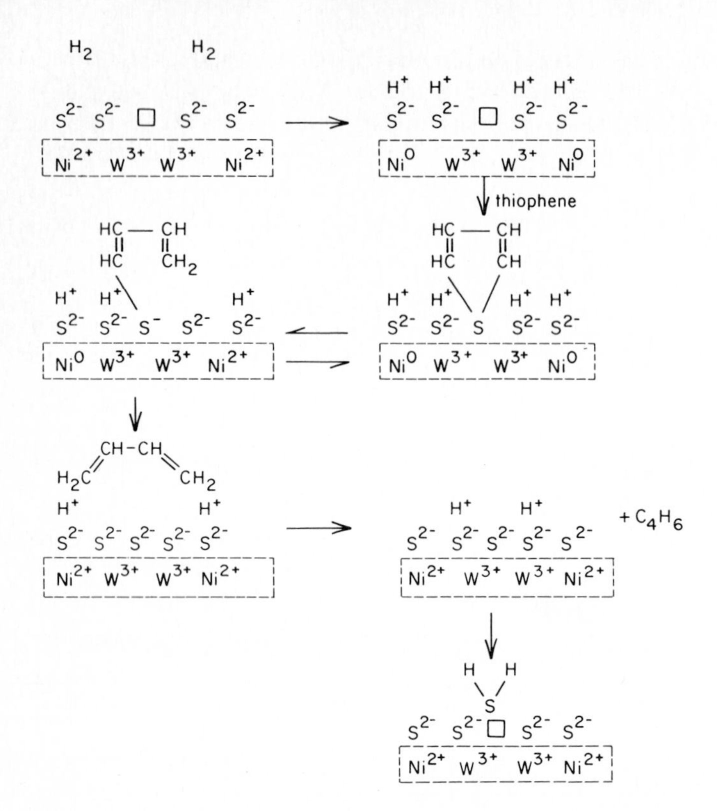

Figure 5-15 Schematic description of the suggested thiophene hydrodesulfurization mechanism: □ = anion vacancy; the rectangle ⌜⌟ indicates that there may be (partial) delocalization of electrons over the enclosed cations.

Incorporation of methyl groups in the ring positions nearest to sulfur could form a structure which is sterically hindered from adsorbing in the manner shown. This steric restriction accounts for the previously mentioned decrease in hydrodesulfurization rate resulting from incorporation of the methyl substituents.

Hydrogenation of aromatics on these sulfide catalysts is even less well understood than hydrodesulfurization [43], but it probably takes place by a similar mechanism, first involving electron donation by Mo^{3+} then proton donation from SH^- groups associated with Co. For symmetrical molecules such as benzene, the first adsorption of the aromatic might be envisaged as forming a flat or "half sandwich" species parallel to the surface and located above Mo cations. The anion vacancy necessary to allow formation of such a species might be too small to accommodate larger aromatic molecules. These would then be adsorbed on edge, perpendicular to the surface, or flat in positions where enough room could be made by removal of more contiguous sulfur atoms, as would result from application of higher hydrogen partial pressures.

Thiophene hydrodesulfurization and benzene hydrogenation both presumably involve formation of primary products having the nature of a (cyclo) diene. It is generally observed that these products are rapidly hydrogenated, but information about the mechanism is lacking. We suggest that two plausible mechanisms are like those shown in Fig. 5-16.

Hydrodenitrogenation of heterocyclics such as pyridine takes place via reaction networks different from those encountered in hydrodesulfurization [20, 44, 45]. First, there is complete hydrogenation preceding removal of nitrogen, e.g.,

$$\text{Pyridine} \xrightarrow{H_2} \text{piperidine}$$

The nitrogen-removal reaction is fundamentally different from that involving sulfur removal, and it was supposed by Goudriaan [44] to result from cracking on the acidic support, presumably requiring hydrogen spillover from the oxide.

The interesting feature common to all these hydrogenation and hydrodenitrogenation reactions is that they occur on both oxides and sulfides (the sulfides being more active) and they all benefit from the promotion of Co and Ni. Accepting the mechanism given above, we expect that H donation should occur by reductive adsorption of H_2. Since Co compounds are generally observed to be somewhat less readily reduced than Ni compounds, with reduction of either Mo^{4+} or W^{4+} compounds being even more difficult, the promotional effect might be assumed to arise from a more or less facile reduction with the rate increasing in the sequence $Mo^{4+} < W^{4+} < Co^{2+} < Ni^{2+}$. On the other hand, there is the electron donation by Mo^{3+} or W^{3+}; Mo^{3+} is known to tend to pair forming because of the formation of Mo—Mo bonds, as exemplified clearly by the compound $CoMo_2S_4$ [36]. With Mo in such a form, electron donation would have to occur from a bonding molecular orbital, and it would therefore be more difficult. (This suggestion might account for the previously mentioned inactivity of $CoMo_2S_4$ as a hydrodesulfurization catalyst.) In general one would expect Mo compounds to be less active than W compounds. Therefore, the expected order in activity is

$$MoS_2 < WS_2 < \text{Co–Mo sulfides} < \text{Ni–Mo sulfides} < \text{Ni–W sulfides}$$

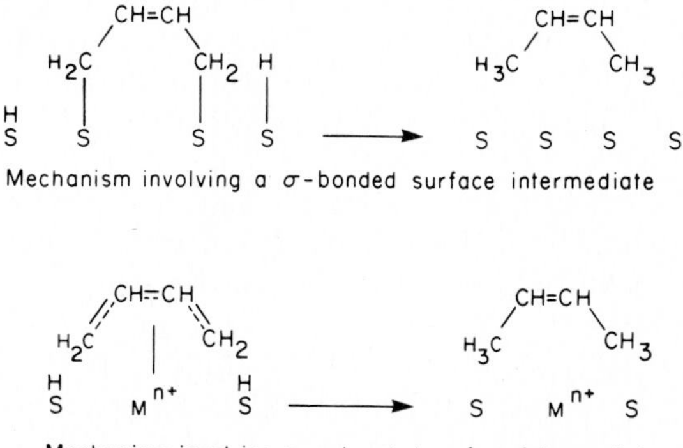

Figure 5-16 Suggested mechanisms of olefin hydrogenation on sulfided catalysts.

The fact that Ni–W sulfide catalysts do not find application as hydrodesulfurization catalysts may be explained by their tendency to hydrogenate aromatics simultaneously, which is undesirable in increasing hydrogen consumption. (But Ni–W catalysts are also more expensive than Co–Mo and Ni–Mo catalysts.)

PROCESS ENGINEERING

INTRODUCTION

The general description of hydrodesulfurization processes given earlier needs little elaboration for light-distillate hydrodesulfurization processes. An account of commercial operating experiences, including details of reactor design, construction, and operation has been given by Lister [46], and some of his data are collected in Table 5-13. Conversions of sulfur were not given, but they were surely somewhat less than conversions in modern units, which are required to provide catalytic reformer feeds containing only a few parts per million of sulfur, corresponding to equilibrium-limited conversions of as much as 99.97 percent. Further details of these processes can be found elsewhere [47, 48].

The following section emphasizes engineering considerations for hydrodesulfurization of the heavier feeds, which are residua and coal-derived liquids. Since these feeds present the greater and more recently recognized engineering challenges, the related literature is far less developed. Our intention is to identify the engineering problems, to evaluate their importance, and to direct the reader toward the literature which provides detailed design methods. Frequently the evaluations are only qualitative, since few quantitative data are available.

RESIDUUM HYDRODESULFURIZATION PROCESSES

A representative process flow diagram for residuum hydrodesulfurization is shown in Fig. 5-17. The properties of several feeds used in such processes are collected in Table 5-14. The important differences between the feeds are the sulfur content and, especially, the Ni + V content, since these, for the most part, determine the required processing severities.

During a run, the average reactor temperature is raised gradually to compensate for catalyst activity loss. Representative run plots are shown in Fig. 5-18. The temperatures are not given explicitly, but we can estimate that the base temperature shown was about 350°C.

Some product properties and yields are summarized in Table 5-15, and hydrogen consumption data are shown in this table and in Figs. 5-19 and 5-20.

Table 5-13 Distillate hydrodesulfurization plants [46][a]

		Process variables					Catalyst-bed details					
Location	Approx. design date	$10^3 \times$ throughput, m^3 oil/s	LHSV	Pressure, atm	Temp., °C	Recycle gas rate, std m^3/m^3 oil	Diam, m	Depth, m	Vol., m^3	Vessel ID, m	Tangent length, m	Type of reactor construction[b]
Kwinana	1952	11.5	8	69	416	720	1.2	4.2	4.6	1.2	5.2	A
Hamburg	1954	13.3	1.4	47	366–382	260	2.0	9.9	31.4	2.0	11.9	B
Kent (nos. 1 and 2)	1955	20.4	3	42	388–416	450	1.8	2.7	7.0	2.0†	3.4	C
Antwerp												
Venice	1957	13.3	4	36	416	360	1.6	5.4	11.0	1.6	6.8	B
Ruhr	1958	20.4	8	69	416	180	1.8	3.6	8.6	1.9	4.3	C
Belfast	1961	7.8	8	69	421	180	1.4	2.3	3.4	1.4	3.2	D
Lavera	1962	30.7	4	36	421	360	2.2	6.6	25.3	2.5	6.7	C

[a] Data of Lister were given in units of barrels per standard day. One standard day was assumed to be 7.76×10^4 s. Three beds, each 3.3 m deep.

† Three identical reactors in series on each unit (total catalyst volume 21 m^3).

[b] A = solid stainless steel (18/8/Ti), B = clad plate: 18/8/Ti on C/$\frac{1}{2}$Mo, C = C/$\frac{1}{2}$Mo vessel with refractory lining + 18/8/Ti shroud, D = clad plate: 18/8/Ti on Cr/$\frac{1}{2}$Mo.

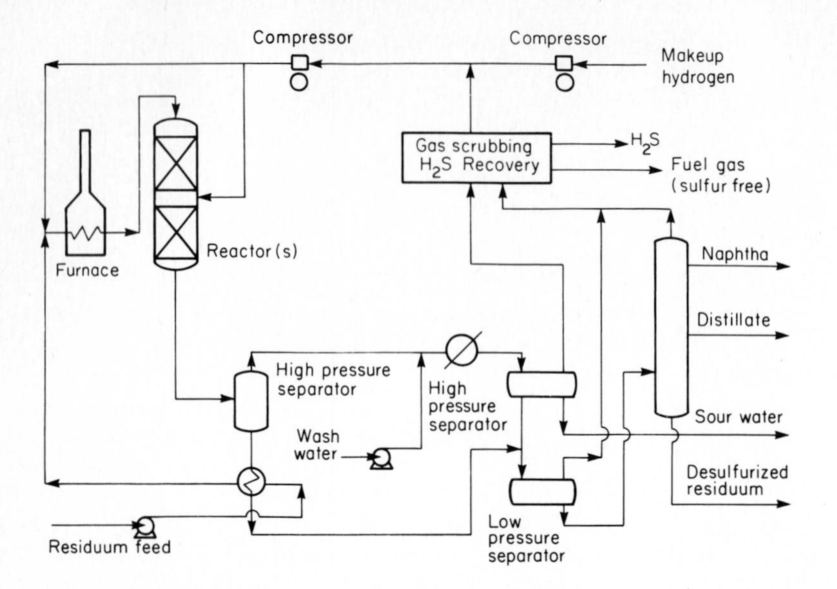

Figure 5-17 Schematic diagram of a residuum hydrodesulfurization process [49]. A guard-bed reactor may be applied upstream of the main reactor.

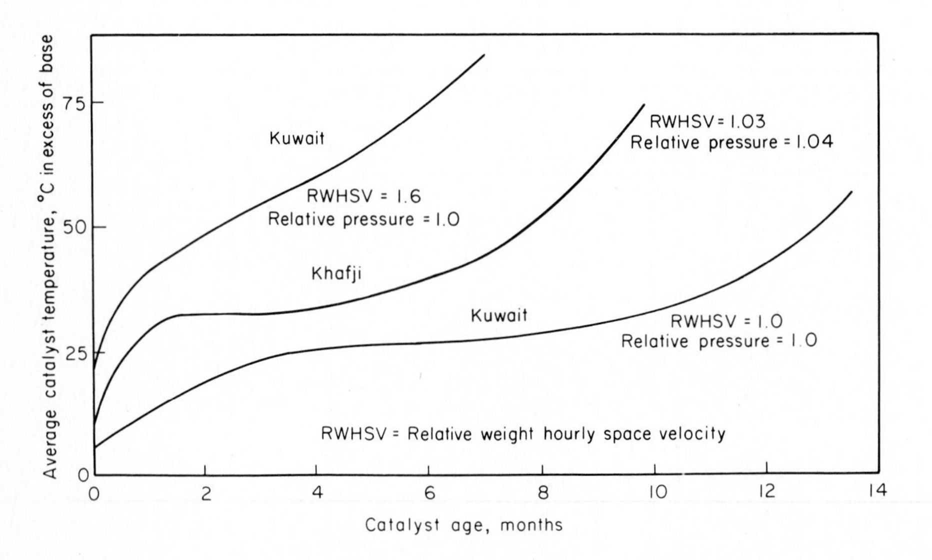

Figure 5-18 Residuum hydrodesulfurization run plots showing increasing catalyst temperature to compensate for catalyst aging and to maintain a constant product sulfur content of 1 wt% [51].

Table 5-14 Analysis of petroleum residua

Residuum	Khafji [18]	Gach Saran [18]	West Texas sour [18]	Venezuelan [16]	West Coast USA [50]	Kuwait [50]
			Elemental analysis			
S, wt %	4.30	2.40	3.65	2.17	1.73	3.66
N, wt %	0.27	0.46	0.23	...	0.90	0.20
H, wt %	11.01	11.44	11.08	11.43	...	...
C, wt %	83.87	86.48	85.04	85.93	...	...
Ni, ppm	32	60	16	37	75	11
V, ppm	93	160	25	290	63	38
			Component analysis			
Saturates, wt %	...	...	...	...	20.6	26.7
Aromatics, wt %	...	...	...	...	41.2	50.4
Polar aromatics, wt %	...	...	...	...	22.9	10.4
Asphaltenes, wt %	...	...	...	...	10.7	6.3
Sulfur, unsaturates, ppm	...	...	...	...	22	34
In aromatics, wt %	...	...	...	...	0.94	2.36
In polar aromatics, wt %	...	...	...	...	0.41	0.59
In asphaltenes, wt %	...	...	...	...	0.22	0.38

Table 5-15 Yields in residuum hydrodesulfurization [18]

Residuum feedstock	Khafji[a]	West Texas sour[a]	
Sulfur content of product boiling at temperatures exceeding 343°C, wt %	1.0	1.0	0.3
Yields of products boiling at:			
< 182°C, wt %	6.35	5.12	6.75
182–343°C, vol %	9.3	5.7	11.1
> 343°C, vol %	91	94	89
Hydrogen consumption, std ft^3/bbl	580	420	600

[a] Feed properties are given in Table 5-14.

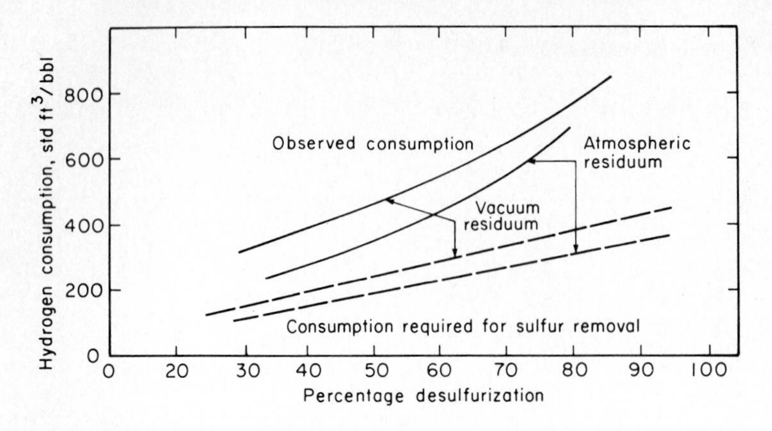

Figure 5-19 Dependence of hydrogen consumption on the degree of desulfurization of Kuwait residuum [17].

These data show that a significant fraction of the hydrogen is used to saturate aromatic rings rather than to remove sulfur. The fraction of hydrogen consumed by reactions other than hydrodesulfurization increases with increasing degree of desulfurization (Fig. 5-19) and with increasing pressure (Fig. 5-20).

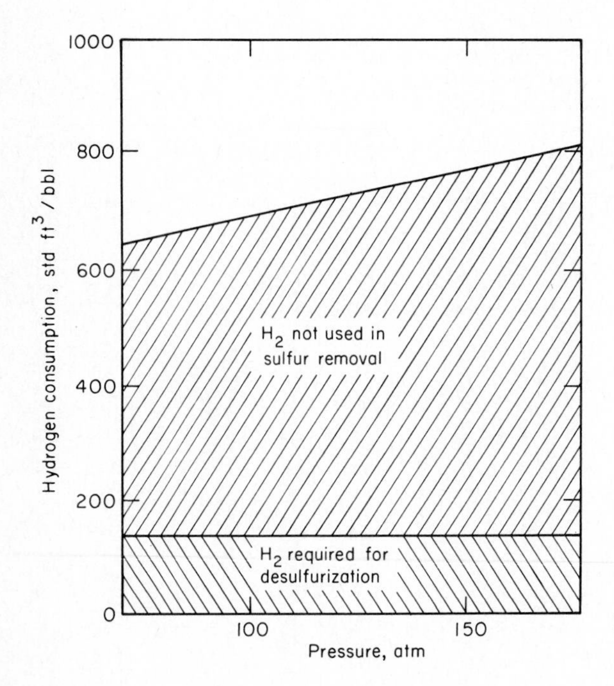

Figure 5-20 Dependence of hydrogen consumption on pressure for hydrodesulfurization of Kuwait residuum to 0.5 wt% sulfur [52].

COAL HYDRODESULFURIZATION PROCESSES

Synthetic liquid can be produced from coal by liquefaction and solids-removal processes, most of which involve pyrolysis of the polymeric coal matrix and hydrogen transfer to give smaller molecules, many of them polycyclic aromatics. The extensive literature of coal-liquefaction processes involving catalysis dating from prewar Germany has been reviewed by Weisser and Landa [43]. In some processes now undergoing development [53] the coal liquefaction takes place in the same reactor as catalytic hydrodesulfurization of the coal-derived liquids; both slurry-bed and fixed-bed reactors have been used, and the process flow diagrams are similar to those mentioned for petroleum residua. The more common processing arrangement, however, allows for liquefaction in one stage, followed by removal of the solids (largely mineral matter) and then catalytic hydrodesulfurization in separate downstream trickle-bed reactors.

Breakup of the organic coal matrix is primarily a pyrolytic (and noncatalytic) process similar to the thermal cracking described in Chap. 1. It may involve hydrogen transfer to the coal fragments from molecules such as Tetralin; after being dehydrogenated, these molecules can be rehydrogenated on a catalyst surface by the mechanisms discussed in the preceding section. The coal loses a small amount of its organic sulfur during breakup of the matrix, but most of the sulfur remains organically bound in product molecules such as dibenzothiophene, benzonaphthothiophene, and related compounds [13]. These compounds are now small enough to diffuse into the pores of a catalyst particle, and on the surface they undergo hydrodesulfurization reactions like the ones just described.

Since coal-derived liquids are highly aromatic, they may require considerable upgrading by hydrocracking to give suitable fuels; consequently, processing conditions may be more severe than those required for hydrodesulfurization of petroleum residuum (Table 5-3), even though hydrodesulfurization of compounds found in coal-derived liquids proceeds more rapidly than hydrodesulfurization of the compounds in petroleum residua, since the latter have more of the highly branched substituent groups which can cause steric hindrance of reaction [54].

SEPARATION PROCESSES

The product purification procedures in commercial hydrodesulfurization processes are well established. Gaseous and liquid products are separated at high pressures in one or two stages, and the oil is stripped to remove remaining dissolved light products. The hydrogen stream from the phase separator is scrubbed with an amine solution to remove hydrocarbon and H_2S, which inhibits reaction.

Standard operation includes scrubbing of H_2S from hydrogen recycle streams. Scrubbing of interstage reactor streams may also be practiced, and an alternative procedure for interstage purification involves selective adsorption of H_2S on zinc oxide [11].

The ease of separation of fluid-phase reaction products from solid catalysts is a strong advantage of fixed-bed reactor designs. In the slurry-bed processes, small particles of catalyst are continuously carried overhead with products. Problems are encountered with filtration, and filter plugging may be difficult to avoid. The solids-removal problem is especially important in developing slurry-bed coal liquefaction/hydrodesulfurization processes, since the mineral content of coal contributes so much particulate matter to the product.

MASS TRANSFER

Experiments with trickle-bed hydrodesulfurization generally confirm the predictions of mass-transfer correlations and show the absence of fluid-phase mass-transfer influence. For example, Cecil et al. [16] performed pilot-plant experiments with several distillates, heavy vacuum gas oils, and residua, finding no effect on reaction rate resulting from changes in mass velocity of oil in the range of about 0.13 to 0.54 $kg/m^2 \cdot s$. A representative particle dimension was probably about 3 mm. Since external-phase mass transfer does not influence rates of hydrodesulfurization in well-designed pilot-scale reactors, and since mass velocities are greater in commercial-scale reactors, as summarized in Table 5-16, the external-phase resistances are expected to be negligible generally.

In contrast, mass-transfer resistance within catalyst pores is significant for feeds ranging from light distillates to residua. Rates of transport of hydrogen or more probably of sulfur-containing oil molecules in liquid-filled pores may be low compared with intrinsic rates of reaction. The effects can usually be accounted for satisfactorily by the Thiele model for diffusion and first-order isothermal reaction in porous catalyst particles. The standard experimental test for evaluating the intraparticle mass-transfer resistance is a series of reaction-rate determinations with catalyst particles of various sizes and otherwise constant operating conditions. A perfectly mixed stirred-tank reactor is appropriate for obtaining such data, since reaction rates can be determined directly at any conversion. The catalyst may best be contained in a spinning basket (Carberry reactor).

Adlington and Thompson [56] estimated that effectiveness factors for hydrodesulfurization at 415°C and 35 atm were about 0.6 for 3.2-mm-diameter pelleted catalyst. Their estimates showed that the effectiveness factor did not change appreciably with feed boiling range, a result suggesting that the decreased reactivities of the heavier sulfur compounds were roughly compensated by their decreased diffusivities. Effects of catalyst particle size on rates of hydrodesulfurization of Middle Eastern residua were measured by Cecil et al. [16], who found an effectiveness factor of 0.4 for cylindrical particles 1.6 mm in diameter; when the average pore diameter was increased from 78 to 103 Å, the estimated effectiveness factor increased to 0.8.

Even though there are not many data available, it appears to be safe to generalize that effectiveness factors in hydrodesulfurization are slightly less than 1 for many feeds and catalysts. The right-hand side of Eq. (12) should be multiplied by an effectiveness factor, the values of which must be determined experimentally for each catalyst and feed component.

The effectiveness factor may change significantly during catalyst aging; aging is caused in part by the demetallization reactions, which are also influenced by intraparticle mass transport, as discussed in a later section on catalyst aging.

FLUID FLOW AND MIXING

The performance of a trickle-bed reactor can be approximated simply by an equation analogous to that for a piston-flow reactor with a single reactant phase, provided that the following assumptions are valid:

1. The liquid flows through the reactor as though it were a piston and completely wets all the catalyst particles.
2. There is no influence of heat or mass transfer on reaction rate.
3. The reaction is pseudo first order and takes place isothermally on the catalyst surface and not in an external phase.
4. There is negligible condensation or vaporization.

For this case of a so-called ideal trickle-bed reactor, the integrated mass-balance equation, written for the reactant about a differential element of reactor volume, takes the form

$$\ln \frac{(C_s)_{\text{in}}}{(C_s)_{\text{out}}} = \frac{3600 k_V}{\text{LHSV}} \tag{23}$$

Here k_V is the pseudo-first-order-reaction rate constant based on the volume of catalyst.

There is a mass of evidence showing that values of k_V determined from trickle-bed data increase as the liquid flow rate is increased. In other words, the fractional conversion increases with liquid flow rate, even though Eq. (23) predicts no change. The deviation from ideality is caused by inefficient contacting resulting, for example, from channeling and nonuniform wetting of the catalyst by the liquid.

To account approximately for the efficiency of contacting, we define the apparent rate constant k_{app} as that determined from trickle-bed data by use of Eq. (23). Following Satterfield [55], we now define the contacting effectiveness as the ratio of k_{app}/k_V. There are few data available, but Satterfield has used them as a basis for suggesting a preliminary correlation predicting the dependence of contacting effectiveness on liquid flow rate, as shown in Fig. 5-21. This correlation predicts that at the liquid flow rates normally encountered in commercial hydrodesulfurization reactors (Table 5-16), the contacting effectiveness is nearly 1. For pilot-scale reactors, however, the efficiency is likely to be substantially less, which suggests that there is usually a safety factor built into scale-up calculations based on equal space velocities in pilot-plant and commercial reactors.

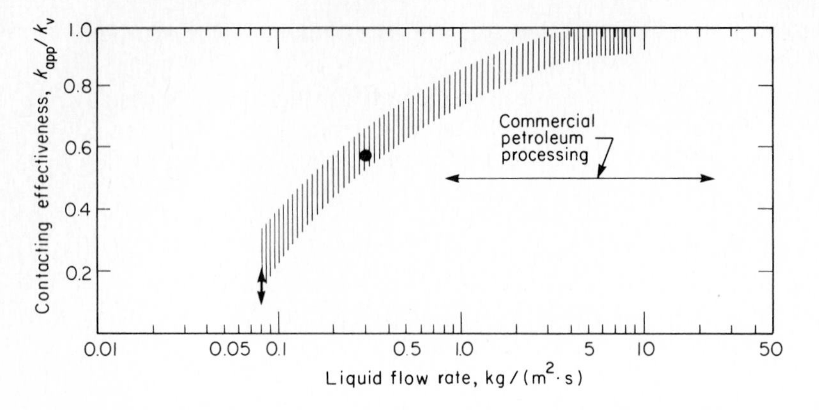

Figure 5-21 Preliminary correlation of contacting effectiveness data in trickle beds [55].

Nonetheless, one or more large-scale units were found to have contacting efficiencies much less than those expected on the basis of pilot-plant experience [57]. These results are considered anomalous, indicating maldistribution of liquid resulting from poor liquid-distributor design. Several industrial workers have emphasized the need for good distributor design and the advantage of redistributing liquid between reactor stages. We infer that ideal trickle-bed performance can be closely approached in large-scale reactors with proper distributor design; proper design is not specified in the open literature.

The foregoing paragraphs do not give more than a superficial representation of the complex fluid-mixing phenomena in trickle beds. Effects of gas flow rate,

Table 5-16 Representative limiting flow rates applied in hydrodesulfurization [55]

	Superficial liquid velocity		Superficial gas velocity			
			ft/h (STP)		kg/m^2 · s	
	ft/h	kg/m^2 · s	A[a]	B[a]	A[a]	B[a]
Pilot plant[b]	1 to 30	0.08 to 2.5	180 to 5400	890 to 27,000	0.0013 to 0.040	0.0066 to 0.20
Commercial reactor	10 to 300	0.8 to 25	1800 to 54,000	8900 to 270,000	0.013 to 0.40	0.066 to 2.0

[a] The values of gas velocity are shown for (A) 1000 and (B) 5000 std ft^3 of H_2 per barrel, assuming that all the oil is in the liquid phase. The linear velocity is calculated as though the gas were at standard temperature and pressure.

[b] The length of the pilot-plant reactor was assumed to be one-tenth the length of the commercial reactor.

catalyst aging, heat of reaction, and vaporization-condensation processes in often supercritical fluids are also important. The poorly developed literature of this subject has been reviewed critically by Satterfield [55].

In the prospective absence of deviations from piston flow and external-phase mass-transport resistance, Eq. (12), the recommended rate equation, can be integrated straightforwardly to give reactor design estimates, provided methods of accounting for catalyst aging are available.

The distribution of flow in slurry-bed hydrodesulfurization reactors is not well characterized. It is probably a good first approximation, however, that the oil and catalyst are perfectly mixed in the reactor. The hydrogen flowing through the slurry as bubbles might be nearly in piston flow.

The consequences of this mixing of reacted and unreacted compounds in the oil are clear. For example, for a pseudo-second-order desulfurization reaction in a perfectly mixed reactor, the required reactor volume is roughly twice that of a piston-flow reactor at 40 percent conversion, and about 5 times that of a piston-flow reactor at 80 percent conversion. Reactor volume in the slurry-bed process can be significantly reduced if multiple stages are used. High-temperature reactors and consequently higher reaction rates can also be used; this option is probably most desirable when processing objectives include hydrocracking of residua, since hydrocracking in the presence of aged catalysts evidently involves primarily thermal (and noncatalytic) reactions [51], and these are favored by the high ratio of reactor volume to catalyst volume of the slurry reactor. The slurry reactor may also be especially attractive for coal hydroprocessing, discussed in the next section.

The design of flow distributors is undoubtedly important in the slurry-bed processes. Even distribution of gas bubbles in the bed and avoidance of slugging are essential to maintain fluidization and even temperature distribution. Mechanical recirculation of oil may be employed to provide sufficient velocity for fluidization of larger catalyst particles [58].

Example 5-1: Contacting efficiency in a pilot-scale trickle-bed hydrodesulfurization reactor Use the conversion data of Paraskos et al. [59] to determine the contacting efficiency in a 2.5-cm-diameter, 100-cm-long trickle-bed reactor for hydrodesulfurization of Kuwait residuum. The catalyst was presumably a 0.16-cm-diameter extrudate.

Solution Empirical results have been used to derive correlations for conversion as a function of liquid hourly space velocity and bed length l in trickle-bed reactors. The results have been summarized by Paraskos et al. in the following equations, which are valid for first-order reactions. If inefficiency of contacting results from liquid holdup or from uneven catalyst wetting, then

$$\ln \frac{(C_s)_i}{(C_s)_o} = \frac{kbl^{\beta}}{(\mathrm{LHSV})^{1-\beta}} \tag{24}$$

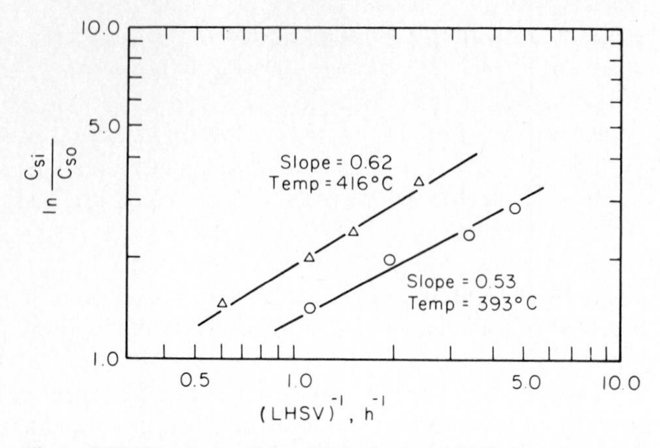

Figure 5-22 Conversion data of Paraskos et al. [59] plotted to test for contacting inefficiency with a first-order reaction. (Reprinted with permission from *Industrial and Engineering Chemistry Process Design and Development*. Copyright by the American Chemical Society.)

where b and β are empirical constants. Alternatively, if inefficiency of contacting results from backmixing, then

$$\ln \frac{(C_s)_i}{(C_s)_o} \approx \frac{k}{\text{LHSV}} - \frac{k^2 d_s}{a(\text{LHSV})^{2+\alpha} l^{1+\alpha}} \tag{25}$$

where a and α are empirical constants and d_s is the equivalent spherical particle diameter of the catalyst. If the reactions are second order, then in each case the left-hand side of the equation is replaced by $1/(C_s)_o - 1/(C_s)_i$.

The data are plotted as $\ln [(C_s)_i/(C_s)_o]$ vs. 1/LHSV on logarithmic coordinates in Fig. 5-22 and as $[1/(C_s)_o - 1/(C_s)_i]$ vs. 1/LHSV on logarithmic coordinates in Fig. 5-23. The linearity of the data in the former plot and the curvature of the data in the latter indicate that the data are better represented by first- than by second-order kinetics. The linearity of the former plot

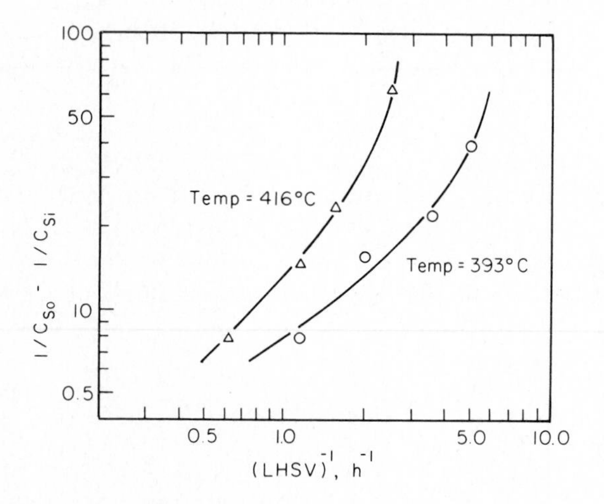

Figure 5-23 Conversion data of Paraskos et al. [59] plotted to test for contacting inefficiency with a second-order reaction. (Reprinted with permission from *Industrial and Engineering Chemistry Process Design and Development*. Copyright by the American Chemical Society.)

is consistent with Eq. (24). Since Eq. (25) predicts a curved plot, it is concluded that the data were not influenced by backmixing. It is also concluded that significant inefficiency in contacting resulted from effects of liquid holdup or incomplete catalyst wetting, since the slopes of the lines are about 0.5 and 0.6, and the value corresponding to piston flow is 1.0 [Eq. (23)].

CATALYST AGING

Deposition of coke and metal sulfides causes loss of catalyst activity by chemical modification of the surface and by physical blocking of the pores and fixed-bed interstices. Coke formation in a residuum desulfurization catalyst is relatively rapid initially, and the deposit may approach a steady-state level after several weeks on stream [18]. Coke formation is retarded by increased hydrogen partial pressure, as discussed in relation to catalytic cracking in Chap. 1.

The unique engineering problems associated with hydrodesulfurization of petroleum residua result from the presence of the organometallic compounds contained primarily in the heaviest (asphaltene) fraction of the residuum. Feedstock inspections of a Middle Eastern residuum, for example, show a Ni + V content of the order of 100 ppm, whereas a gas oil distilled from this contains only about 1 ppm of Ni + V. Although the metals are almost completely contained in the residual fraction, the sulfur is almost evenly distributed over the entire boiling range.

The nature of V and Ni compounds in petroleum has been discussed by Larson and Beuther [60] and by Dickie and Yen [61]. Metals in the asphaltene fraction are believed to be present as organometallic compounds associated in the form of structures like micelles. Asphaltenes structures are poorly known, but a schematic representation is suggested in Fig. 5-24. Asphaltenes are also formed in

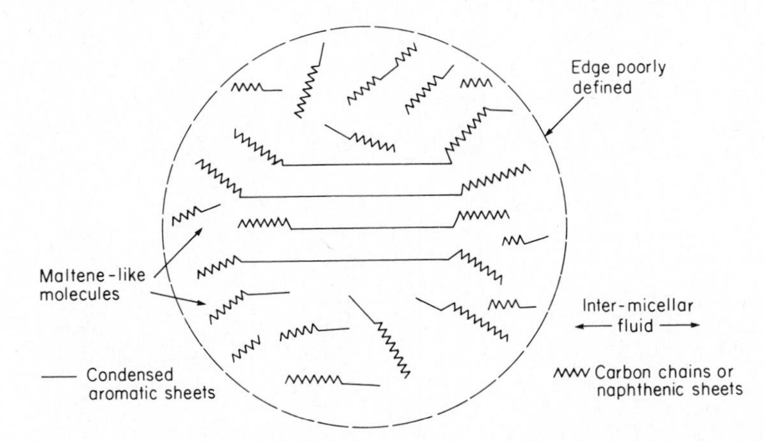

Figure 5-24 Schematic depiction of an asphaltene [60]. (Reprinted with permission from *Preprints, American Chemical Society Division of Petroleum Chemistry*. Copyright by the American Chemical Society.)

Table 5-17 Catalysts for hydrotreating residua [62]

Manufacturer	Preferred metals	SiO_2 in silica-alumina, wt % (plus other components)	Bulk density, g/cm^3	Average pore volume, cm^3/g	Average pore diameter, Å	Specific surface area m^2/g
Esso	Ni + Mo or Co + Mo	1–6	< 0.70	> 0.25	Mainly 30–70	> 150
UOP	Ni + Mo	10–40, 10–25[a]	0.625–0.875	0.3–0.5	60–100	150–250
Standard Oil Indiana	Co–Mo[b]	1–10[a]	...	> 0.5	100–200	150–500 300–350[a]
Gulf	Mo + Co + Ni	0	...	0.46	Regular distribution 0–240; average diameter 140–180	165–220
Texaco	Ni or Co + Mo or Ni + W	2–30	0.5–0.7	0.6–0.8	...	300–800
UOP	Ni + Mo	10–90 (boron phosphate[c])	0.15–0.35	1.23	125	
Shell	Ni–Co–W–Mo	70–90[a]	0.71	...	90	292
Chevron	Groups VI and VIII	Practically none (+ metal phosphates)	...	Porosity > 60%	> 60	> 100
Nippon Oil Co.	Co Mo or Ni W	0	...	0.3 of pores > 75 Å	Many pores of 1000–50,000	...
Hydrocarbon Research	Co Mo (+Ni)	0–100	...	0.45–0.5	60–70 plus channels > 1000	260–355
			Commercial samples			
Girdler G-35B	Co–Mo	...	0.96	0.22	Mainly < 200	270
Pro Catalyse Co. IFD HR304	Co–Mo	...	0.5	0.78	Regular distribution up to 10,000 plus larger pores	283
Davison Chemical Co.	Ni–Co–Mo	...	0.64	0.39	Peak at 400	222

[a] Preferred. [b] May contain zeolites. [c] Boron phosphate probably provides mechanical strength to catalysts of low density.

liquefaction of coal, and their properties are distinct from those of asphaltenes found in petroleum. An asphaltene may be 40 to 50 Å in diameter, too large to pass through many of the pores in practical hydrodesulfurization catalysts, for which representative physical properties are indicated in the compilation of Table 5-17.

Removal of V is generally found to be more rapid than removal of Ni, as shown by the kinetics data of Fig. 5-25. These data show that the demetallization reaction rates for both V and Ni compounds increase with decreasing catalyst particle size. They therefore confirm the earlier results of Hiemenz [64], who found that inorganic V deposited within the pores of catalyst particles used to hydrodesulfurize an Iranian residuum was concentrated near the particle peripheries. The outer layer containing V included only about 8 percent of the catalyst pore volume. Hiemenz found that deposited Ni was more nearly uniform in its distribution in the catalyst particle, occupying the outer 18 percent of the particle volume; the distribution of deposited carbon (coke) was uniform in the particle.

Quantitative data generally confirming these observations have become available with application of the electron microprobe, an instrument in which a beam of electrons impinges on a roughly 1-μm^2 area of solid, producing x-rays and giving an analysis of the sample to a depth of about 1 μm. Data obtained in this way, shown in Fig. 5-26, indicate massive deposits of V and Ni (as sulfides) in the catalyst pores, especially near the particle periphery.

These results show clearly that deposits of Ni and especially V form within the pores of hydrodesulfurization catalysts; these deposits lead to restriction of the

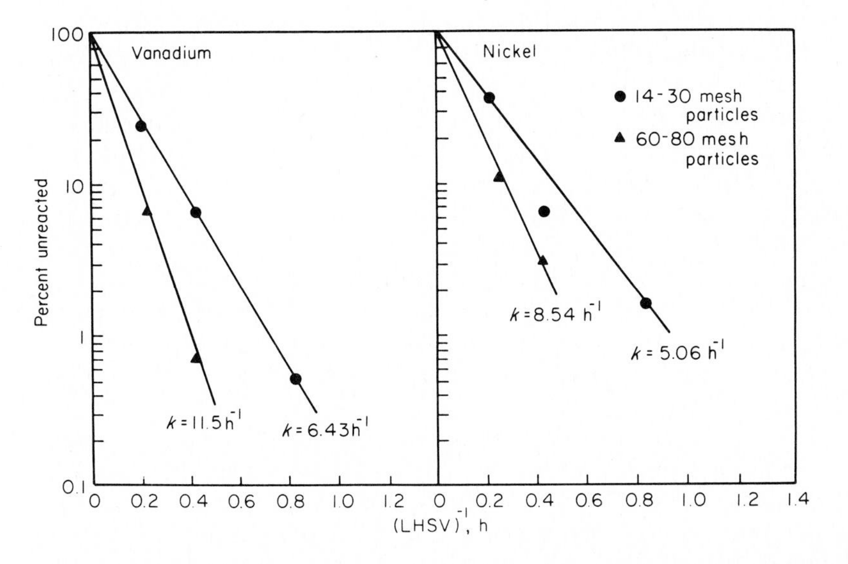

Figure 5-25 First-order demetallization of Agha Jari residuum in the presence of manganese nodules at 398°C and 137 atm [63]. (Reprinted with permission from *Industrial and Engineering Chemistry Process Design and Development*. Copyright by the American Chemical Society.)

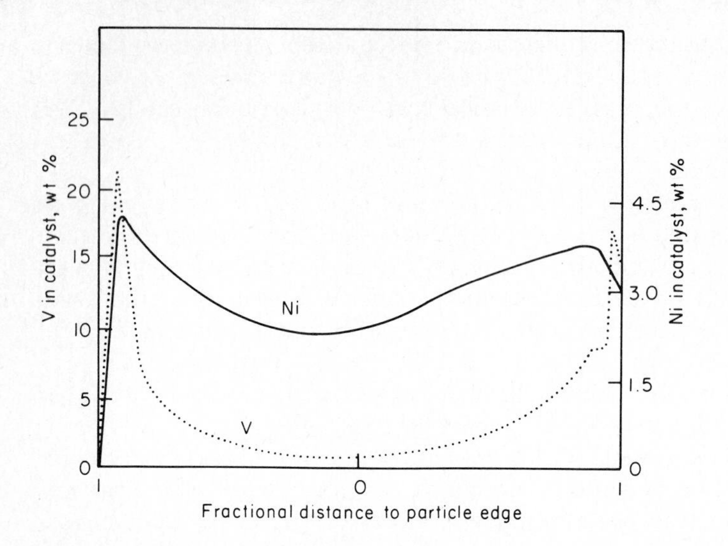

Figure 5-26 Electron microprobe analysis of the circular cross section of an aged residuum hydrodesulfurization catalyst [49]. The profiles show that V is deposited in the pores preferentially near the particle periphery, whereas Ni is more nearly uniformly deposited. The dimensions of the particle were not given, but a typical particle is probably a cylinder with a diameter of about 1.5 mm.

diffusion paths of reactant molecules, ultimately blocking the paths and deactivating the catalyst irreversibly.

The occurrence of this pore-blocking phenomenon indicates the advantage of catalysts with large pores and increased pore volume to provide easy access and high capacity for the inorganic metal compounds. Trade-offs in the catalyst design are easily recognized: larger pores and increased pore volume can be obtained at the expense of small pores and surface area, hence at the expense of intrinsic catalytic activity. Pore-volume increases obtained through increased catalyst porosity produce particles of decreasing mechanical strength. It is not yet known what the optimum pore-size distribution is for any given application; the patent literature contains many conflicting reports [50], and the information in Table 5-17 confirms the wide variations applied industrially. It is clear, however, that catalysts with large pore volumes are applied to accommodate large amounts of deposits; Mosby's catalyst accumulated almost $1\frac{1}{2}$ times its fresh weight in deposits, mostly metal sulfides [18].

Metal deposition in catalyst pores may also take place in liquefaction/hydroprocessing of coal. The problem is not defined quantitatively, but deposits of Ti have been found in the pores of used catalysts [65], and they were presumably formed from organotitanium compounds derived from coal.

The fractional removal of V + Ni was reported to be about the same as the fractional removal of sulfur with an unidentified "conventional" catalyst used in hydrodesulfurization of Kuwait atmospheric residuum [66]. With an unspecified improved catalyst, however, only about half this degree of metal removal was observed; consequently catalyst aging was less rapid, and the amount of sulfur removed before catalyst replacement was increased. These and more recent results [50] demonstrate the application of a catalyst with molecular-sieving properties. Most of the pores had diameters small enough to exclude the metal-containing asphaltenes, while most of the sulfur-containing molecules could enter the catalyst pores and penetrate toward the particle interior.

These results are suggestive of the shape-selective zeolite cracking catalysts discussed in Chap. 1. It is most likely that the shape-selective hydrodesulfurization catalysts are based on aluminas with tailor-made pore structures, but there is a possibility that zeolites have been applied. The patent literature [67] describes a two-stage residuum hydrocracking process using zeolites to serve the desired purpose. In the first stage metals and other contaminants are partially removed by reaction catalyzed by Co–Mo/Al_2O_3 having an average pore diameter of 80 Å; in the second stage a palladium-loaded zeolite (such as type Y) is used for further selective reaction of the smaller molecules, sieving out large metal-containing species which otherwise would react to form deposits of inorganic metals and coke.

Although accumulation of inorganic material within catalyst pores causes an irreversible loss of catalyst activity, the simultaneous accumulation of coke does not, and regeneration by controlled oxidation is standard practice in operation with light feeds, which contain very little metal, provided they are handled properly and corrosion products are excluded. For example, van Deemter [68] observed a rapid decline in activity of Co–Mo/Al_2O_3 after about 1.5×10^7 kg of gas oil had been hydrodesulfurized per cubic meter of catalyst bed. The catalyst regained full activity when the carbon was burned off under carefully controlled conditions. The rapid activity loss was perhaps indicative of pore blocking by coke.

Significant engineering problems result not only from deposition in catalyst pores but also from deposition in the interstices of fixed beds. Lister [46] cited uneven distribution of flow and excessive pressure drop resulting from deposition of solids at the upstream end of a fixed-bed reactor used in distillate processing. Feed contamination with sodium chloride and corrosion products must be avoided, since they rapidly form deposits near the reactor entrance. Lister recommended the placement of mesh baskets at the top of a fixed-bed reactor. These baskets should contain loosely packed solids, providing high interstitial capacity for deposited material. The baskets can be replaced periodically.

A separate guard chamber operated as a swing reactor upstream of the main reactor serves the same purpose, and this design has found application for residuum hydrodesulfurization processes [69]. The guard-bed catalyst should have high demetallization activity [such as that evidenced by manganese nodules (Fig. 5-25)] and large interstitial volume so that it can accept a maximum amount of metal-containing deposits. Data of Kubička [70] and Mosby et al. [18] demonstrate the usefulness of a guard-bed reactor. The solid deposits (especially iron and scale) were found to be concentrated near the bed entrance; V was more strongly concentrated near the entrance than Ni, corresponding to the greater reactivity of organovanadium compounds.

An advantage of the slurry-bed reactor is the continuous removal of solid material formed in reaction. This advantage may be especially strong when coal is liquefied and hydroprocessed in a single reactor, since large amounts of deposits are formed from the mineral content of coal and it is not clear that a fixed-bed reactor can be designed to accommodate the material [65]. As mentioned earlier, separation of solids from liquid products also presents a significant engineering problem, especially for coal.

REACTOR STABILITY

The heat of reaction of residuum hydrodesulfurization is sufficient to raise the reactant temperature about 20 to 80°C at typical operating conditions [58]. To compensate for the temperature rise in fixed-bed reactors, cold hydrogen is added between stages. Nonetheless, the occurrence of hot spots like those described in Chap. 4 has been repeatedly cited, and the hot spots have been identified as a symptom of catalyst aging causing poor flow distribution [46]. It is possible that uncontrollable temperature excursions could occur in fixed-bed reactors, since highly exothermic side reactions like hydrocracking begin to take place rapidly at temperatures not much higher than those normally encountered.

Instability would seem to be less likely for the backmixed slurry-bed reactors since heat transfer is rapid; the maximum temperature difference between any two points in a reactor is reported to be normally less than 3°C [58]. Yet a commercial-scale slurry-bed hydrodesulfurization reactor is known to have exploded [71], and we may speculate that the explosion resulted from localized overheating in the reactor. Such overheating would have resulted if fluidization had not been maintained at some location. The potential hazard points to a clear need for operating data and analysis of the instability phenomena.

NOTATION

a	empirical constant in Eq. (25)
b	empirical constant in Eq. (24)
c	concentration, weight percent
d_s	equivalent spherical diameter of a catalyst particle
K	rate-equation parameter, l^2/F
k	reaction rate constant, variable dimensions
LHSV	liquid hourly space velocity, volume of liquid feed per hour per volume of catalyst
l	length of catalyst bed
P	pressure; partial pressure, F/l^2
r	reaction rate, variable dimensions
T	temperature, °C or K
WHSV	weight hourly space velocity, mass of liquid feed per mass of catalyst per hour

Dimensional

E = energy
F = force
l = length
m = mass
t = time

Greek

α	empirical constant in Eq. (25)
α_1	fraction of sulfur which is relatively reactive
α_2	fraction of sulfur which is relatively unreactive
β	empirical constant in Eq. (24)
σ	void fraction of reactor

Subscripts

app	apparent
ar	aromatic
B	butene
HC	hydrocarbon
HDS	hydrodesulfurization
hyd	hydrogenation
i	in
o	out
S	sulfur-containing compound or sulfur
T	thiophene
V	based on unit volume of catalyst

REFERENCES

1. Owens, P. J., and C. H. Amberg, *Adv. Chem. Ser.* **33,** 182 (1961).
2. Owens, P. J., and C. H. Amberg, *Can. J. Chem.* **40,** 941 (1962).
3. Owens, P. J., and C. H. Amberg, *Can. J. Chem.* **40,** 947 (1962).
4. Desikan, P., and C. H. Amberg, *Can. J. Chem.* **41,** 1966 (1963).
5. Desikan, P., and C. H. Amberg, *Can. J. Chem.* **42,** 843 (1964).
6. Kolboe, S., and C. H. Amberg, *Can. J. Chem.* **44,** 2623 (1966).
7. Givens, E. N., and P. B. Venuto, *Prepr. Am. Chem. Soc. Div. Pet. Chem.*, **15**(4), A183 (1970).
8. Bartsch, R., and C. Tanielian, *J. Catal.*, **35,** 353 (1974).

8*a*. Houalla, M., N. K. Nag, A. V. Sapre, D. H. Broderick, and B. C. Gates, to be published.

9. Satterfield, C. N., and G. W. Roberts, *Am. Inst. Chem. Eng. J.*, **14,** 159 (1968).
10. Frye, C. G., and J. F. Mosby, *Chem. Eng. Prog.*, **63**(9), 66 (1967).
11. Phillipson, J. J., paper presented at *Am. Inst. Chem. Eng. Meet., Houston, 1971.*
12. Schuit, G. C. A., and B. C. Gates, *Am. Inst. Chem. Eng. J.*, **19,** 417 (1973).
13. Kilanowski, D. R., H. Teeuwen, V. H. J. de Beer, B. C. Gates, G. C. A. Schuit, and H. Kwart, *J. Catal.*, to be published.
14. Houalla, M., D. Broderick, V. H. J. de Beer, B. C. Gates, and H. Kwart, *Prepr. Am. Chem. Soc. Div. Pet. Chem.* **22**(3) 941 (1977).
15. Metcalfe, T. B., *Chim. Ind. Gen. Chim.*, **102,** 1300 (1969).
16. Cecil, R. R., F. Z. Mayer, and E. N. Cart, paper presented at *Am. Inst. Chem. Eng. Meet., Los Angeles, 1968.*
17. Beuther, H., and B. K. Schmid, *Proc. 6th World Pet. Cong.*, sec. III, p. 297, Verein für Forderung des 6. Welt-Erdöl Kongresses, Hamburg, 1964.

18. Mosby, J. F., G. B. Hockstra, T. A. Kleinhanz, and J. M. Sroka, *Hydrocarbon Process.*, **52**(5), 93 (1973).
19. Richardson, J. T., *Ind. Eng. Chem. Fundam.*, **3,** 154 (1964).
20. Sonnemans, J., and P. Mars, *J. Catal.*, **31,** 209 (1973).
21. Dufaux, M., M. Che, and C. Naccache, *J. Chim. Phys.*, **67,** 527 (1970).
22. Kittel, C., "Introduction to Solid State Physics," p. 550, Wiley, New York, 1971.
23. Lippens, B. C., Ph.D. thesis, Technical University of Delft, The Netherlands, 1961.
24. Knözinger, H., and P. Ratnasamy, *Catal. Rev.-Sci. Eng.*, in press.
25. Ashley, J. H., and P. C. H. Mitchell, *J. Chem. Soc.*, **A1968,** 2821; **A1969,** 2730.
26. Stork, W. H. J., J. G. F. Coolegem, and G. T. Pott, *J. Catal.*, **32,** 497 (1974).
27. Lipsch, J. M. J. G., and G. C. A. Schuit, *J. Catal.*, **15,** 163, 174, 179 (1969).
28. deBeer, V. H. J., T. H. M. van Sint Fiet, G. H. A. M. van der Steen, A. C. Zwaga, and G. C. A. Schuit, *J. Catal.*, **35,** 297 (1974).
29. LoJacono, M., A. Cimino, and G. C. A. Schuit, *Gazz. Chim. Ital.*, **103,** 1281 (1973).
30. Kiviat, F. E., and L. Petrakis, *J. Phys. Chem.*, **77,** 1232 (1973).
31. Mone, R., and L. Moscou, *Prepr. Am. Chem. Soc. Div. Pet. Chem.*, **20**(2), 564 (1975).
32. Voorhoeve, R. J. H., *J. Catal.*, **23,** 236 (1971).
32*a*. Voorhoeve, R. J. H., and J. C. M. Stuiver, *J. Catal.*, **23,** 228 (1971).
32*b*. Voorhoeve, R. J. H., and J. C. M. Stuiver, *J. Catal.*, **23,** 243 (1971).
33. Farragher, A. L., and P. Cossee, *Proc. 5th Int. Cong. Catal.*, p. 1301, North-Holland, Amsterdam, 1973.
34. Huisman, R., J. deJonge, C. Haas, and F. Jellinek, *J. Solid State Chem.*, **3,** 56 (1971).
35. Wilson, J. A., and A. D. Yoffe, *Adv. Phys.*, **18,** 193 (1969).
36. Berg, J. M. van den, *Inorg. Chim. Acta*, **2,** 216 (1968).
37. Furimsky, E., and C. H. Amberg, *Can. J. Chem.*, **53,** 2542 (1975).
38. Hagenbach, G., P. Courty, and B. Delmon, *J. Catal.*, **23,** 295 (1971); **31,** 264 (1973); G. Hagenbach, and B. Delmon, *C. R. Acad. Sci. Paris Ser. C.*, **273,** 1489 (1971).
39. deBeer, V. H. J., and G. C. A. Schuit, *Ann. N.Y. Acad. Sci.*, **272,** 61 (1976).
40. Smith, G. V., C. C. Hinckley, and F. Behbahmy, *J. Catal.*, **30,** 218 (1973).
41. Mikovsky, R. J., A. J. Silvestri, and H. Heinemann, *J. Catal.*, **34,** 324 (1974).
42. LoJacono, M., J. L. Verbeek, and G. C. A. Schuit, *J. Catal.*, **29,** 463 (1973).
43. Weisser, O., and S. Landa, "Sulphide Catalysts: Their Properties and Applications," Pergamon, London, 1973.
44. Goudriaan, F., Ph.D. thesis, Technical University of Twente, The Netherlands, 1974.
45. Satterfield, C. N., M. Modell, and J. F. Mayer, *Am. Inst. Chem. Eng. J.*, **21,** 1100 (1975).
46. Lister, A., *3d Eur. Symp. Chem. React. Eng.*, p. 225, Pergamon, Oxford, 1965.
47. Docksey, P., and R. J. H. Gilbert, *Proc. 7th World Pet. Cong.*, vol. 4, p. 153, Elsevier, Barking, Essex, England, 1967.
48. Schuman, S. C., and H. Shalit, *Catal. Rev.*, **4,** 245 (1970).
49. Oxenreiter, M. F., C. G. Frye, G. B. Hockstra, and J. M. Sroka, paper presented at *Jap. Pet. Inst. Meet., 1972.*
50. Richardson, R. L., and S. K. Alley, *Prepr. Am. Chem. Soc. Div. Pet. Chem.*, **20**(2), 554 (1975).
51. Paraskos, J. A., A. A. Montagna, and L. W. Brunn, paper presented at *Am. Inst. Chem. Eng. Meet., Washington, 1974.*
52. Fant, B. T., cited in J. H. Krasuk, P. Andreu, and N. Barroeta, *Acta Cientif. Venez.*, **25,** 49 (1974).
53. National Academy of Sciences, "Assessment of Technology for the Liquefaction of Coal," National Academy of Sciences, Washington, D. C., 1977.
54. Larson, O. A., *Prepr. Am. Chem. Soc. Div. Pet. Chem.*, **19**(3), 417 (1974).
55. Satterfield, C. N., *Am. Inst. Chem. Eng. J.*, **21,** 209 (1975).
56. Adlington, D., and E. Thompson, *3d Eur. Symp. Chem. React. Eng.*, p. 203, Pergamon, Oxford, 1965.
57. Ross, L. D., *Chem. Eng. Prog.*, **61**(10), 77 (1965).
58. Mounce, E., and R. S. Rubin, *Chem. Eng. Prog.*, **67**(8), 81 (1971).
59. Paraskos, J. A., J. A. Frayer, and Y. T. Shah, *Ind. Eng. Chem. Process Des. Dev.*, **14,** 315 (1975).

60. Larson, O. A., and H. Beuther, *Prepr. Am. Chem. Soc. Div. Pet. Chem.* **11**(2), B95 (1966).
61. Dickie, J. P., and T. F. Yen, *Anal. Chem.*, **39,** 1847 (1967).
62. Vlugter, J. C., and P. van't Spijker, *Proc. 8th World Pet. Cong.*, vol. 4, p. 159, Applied Science, London, 1971.
63. Chang, C. D., and A. J. Silvestri, *Ind. Eng. Chem. Process Des. Dev.*, **13,** 315 (1974).
64. Hiemenz, E., *Proc. 6th World Pet. Cong.*, sec. III, p. 307, Verein für Forderung des 6. Welt-Erdöl Kongresses, Hamburg, 1964.
65. Stanulonis, J. J., B. C. Gates, and J. H. Olson, *Am. Inst. Chem. Eng. J.*, **22,** 576 (1976).
66. Moritz, K. H., H. R. Savage, D. Traficante, W. Weissman, and B. J. Young, *Chem. Eng. Prog.* **67**(8), 63 (1971).
67. Arey, W. F., R. B. Mason, and R. C. Paule, U.S. Patent 3,254,017 (1966).
68. Deemter, J. J. van, *3d Eur. Symp. Chem. React. Eng.*, p. 215, Pergamon, Oxford, 1965.
69. Ginneken, A. J. J. van, M. M. van Kessel, K. M. A. Pronk, and G. Renstrom, *Oil Gas J.*, April 28, 1975, p. 59.
70. Kubička, R., J. Čir, V. Novák, and J. Vepřek, *Brennst. Chem.*, **49**(10), 308 (1968).
71. Davis, J. C., *Chem. Eng.*, July 10, 1972, p. 36.

PROBLEMS

5-1 Paraskos et al. [51] reported an experiment in which residuum was hydrodesulfurized to give a product sulfur concentration which was held constant as catalyst temperature was increased to compensate for catalyst aging. They observed that the yield of distillate product in a certain boiling range decreased for a relatively short period of operation, passed through a minimum, and then increased as the temperature was increased. What do these results imply about the chemistry of hydrocracking?

5-2 Equation (9) has been described as difficult to justify on theoretical grounds. Explain why specifically.

5-3 Some residuum hydrodesulfurization conversion data of Mosby et al. [18] are given below. Show that a pseudo-second-order reaction in total sulfur is not a good representation of these data. Determine an equation to account for the observed dependence of conversion on space velocity.

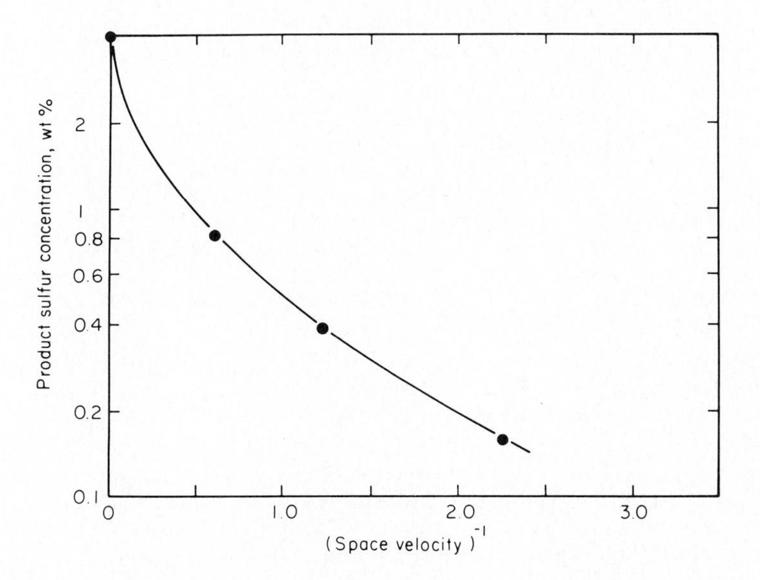

5-4 Design a guard-bed demetallization reactor to operate initially at 400°C to remove 50 percent of the V + Ni from Khafji residuum. Use the data of Fig. 5-25 as a basis for the estimate.

5-5 The data of Fig. 5-25 refer to catalyst particles in two size ranges; the average diameters are approximately 1.0 and 0.2 mm for these ranges. Estimate the effectiveness factors of these particles for removal of V and Ni, respectively. For comparison, what are the approximate time-averaged effectiveness factors for the particle referred to in Fig. 5-26?

5-6 The life of a residuum hydrodesulfurization catalyst may be limited by its capacity for deposited metal sulfides. According to a rule of thumb, the lifetime of a catalyst is determined by the V + Ni content of the feedstock, provided that this value exceeds 100 ppm; otherwise, the lifetime is determined by the coke deposition. Estimate the lifetime of a catalyst used in hydrodesulfurization of Khafji residuum to give product with 1.0 percent S. Assume that the effectiveness factors indicated by Fig. 5-26 are representative of the whole reactor for all times. Also assume that 50 percent of the V + Ni are removed by a guard bed and that the fractional removal of V + Ni in the main reactor is the same as the fractional removal of sulfur. Estimate the catalyst physical properties from Table 5-17. Assume that the value of LHSV is 1.

5-7 Consider hydrodesulfurization of thiophene at 250°C and 1 atm catalyzed by Co–Mo/Al_2O_3. Assume a feed composition equimolar in heptane, H_2, and thiophene, and compare the initial rates of hydrodesulfurization, i.e., the rates at the reactor entrance, predicted by the equations of Satterfield and Roberts [Eq. (6)] and Phillipson [Eq. (9)].

5-8 The hydrocracking activity of catalysts is strongly reduced in the presence of feedstocks containing high concentrations of compounds like quinoline. Explain why. How could a hydrocracking process be designed to minimize this effect?

5-9 If the Mo in a typical Co–Mo/Al_2O_3 hydrodesulfurization catalyst is all near the surface in the form of (intercalated) MoS_2 crystallites, estimate roughly the average crystallite size.

5-10 Residuum hydrodesulfurization catalysts extruded with a cross section shaped like a cloverleaf have found industrial application. What are the advantages of this shape?

5-11 Suggest a mechanism to account specifically for the result of Mikovsky et al. [41] showing that H_2S was produced initially in the reaction between thiophene and D_2.

5-12 Since catalyst aging is caused by metals, suggest an alternative processing route to give low-sulfur fuel oil from a crude oil with a high metals content.

5-13 Deviations from piston flow (represented as axial dispersion) have the greatest effect on reactor performance at high conversions. Estimate the range of conversions for a light-distillate hydrodesulfurization reactor for which the deviation from piston flow can be considered negligible. Use the criterion of Mears [*Chem. Eng. Sci.*, **26**, 1361 (1971)]:

$$\frac{H}{d_p} > \frac{20n}{\text{Pe}} \ln \frac{(C_s)_i}{(C_s)_o}$$

where H = reactor height
d_p = catalyst particle diameter
n = reaction order

Assume that the Peclet number Pe is about 0.1, and estimate the other required values from information given in the chapter.

5-14 In processes for the hydrodesulfurization of distillates as feeds to catalytic reforming reactors, it has been observed that increases in reactor temperature, intended to give increased conversions, have actually led to decreased conversions even though catalyst aging was not markedly accelerated. Explain these observations.

5-15 In Fig. 5-15 the reaction of thiophene on a hydrodesulfurization catalyst has been suggested to proceed through an "end-on" adsorption involving interaction of the S atom with a surface-anion vacancy. Recognizing that aromatic compounds such as benzene are inhibitors of the hydrodesulfurization reactions and also undergo hydrogenation, suggest an alternate mode of adsorption of thiophene and other aromatic sulfur-containing compounds. What is the evidence in favor of each of the two modes of adsorption?

5-16 How is the surface structure of a hydrodesulfurization catalyst expected to change with variations in the ratio of hydrogen to H_2S partial pressures?

5-17 At subatmospheric pressures the hydrodesulfurization reactivities of thiophene, benzothiophene, and dibenzothiophene have been observed to be about equal, whereas at 100 atm thiophene and benzothiophene have been observed to be much more reactive than dibenzothiophene. Suggest an explanation for these observations.

5-18 Some authors contend that a monolayer model best describes the hydrodesulfurization catalyst. Prepare a summary of the evidence supporting this contention, consulting ref. [39] and the paper by Massoth [*J. Catal.*, **50,** 190 (1977)].

5-19 Design a reactor for hydrodesulfurization of a light distillate containing 50 ppm of thiophene and 150 ppm of mercaptans, sulfides, and disulfides. Assume that vapor-phase reactants contact the catalyst at 370°C and 20 atm with a hydrogen/hydrocarbon mole ratio of 1. The product sulfur concentration should be 0.5 ppm. Catalyst aging is negligible under these conditions.

NAME INDEX

SUBJECT INDEX